SELECTED SOLUTIONS MANUAL

C. ALTON HASSELL
BAYLOR UNIVERSITY

GENERAL CHEMISTRY

AN INTEGRATED APPROACH

THIRD EDITION

HILL
PETRUCCI

Prentice Hall
Upper Saddle River, NJ 07458

Editor in Chief: John Challice
Senior Editor: Kent Porter-Hamann
Project Manager: Kristen Kaiser
Executive Managing Editor: Kathleen Schiaparelli
Assistant Managing Editor: Dinah Thong
Production Editor: Natasha Wolfe
Supplement Cover Designer: Jonathan Boylan
Buyer: Ilene Kahn
Cover photo credit: Pencho B. Dimitroff, Aurora, CO.

Prentice-Hall, Inc.
Upper Saddle River, NJ 07458

Printed in the United States of America

10 9 8 7 6 5 4 3 2 1

ISBN 0-13-062004-1

Pearson Education Ltd., London
Pearson Education Australia Pty. Ltd., Sydney
Pearson Education Singapore, Pte. Ltd.
Pearson Education North Asia Ltd., Hong Kong
Pearson Education Canada, Inc., Toronto
Pearson Educacíon de Mexico, S.A. de C.V.
Pearson Education—Japan, Tokyo
Pearson Education Malaysia, Pte. Ltd.

Table of Contents

To the Student

This manual is meant to be an aid in your learning chemistry, especially to your learning to solve chemical problems. It can be misused and not be helpful, but used correctly, the manual can be a great help.

Contained in these pages are the worked out solutions for the in-chapter exercises, selected review questions, and the odd numbered end-of-chapter problems for General Chemistry, 3nd ed. by John W. Hill and Ralph H. Petrucci, Prentice Hall (2002). In working the mathematical problems, values of constants, such as the universal gas constant or the atomic weights, were used that included one more significant figure than the least well-known input data. The reason was to insure that the input data would be the limiting factor in the accuracy of the answer. When the answer to an intermediate step was listed, it was usually written with one more significant figure than should be listed in the final answer. Again, this was done so that the input data, not the intermediate answer, should be the limiting value. The final answer was written with the correct number of significant figures.

Learning problem solving is much like learning to play a sport or a musical instrument. You must do it yourself; you cannot learn by watching. Skill is developed by practice and more practice. Practice time should be planned so that you are at your best. Very little will be gained during practice time when you are very tired or distracted.

Before attempting the problems, read the text chapter carefully, reading the example problems and working out the exercises. If you work on an exercise for ten or fifteen minutes and still do not have an answer, only then should you look in the solutions manual to get an idea of how to start working the problem. After the text is read, work through the review questions and the odd numbered problems. Working a problem may require looking back in the text to find a similar example. Only after ten or fifteen minutes of attempting a problem without result should the solutions manual be consulted. Any problem that requires use of the solutions manual should be reworked a few days later to see if the problem solving skill was really acquired or just the answer read from the manual.

When you refer to the manual, the solution method may be different from yours as there is usually more than one way to work a problem. There are some problems in the manual that actually show two different methods. Sometimes these solutions differ in the last significant figure because of the round off of numbers. If your answer is within 1 or 2 in the last significant figure, then it is probably the same number.

Don't work these problems just to get an answer. Work these problems to develop the skills to be able to solve other problems, such as those on the next test or those in your future job. While practicing these skills, it is better if you work in a quiet, undisturbed atmosphere while you are as fresh as possible. It is also better to work some every day than to cram all of your studying into one day.

Baylor University
Waco, Texas

C. Alton Hassell
Alton_Hassell@baylor.edu

Dedication

I dedicate this manual to the memory of two wonderful men.

The first was my father, Clinton A. "Brit" Hassell (1914–1995). My greatest inheritance was the education that he provided for me.

The second was my first college chemistry professor, Dr. Thomas C. Franklin (1923–1997). He became a mentor and dear friend. He is deeply missed.

Acknowledgements

This manual is the compilation of the work of many people. I owe a great debt to each one. They have combined to keyboard, proof, standardize, edit, double-check, clarify and even beautify the initial rough draft. The errors that remain are my sole responsibility.

Adonna Cook, Barbara Rauls, Andrew Garner, and Erin Saenz keyboarded the manuscript. That is, they turned illegible scrawling and crude drawings into printed text and illustrative figures. Lee Ann Marshall more than once gave us the expert guidance that was needed in the use of Word.

Proofreaders and/or accuracy-checkers included Michael Wismer, Tony Yiannakos, and Matt Johl. These wonderful, hard-working people have pored over the manual and found my many mistakes.

The Department of Chemistry at Baylor University and especially the department chairman, Dr. Marianna Busch, have given me great support and encouragement.

John Hill and Ralph Petrucci wrote a wonderful text without which this manual would be unnecessary. Terry McCreary helped with some of the solutions. Ralph went the extra mile (miles) to proof and edit the manual. Robert Wismer wrote solutions for some problems in another book which are being reused in this manual.

To work with the editors and marvelous staff at Prentice Hall is to work with the best. Paul Corey signed me to the original project. Mary Hornby became a dear friend and the glue that held us all together for earlier editions. Kristen Kaiser has done the same for this edition.

My wife, Patricia, and my children, Clint and Sharina, put up with me or with my absence during the entire process.

Chapter 1

Chemistry: Matter and Measurement

Exercises

1.1 (a) $7.42 \times 10^{-3}\ \text{s} \times \dfrac{\text{ms}}{10^{-3}\ \text{s}} = 7.42\ \text{ms}$

(b) $5.41 \times 10^{-4}\ \text{m} \times \dfrac{\mu\text{m}}{10^{-6}\ \text{m}} = 5.41\ \mu\text{m}$

(c) $1.19 \times 10^{-9}\ \text{g} \times \dfrac{\text{ng}}{10^{-9}\ \text{g}} = 1.19\ \text{ng}$

(d) $5.98 \times 10^{3}\ \text{m} \times \dfrac{\text{km}}{10^{3}\ \text{m}} = 5.98\ \text{km}$

1.2 (a) $475\ \text{nm} = 475 \times 10^{-9}\ \text{m} = 4.75 \times 10^{2} \times 10^{-9}\ \text{m} = 4.75 \times 10^{-7}\ \text{m}$
(b) $225\ \text{ns} = 225 \times 10^{-9}\ \text{s} = 2.25 \times 10^{2} \times 10^{-9}\ \text{s} = 2.25 \times 10^{-7}\ \text{s}$
(c) $1415\ \text{km} = 1415 \times 10^{3}\ \text{m} = 1.415 \times 10^{3} \times 10^{3}\ \text{m} = 1.415 \times 10^{6}\ \text{m}$
(d) $2.26 \times 10^{6}\ \text{g} = 2.26 \times 10^{6} \times 10^{-3}\ \text{kg} = 2.26 \times 10^{3}\ \text{kg}$

1.3 (a) $t_F = (1.8 \times 85.0\ ^\circ\text{C}) + 32 = 185\ ^\circ\text{F}$
(b) $t_F = (1.8 \times -12.2\ ^\circ\text{C}) + 32.0 = 10.0\ ^\circ\text{F}$
(c) $t_C = (355\ ^\circ\text{F} - 32)/1.8 = 179\ ^\circ\text{C}$
(d) $t_C = (-20.8\ ^\circ\text{F} - 32.0)/1.8 = -29.3\ ^\circ\text{C}$

1.4 $6.4\ \text{mm} = 6.4 \times 10^{-3}\ \text{m}$
$1.827\ \text{m} \times 0.762\ \text{m} \times 6.4 \times 10^{-3}\ \text{m} = 8.9 \times 10^{-3}\ \text{m}^3$
or
$1.39\ \text{m}^2 \times 6.4 \times 10^{-3}\ \text{m} = 8.9 \times 10^{-3}\ \text{m}^3$

1.5 $21.60\ \text{g} \times 2.04 \times 21 = 925\ \text{g Zn}$

1.6A (a)
```
   48.2 m
    3.82 m
   48.4394 m
 ___________
  100.4594 m   rounds to 100.5 m
```
(b)
```
  148 g
    2.39 g
    0.0124 g
 ___________
  150.4024 g   rounds to 1.50 × 10^3 g
```

(c) 451 g
−15.46 g
−20.3 g
415.24 g rounds to 415 g

(d) 15.436 L
5.3 L
−6.24 L
−8.177 L
6.319 L rounds to 6.3 L

1.6B (a) 51.5 + 2.67 = 54.17 m; 33.42 − 0.124 = 33.296 m

$54.2 \text{ m} \times 33.30 \text{ m} = 1804.86$ rounds to $1.80 \times 10^3 \text{ m}^2$

(b) 125.1 − 1.22 = 123.88; 52.5 + 0.63 = 53.13

$$\frac{123.9 \text{ g}}{53.1 \text{ mL}} = 2.33 \text{ g/mL}$$

(c) 47.5 − 1.44 = 46.06

$$\frac{46.1 \text{ kg}}{10.5 \text{ m} \times 0.35 \text{ m} \times 0.175 \text{ m}} = 72 \text{ kg/ m}^3$$

(d)

$14.2 \text{ mg} = 14.2 \times 10^{-3} \text{ g} = 0.0142 \text{ g}$

$3.52 \text{ mg} = 3.52 \times 10^{3} \text{ g} = 0.00352 \text{ g}$

0.307 g
− 0.0142 g
− 0.00352 g
0.28928 g

$0.28 \text{ mm} = 0.28 \times 10^{-3} \text{ m} = 0.28 \times 10^{-3} \times 10^{2} \text{ cm} = 0.028 \text{ cm}$

1.22 cm
− 0.028 cm
1.192 cm

$$\frac{0.289 \text{ g}}{1.19 \text{ cm} \times 0.752 \text{ cm} \times 0.51 \text{ cm}} = 0.63 \text{ g/cm}^3$$

1.7 (a) $? \text{ m} = 76.3 \text{ mm} \times \frac{10^{-3} \text{ m}}{\text{mm}} = 0.0763 \text{ m}$

(b) $? \text{ mg} = 0.0856 \text{ kg} \times \frac{10^3 \text{ g}}{\text{kg}} \times \frac{\text{mg}}{10^{-3} \text{ g}} = 8.56 \times 10^4 \text{ mg}$

(c) $? \text{ ft} = 0.556 \text{ km} \times \frac{0.6214 \text{ mi}}{\text{km}} \times \frac{5280 \text{ ft}}{\text{mi}} = 1.82 \times 10^3 \text{ ft}$

(d) $? \text{ kg} = 48.8 \text{ oz} \times \frac{28.35 \text{ g}}{\text{oz}} \times \frac{\text{kg}}{10^3 \text{ g}} = 1.38 \text{ kg}$

(e) $? \text{ fl oz} = 3.50 \text{ gal} \times \frac{3.785 \text{ L}}{1 \text{ gal}} \times \frac{\text{mL}}{10^{-3} \text{ L}} \times \frac{\text{fl oz}}{29.57 \text{ mL}} = 448 \text{ fl oz}$

1.8A (a) $? \text{ in}^2 = 476 \text{ cm}^2 \times \left(\frac{1 \text{ in.}}{2.54 \text{ cm}}\right)^2 = 73.8 \text{ in.}^2$

(b) $? \text{ m}^3 = 1.56 \times 10^4 \text{ in.}^3 \times \left(\frac{1 \text{ m}}{39.37 \text{ in.}}\right)^3 = 0.256 \text{ m}^3$

1.8B $? \text{ kg/m}^2 = \frac{14.70 \text{ lb}}{\text{in.}^2} \times \left(\frac{1 \text{ in.}}{2.54 \text{ cm}}\right)^2 \times \left(\frac{\text{cm}}{10^{-2} \text{ m}}\right)^2 \times \frac{453.6 \text{ g}}{\text{lb}} \times \frac{\text{kg}}{10^3 \text{ g}} = \frac{1.034 \times 10^4 \text{ kg}}{\text{m}^2}$

1.9 (a) $? \text{ m/s} = \frac{90.0 \text{ km}}{\text{h}} \times \frac{\text{h}}{60 \text{ min}} \times \frac{\text{min}}{60 \text{ s}} \times \frac{10^3 \text{ m}}{\text{km}} = \frac{25.0 \text{ m}}{\text{s}}$

(b) $? \text{ km/h} = \frac{1.39 \text{ ft}}{\text{s}} \times \frac{60 \text{ s}}{\text{min}} \times \frac{60 \text{ min}}{\text{h}} \times \frac{12 \text{ in.}}{\text{ft}} \times \frac{2.54 \text{ cm}}{\text{in}} \times \frac{10^{-2} \text{ m}}{\text{cm}} \times \frac{\text{km}}{10^3 \text{ m}}$

$= \frac{1.53 \text{ km}}{\text{h}}$

(c) $? \text{ kg/h} = \frac{4.17 \text{ g}}{\text{s}} \times \frac{\text{kg}}{10^3 \text{ g}} \times \frac{60 \text{ s}}{\text{min}} \times \frac{60 \text{ min}}{\text{h}} = \frac{15.0 \text{ kg}}{\text{h}}$

1.10A $d = \frac{m}{V} = \frac{1.25 \text{ kg}}{10.5 \text{ cm x } 11.2 \text{ cm x } 18.7 \text{ cm}} \times \frac{10^3 \text{ g}}{\text{kg}} = \frac{0.568 \text{ g}}{\text{cm}^3}$

1.10B $d = \frac{m}{V} = \frac{76.0 \text{ lb}}{2.54 \text{ L}} \times \frac{453.6 \text{ g}}{\text{lb}} \times \frac{10^{-3} \text{ L}}{\text{mL}} = 13.6 \text{ g/mL}$

1.11A $? \text{ gal} = 10.0 \text{ kg} \times \frac{10^3 \text{ g}}{\text{kg}} \times \frac{\text{mL}}{0.791 \text{ g}} \times \frac{10^{-3} \text{L}}{\text{mL}} \times \frac{1.057 \text{ qt}}{1.00 \text{ L}} \times \frac{\text{gal}}{4 \text{ qt}} = 3.34 \text{ gal}$

1.11B $? \text{ mL} = 10.00 \text{ gal} \times \frac{4 \text{ qt}}{\text{gal}} \times \frac{1 \text{ L}}{1.057 \text{ qt}} \times \frac{\text{mL}}{10^{-3} \text{ L}} \times \frac{0.690 \text{ g}}{\text{mL}} \times \frac{\text{mL}}{0.791 \text{ g}}$

$= 3.30 \times 10^4 \text{ mL}$

OR

$? \text{ gal} = 10.00 \text{ gal} \times \frac{4\text{qt}}{\text{gal}} \times \frac{1.\text{L}}{1.057 \text{ qt}} \times \frac{\text{mL}}{10^{-3} \text{L}} \times \frac{0.690 \text{ g}}{\text{mL}} \times \frac{\text{mL}}{0.791 \text{ g}}$

$\times \frac{10^{-3} \text{ L}}{\text{mL}} \times \frac{1.057 \text{ qt}}{1\text{L}} \times \frac{\text{gal}}{4 \text{ qt}} = 8.723 \text{ gal}$

This simplifies to:

$? \text{ gal} = 10.00 \text{ gal} \times \frac{0.690 \text{ g}}{\text{mL}} \times \frac{\text{mL}}{0.791 \text{ g}} = 8.723 \text{ gal}$

1.12 Water is about 1 g/mL

$$20\text{ qt} \times \frac{\text{L}}{1.06\text{ qt}} \times \frac{\text{mL}}{10^{-3}\text{ L}} \times \frac{1\text{ g}}{\text{mL}} \times \frac{\text{kg}}{10^{3}\text{ g}} \text{ simplifies to } 20\text{ qt} \times \frac{\text{L}}{1.1\text{ qt}} \times \frac{\text{kg}}{\text{L}}$$

$\frac{20}{1.1}$ is closer to 20 than 15, so 20 kg is a reasonable answer.

1.13 mass of Saturn $= 95.2 \times 6.0 \times 10^{24}\text{ kg} \times \frac{10^{3}\text{ g}}{\text{kg}} = 5.7 \times 10^{29}\text{ g}$

$$\text{volume of Saturn} = \frac{4}{3}\pi r^3 = \frac{4}{3} \times 3.1416 \times (5.82 \times 10^4\text{ km})^3 \times \left(\frac{10^3\text{ m}}{\text{km}}\right)^3 \times \left(\frac{\text{cm}}{10^{-2}\text{ m}}\right)^3 = 8.3\text{ cm}^3$$

$$\text{density of Saturn} = \frac{m}{V} = \frac{5.7\times10^{29}\text{ g}}{8.27\times10^{29}\text{ cm}^3} = 0.69\text{ g/ cm}^3$$

Since water is about 1 g/cm^3, Saturn would float in water.

1.14 An ocean liner constructed of several separate compartments would stay afloat even if some of the compartments were 'holed' and contained water. *Titanic* sank because the compartments were not truly separate but open at the top. When she took on enough water to list, the water flowed over the top of the compartments to fill the next compartment and the next, until she was heavier than water. For this reason, naval warships close off all compartments as they go to battle.

1.15 One essential assumption is that all of the invitations weigh the same amount. Thus, we can weigh 100 to get the weight of each. Another assumption is that 10-g differences are enough of a difference to determine if the 100 invitations weigh more or less than 28.39 g per invitation. The method does not work if extra notes have been added to some invitations, if there are less than 20 invitations, if the masses of envelopes are too close to 28.35 g, or if the scale is incorrectly calibrated.

Review Questions

2. (a) Iron, (b) air, (c) the human body, and (e) gasoline are matter.

11. (c) Combustion of natural gas is a chemical property; it reacts with air.

12. (a) physical (b) chemical (c) chemical (d) chemical (e) physical

13. C, Cl, and Na are symbols representing elements. CO, $CaCl_2$, and KI are combinations of symbols for two elements (formulas) and represent compounds.

14. Helium and salt are substances. Lemon juice and wine are mixtures. Helium and salt are composed of only helium atoms and sodium and chloride ions, respectively. Lemon juice and wine are composed of a number of substances.

15. Gasoline is the same throughout and thus is homogeneous. Raisin pudding has raisins as a separate component. The large bits of spice in Italian salad dressing are a separate component. The bubbles of carbon dioxide gas are a separate component. Thus, raisin pudding, Italian salad dressing, and an open cola drink are heterogeneous. The cola drink will be homogeneous until it is opened.

Problems

27. (a) $? \text{ ng} = 4.54 \times 10^{-9} \text{ g} \times \frac{1 \text{ ng}}{10^{-9} \text{ g}} = 4.54 \text{ ng}$

(b) $? \text{ km} = 3.76 \times 10^{3} \text{ m} \times \frac{1 \text{ km}}{10^{3} \text{ m}} = 3.76 \text{ km}$

(c) $? \text{ µg} = 6.34 \times 10^{-6} \text{ g} \times \frac{1 \text{ µg}}{10^{-6} \text{ g}} = 6.34 \text{ µg}$

29. (a) $t_F = 1.8\ (23.5\ °C) + 32.0 = 74.3\ °F$

(b) $t_C = \frac{173.9°F - 32.0}{1.8} = 78.83\ °C$

(c) $t_F = 1.8\ (-98.0\ °C) + 32 = -144\ °F$

31. $t_F = 1.8\ (37\ °C) + 32 = 99\ °F$

$t_C = \frac{98.2°F - 32.0}{1.8} = 36.8\ °C$

33. (a) $t_C = \frac{136°F - 32}{1.8} = 57.8\ °C$

(b) $t_F = 1.8\ (-120\ °C) + 32 = -184\ °F$

35. (a) $? \text{ m} = 50.0 \text{ km} \times \frac{10^3 \text{ m}}{\text{km}} = 5.00 \times 10^4 \text{ m}$

(b) $? \text{ L} = 47.9 \text{ mL} \times \frac{10^{-3} \text{ L}}{\text{mL}} = 4.79 \times 10^{-2} \text{ L}$

(c) $? \text{ ms} = 578 \text{ µs} \times \frac{10^{-6} \text{ s}}{\text{µs}} \times \frac{\text{ms}}{10^{-3} \text{ s}} = 0.578 \text{ ms}$

(d) $? \text{ mg} = 1.55 \times 10^{2} \text{ kg} \times \frac{10^3 \text{ g}}{\text{kg}} \times \frac{\text{mg}}{10^{-3} \text{ g}} = 1.55 \times 10^{8} \text{ mg}$

(e) $? \text{ mm}^2 = 87.4 \text{ cm}^2 \times \left(\frac{10^{-2} \text{ m}}{\text{cm}}\right)^2 \times \left(\frac{\text{mm}}{10^{-3} \text{ m}}\right)^2 = 8.74 \times 10^3 \text{ mm}^2$

(f) $? \frac{\text{m}}{\text{s}} = \frac{0.0962 \text{ km}}{\text{min}} \times \frac{\text{min}}{60 \text{ s}} \times \frac{10^3 \text{ m}}{\text{km}} = 1.60 \frac{\text{m}}{\text{s}}$

37. The yardstick (3 ft.) is shorter than the 3-ft 5-in. rattlesnake. The rattlesnake is 41 in. [(3 × 12) + 5], which is shorter than the chain, because 1 meter is about 40 in., so the chain is about 48 (40 × 1.2) in.. The rope is longest at 75 in.
(4) < (3) < (1) < (2)
yardstick < rattlesnake < chain < rope

39. $? \text{ in.} = 1 \text{ link} \times \frac{\text{chain}}{100 \text{ links}} \times \frac{\text{furlong}}{10 \text{ chain}} \times \frac{\text{mi}}{8 \text{ furlongs}} \times \frac{5280 \text{ ft}}{\text{mi}} \times \frac{12 \text{ in.}}{\text{ft}} = 7.92 \text{ in.}$

41. (a) 4 (b) 2 (c) 5 (d) 3 (e) 4 (f) 4

43. (a) 2.804×10^3 m (b) 9.01×10^2 s
(c) 9.0×10^{-4} cm (d) 2.210×10^2 s

45. (a) 505.5 m (b) 2120, zero not significant
(c) 0.00610 (d) 40000 mL, last two zeros are not significant

47. (a)
```
  36.5    m
 –2.16    m
  3.452   m
 37.792   m => 37.8 m
```
(b)
```
  151        g
    4.16     g
 –  0.0220   g
  155.1380 => 155 g
```
(c)
```
  15.44  mL
 – 9.1   mL
 105     mL
 111.34  => 111 mL
```
(d)
```
  12.52 cm
 + 5.1  cm
 – 3.18 cm
 –12.02 cm
   2.42 cm => 2.4 cm
```

49. (a) 73.0 mm × 1.340 mm × (25.31 – 1.6) mm = 73.0 mm × 1.340 mm × 23.7 mm
$= 2318 \text{ mm}^3 \Rightarrow 2.32 \times 10^3 \text{ mm}^3$

(b) $\frac{33.58 \text{ cm} \times 1.007 \text{ cm}}{0.00705 \text{ g}} = 4796.4624 \frac{\text{cm}^2}{\text{g}} \Rightarrow 4.80 \times 10^3 \frac{\text{cm}^2}{\text{g}}$

(c) $\frac{418.7 \text{ mm} \times 31.8 \text{ mm}}{(19.27 \text{ mg} - 18.98 \text{ mg})} = \frac{418.7 \text{ mm} \times 31.8 \text{ mm}}{(0.29 \text{ mg})} = 45913 \frac{\text{mm}^2}{\text{mg}}$

$\Rightarrow 4.6 \times 10^4 \frac{\text{mm}^2}{\text{mg}}$

(d) $\dfrac{2.023\,\text{g}-\left(1.8\times10^{-3}\ \text{g}\right)}{1.05\times10^{4}\ \text{mL}}=\dfrac{2.021\,\text{g}}{1.05\times10^{4}\ \text{mL}}=1.92495\times10^{-4}\dfrac{\text{g}}{\text{mL}}$

$\Rightarrow 1.9\times10^{-4}\dfrac{\text{g}}{\text{mL}}$

51. $d=\dfrac{m}{V}=\dfrac{78.0\ \text{g}}{25.0\ \text{mL}}=3.12\ \text{g/mL}$

53.

18.432 g	metal + paper
– 1.214 g	paper
17.218 g	metal

$d=\dfrac{m}{V}=\dfrac{17.218\ \text{g}}{3.29\ \text{cm}^3}=5.23\ \text{g/cm}^3$

55. $d=\dfrac{m}{V}=\dfrac{5901\ \text{g}}{1.20\ \text{cm}\times2.41\ \text{cm}\times1.80\ \text{cm}}=\dfrac{59.01\ \text{g}}{5.206\ \text{cm}^3}=11.3\ \text{g/cm}^3$

57. $V=\dfrac{m}{d}=5.79\ \text{mg}\times\dfrac{10^{-3}\ \text{g}}{\text{mg}}\times\dfrac{\text{cm}^3}{19.3\,\text{g}}=3.00\times10^{-4}\ \text{cm}^3$

$l=\dfrac{V}{A}=\dfrac{3.00\text{x}10^{-4}\ \text{cm}^3}{44.6\,\text{cm}^2}=6.73\times10^{-6}\ \text{cm}$

59. $d=\dfrac{m}{V}=\dfrac{3.2\ \text{kg}}{0.80\ \text{m x }0.80\ \text{m x }1.20\ \text{m}}\times\dfrac{10^3\,\text{g}}{\text{kg}}\times\left(\dfrac{10^{-2}\ \text{m}}{\text{cm}}\right)^3=\dfrac{3.2\text{x}10^3\ \text{g}}{7.68\text{x}10^5\ \text{cm}^3}$

$=4.2\times10^{-3}\ \text{g/cm}^3$

61. $15\ \text{gal}\times\dfrac{4\ \text{qt}}{\text{gal}}\times\dfrac{\text{L}}{1.06\ \text{qt}}\times\dfrac{\text{mL}}{10^{-3}\ \text{L}}\times\dfrac{1\ \text{g}}{\text{mL}}\times\dfrac{\text{lb}}{454\ \text{g}}$

estimate $\dfrac{15\text{ x }4\text{ x }1000}{1\text{ x }500}\approx$ 120 lb (actually 125 lb) water

$3.0\ \text{L}\times\dfrac{\text{mL}}{10^{-3}\ \text{L}}\times\dfrac{\text{cm}^3}{\text{mL}}\times\dfrac{13.6\,\text{g}}{\text{cm}^3}\times\dfrac{\text{lb}}{454\ \text{g}}$ = leads to

estimate $\dfrac{3\text{ x }1000\text{ x }14}{500}\approx$ 84 lb (actually 90 lb) mercury

(2) The water is most difficult to lift.

63. $d=\dfrac{m}{V}=\dfrac{30.0\ \text{g}}{(22.2-18.0)\ \text{mL}}=7.143\ \text{g/mL}\Rightarrow7.1\ \text{g/mL}$

Additional Problems

65. Sample B weighs less on the moon than Sample A weighs on Earth because the moon has a smaller gravitational pull. Sample C has more mass than Sample A.

For Sample C to weigh the same in the smaller gravitational field, it would have to have more mass than Sample A.

67. The meter stick, with 1000-mm markings, is actually 1005 mm long. The true poster area is $1.827\ m \times 0.763\ m = 1.39\ m^2$. The measured size is 1.827 m

$$\times \frac{1.005\ m}{1.000\ m} \times 0.762\ m \times \frac{1.005\ m}{1.000\ m} = 1.836\ m \times .766\ m = 1.41\ m^2.$$

$1.41\ m^2 - 1.39\ m^2 = 0.02\ m^2$ error.

69. $$?\frac{\text{furlongs}}{\text{fortnight}} = \frac{3.00\times10^8\ m}{s} \times \frac{60\ s}{min} \times \frac{60\ min}{h} \times \frac{24\ h}{d} \times \frac{14\ d}{\text{fortnight}} \times \frac{cm}{10^{-2}\ m} \times \frac{in.}{2.54\ cm} \times \frac{ft}{12\ in.} \times \frac{mile}{5280\ ft} \times \frac{8\ \text{furlongs}}{mile} = 1.80\times10^{12}\ \frac{\text{furlongs}}{\text{fortnight}}$$

71. $$?\ \text{acres} = 1\ \text{hectare} \times \frac{1\ hm^2}{\text{hectare}} \times \left(\frac{100\ m}{hm}\right)^2 \times \left(\frac{39.37\ in.}{1\ m}\right)^2 \times \left(\frac{ft}{12\ in.}\right)^2 \times \left(\frac{mi}{5280\ ft}\right)^2 \times \frac{640\ \text{acre}}{mi^2} = 2.47\ \text{acres}$$

The hectare is larger.
1 hectare = 2.47 acres

73. 4.72 kg bottle and wine
−1.70 kg bottle
3.02 kg wine

$$d = \frac{3.02\ kg}{3.00\ L} \times \frac{10^3\ g}{kg} \times \frac{10^{-3}\ L}{mL} = 1.007\ g/mL$$

$$?\ oz = 275\ mL \times \frac{1.007\ g\ \text{wine}}{mL} \times \frac{11.0\ g\ \text{alcohol}}{100\ g\ \text{wine}} \times \frac{oz}{28.35\ g} = 1.07\ oz$$

75. $$?\ g = 125\ cm^3 \times \frac{2.2\ g}{cm^3} = 275\ g$$

$$?\ cm^3 = 275\ g \times \frac{cm^3}{0.015\ g} = 1.8 \times 10^4\ cm^3$$

77. $$?\ mg/m^2\ hr = \frac{10\ tons}{mi^2\ month} \times \left(\frac{mi}{5280\ ft}\right)^2 \times \left(\frac{ft}{12\ in.}\right)^2 \times \left(\frac{in.}{2.54\ cm}\right)^2 \times \left(\frac{cm}{10^{-2}\ m}\right)^2 \times \left(\frac{2000\ lb}{tons}\right) \times \left(\frac{454\ g}{lb}\right) \times \left(\frac{mg}{10^{-3}\ g}\right) \times \left(\frac{month}{30\ days}\right) \times \left(\frac{day}{24\ hr}\right) = 4.9\ mg/m^2\ hr$$

Density is not needed.

79. $V_{brass} = (2.0 \text{ cm})^3 = 8.0 \text{ cm}^3$

The brass cube will sink to the bottom, displacing 8.0 cm^3, or 8.0 mL of water.

$V_{cork} = 5.0 \text{ cm} \times 4.0 \text{ cm} \times 2.0 \text{ cm} = 40 \text{ cm}^3$

$$m_{cork} = dV = \frac{0.22 \text{ g}}{\text{cm}^3} \times 40 \text{ cm}^3 = 8.8 \text{ g}$$

The cork will displace 8.8 g of water, which is about 8.8 mL of water. More water will overflow from the vessel of water in which the cork is floated (right).

81. (a) density of object $= \dfrac{5.15 \text{ kg}}{30.2 \text{ cm} \times 12.9 \text{ cm} \times 11.5 \text{ cm} - \pi (2.5 \text{ cm})^2 \times 11.5 \text{ cm}}$

$$\times \left(\frac{10^3 \text{ g}}{\text{kg}}\right) = \frac{5150 \text{ g}}{4.25\text{x}10^3 \text{ cm}^3} = 1.21 \text{ g/cm}^3$$

by estimation $\dfrac{5 \times 1000 \text{ g}}{30 \text{ cm} \times 13 \text{ cm} \times 12 \text{ cm} - (3 \text{ cm})^2 \times 12 \text{ cm}}$

$$= \frac{5000 \text{ g}}{4680 \text{ cm}^3 - 324 \text{ cm}^3} = \frac{5000 \text{ g}}{4356 \text{ cm}^3}$$

The density is greater than 1 g/cm^3. The object will not float.

(b) ? g of balsa = $\pi (2.5 \text{ cm})^2 \times 11.5 \text{ cm} \times \dfrac{0.1 \text{ g}}{\text{cm}^3} = 23$ g balsa

density of object $= \dfrac{5150 \text{ g} + 23 \text{ g}}{30.2 \text{ cm x } 12.9 \text{ cm x } 11.5 \text{ cm}} = \dfrac{5173 \text{ g}}{4.48 \times 10^3 \text{ cm}^3}$

$$= 1.15 \text{ g/cm}^3$$

The object still will not float.

(c) Weight of material of object removed:

? g = $\pi (2.5 \text{ cm})^2 \times 11.5 \text{ cm} \times 1.21 \text{ g/cm}^3 = 273$ g

Weight loss when replaced by balsa:

? g = 273 g – 23 g = 250 g

Object must weigh less than 4.48×10^3 g to float.

? g = 5150 g – 4480 g = 670 g

? holes = 670 g $\times \left(\dfrac{\text{hole}}{250 \text{ g}}\right)$ = 2.7 holes

At least three holes must be drilled and filled with balsa.

83. The ground crew called pounds kilograms. The fuel needed was

(13,597 + 8703) kg = 22,300 kg. They had 7682 L $\times \dfrac{1.77 \text{ lb}}{\text{L}} \times \dfrac{1.000 \text{ kg}}{2.205 \text{ lb}}$

= 6167 kg fuel.

? kg fuel needed = 22,300 kg – 6167 kg = 16133 kg

? L fuel needed = 16133 kg $\times \dfrac{2.205 \text{ lb}}{1.00 \text{ kg}} \times \dfrac{\text{L}}{1.77 \text{ lb}} = 2.10 \times 10^4$ L

They should have added at least 2.10×10^4 L of fuel.

85. $$?\frac{\text{mg}}{\text{L}} = \frac{10\,\mu\text{g}}{\text{dL}} \times \frac{\text{dL}}{10^{-1}\text{L}} \times \frac{10^{-6}\text{g}}{\mu\text{g}} \times \frac{\text{mg}}{10^{-3}\text{g}} = 0.10\frac{\text{mg}}{\text{L}}$$

$$?\frac{\text{mg}}{\text{L}} = \frac{10\,\text{mg}}{10\,\text{L}} = 1\frac{\text{mg}}{\text{L}}$$

They are not the same. They differ by a factor of 10. The reporter may have used the conversion for cL instead of dL, but probably he used mg for μg and daL (10 L) for dL (0.1 L). That would make a factor of 10 difference.

Apply Your Knowledge

87 (a) $$\text{microballoon volume} = 1.00\text{L} \times \frac{68\,\text{L}}{100\,\text{L}} \times \frac{\text{mL}}{10^{-3}\,\text{L}} \times \frac{\text{cm}^3}{\text{mL}} = 6.80 \times 10^2\ \text{cm}^3$$

$$?\frac{\text{volume}}{\text{microballoon}} = \frac{4}{3}\pi r^3 = \frac{4}{3} \times 3.1416 \times \left(\frac{0.070\,\text{mm}}{2}\right)^3 \times \left(\frac{10^{-3}\ \text{m}}{\text{mm}}\right)^3 \times \left(\frac{\text{cm}}{10^{-2}\ \text{m}}\right)^3 = 1.80 \times 10^{-7}\ \text{cm}^3$$

$$?\text{microballoons} = 6.80 \times 10^2\ \text{cm}^3 \times \frac{\text{microballoon}}{1.80 \times 10^{-7}\ \text{cm}^3} = 3.78 \times 10^9\ \text{microballoons}$$

$$?\frac{\text{g}}{\text{microballoons}} = \frac{62\,\text{g}}{3.78 \times 10^9\ \text{microballoons}} = 1.64 \times 10^{-8}\frac{\text{g}}{\text{microballoon}}$$

surface area = $4\pi r^2$

$$= 4 \times 3.1416\left(\frac{0.070\,\text{mm}}{2}\right)^2 \times \left(\frac{10^{-3}\ \text{m}}{\text{mm}}\right)^2 \times \left(\frac{\text{cm}}{10^{-2}}\right)^2 = 1.54 \times 10^{-4}\ \text{cm}^2$$

$$\text{volume of glass} = \frac{1.64 \times 10^{-8}\ \text{g}}{\text{microballoon}} \times \frac{\text{cm}^3}{2.2\,\text{g}} = 7.45 \times 10^{-9}\ \text{cm}^3$$

$$\text{thickness} = \frac{\text{volume}}{\text{surface area}} = \frac{7.45 \times 10^{-9}\ \text{cm}^3}{1.54 \times 10^{-4}\ \text{cm}^2} = 4.84 \times 10^{-5}\ \text{cm}$$

(b) $$\text{mass glue} = \frac{32\,\text{L}}{100\,\text{L}} \times 1.00\,\text{L} \times \frac{\text{mL}}{10^{-3}\ \text{L}} \times \frac{1.67\,\text{g}}{\text{mL}} = 534\,\text{g}$$

total mass = mass glue + mass microballoon = 534 g + 62 g = 596 g

$$\text{density} = \frac{\text{mass}}{\text{volume}} = \frac{596\,\text{g}}{\text{L}} \times \frac{10^{-3}}{\text{mL}} = \frac{0.596\,\text{g}}{\text{mL}}$$

89. (a) ? g = 42.062 g – 32.105 g = 9.957 g

$$\text{V pyc} = 9.957\ \text{g} \times \frac{\text{mL}}{0.9982\ \text{g}} = 9.975\ \text{mL}$$

$? \text{ g} = 40.873 \text{ g} - 32.105 \text{ g} = 8.768 \text{ g}$

$$d = \frac{m}{V} = \frac{8.768 \text{ g}}{9.975 \text{ mL}} = 0.8790 \text{ g/mL}$$

(b) $? \text{ g} = 38.1055 \text{ g} - 36.2142 \text{ g} = 1.8913 \text{ g Zn}$ mass Zn

$$? \text{ mL} = (46.1894 \text{ g} - 36.2142 \text{ g}) \times \frac{\text{mL}}{0.99823 \text{ g}} = 9.9929 \text{ mL volume } H_2O \text{ in pyc without zinc}$$

$$? \text{ mL} = (47.8161 \text{ g} - 1.8913 \text{ g Zn} - 36.2142 \text{ g pyc}) \times \frac{\text{mL}}{0.99823 \text{ g}} = 9.7278 \text{ mL}$$

volume H_2O in pyc with zinc

$? \text{ mL} = 9.9929 \text{ mL} - 9.7278 \text{ mL} = 0.2651 \text{ mL}$ volume Zn

$$d = \frac{m}{V} = \frac{1.8913 \text{ g}}{0.2651 \text{ mL}} = 7.134 \text{ g/mL}$$

91. (a) surface area $= 4\pi r^2 = 4 \times 3.1416 \times \left(\frac{7930 \text{ miles}}{2}\right)^2 \times \left(\frac{5280 \text{ ft}}{\text{mile}}\right)^2 \times \left(\frac{12 \text{ in.}}{\text{ft}}\right)^2$

$= 7.931 \times 10^{17} \text{ in.}^2$

$$\text{mass} = \text{surface area} \times \frac{14.7 \text{ pounds}}{\text{in.}^2} = 7.931 \times 10^{17} \text{ in.}^2 \times \frac{14.7 \text{ lbs}}{\text{in.}^2} \times \frac{454 \text{ g}}{\text{lb}} \times \frac{\text{kg}}{10^3 \text{ g}} = 5.29 \times 10^{18} \text{ kg}$$

(b) volume of Earth $= \frac{4}{3}\pi r^3 = \frac{4}{3} \times 3.1416 \times \left(\frac{7930 \text{ miles}}{2}\right)^3 \times \left(\frac{5280 \text{ ft}}{\text{mile}}\right)^3 \times \left(\frac{12 \text{ in.}}{\text{ft}}\right)^3 \times \left(\frac{2.54 \text{ cm}}{\text{in.}}\right)^3 \times \frac{\text{mL}}{\text{cm}^3} = 1.088 \times 10^{27} \text{ mL}$

$$d = \frac{\text{mass}}{\text{volume}} = \frac{5.98 \times 10^{24} \text{ kg}}{1.088 \times 10^{27} \text{ mL}} \times \frac{10^3 \text{ g}}{\text{kg}} = 5.49 \frac{\text{g}}{\text{mL}}$$

Chapter 2

Atoms, Molecules, and Ions

Exercises

2.1 The bulb will still weigh 7.500 g. All of the reactants and products are sealed inside the bulb; nothing can escape or be added.

2.2A $? \text{ g magnesium oxide} = 1.500 \text{ g oxygen} \times \left(\frac{3.317 \text{ g magnesium oxide}}{1.317 \text{ g oxygen}}\right)$
$= 3.778 \text{ g magnesium oxide}$

2.2B $? \text{ g oxygen reacted} = 3.250 \text{ g magnesium} \times \frac{0.6583 \text{ g oxygen}}{1.000 \text{ g magnesium}} = 2.139 \text{ g oxygen}$
$? \text{ g oxygen excess} = 12.500 \text{ g} - 2.139 \text{ g} = 10.361 \text{ g oxygen}$

2.3A $A = Z + 66 = 50 + 66 = 116$ $\qquad {}^{116}_{50}\text{Sn}$

2.3B Cadmium has 48 protons and 48 electrons
$A = 48 + 66 = 114$ mass number 114.

2.4A
$19.99244 \times 0.9051 = 18.095$
$20.99395 \times 0.0027 = 0.057$
$21.99138 \times 0.0922 = 2.028$
$20.180 \Rightarrow 20.18$

2.4B X = fractional abundance of copper 63
$X \times 62.9298 + (1.000 - X) \times 64.9278 = 63.546$
$62.9298\,X + 64.9278 - 64.9278\,X = 63.546$
$X = \frac{-1.382}{-1.998} = 0.6917$
69.17% copper 63
30.83% copper 65

2.5 The most abundant isotope is ^{24}Mg. Since the average atomic mass is 24.3050, the atomic mass is closer to the mass of magnesium 24.
It is difficult to determine the second most abundant unless we know the % abundance of the magnesium 24. There are many combinations of the three percentages that would work. Since both ^{25}Mg and ^{26}Mg are heavier than the average atomic mass, it is not obvious which is second most abundant. A study of isotopes indicates that the percentage usually decreases the farther away from the average atomic mass [implying ^{25}Mg] but that isotopes with an odd number of neutrons often have a lesser percent abundance [implying ^{26}Mg].

The actual percentages are ^{24}Mg 78.70%, ^{25}Mg 10.13%, and ^{26}Mg 11.17%.

2.6 N_2F_4 dinitrogen tetrafluoride

2.7 (a) P_4O_{10}
(b) heptasulfur dioxide

2.8 (a) AlF_3 (b) K_2S (c) Ca_3N_2 (d) Li_2O

2.9A (a) calcium bromide
(b) lithium sulfide
(c) iron(II) bromide
(d) copper(I) iodide

2.9B Copper(I) sulfide Cu_2S

2.10 (a) $(NH_4)_2CO_3$ (b) $Ca(ClO)_2$ (c) $Cr_2(SO_4)_3$

2.11 (a) potassium hydrogen carbonate
(b) iron(III) phosphate
(c) magnesium dihydrogen phosphate

2.12 (a) No, the formulas are different: C_7H_{16} for the left structure and C_7H_{14} for the cyclic structure on the right.
(b) Yes, both hydrocarbons have the formula C_9H_{20}. One has methyl groups on the second and fourth carbon atoms, and the other, on the second and fifth carbon atoms.

2.13

	BP °C	increase per CH_2 emit (°C)
C_8H_8	125.7	≈ 25
C_9H_{20}		≈ 23
$C_{10}H_{22}$	174.1	≈ 22
$C_{11}H_{24}$	196	≈ 21
$C_{12}H_{26}$	217	≈ 19
$C_{13}H_{28}$	236	≈ 18
$C_{14}H_{30}$	254	≈ 17
$C_{15}H_{32}$	271	≈ 16
$C_{16}H_{34}$	286.8	≈ 15

Kerosene probably has the alkanes $C_{11}H_{24}$, $C_{12}H_{26}$, $C_{13}H_{28}$, $C_{14}H_{30}$, and $C_{15}H_{32}$.

Review Questions

1. (a) 104.3 g before and after reaction.

(b) 40.3 g magnesium oxide
(c) Law of conservation of mass, and law of definite proportions
(d) 80.6 g magnesium oxide is formed. There is twice the amount of magnesium to react, as well as excess oxygen (more than 32.0 g).

2. Compound B has an oxygen-to-sulfur mass ratio of 3:2. The law of multiple proportions is illustrated.

7. (a) isobars (b) isotopes (c) identical atoms (d) different elements
(e) different elements

8. (a) ${}^{8}_{5}B$ (b) ${}^{14}_{6}C$ (c) ${}^{235}_{92}U$ (d) ${}^{60}_{27}Co$

11. $FeCl_2$ is called iron(II) chloride, or ferrous chloride. $FeCl_3$ is called iron(III) chloride, or ferric chloride. The Stock system using the Roman numerals is the preferred method.

12. (a) Magnesium is an element.
(b) Methyl is a prefix, meaning a CH_3 group.
(c) Chloride is a negatively charged atom of chlorine, the ion Cl^-.
(d) Ammonia is a compound, NH_3.
(e) Ammonium is an ion, NH_4^+.
(f) Ethane is a compound, CH_3CH_3.
The only substances that could be found on a stockroom shelf are (a), (d), and (f).

19. (a) Usually it is organic if it contains C, but there are a few exceptions, so a molecular formula is sufficient.
(b) It contains only H and C, so a molecular formula is sufficient.
(c) An alcohol requires a structural formula.
(d) It contains only H and C, and the ratio is 2n + 2 hydrogens for each carbon, so a molecular formula is sufficient.
(e) A carboxylic acid requires a structural formula.

Problems

21. $Zn + S \rightarrow ZnS$
Before reaction: 1.000 g Zn + 0.200 g S = 1.200 g
After reaction: 0.608 g ZnS + 0.592 g Zn = 1.200 g
The mass of substances after the reaction equals the mass of substances before the reaction. The law of conservation of mass is confirmed.

23. $\frac{0.625\text{ g}}{1.000\text{ g}} \times 100\% = 62.5\%\text{ C}$ $\quad \frac{0.0419\text{ g}}{1.000\text{g}} \times 100\% = 4.19\%\text{ H}$
$\frac{0.968\text{ g}}{1.549\text{ g}} \times 100\% = 62.5\%\text{ C}$ $\quad \frac{0.0649\text{ g}}{1.549\text{ g}} \times 100\% = 4.19\%\text{ H}$

$\frac{0.618\ g}{0.988\ g} \times 100\% = 62.6\%\ C \qquad \frac{0.0414\ g}{0.988\ g} \times 100\% = 4.19\%\ H$

Yes, each sample has the same percent carbon and percent hydrogen.

25. $?\ g\ SO_2 = 1.305\ g\ S \times \frac{0.632\ g\ SO_2}{0.312\ g\ S} = 2.61\ g\ SO_2$ produced

27. Using the $\frac{1.142\ g\ oxygen}{1.000\ g\ nitrogen}$ ratio but changing the relative number of atoms produces new ratios.

Twice the number of oxygen $\frac{2.284\ g\ oxygen}{1.000\ g\ nitrogen}$; (c) is possible.

Twice the number of nitrogen $\frac{1.142\ g\ oxygen}{2.000\ g\ nitrogen} = \frac{0.571\ g\ oxygen}{1.000\ g\ nitrogen}$; (a) is possible.

(b) and (d) are not possible; no combination of atoms would produce those ratios.

29.

	Protons	Neutrons	Electrons
(a) ^{62}Zn	30	32	30
(b) ^{241}Pu	94	147	94
(c) ^{99}Tc	43	56	43
(d) ^{99}Mo	42	57	42

31. #2 $^{40}_{20}Ca$ #6 $^{48}_{22}Ti$ #7 $^{48}_{20}Ca$

#2 and #7 are isotopes.

33. There are multiple isotopes whose masses average the atomic mass of 79.904 u.

35. $68.926 \times 0.601 = 41.42$

$70.925 \times 0.399 = \underline{28.30}$

69.72 => 69.7 u

37. $19.9924 \times 0.9051 = 18.095$

$20.9940 \times 0.0027 = 0.057$

$21.9914 \times 0.0922 = \underline{2.028}$

20.180 => 20.18 u

39. $84.91179\,X + (1.000 - X) \times 86.90919 = 85.4678$

$84.91179\,X + 86.90919 - 86.90919\,X = 85.4678$

$X = \frac{-1.4414}{-1.9974} = 0.72164$

72.164% rubidium 85 27.836% rubidium 87

41.

	Group	Period	Type
(a) C	4A	2	nonmetal

(b) Ca	2A	4	metal
(c) S	6A	3	nonmetal
(d) Ti	4B	4	metal
(e) Br	7A	4	nonmetal
(f) Bi	5A	6	metal
(h) Au	1B	6	metal

43. (a) O_2 (b) Br_2 (c) H_2 (d) N_2

45. ClF_3 and H_2O are binary molecular compounds. They have two elements and are molecular compounds. BaI_2 is an ionic compound. HCN and ONF are ternary molecular compounds.

47. (a) N_2O (b) SF_6 (c) P_4S_3(d) NCl_3
    (e) carbon disulfide
    (f) diboron tetrachloride
    (g) dichlorine heptoxide
    (h) oxygen difluoride

49. (a) potassium ion (b) oxide ion (c) copper(II) ion
    (d) Al^{3+} (e) N^{3-} (f) Cr^{3+}

51. (a) carbonate ion (b) sulfate ion (c) hydroxide ion
    (d) dihydrogen phosphate ion (e) NH_4^+ (f) NO_2^-
    (g) CN^- (h) HCO_3^-

53. (a) potassium bromide
    (b) calcium chloride
    (c) iron(III) oxide
    (d) copper(I) hydroxide
    (e) ammonium hydrogen sulfate or ammonium bisulfate
    (f) titanium(II) nitrate
    (g) barium chloride dihydrate
    (h) lithium oxalate
    (i) sodium sulfate decahydrate

55. (a) KCl (b) $CaCO_3$ (c) Cr_2O_3
    (d) $KClO_4$ (e) $NaClO_3$ (f) $FeSO_4{\cdot}7H_2O$

57. (a) Ammonium chlorate is NH_4ClO_3; NH_4Cl is ammonium chloride
    (b) Potassium nitrate is KNO_3; KNO_2 is potassium nitrite.
    (c) Sodium sulfate is Na_2SO_4; $NaSO_4$ is not a correct formula.
    (d) Barium hydroxide is $Ba(OH)_2$; BaOH is not a correct formula.
    (e) Zinc oxalate is ZnC_2O_4; ZnO is zinc oxide.

(f) Manganese(IV) oxide is MnO_2, not MnO_4. MnO_4^- is the permanganate ion.
(g) Strontium chromate is $SrCrO_4$; $Sr(CrO_4)_2$ is not a correct formula.
(h) Copper(II) phosphate is $Cu_3(PO_4)_2$; Cu_2PO_4 is not a correct formula.

59. (a) HI (b) LiOH
(c) HNO_2 (d) $Ba(OH)_2$
(e) hypochlorous acid (f) periodic acid
(g) calcium hydroxide (h) phosphorous acid

61. SeO_3^{2-} is the selenite ion. H_2SeO_3 is selenous acid.
SeO_4^{2-} is the selenate ion. H_2SeO_4 is selenic acid.

63.

(a) 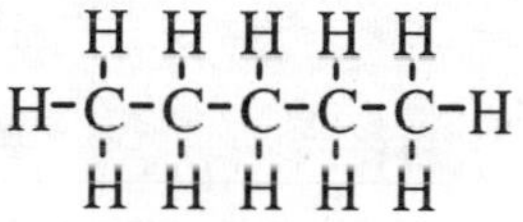
$CH_3(CH_2)_3CH_3$

(b) 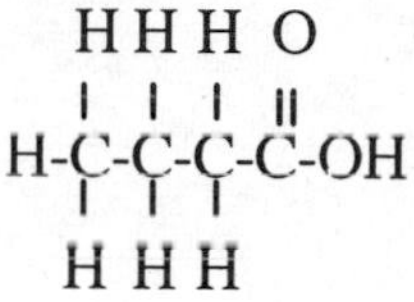
$CH_3(CH_2)_2COOH$

(c) 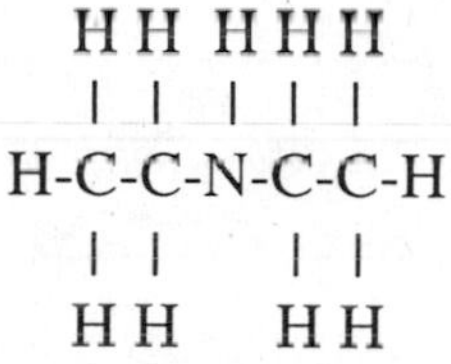
$(CH_3CH_2)_2NH$

(d) 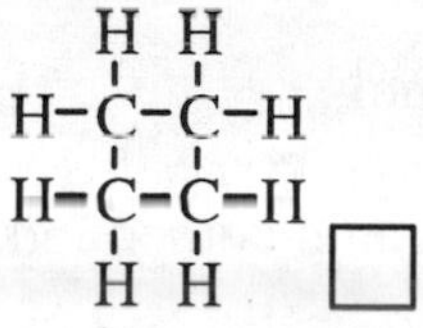

(e) 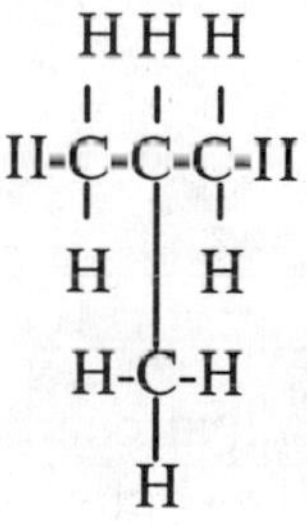
$CH_3CH(CH_3)CH_3$

(f) 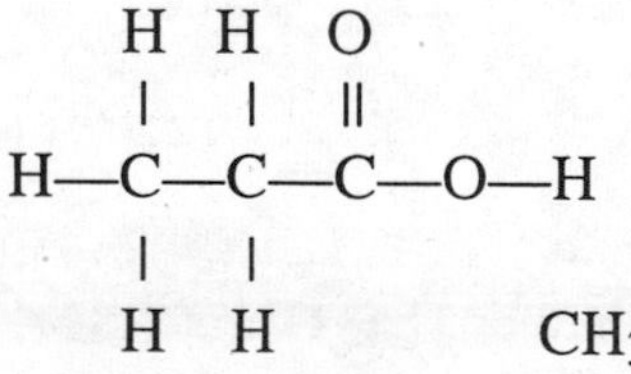
CH_3CH_2COOH

(g) 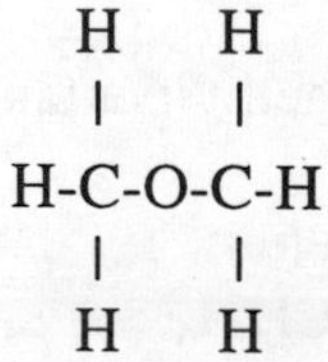
CH_3OCH_3

(h) 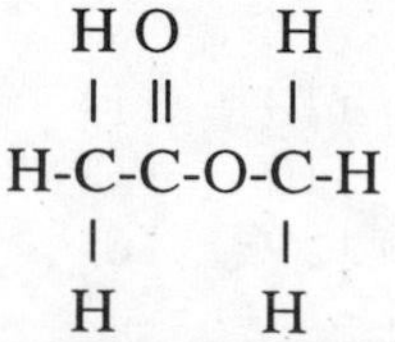
CH_3COOCH_3

(i) 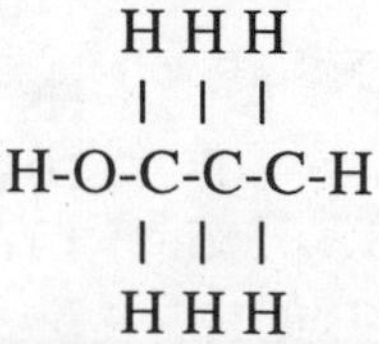
$HOCH_2CH_2CH_3$

(j) 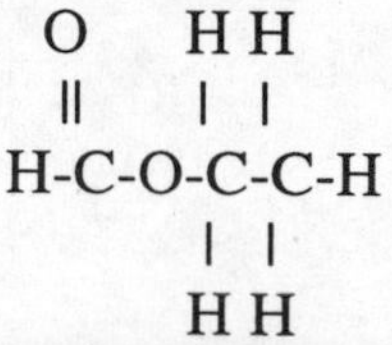
$HCOOCH_2CH_3$

65. (a)

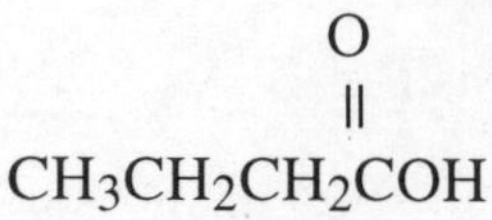

$$\overset{\overset{\displaystyle O}{\|}}{CH_3CH_2CH_2COH}$$

butyric acid

$$\overset{\overset{\displaystyle O}{\|}}{CH_3CH(CH_3)COH}$$

isobutyric acid

(b)

$$\overset{\overset{\displaystyle O}{\|}}{CH_3CH_2CH_2COCH_2CH_2CH_3}$$

propyl butyrate

$$\overset{\overset{\displaystyle O}{\|}}{CH_3CH(CH_3)COCH_2CH_2CH_3}$$

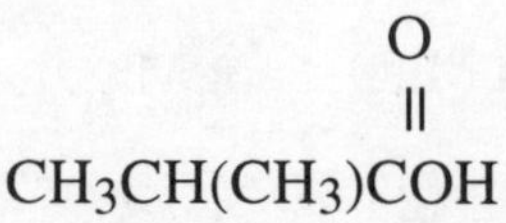

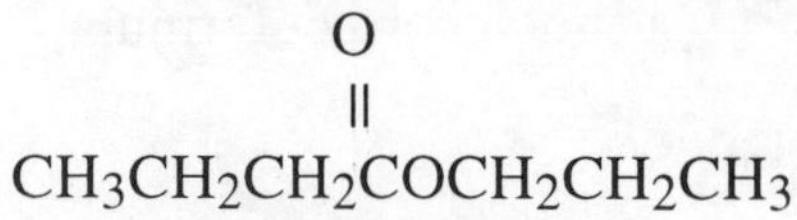

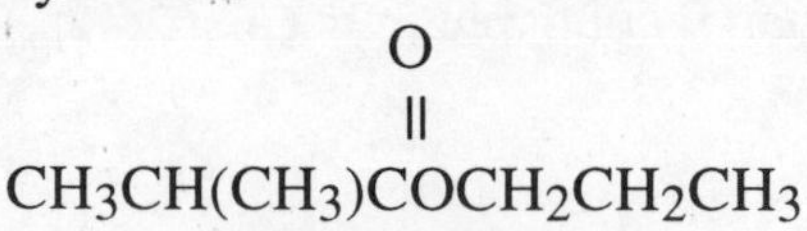

propyl isobutyrate

67. $C_7H_{16}O$

(a) $C_7H_{16}O$ isomer
(b) $C_7H_{16}O$ isomer
(c) identical to b
(d) $C_8H_{18}O$ not isomer

69.

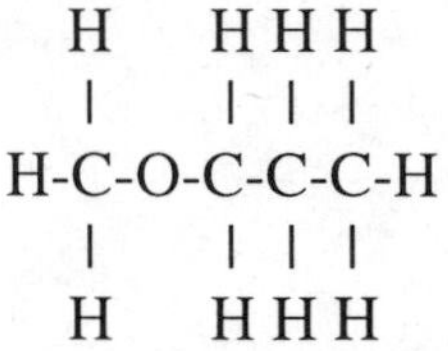

```
  H   H H H
  |   | | |
H-C-O-C-C-C-H
  |   | | |
  H   H H H
```
methyl propyl ether

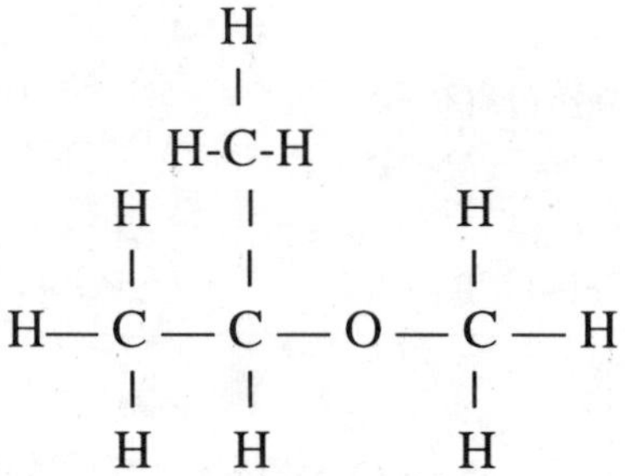

```
          H
          |
        H-C-H
 H      |        H
 |      |        |
H— C — C — O — C — H
 |      |        |
 H      H        H
```
isopropyl methyl ether

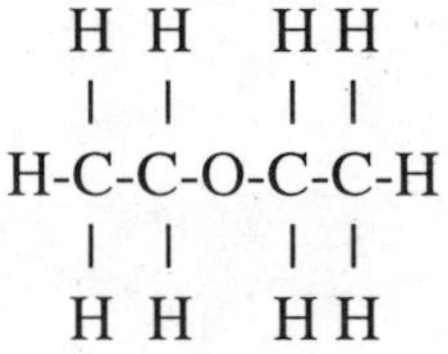

```
H H   H H
| |   | |
H-C-C-O-C-C-H
| |   | |
H H   H H
```
diethyl ether

71. (a) straight-chain alkane hydrocarbon
    (b) alcohol
    (c) hydrocarbon (could be cyclic alkane)
    (d) hydrocarbon
    (e) carboxylic acid
    (f) inorganic compound
    (g) ester
    (h) ether

73. b: 2,2,4-trimethylpentane

Additional Problems

75. $Ge[S(CH_2)_4CH_3]_4$ $1\,Ge + 4\,S + 4 \times (4+1)\,C + \{(4 \times 2) + 3\} \times 4\,H = 69$ atoms

77.

element	Rf	Db	Sg	Bh	Hs	Mt	110	111	112
protons	104	105	106	107	108	109	110	111	112
neutrons	157	157	157	155	157	157	159	161	165

79. Phosphate ion has to have some positive ion with it for it to be a solid. A salt like sodium phosphate would produce the phosphate ion in solution.

81. Except for carbon-12, isotopes don't have integral masses because the proton and neutron are not an exact integral mass. Carbon-12 is defined as 12.000 u.

83. Before reaction:

$$? \text{ g} = 100.0 \text{ mL} \times \frac{1.148\text{g}}{\text{mL}} = 114.8 \text{ g HCl solution}$$

114.8 g HCl solution
+10.00 g $CaCO_3$
124.8 g total

After reaction:

$$? \text{ g} = 2.22 \text{ L} \times \frac{0.0019769 \text{ g}}{\text{mL}} \times \frac{\text{mL}}{10^{-3} \text{ L}} = 4.39 \text{ g } CO_2$$

4.39 g CO_2 gas
+120.40 g solution
124.79 g total

The mass of substances after the reaction equals the mass of substances before the reaction. The law of conservation of mass is confirmed.

85.

	ratio C : sample	ratio H : sample
sample 1	$\frac{0.1141 \text{ g}}{0.2450 \text{ g}} = 0.4657$	$\frac{0.0216 \text{ g}}{0.2450 \text{ g}} = 0.0882$
sample 2	$\frac{0.1400 \text{ g}}{0.3005 \text{ g}} = 0.4659$	$\frac{0.0264 \text{ g}}{0.3005 \text{ g}} = 0.0879$
sample 3	$\frac{0.0639 \text{ g}}{0.1371 \text{ g}} = 0.4661$	$\frac{0.0121 \text{ g}}{0.1371\text{g}} = 0.0883$

The composition of carbon and hydrogen is constant, and the composition of the total oxygen and nitrogen together is consistent, but it is impossible to determine if the ratio of oxygen to nitrogen changes.

87. In a 100-g sample,

$$\frac{96.2 \text{ g Hg}}{3.8 \text{ g O}} = 25{:}1 \qquad \frac{92.6 \text{ g Hg}}{7.4 \text{ g O}} = 12.5{:}1 \qquad \frac{25}{12.5} = 2{:}1$$

Thus, there are two compounds of different proportions.

89. First compound:

$$\frac{77.64 \text{ g Fe}}{22.36 \text{ g O}} = 3.472$$

(a) $\frac{72.36 \text{ g Fe}}{27.64 \text{ g O}} = 2.618$ $\quad \frac{3.472}{2.618} = 1.33 = 4{:}3$, a ratio of small whole numbers.

(b) $\frac{55.28 \text{ g Fe}}{44.72 \text{ g O}} = 1.236$ $\quad \frac{3.472}{1.236} = 2.81$, not a ratio of small whole numbers.

(c) $\frac{50.00 \text{ g Fe}}{50.00 \text{ g O}} = 1.0000$ $\quad \frac{3.472}{1.000} = 3.47$, not a ratio of small whole numbers.

(d) $\frac{32.92 \text{ g Fe}}{67.08 \text{ g O}} = 0.4908$ $\quad \frac{3.472}{0.4908} = 7.07$, not a ratio of small whole numbers.

The probable second compound is compound (a)

91. y = fractional abundance of Mg 25

$23.98504 \times 0.7899 + 24.98584\,y + 25.98259 \times (0.2101 - y) = 24.3050$

$18.9458 + 24.98584\,y + 5.4589 - 25.98259\,y = 24.3050$

$-0.99675\,y = -0.0997$

$y = \dfrac{-0.0997}{-0.99675} = .1000$

$0.2101 - 0.1000 = 0.1101$

Mg 24 78.99% Mg 25 10.00% Mg26 11.01%

93.

	BP °C	increase per CH_2 unit (°C)
C_8H_{18}	125.7	≈ 25
C_9H_{20}	151	≈ 23
$C_{10}H_{22}$	174	≈ 22
$C_{11}H_{24}$	196	≈ 21
$C_{12}H_{26}$	217	≈ 19
$C_{13}H_{28}$	236	≈ 18
$C_{14}H_{30}$	254	≈ 17
$C_{15}H_{32}$	271	≈ 16
$C_{16}H_{34}$	286.8	≈ 15
$C_{17}H_{36}$	301.82	≈ 14
$C_{18}H_{38}$	316.12	≈ 13.6
$C_{19}H_{40}$	329.7	≈ 13
$C_{20}H_{42}$	342.7	≈ 12.6
$C_{21}H_{44}$	355	≈ 12
$C_{22}H_{46}$	367	

$?T = 287\ °C - 147\ °C = 113\ °C$

$\dfrac{113\ °C}{6} = 18.8\ °C$

Each successive step needs to be smaller.

$22 + 21 + 19 + 18 + 17 + 16 = 113$

That would make the BPs for:

$C_{11}H_{24}$	196 °C
$C_{12}H_{26}$	217 °C
$C_{13}H_{28}$	236 °C
$C_{14}H_{30}$	254 °C
$C_{15}H_{32}$	271 °C

With a smaller increase at the higher alkanes,

$C_{21}H_{44}$	355 °C
$C_{22}H_{46}$	367 °C

95. (a) No. One structure is simply a flipped-over version of the other.
(b) Yes. The structures have the same formula (C_9H_{20}), but the positions of the CH_3 group are different.
(c) No. The structures have different formulas.
(d) No. The structures and formulas are identical.

97. (b) methylcyclohexane is a cyclic alkane.

99. (f) lactic acid is $CH_3CH(OH)COOH$

101. The molecular formula of (e) 1,4-butanediamine is $C_4H_{12}N_2$. The empirical formula is C_2H_6N.

103. (a) $(CH_3(CH_2)_{10}COO)NH_4$
(b) $(CH_3(CH_2)_{16}COO)_2Ca$
(c) $(CH_3(CH_2)_7CH{=}CH(CH_2)_7COO)K$

105.
```
C-C-C-O-C-C-C          C-C-C-O-C-C          C-C-O-C-C
                               |              |   |
                               C              C   C
```
There are other ether isomers with the formula $C_6H_{14}O$, such as $CH_3OC_5H_{11}$ (6 isomers) and $C_2H_5OC_4H_9$ (4 isomers). There are also alcohols (21 isomers), such as $C_6H_{13}OH$.

107. Organic food is food grown without manufactured fertilizers or pesticides. Organic fertilizer is made from natural components such as manure or by using a compost pile. Organic medicines are not inorganic. That is, the medicine is a carbon-containing compound. Organic cosmetics are natural products.
(a) Cane sugar is a carbon-containing compound.
(b) *E.* coli is carbon-containing but also comes from living tissue.
(c) Organic cabbage means grown without synthetic fertilizers or pesticides.
(d) Cow manure is from a living creature and contains carbon.
(e) Shampoo is carbon containing.
(f) Cocaine is produced from a living plant and contains carbon.

Apply Your Knowledge

109. (a) The atomic number is 20, that is, calcium. The mass number is 40, so the isotope is ${}^{40}_{20}\text{Ca}$.
(b) $Z + 1.6\,Z = 234$
$2.6\,Z = 234$
$Z = 90$ ${}^{234}_{90}\text{Th}$
(c) $A_1 = 16 \times 7.50 = 120$
$A_2 = 4 \times 12 = 48$
$Z_1 = 2\,Z_2$

$A_1 - Z_1 = 3(A_2 - Z_2)$
$120 - Z_1 = 3(48 - Z_2) = 144 - Z_2$
$120 - 2\,Z_2 = 144 - Z_2$
$Z_2 = 24$ $\qquad Z_1 = 48$

The isotope is $^{120}_{48}Cd$. (Isotope 2 is $^{48}_{24}Cr$.)

(d) $A = (Z + n) \times 2$
Z = 10, 12, 14, 16, 18, 20, 28, 30, 32, 34, 36, 66, 68, 70, 73, 104, 106, 108, 170, 172
$\#e^- = Z - n$ = 2, 10, 18, 36, 54, 86, 12, 20, 38, 56, 88, 28, 46, 64, 96, 54, 72, 104, 90, 122, 140
for Z 10, 18, 36 are noble gases and could not be an ion.
12, 14, 16, 20 have A = 2Z not 2(Z + n)
170, 172, 104, 106, 108 are too large for common ions.
A is too large for 66, 68, 70, 72.
The only Z – n that is close to 28, 30, 32, or 34 is 28, or if Z – n = 28, then Z can be 30(n = 2) or 32(n = 4). If Z = 30, A = 2 × 32 = 64. One choice is $^{64}_{30}Zn^{2+}$. If Z = 32 , A = 2 × 36 = 72. Another choice is $^{72}_{32}Ge^{4+}$. The 4+ ion is probably less likely than a 2+ ion.

111. (a) There are two stable isotopes of bromine, masses 79 and 81. The peaks at 79 and 81 are monoatomic ions of bromine. The peak at 158 is diatomic bromine consisting of two atoms of mass 79. The peak at 160 is diatomic bromine consisting of atom of mass 79 and one atom of 81. The peak at 162 is diatomic bromine consisting of two atoms of mass 81. There are four ways to make diatomic bromine from two atoms. Two of these ways are one atom of mass 81 and one atom of mass 79. Thus, the mass peak at 160 is twice the size of 158 and 160.

(b) The peak at 86 is that of the molecular ion. Breakage of a C–C bond between the first and second C atoms of 2-methylpentane yields the ion with m/z =71; between the second and third C atoms, the ion with m/z = 43(the most abundant); and between the third and fourth C atoms, the ion with m/z =57. In 3-methylpentane, breakage of a C–C bond between the second and third C atoms and between the third and fourth C atoms both yield the ion with m/z= 57, producing the highest peak.

Ions from 2-methylpentane:

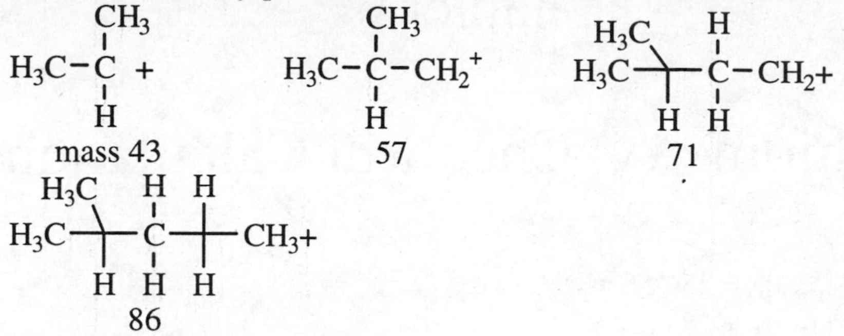

Ions from 3-methylpentane:

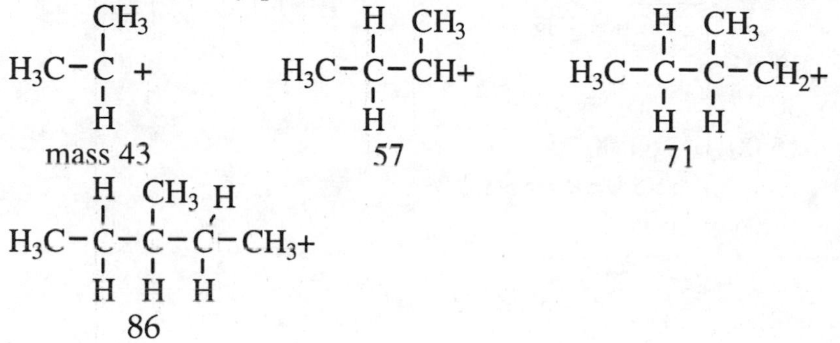

(c) BrCl can be made four ways from the two chlorine isotopes (35 + 37) and the two bromine isotopes (79 + 81). The mass numbers are 114, 116, 116, and 118. The weighted average molecular mass of BrCl from problem data-

for 79 and 35	$(0.5069 \times 0.7577) \times (78.9183 + 34.9689)\ u = 43.74\ u$
for 79 and 37	$(0.5069 \times 0.2423) \times (78.9183 + 36.9659)\ u = 14.23\ u$
for 81 and 35	$(0.4931 \times 0.7577) \times (80.9163 + 34.9689)\ u = 43.30\ u$
for 81 and 37	$(0.4931 \times 0.2423) \times (80.9163 + 36.9659)\ u = \underline{14.08\ u}$
	$115.35\ u$

The molecular mass of BrCl from periodic table-
$79.904\ u + 35.4527\ u = 115.36\ u$

Chapter 3

Stoichiometry: Chemical Calculations

Exercises

3.1A (a)	S	$8 \times 32.07 = 256.56$ u => 257 u	
(b)	N	$2 \times 14.01 = 28.02$	
	H	$4 \times 1.01 = \underline{4.04}$	
		32.06 u => 32.1	
(c)	H	$3 \times 1.01 = 3.03$	
	P	$1 \times 30.97 = 30.97$	
	O	$4 \times 16.00 = \underline{64.00}$	
		98.00 u => 98.0	
(d)	C	$5 \times 12.01 = 60.05$	
	H	$12 \times 1.01 = \underline{12.12}$	
		72.17 u => 72.2 u	

3.1B (a)	PF_5	P	$1 \times 30.9738 = 30.9738$
		F	$5 \times 18.9984 = \underline{94.9920}$
			125.9648 u => 125.97 u
(b)	N_2O_4	N	$2 \times 14.0067 = 28.0134$
		O	$4 \times 15.9994 = \underline{63.9976}$
			92.0110 u => 92.011 u
(c)	C_6H_{14}	C	$6 \times 12.011 = 72.066$
		H	$14 \times 1.0079 = \underline{14.1106}$
			86.1766 u => 86.177 u
(d)	CH_3CH_2COOH	C	$3 \times 12.011 = 36.033$
		H	$6 \times 1.008 = 6.048$
		O	$2 \times 15.999 = \underline{31.998}$
			74.079 u

3.2A (a)	Li	$2 \times 6.94 = 13.88$	
	O	$1 \times 16.00 = \underline{16.00}$	
		22.88 u => 29.9 u	
(b)	Mg	$1 \times 24.3 = 24.3$	
	N	$2 \times 14.0 = 28.0$	
	O	$6 \times 16.0 = \underline{96.0}$	
		148.3 u = 148	
(c)	Ca	$1 \times 40 = 40$	
	H	$4 \times 1.0 = 4$	
	P	$2 \times 31 = 62$	
	O	$8 \times 16 = \underline{128}$	
		234 u	

(d) K $2 \times 39 = 78$
Sb $1 \times 122 = 122$
F $5 \times 19 = \underline{95}$
295 u

3.2B (a) $NaHSO_3$
Na $1 \times 22.990 = 22.990$
H $1 \times 1.008 = 1.008$
S $1 \times 32.066 = 32.066$
O $3 \times 15.999 = \underline{47.997}$
104.061 u => 104.06 u

(b) NH_4ClO_4
N $1 \times 14.007 = 14.007$
H $4 \times 1.008 = 4.032$
Cl $1 \times 35.453 = 35.453$
O $4 \times 15.999 = \underline{63.996}$
117.488 => 117.49 u

(c) $Cr_2(SO_4)_3$
Cr $2 \times 51.996 = 103.992$
S $3 \times 32.066 = 96.198$
O $12 \times 15.999 = \underline{191.988}$
392.178 => 392.18 u

(d) $CuSO_4{\cdot}5H_2O$
Cu $1 \times 63.546 = 63.546$
S $1 \times 32.066 = 32.066$
O $9 \times 15.9994 = 143.995$
H $10 \times 1.0079 = \underline{10.079}$
249.686 u => 249.69 u

3.3A (a) C $3 \times 12.01 = 36.03$
H $8 \times 1.01 = \underline{8.08}$
44.11 u

$$? \text{ g } C_3H_8 = 0.155 \text{ mol } C_3H_8 \times \frac{44.11 \text{ g } C_3H_8}{\text{mol } C_3H_8} = 6.84 \text{ g } C_3H_8$$

(b) C $2 \times 12.01 = 24.02$
H $6 \times 1.01 = \underline{6.06}$
30.08 u

$$? \text{ mg } C_2H_6 = 2.45 \times 10^{-4} \text{ mol } C_2H_6 \times \frac{30.08 \text{ g } C_2H_6}{\text{mol } C_2H_6} \times \frac{\text{mg}}{10^{-3} \text{ g}} = 7.37 \text{ mg } C_2H_6$$

(c) C $4 \times 12.01 = 48.04$
H $10 \times 1.008 = \underline{10.08}$
58.12 u

$$? \text{ mol } C_4H_{10} = 165 \text{ kg } C_4H_{10} \times \frac{10^3 \text{ g}}{\text{kg}} \times \frac{\text{mol } C_4H_{10}}{58.12 \text{ g } C_4H_{10}}$$

$$= 2.84 \times 10^3 \text{ mol } C_4H_{10}$$

(d) H $3 \times 1.01 = 3.03$
P $1 \times 30.97 = 30.97$
O $4 \times 16.00 = \underline{64.00}$
98.00 u

$$? \text{ mol } H_3PO_4 = 76.0 \text{ mg} \times \frac{10^{-3}\text{ g}}{\text{mg}} \times \frac{\text{mol } H_3PO_4}{98.00 \text{ g } H_3PO_4} = 7.76 \times 10^{-4} \text{ mol } H_3PO_4$$

3.3B (a) $$? \text{ g Al} = 5.5 \text{ cm} \times 5.5 \text{ cm} \times 5.5 \text{ cm} \times \frac{2.70 \text{ g}}{\text{cm}^3} = 4.49 \times 10^2 \text{ g Al}$$

$$? \text{ mol Al} = 4.49 \times 10^2 \text{ g Al} \times \frac{\text{mol Al}}{27.0 \text{ g Al}} = 17 \text{ mol Al}$$

(b) CCl_4 C $1 \times 12.01 = 12.01$
Cl $4 \times 35.45 = \underline{141.80}$
153.81

$$? \text{ g } CCl_4 = 1.38 \text{ mol} \times \frac{153.8 \text{ g } CCl_4}{\text{mol } CCl_4} = 212 \text{ g } CCl_4$$

$$? \text{ mL } CCl_4 = 212 \text{ g } CCl_4 \times \frac{\text{mL}}{1.59 \text{ g}} = 133 \text{ mL } CCl_4$$

3.4A (a) $$? \frac{\text{g}}{\text{Bi atom}} = \frac{208.980 \text{ g}}{\text{mol}} \times \frac{\text{mol Bi atoms}}{6.022 \times 10^{23} \text{ Bi atoms}} = \frac{3.470 \times 10^{-22} \text{ g}}{\text{Bi atom}}$$

(b) C $3 \times 12.011 = 36.033$
H $8 \times 1.008 = 8.064$
O $3 \times 15.999 = \underline{47.997}$
92.094

$$? \text{ mg } C_3H_8O_3 \text{ molecule} = \frac{92.094 \text{ g}}{\text{mol}} \times \frac{\text{mol } C_3H_8O_3}{6.022 \times 10^{23}\ C_3H_8O_3}$$
$$= 1.529 \times 10^{-22} \text{ mg } C_3H_8O_3 \text{ molecules}$$

(c) $$? \text{ mol } N_2 = 0.0100 \text{ g } N_2 \times \frac{\text{mol } N_2}{28.02 \text{ g } N_2} \times \frac{6.022 \times 10^{23} \text{ molecules } N_2}{\text{mol } N_2}$$
$$= 2.15 \times 10^{20} \text{ molecules } N_2$$

(d) C $12 \times 12.01 = 144.12$
H $22 \times 1.01 = 22.22$
O $\underline{11} \times 16.00 = \underline{176.00}$
45 342.34 u

$$? \text{ atoms} = 215 \text{ g } C_{12}H_{22}O_{11} \times \frac{\text{mol } C_{12}H_{22}O_{11}}{342.34 \text{ g } C_{12}H_{22}O_{11}}$$
$$\times \frac{6.022 \times 10^{23} \text{ molecules}}{\text{mol}} \times \frac{45 \text{ atoms}}{\text{molecule}} = 1.70 \times 10^{25} \text{ atoms}$$

3.4B (a) $? \text{ molecules} = 125 \text{ mL Br}_2 \times \frac{3.12 \text{ g}}{\text{mL}} \times \frac{\text{mol}}{159.8 \text{ g}} \times \frac{6.022 \times 10^{23} \text{ molecules}}{\text{mol}}$

$= 1.47 \times 10^{24} \text{ molecules Br}_2$

(b) C_2H_5OH

C	$2 \times 12.01 =$	24.02
H	$6 \times 1.01 =$	6.06
O	$1 \times 16.00 =$	16.00
		46.08 u

$$? \text{ L} = 1.00 \times 10^{25} \text{ molecules} \times \frac{\text{mol}}{6.022 \times 10^{23} \text{ molecules}} \times \frac{46.08 \text{ g}}{\text{mol}} \times \frac{\text{mL}}{0.789 \text{ g}} \times \frac{10^{-3} \text{ L}}{\text{mL}} = 0.970 \text{ L}$$

3.5 6×10^{23} atoms is 24 g. Thus, $1 \times 10^{23} = 4.0$ g.

$\frac{24.31}{6.022} = 4.04$ actual.

3.6A (a) $(NH_4)_2SO_4$

$2 \times 14.01 + 8 \times 1.008 + 1 \times 32.07 + 4 \times 16.00 = 132.15$ u

$$\%N = \frac{28.02 \text{ g N}}{132.15 \text{ g (NH}_4)_2\text{SO}_4} \times 100\% = 21.20\% \text{ N}$$

$$\%H = \frac{8.064 \text{ g H}}{132.15 \text{ g (NH}_4)_2\text{SO}_4} \times 100\% = 6.10\% \text{ H}$$

$$\%S = \frac{32.07 \text{ g S}}{132.15 \text{ g (NH}_4)_2\text{SO}_4} \times 100\% = 24.27\% \text{ S}$$

$$\%O = \frac{64.00 \text{ g O}}{132.15 \text{ g (NH}_4)_2\text{SO}_4} \times 100\% = \underline{48.43\% \text{ O}}$$

100.00%

(b) $CO(NH_2)_2$ $1 \times 12.01 + 1 \times 16.00 + 2 \times 14.01 + 4 \times 1.008 = 60.06$ u

$$\%N = \frac{28.02 \text{ g N}}{60.06 \text{ g CO(NH}_2)_2} \times 100\% = 46.65\% \text{ N}$$

$$\%C = \frac{12.01 \text{ g C}}{60.06 \text{ g CO(NH}_2)_2} \times 100\% = 20.00\% \text{ C}$$

$$\%O = \frac{16.00 \text{ g O}}{60.06 \text{ g CO(NH}_2)_2} \times 100\% = 26.64\% \text{ O}$$

$$\%H = \frac{4.032 \text{ g H}}{60.06 \text{ g CO(NH}_2)_2} \times 100\% = \underline{6.71\% \text{ H}}$$

100.00%

$$\%N = \frac{28.02 \text{ g N}}{2 \text{ x } 14.01 + 4 \text{ x } 1.008 + 3 \text{ x } 16.00} = 35.00\% \text{ N in NH}_4\text{NO}_3$$

Urea $CO(NH_2)_2$ has the highest %N.

3.6B (a) $Ca(NO_3)_2$

$$\%\,N = \frac{(2\times 14.02)\text{ g N}}{40.08+2\times 14.01+6\times 16.00}\times 100\% = 17.09\%\text{ N}$$

(b) $(NH_4)_2S$

$$\%\,N = \frac{(2\text{ x }14.02)\text{ g N}}{2\text{ x }14.02 + 8\text{ x }1.008 + 32.07}\times 100\% = 41.13\%\text{ N}$$

(c) $N(CH_2CH_2OH)_3$

$(1 \times 14.01) + (6 \times 12.01) + (3 \times 16.00) + (15 \times 1.01) = 149.2$

$$\%N = \frac{14.01}{149.2}\times 100\% = 9.38\%\text{ N}$$

3.7A $NaHCO_3$ $1 \times 22.99 + 1 \times 1.008 + 1 \times 12.01 + 3 \times 16.00 = 84.01$ u

$$?\text{ mg Na} = 5.00\text{g NaHCO}_3 \times \frac{\text{mol NaHCO}_3}{84.01\text{ g NaHCO}_3} \times \frac{\text{mol Na}}{\text{mol NaHCO}_3} \times \frac{22.99\text{ g Na}}{\text{mol Na}}$$

$$\times \frac{\text{mg}}{10^{-3}\text{ g}} = 1.37 \times 10^3\text{ mg Na}$$

3.7B NH_4NO_3

$$?\frac{\text{molar mass N}}{\text{molar mass NH}_4\text{NO}_3} = \frac{2\text{ x }14.007}{2\text{ x }14.007 + 3\text{ x }15.999 + 4\text{ x }1.0079} = 0.3500$$

$(NH_4)_2SO_4$

$$?\frac{\text{molar mass N}}{\text{molar mass (NH}_4)_2\text{SO}_4} = \frac{2\text{ x }14.007}{2\text{ x }14.007 + 4\text{ x }15.999 + 8\text{ x }1.0079 + 1\text{ x }32.066}$$

$$= 0.2120$$

$$?\text{ g N} = 1.00\text{ kg} \times \frac{10^3\text{ g}}{\text{kg}} \times \left(\frac{(0.3500\text{ x }12.5)\text{ g N} + (0.2120\text{ x }35.3)\text{ g N}}{100\text{ g FERT}}\right) = 119\text{ g N}$$

3.8A LiH_2PO_4 $?\%\,P = \frac{31\text{ x }100\%}{7+2+31+64} = \frac{3100}{104} \approx 31\%$ Actual 30%

$Ca(H_2PO_4)_2$ $?\%\,P = \frac{2\text{ x }31\text{ x }100\%}{40+4+62+128} = \frac{6200}{234} \approx 25\%$ Actual 26%

$(NH_4)_2HPO_4$ $?\%\,P = \frac{31\text{ x }100\%}{28+8+1+31+64} = \frac{3100}{132} \approx 23\%$ Actual 23%

Once the equations are written out, it is easy to see that the denominator is the smallest for LiH_2PO_4— that is 7 + 2 is less than 28 + 8 + 1 and less than 1/2 of 40 + 4. So the largest percentage is LiH_2PO_4.

3.8B methanol CH_3OH; acetic acid CH_3COOH; butane C_4H_{10}; octane C_8H_{18}
Methane and acetic acid have a lower percent of carbon, because an oxygen is included in the formula. Octane has a greater carbon mass percent, because octane has 9 hydrogens per 4 carbons compared to 10 hydrogens per 4 carbons for butane.

actual C_8H_{16} $\frac{8\text{ x }12.01}{8\text{ x }12.01 + 16\text{ x }1.008} \times 100\% = 85.63\%\text{ C}$

C_4H_{10} $\quad \dfrac{4 \text{ x } 12.01}{4 \text{ x } 12.01 + 10 \text{ x } 1.008} \times 100\% = 82.66\%\ C$

CH_3COOH $\quad \dfrac{2 \text{ x } 12.01}{2 \text{ x } 12.01 + 2 \text{ x } 16.00 + 4 \text{ x } 1.008} \times 100\% = 40.05\%\ C$

CH_3OH $\quad \dfrac{12.01}{12.01 + 16.00 + 4 \text{ x } 1.008} \times 100\% = 37.48\%\ C$

3.9A $? \text{ mol C} = 71.95 \text{ g C} \times \dfrac{\text{mol C}}{12.01 \text{ g C}} = 5.991 \text{ mol C} \times \dfrac{1}{0.998 \text{ mol O}} = \dfrac{6.00 \text{ mol C}}{\text{mol O}}$

$? \text{ mol H} = 12.08 \text{ g H} \times \dfrac{\text{mol H}}{1.008 \text{ g H}} = 11.98 \text{ mol H} \times \dfrac{1}{0.998 \text{ mol O}} = \dfrac{12.00 \text{ mol H}}{\text{mol O}}$

$? \text{ mol O} = 15.97 \text{ g O} \times \dfrac{\text{mol O}}{16.00 \text{ g O}} = 0.998 \text{ mol O} \times \dfrac{1}{0.998 \text{ mol O}} = \dfrac{1.00 \text{ mol O}}{\text{mol O}}$

$C_6H_{12}O$

3.9B $? \text{ mol C} = 51.70 \text{ g C} \times \dfrac{\text{mol C}}{12.01 \text{ g C}} = 4.305 \text{ mol C} \times \dfrac{1}{0.8608 \text{ mol N}} = \dfrac{5.00 \text{ mol C}}{\text{mol N}}$

$? \text{ mol H} = 8.68 \text{ g H} \times \dfrac{\text{mol H}}{1.008 \text{ g H}} = 8.61 \text{ mol H} \times \dfrac{1}{0.8608 \text{ mol N}} = \dfrac{10.0 \text{ mol H}}{\text{mol N}}$

$? \text{ mol N} = 12.06 \text{ g N} \times \dfrac{\text{mol N}}{14.01 \text{ g N}} = 0.8608 \text{ mol N} \times \dfrac{1}{0.8608 \text{ mol N}} = \dfrac{1.00 \text{ mol N}}{\text{mol N}}$

$? \text{ mol O} = 27.55 \text{ g O} \times \dfrac{\text{mol O}}{16.00 \text{ g O}} = 1.722 \text{ mol O} \times \dfrac{1}{0.8608 \text{ mol N}} = \dfrac{2.00 \text{ mol O}}{\text{mol N}}$

$C_5H_{10}NO_2$

3.10A $? \text{ mol C} = 94.34 \text{ g C} \times \dfrac{\text{mol C}}{12.01 \text{ g C}} = 7.855 \text{ mol C} \times \dfrac{1}{5.615 \text{ mol H}} = \dfrac{1.399 \text{ mol C}}{\text{mol H}}$

$? \text{ mol H} = 5.66 \text{ g H} \times \dfrac{\text{mol H}}{1.008 \text{ g H}} = 5.615 \text{ mol H} \times \dfrac{1}{5.615 \text{ mol H}} = \dfrac{1.00 \text{ mol H}}{\text{mol H}}$

$C_{(1.4 \times 5)}H_{(1 \times 5)} = C_7H_5$

3.10B(a) $? \text{ mol Fe} = 72 \text{ g Fe} \times \dfrac{\text{mol Fe}}{55.85 \text{ g Fe}} = 1.3 \text{ mol Fe} \times \dfrac{1}{1.3 \text{ mol Fe}} = 1 \dfrac{\text{mol Fe}}{\text{mol Fe}}$

$? \text{ mol O} = 27.6 \text{ g O} \times \dfrac{\text{mol O}}{16.00 \text{ g O}} = 1.73 \text{ mol O} \times \dfrac{1}{1.3 \text{ mol Fe}} = 1.3 \dfrac{\text{mol O}}{\text{mol Fe}}$

$Fe_{(1\times3)}O_{(1.3 \times 3)} = Fe_3O_4$

(b) $? \text{ mol C} = 9.93 \text{ g C} \times \dfrac{\text{mol C}}{12.01 \text{ g C}} = 0.827 \text{ mol C} \times \dfrac{1}{0.827 \text{ mol C}} = 1 \dfrac{\text{mol C}}{\text{mol C}}$

$? \text{ mol Cl} = 58.6 \text{ g Cl} \times \dfrac{\text{mol Cl}}{35.45 \text{ g Cl}} = 1.65 \text{ mol Cl} \times \dfrac{1}{0.827 \text{ mol C}} = \dfrac{2 \text{ mol Cl}}{\text{mol C}}$

$? \text{ mol F} = 31.4 \text{ g F} \times \dfrac{\text{mol F}}{19.00 \text{ g F}} = 1.65 \text{ mol F} \times \dfrac{1}{0.827 \text{ mol C}} = \dfrac{2 \text{ mol F}}{\text{mol C}}$

CCl_2F_2

(c) $? \text{ mol C} = 37.01 \text{ g C} \times \dfrac{\text{mol C}}{12.01 \text{ g C}} = 3.082 \text{ mol C} \times \dfrac{1}{1.320 \text{ mol N}} = \dfrac{2.335 \text{ molC}}{\text{molN}}$

$$? \text{ mol H} = 2.22 \text{ g H} \times \frac{\text{mol H}}{1.008 \text{ g H}} = 2.202 \text{ mol H} \times \frac{1}{1.320 \text{ mol N}} = \frac{1.667 \text{ molH}}{\text{molN}}$$

$$? \text{ mol N} = 18.50 \text{ g N} \times \frac{\text{mol N}}{14.01 \text{ g N}} = 1.320 \text{ mol N} \times \frac{1}{1.320 \text{ mol N}} = \frac{1.000 \text{ molN}}{\text{molN}}$$

$$? \text{ mol O} = 42.27 \text{ g O} \times \frac{\text{mol O}}{16.00 \text{ g O}} = 2.642 \text{ mol O} \times \frac{1}{1.320 \text{ mol N}} = \frac{2.000 \text{ molO}}{\text{molN}}$$

$C_{2.335}H_{1.667}NO_2$ (multiplication by 3 to obtain whole numbers) $C_7H_5N_3O_6$

3.11A CH_2 $1 \times 12.01 + 2 \times 1.008 = 14.0$ u

ethylene $\frac{28.0 \text{ u}}{\text{molecule}} \times \frac{\text{formula}}{14.0 \text{ u}} = \frac{2 \text{ formula}}{\text{molecule}}$

C_2H_4

cyclohexane $\frac{84.0 \text{ u}}{\text{molecule}} \times \frac{\text{formula}}{14.0 \text{ u}} = \frac{6 \text{ formula}}{\text{molecule}}$

C_6H_{12}

1-pentene $\frac{70.0 \text{ u}}{\text{molecule}} \times \frac{\text{formula}}{14.0 \text{ u}} = \frac{5 \text{ formula}}{\text{molecule}}$

C_5H_{10}

3.11B (a) P_2O_3 (b) C_2H_3 (c) $C_3H_4O_3$ (d) $C_2H_4O_3$ (e) CH_3O (f) $CuCO_2$

3.12A (a) $? \text{ g C} = 0.561 \text{ g } CO_2 \times \frac{12.01 \text{ g C}}{44.01 \text{ g } CO_2} = 0.153 \text{ g C}$

$$? \text{ g H} = 0.306 \text{ g } H_2O \times \frac{2 \text{ x } 1.0079 \text{ g H}}{18.015 \text{ g } H_2O} = 0.0342 \text{ g H}$$

$$?\% \text{ C} = \frac{0.153 \text{ g C}}{0.255 \text{ g sample}} \times 100\% = 60.0\% \text{ C}$$

$$?\% \text{ H} = \frac{0.0342 \text{ g H}}{0.255 \text{ g sample}} \times 100\% = 13.4\% \text{ H}$$

$?\% \text{ O} = 100.0\% - 60.0\% \text{ C} - 13.4\% \text{ H} = 26.6\% \text{ O}$

(b) $? \text{ mol C} = 60.0 \text{ g C} \times \frac{\text{mol C}}{12.011 \text{ g C}} = 5.00 \text{ mol C} \times \frac{1}{1.66 \text{ mol O}} = \frac{3.01 \text{ mol C}}{\text{mol C}}$

$$? \text{ mol H} = 13.4 \text{ g H} \times \frac{\text{mol H}}{1.0079 \text{ g H}} = 13.3 \text{ mol H} \times \frac{1}{1.66 \text{ mol O}} = \frac{8.01 \text{ mol H}}{\text{mol O}}$$

$$? \text{ mol O} = 26.6 \text{ g O} \times \frac{\text{mol O}}{15.999 \text{ g O}} = 1.66 \text{ mol O} \times \frac{1}{1.66 \text{ mol O}} = \frac{1.00 \text{ mol O}}{\text{mol O}}$$

C_3H_8O

3.12B (a) $?\% \text{ C} = 1.0666 \text{ g } CO_2 \times \frac{12.011 \text{ g C}}{44.010 \text{ g } CO_2} \times \frac{100\%}{0.3629 \text{ g sample}} = 80.213\% \text{ C}$

$$?\% \text{ H} = 0.3120 \text{ g } H_2O \times \frac{2 \text{ x } 1.0079 \text{ g H}}{18.015 \text{ g } H_2O} \times \frac{100\%}{0.3629 \text{ g sample}} = 9.620\% \text{ H}$$

$100.00\% - 80.213\% \text{ C} - 9.620\% \text{ H} = 10.167\% \text{ O}$

(b) $? \text{ mol C} = 80.213 \text{ g C} \times \frac{\text{mol C}}{12.011 \text{ g C}} = 6.681 \text{ mol C} \times \frac{1}{0.6355 \text{ mol O}} = \frac{10.51 \text{ mol C}}{\text{mol C}}$

$? \text{ mol H} = 9.620 \text{ g H} \times \frac{\text{mol H}}{1.0079 \text{ g H}} = 9.545 \text{ mol H} \times \frac{1}{0.6355 \text{ mol O}} = \frac{15.02 \text{ mol H}}{\text{mol O}}$

$? \text{ mol O} = 10.14 \text{ g O} \times \frac{\text{mol O}}{15.999 \text{ g O}} = 0.6355 \text{ mol O} \times \frac{1}{0.6355 \text{ mol O}} = \frac{1.00 \text{ mol O}}{\text{mol O}}$

$C_{10.54}H_{15.02}O_{1.00} \Rightarrow C_{21}H_{30}O_2$

3.13 (a) $SiCl_4 + 2\ H_2O \rightarrow SiO_2 + 4\ HCl$
(b) $PCl_5 + 4\ H_2O \rightarrow H_3PO_4 + 5\ HCl$
(c) $6\ CaO + P_4O_{10} \rightarrow 2\ Ca_3(PO_4)_2$

3.14 (a) $2\ C_4H_{10} + 13\ O_2 \rightarrow 8\ CO_2 + 10\ H_2O$
(b) $C_5H_{12}O_2 + 7\ O_2 \rightarrow 5\ CO_2 + 6\ H_2O$

3.15 (a) $FeCl_3 + 3NaOH \rightarrow Fe(OH)_3 + 3\ NaCl$
(b) $3\ Ba(NO_3)_2 + Al_2(SO_4)_3 \rightarrow 3\ BaSO_4 + 2\ Al(NO_3)_3$
(c) $3\ Ca(OH)_2 + 2\ H_3PO_4 \rightarrow Ca_3(PO_4)_2 + 6\ H_2O$

3.16 $? \text{ mol C} = 47.99 \text{ g C} \times \frac{\text{mol C}}{12.011 \text{ g C}} = 3.996 \text{ mol C} \times \frac{1}{2.664 \text{ mol O}} = \frac{1.500 \text{ mol C}}{\text{mol O}}$

$? \text{ mol H} = 9.40 \text{ g H} \times \frac{\text{mol H}}{1.008 \text{ g H}} = 9.325 \text{ mol H} \times \frac{1}{2.664 \text{ mol O}} = \frac{3.500 \text{ mol H}}{\text{mol O}}$

$? \text{ mol O} = 42.62 \text{ g O} \times \frac{\text{mol O}}{15.996 \text{ g O}} = 2.664 \text{ mol O} \times \frac{1}{2.664 \text{ mol O}} = \frac{1.000 \text{ mol O}}{\text{mol O}}$

$C_{(1.5\times2)}H_{(3.5\times2)}O_{(1.0\times2)} \Rightarrow C_3H_7O_2$

formula mass = $(3 \times 12.01 \text{ u}) + (7 \times 1.008 \text{ u}) + (2 \times 16.00 \text{ u}) = 75.09 \text{ u}$

$\frac{? \text{ formula units}}{\text{molecule}} = \frac{150.2 \text{ u}}{\text{molecule}} \times \frac{\text{formula units}}{75.09 \text{ u}} = \frac{2 \text{ formula units}}{\text{molecule}}$ $C_6H_{14}O_4$

$2\ C_6H_{14}O_4 + 15\ O_2 \rightarrow 12\ CO_2 + 14\ H_2O$

3.17 (a) $C_3H_8 + 5\ O_2 \rightarrow 3\ CO_2 + 4\ H_2O$

$? \text{ mol } CO_2 = 0.529 \text{ mol } C_3H_8 \times \frac{3 \text{ mol } CO_2}{\text{mol } C_3H_8} = 1.59 \text{ mol } CO_2$

(b) $? \text{ mol } H_2O = 76.2 \text{ mol } C_3H_8 \times \frac{4 \text{ mol } H_2O}{\text{mol } C_3H_8} = 305 \text{ mol } H_2O$

(c) $? \, CO_2 = 1.010 \text{ mol } O_2 \times \frac{3 \text{ mol } CO_2}{5 \text{ mol } O_2} = 0.6060 \text{ mol } CO_2$

3.18A $2\,Mg\,(s) + TiCl_4 \rightarrow 2\,MgCl_2 + Ti$

$$? \text{ g Mg} = 83.6 \text{ g } TiCl_4 \times \frac{\text{mol } TiCl_4}{189.68 \text{ g } TiCl_4} \times \frac{2 \text{ mol Mg}}{\text{mol } TiCl_4} \times \frac{24.31 \text{ g Mg}}{\text{mol Mg}}$$
$$= 21.4 \text{ g Mg}$$

3.18B $2\,NH_4NO_3 \rightarrow 2\,N_2 + O_2 + 4\,H_2O$

$$? \text{ g } N_2 = 75.5 \text{ g } NH_4NO_3 \times \frac{\text{mol } NH_4NO_3}{80.06 \text{ g } NH_4NO_3} \times \frac{2 \text{ mol } N_2}{2 \text{ mol } NH_4NO_3}$$
$$\times \frac{28.02 \text{ g } N_2}{\text{mol } N_2} = 26.4 \text{ g } N_2$$

$$? \text{ g } O_2 = 75.5 \text{ g } NH_4NO_3 \times \frac{\text{mol } NH_4NO_3}{80.06 \text{ g } NH_4NO_3} \times \frac{1 \text{ mol } O_2}{2 \text{ mol } NH_4NO_3} \times \frac{32.00 \text{ g } O_2}{\text{mol } O_2}$$
$$= 15.1 \text{ g } O_2$$

3.19 $2\,C_8H_{18} + 25\,O_2 \rightarrow 16\,CO_2 + 18\,H_2O$

$$? \text{ mL} = 775 \text{ mL } C_8H_{18} \times \frac{0.7025 \text{ g } C_8H_{18}}{\text{mL } C_8H_{18}} \times \frac{\text{mol } C_8H_{18}}{114.22 \text{ g } C_8H_{18}} \times \frac{18 \text{ mol } H_2O}{2 \text{ mol } C_8H_{18}}$$
$$\times \frac{18.015 \text{ g } H_2O}{\text{mol } H_2O} \times \frac{\text{mL}}{0.9982 \text{ g}} = 774 \text{ mL}$$

3.20A $? \text{ g } H_2S = 10.2 \text{ g HCl} \times \frac{\text{mol HCl}}{36.46 \text{ g HCl}} \times \frac{\text{mol } H_2S}{2 \text{ mol HCl}} \times \frac{34.08 \text{ g } H_2S}{\text{mol } H_2S} = 4.77 \text{g } H_2S$

$$? \text{ g } H_2S = 13.2 \text{ g FeS} \times \frac{\text{mol FeS}}{87.92 \text{ g FeS}} \times \frac{\text{mol } H_2S}{\text{mol FeS}} \times \frac{34.08 \text{ g } H_2S}{\text{mol } H_2S} = 5.12 \text{ g } H_2S$$

4.77 g H_2S is formed.

$$4.77 \text{ g } H_2S \times \frac{\text{mol } H_2S}{34.08 \text{ g } H_2S} \times \frac{\text{mol FeS}}{\text{mol } H_2S} \times \frac{87.92 \text{ g FeS}}{\text{mol FeS}} = 12.3 \text{ g FeS used}$$

? g FeS excess = 13.2 g – 12.3 g = 0.9 g FeS excess

3.20B $2\,Al + 6\,HCl \rightarrow 3\,H_2 + 2\,AlCl_3$

$$? \text{ g } H_2 = 12.5 \text{ g Al} \times \frac{\text{mol Al}}{26.98 \text{ g Al}} \times \frac{3 \text{ mol } H_2}{2 \text{ mol Al}} \times \frac{2.02 \text{ g } H_2}{\text{mol } H_2} = 1.40 \text{ g } H_2$$

$$? \text{ g } H_2 = 250.0 \text{ mL solution} \times \frac{1.13 \text{ g}}{\text{mL}} \times \frac{25.6 \text{ g HCl}}{100 \text{ g solution}} \times \frac{\text{mol HCl}}{36.46 \text{ g HCl}}$$
$$\times \frac{3 \text{ mol } H_2}{6 \text{ mol HCl}} \times \frac{2.02 \text{ g } H_2}{\text{mol } H_2} = 2.00 \text{ g } H_2$$

1.40 g H_2 are produced.

3.21A $? \text{ g acet} = 20.0 \text{ g alcohol} \times \frac{\text{mol alch}}{88.15 \text{ g alch}} \times \frac{1 \text{ mol acetate}}{1 \text{ mol alch}} \times \frac{130.18 \text{ g acet}}{\text{mol acet}}$

$= 29.5 \text{ g acet}$

$? \text{ g acet} = 25.0 \text{ g acid} \times \frac{\text{mol aa}}{60.05 \text{ g aa}} \times \frac{1 \text{ mol acet}}{1 \text{ mol aa}} \times \frac{130.18 \text{ g acet}}{\text{mol acet}} = 54.20 \text{ g acet}$

Theoretical yield is 29.5 g isopentyl acetate.

$\text{Actual Y} = \frac{\% \text{ Y theoret. Y}}{100\%} = \frac{29.5 \text{ g} \times 90.0\%}{100\%} = 26.6 \text{ g}$

3.21B $? \text{ g} = \text{theor Y} = \frac{\text{actual Y x } 100\%}{\% \text{ Y}} = \frac{433 \text{ g} \times 100\%}{78.5\%} = 552 \text{ g}$

$? \text{ g} = 552 \text{ g acet} \times \frac{\text{mol acet}}{130.18 \text{ g acet}} \times \frac{\text{mol alch}}{\text{mol acet}} \times \frac{88.15 \text{ g alch}}{\text{mol alch}}$

$= 374 \text{ g isopentyl alcohol}$

3.22 $H_3PO_4 + 2\ NH_3 \rightarrow (NH_4)_2HPO_4$

$? \text{ kg } (NH_4)_2HPO_4 = 1.00 \text{ kg } H_3PO_4 \times \frac{\text{mol } H_3PO_4}{98 \text{ g } H_3PO_4} \times \frac{\text{mol } (NH_4)_2HPO_4}{\text{mol } H_3PO_4}$

$\times \frac{132 \text{ g } (NH_4)_2HPO_4}{\text{mol } (NH_4)_2HPO_4} = 1.35 \text{ kg } (NH_4)_2HPO_4$

3.23A (a) $? \text{ M} = \frac{3.00 \text{ mol KI}}{2.39 \text{ L solution}} = 1.26 \text{ M KI}$

(b) $? \text{ M} = \frac{0.522 \text{ g HCl}}{0.592 \text{ L solution}} \times \frac{\text{mol HCl}}{36.46 \text{ g HCl}} = 0.0242 \text{ M HCl}$

(c) $? \text{ M} = \frac{2.69 \text{ g } C_{12}H_{22}O_{11}}{225 \text{ mL}} \times \frac{\text{mL}}{10^{-3} \text{ L}} \times \frac{\text{mol } C_{12}H_{22}O_{11}}{342.3 \text{ g } C_{12}H_{22}O_{11}}$

$= 3.49 \times 10^{-2} \text{ M } C_{12}H_{22}O_{11}$

3.23B (a) $? \text{ M} = \frac{126 \text{ mg}}{100.0 \text{ mL}} \times \frac{\text{mol } C_6H_{12}O_6}{180.2 \text{ g}} = 6.99 \times 10^{-3} \text{ M } C_6H_{12}O_6$

(b) $? \text{ M} = \frac{10.5 \text{ mL } C_2H_5OH}{25.0 \text{ mL}} \times \frac{0.789 \text{ g } C_2H_5OH}{\text{mL } C_2H_5OH} \times \frac{\text{mol } C_2H_5OH}{46.08 \text{ g } C_2H_5OH}$

$\times \frac{\text{mL}}{10^{-3} \text{ L}} = 7.19 \text{ M } C_2H_5OH$

(c) $? \text{ M} = \frac{9.5 \text{ mg N}}{\text{mL}} \times \frac{60.07 \text{ g } CO(NH_2)_2}{28.02 \text{ g N}} \times \frac{\text{mol } CO(NH_2)_2}{60.07 \text{ g } CO(NH_2)_2}$

$= 0.34 \text{ M } CO(NH_2)_2$

3.24A (a) $? \text{ g} = 2.00 \text{ L} \times 6.00 \text{ M KOH} \times \frac{56.11 \text{ g KOH}}{\text{mol KOH}} = 673 \text{ g KOH}$

(b) $? \text{ g} = 10.0 \text{ mL} \times \frac{10^{-3}\text{ L}}{\text{mL}} \times 0.100 \text{ M KOH} \times \frac{56.11 \text{ g KOH}}{\text{mol KOH}} = 0.0561 \text{ g KOH}$

(c) $? \text{ g} = 35.0 \text{ mL} \times \frac{10^{-3}\text{ L}}{\text{mL}} \times 2.50 \text{ M KOH} \times \frac{56.11 \text{ g KOH}}{\text{mol KOH}} = 4.91 \text{ g KOH}$

3.24B $? \text{ mL} = 725 \text{ mL} \times 0.350 \text{ M} \times \frac{10^{-3}\text{ mol}}{\text{mmol}} \times \frac{74.14 \text{ g } C_4H_{10}O}{\text{mol } C_4H_{10}O} \times \frac{\text{mL}}{0.810 \text{ g}}$

$= 23.2 \text{ mL } C_4H_{10}O$

3.25A $? \text{ M} = \frac{90.0 \text{ g HCOOH}}{100 \text{ g solution}} \times \frac{1.20 \text{ g}}{\text{mL}} \times \frac{\text{mL}}{10^{-3}\text{ L}} \times \frac{\text{mol HCOOH}}{46.03 \text{ g HCOOH}} = 23.5 \text{ M HCOOH}$

3.25B $?\% \ HClO_4 = \frac{11.7 \text{ mol } HClO_4}{\text{L solution}} \times \frac{10^{-3}\text{ L}}{\text{mL}} \times \frac{\text{mL}}{1.67 \text{ g}} \times \frac{100.5 \text{ g}}{\text{mol}} \times 100\%$

$= 70.4\% \ HClO_4$, by mass

3.26A $? \text{ mL} = \frac{15.0 \text{ L x } 0.315 \text{ M}}{10.15 \text{ M}} \times \frac{\text{mL}}{10^{-3}\text{ L}} = 466 \text{ mL}$

3.26B $? \text{ mL} = \frac{7.50 \text{ mg}}{\text{mL}} \times 375 \text{ mL} \times \frac{\text{mol } CH_3OH}{32.05 \text{ g } CH_3OH} \times \frac{\text{L}}{5.15 \text{ mol}} = 17.0 \text{ mL } CH_3OH$

3.27A $? \text{ mL} = 750.0 \text{ mL} \times 0.0250 \text{ M } Na_2CrO_4 \times \frac{2 \text{ mol } AgNO_3}{\text{mol } Na_2CrO_4}$

$\times \frac{\text{L}}{0.100 \text{ mol } AgNO_3} = 375 \text{ mL } AgNO_3$

3.27B (a) $? \text{ g } CO_2 = 175 \text{ mL} \times 1.55 \text{ M } NaHCO_3 \times \frac{\text{mol } CO_2}{\text{mol } NaHCO_3} \times \frac{10^{-3}\text{ L}}{\text{mL}}$

$\times \frac{44.01 \text{ g } CO_2}{\text{mol } CO_2} = 11.9 \text{ g } CO_2$

$? \text{ g } CO_2 = 235 \text{ mL} \times 1.22 \text{ M HCl} \times \frac{\text{mol } CO_2}{\text{mol HCl}} \times \frac{10^{-3}\text{ L}}{\text{mL}}$

$\times \frac{44.01 \text{ g } CO_2}{\text{mol } CO_2} = 12.6 \text{ g } CO_2$

$NaHCO_3$ is limiting; 11.9 g CO_2 is produced.

(b) $? \text{ mol NaCl} = 175 \text{ mL} \times 1.55 \text{ M } NaHCO_3 \times \frac{\text{mol NaCl}}{\text{mol } NaHCO_3} \times \frac{10^{-3}\text{ L}}{\text{mL}}$

$= 0.271 \text{ mol NaCl}$

$$? \text{ M} = \frac{0.271 \text{ mol NaCl}}{410 \text{ mL}} \times \frac{\text{mL}}{10^{-3} \text{ L}} = 0.661 \text{ M NaCl}$$

Review Questions

2. $1.00 \text{ mol} \times \frac{6.022 \text{ x } 10^{23} \text{ molecules}}{\text{mol}} = 6.02 \times 10^{23} \text{ molecules}$

$1.00 \text{ mol} \times \frac{6.022 \text{ x } 10^{23} \text{ molecules}}{\text{mol}} \times \frac{2 \text{ atoms}}{\text{molecule}} = 1.20 \times 10^{24} \text{ atoms}$

3. (a) $1.00 \text{ mol Ca(NO}_3)_2 \times \frac{6.022 \text{ x } 10^{23} \text{ formula units}}{\text{mol}} \times \frac{1 \text{ Ca}^{2+}}{\text{formula unit}}$
$= 6.02 \times 10^{23} \text{ Ca}^{2+}$

$1.00 \text{ mol Ca(NO}_3)_2 \times \frac{6.022 \text{ x } 10^{23} \text{ formula units}}{\text{mol}} \times \frac{2 \text{ NO}_3^-}{\text{formula unit}}$
$= 1.20 \times 10^{24} \text{ NO}_3^-$

(b) $1.00 \text{ mol Ca(NO}_3)_2 \times \frac{6.022 \text{ x } 10^{23} \text{ formula units}}{\text{mol}} \times \frac{2 \text{ N atoms}}{\text{formula unit}}$
$= 1.20 \times 10^{24} \text{ N atoms}$

$1.00 \text{ mol Ca(NO}_3)_2 \times \frac{6.022 \text{ x } 10^{23} \text{ formula units}}{\text{mol}} \times \frac{6 \text{ O atoms}}{\text{formula unit}}$
$= 3.61 \times 10^{24} \text{ O atoms}$

5. (a) HO (b) CH_2 (c) C_5H_4 (d) $C_6H_{16}O$

9. (a) $NH_4NO_3(s) \rightarrow N_2O(g) + 2\ H_2O(l)$
(b) $C_7H_{16}(l) + 11\ O_2(g) \rightarrow 7\ CO_2(g) + 8\ H_2O(l)$
One mole of ammonium nitrate is chemically equivalent to one mole of dinitrogen oxide.
One mole of heptane is chemically equivalent to seven moles of carbon dioxide.

14. These are the same values. By multiplying both numerators and denominators by 10^{-3}, grams are changed to milligrams, and liters are changed to milliliters.

Problems

17. (a)

C	6 × 12.01 u =	72.06
H	5 × 1.008 u =	5.04
Br	1 × 79.90 u =	79.90
		157.00 u

molecular

(b)

Ca	1 × 40.08 =	40.08
H	2 × 1.01 =	2.02
C	2 × 12.01 =	24.02
O	6 × 16.00 =	96.00
		162.12 u

formula

(c)

H	3 × 1.008 u =	3.024
P	1 × 30.974 u =	30.974
O	4 × 15.999 u =	63.996
		97.994 u

(d)

K	2 × 39.10 u =	78.20
Cr	2 × 52.00 u =	104.00
O	7 × 16.00 u =	112.00
		294.20 u

molecular

(e) Al 2×26.98 u = 53.96
S 3×32.07 u = 96.21
O 12×16.00 u = 192.00
H 36×1.008 u = 36.29
O 18×16.00 u = 288.00
666.46 u

formula

(f) formula
Na 2×22.99 u = 45.98
Pt 1×195.08 u = 195.08
C 4×12.01 u = 48.04
N 4×14.01 u = 56.04
H 6×1.01 u = 6.06
O 3×16.00 u = 48.00
399.20 u

formula

(g) C 5×12.01 u = 60.05
H 12×1.01 u = 12.12
O 2×16.00 u = 32.00
104.17 u

molecular

(h) C 3×12.01 u = 36.03
H 9×1.01 u = 9.09
N 1×14.01u = 14.01
59.13 u

molecular

19. (a) C 11×12.01 u = 132.11
H 17×1.01 u = 17.17
O 5×16.00 u = 80.00
P 1×30.97 u = 30.97
S 2×32.07 u = 64.14
324.39 u

(b) C 8×12.01 u = 96.08
H 8×1.01 u = 8.08
O 3×16.00 u = 48.00
152.16 u

21. (a) Ca 1×40.08 u = 40.08
H 2×1.008 u = 2.02
42.10 u

$$? \text{ g } CaH_2 = 1.12 \text{ mol } CaH_2 \times \frac{42.10 \text{ g}}{\text{mol}} = 47.2 \text{ g } CaH_2$$

(b) C $4 \times 12.01 = 48.04$
H $8 \times 1.01 = 8.08$
O $2 \times 16.00 = 32.00$
88.12

$$? \text{ g } C_4H_8O_2 = 0.250 \text{ mol } C_4H_8O_2 \times \frac{88.12 \text{ g}}{\text{mol}} = 22.03 \text{ g } C_4H_8O_2$$

(c) IF_5 I $1 \times 126.90 = 126.90$
F $5 \times 19.00 = 95.00$
221.90

$$? \text{ g } IF_5 = 0.158 \text{ mol } IF_5 \times \frac{221.90 \text{ g}}{\text{mol}} = 35.1 \text{ g } IF_5$$

(d) $Ba(HCO_3)_2$
Ba $1 \times 137.33 = 137.33$
H $2 \times 1.01 = 2.02$
C $2 \times 12.01 = 24.02$
O $6 \times 16.00 = 96.00$
259.37

$$? \text{ g } Ba(HCO_3)_2 = 0.325 \text{ mol} \times \frac{259.37 \text{ g}}{\text{mol}} = 84.3 \text{ g } Ba(HCO_3)_2$$

23. (a) H $1 \times 1.008 = 1.01$
N $1 \times 14.01 = 14.01$
O $3 \times 16.00 = \underline{48.00}$
63.02

$$? \text{ mol } HNO_3 = 98.6 \text{ g } HNO_3 \times \frac{\text{mol}}{63.02 \text{ g}} = 1.56 \text{ mol } HNO_3$$

(b) SF_6
S $1 \times 32.06 = 32.07$
F $6 \times 19.00 = \underline{114.00}$
146.07

$$? \text{ mol } SF_6 = 16.3 \text{ g } SF_6 \times \frac{\text{mol}}{146.07 \text{ g}} = 0.112 \text{ mol } SF_6$$

(c) $Fe_2(SO_4)_3{\cdot}7H_2O$
Fe $2 \times 55.85 = 111.70$
S $3 \times 32.07 = 96.21$
H $14 \times 1.008 = 14.112$
O $19 \times 16.00 = \underline{304.00}$
526.02

$$? \text{ mol } Fe_2(SO_4)_3{\cdot}7H_2O = 35.6 \text{ g } Fe_2(SO_4)_3{\cdot}7H_2O \times \frac{\text{mol}}{526.02 \text{ g}} = 0.0677 \text{ mol } Fe_2(SO_4)_3{\cdot}7H_2O$$

(d) C $1 \times 12.01 = 12.01$
H $4 \times 1.01 = 4.04$
O $1 \times 16.00 = \underline{16.00}$
32.05 u

$$? \text{ mol} = 218 \text{ mg} \times \frac{10^{-3} \text{ g}}{\text{mg}} \times \frac{\text{mol } CH_3OH}{32.05 \text{ g } CH_3OH} = 6.80 \times 10^{-3} \text{ mol } CH_3OH$$

(e) Fe $2 \times 55.85 = 111.70$
O $3 \times 16.00 = \underline{48.00}$
159.70

$$? \text{ mol } Fe_2O_3 = 3.32 \times 10^7 \text{ g} \frac{\text{mol}}{159.70 \text{ g}} = 2.08 \times 10^5 \text{ mol } Fe_2O_3.$$

25. (a) $$? \text{ molecules} = 4.68 \text{ mol} \times \frac{6.022 \times 10^{23} \text{ molecules}}{\text{mol}} = 2.82 \times 10^{24} \text{ molecules}$$

(b) $$? \text{ sulfate ions} = 86.2 \text{ g} \times \frac{\text{mol } Al_2(SO_4)_3}{342 \text{ g } Al_2(SO_4)_3} \times \frac{6.022 \times 10^{23} \text{ molecules}}{\text{mol } Al_2(SO_4)_3} \times \frac{3\, SO_4^{2-} \text{ ions}}{\text{molecules } Al_2(SO_4)_3} = 4.55 \times 10^{23}\ SO_4^{2-} \text{ ions}$$

(c) $$? \text{ g/atom Cu} = \frac{63.546 \text{ g}}{\text{mol}} \times \frac{\text{mol}}{6.022 \times 10^{23} \text{ molecules}} = 1.055 \times 10^{-22} \text{ g/atom Cu}$$

(d) $? \text{ g/ion } CO_3^{-2} = \frac{60.01 \text{ g}}{\text{mol}} \times \frac{\text{mol}}{6.022 \text{ x } 10^{23} \text{ ions}} = 9.97 \times 10^{-23} \text{ g/ion } CO_3^{2-}$

27. 6.1×10^{23} molecules of O_2 is about $2 \times$ Avogadro's number of atoms. 0.76 atoms of HCl is about 1.5 times Avogadro's number of atoms. 250.0 g of U is slightly more than Avogadro's number of atoms, and 0.86 mol Al is less than Avogadro's number of atoms.

$$Al < U < HCl < O_2 \quad (a) < (c) < (d) < (b)$$

actual

(a) $0.86 \text{ mol Al} \times \frac{6.022 \text{ x } 10^{23} \text{ atoms}}{\text{mol}} = 5.18 \times 10^{23} \text{ atoms}$

(b) $6.1 \times 10^{23} \text{ molecules } O_2 \times \frac{2 \text{ atoms O}}{\text{molecules } O_2} = 12.2 \times 10^{23} \text{ atoms}$

(c) $250.0 \text{ g U} \times \frac{\text{mol}}{238.0 \text{ g}} \times \frac{6.022 \text{ x } 10^{23} \text{ atoms}}{\text{mol}} = 6.33 \times 10^{23} \text{ atoms}$

(d) $0.76 \text{ mol HCl} \times \frac{6.022\ 10^{23} \text{ molecules}}{\text{mol}} \times \frac{2 \text{ atoms}}{\text{molecule}} = 9.15 \times 10^{23} \text{ atoms}$

29. (a)

Ba	$1 \times 137.33 =$	137.33
Si	$1 \times 28.09 =$	28.09
O	$3 \times 16.00 =$	48.00
		213.42

$?\% \text{ Ba} = \frac{137.33 \text{ g}}{213.42 \text{ g}} \times 100\% = 64.35\% \text{ Ba}$

$?\% \text{ Si} = \frac{28.09 \text{ g}}{213.42 \text{ g}} \times 100\% = 13.16\% \text{ Si}$

$?\% \text{ O} = \frac{48.00 \text{ g}}{213.42 \text{ g}} \times 100\% = 22.49\% \text{ O}$

(b)

C	$6 \times 12.01 =$	72.06
H	$5 \times 1.008 =$	5.040
N	$1 \times 14.01 =$	14.01
O	$2 \times 16.00 =$	32.00
		123.11

$?\% \text{ C} = \frac{72.06 \text{ g}}{123.11 \text{ g}} \times 100\% = 58.53\% \text{ C}$

$?\% \text{ H} = \frac{5.040 \text{ g}}{123.11 \text{ g}} \times 100\% = 4.094\% \text{ H}$

$?\% \text{ N} = \frac{14.01 \text{ g}}{123.11 \text{ g}} \times 100\% = 11.38\% \text{ N}$ $?\% \text{ O} = \frac{32.00 \text{ g}}{123.11 \text{ g}} \times 100\% = 25.99\% \text{ O}$

(c)

Fe	$2 \times 55.85 =$	111.70
H	$3 \times 1.01 =$	3.03
P	$3 \times 30.97 =$	92.91
O	$12 \times 16.00 =$	192.00
		399.64

$$? \% \text{ Fe} = \frac{111.70 \text{ g}}{399.64 \text{ g}} \times 100 \% = 27.95 \% \text{ Fe}$$

$$? \% \text{ H} = \frac{3.03 \text{ g}}{399.64 \text{ g}} \times 100 \% = 0.76 \% \text{ H}$$

$$? \% \text{ P} = \frac{92.91 \text{ g}}{399.64 \text{ g}} \times 100 \% = 23.25 \% \text{ P}$$

$$? \% \text{ O} = \frac{192.00 \text{ g}}{399.64 \text{ g}} \times 100 \% = 48.04 \% \text{ O}$$

(d) Al $1 \times 26.98 = 26.98$
Br $3 \times 79.90 = 239.70$
O $18 \times 16.00 = 288.00$
H $18 \times 1.008 = \underline{18.14}$
572.82

$$?\% \text{ Al} = \frac{26.98 \text{ g}}{572.82 \text{ g}} \times 100\% = 4.71\% \text{ Al}$$

$$?\% \text{ Br} = \frac{239.70 \text{ g}}{572.82 \text{ g}} \times 100\% = 41.85\% \text{ Br}$$

$$?\% \text{ O} = \frac{288.00 \text{ g}}{572.82 \text{ g}} \times 100\% = 50.28\% \text{ O}$$

$$?\% \text{ H} = \frac{18.14 \text{ g}}{572.82 \text{ g}} \times 100\% = 3.17\% \text{ H}$$

31. (a) $?\% \text{ O} = \dfrac{4 \times 16.00}{132.13} \times 100\% = 48.44\% \text{ O}$

(b) $?\% \text{ N} = \dfrac{14.01}{101.17} \times 100\% = 13.85\% \text{ N}$

33. (a) C $3 \times 12 = 36$
H $2 \times 1 = 2$
Cl $1 \times 35 = \underline{35}$
73

$$\frac{147 \text{ u}}{\text{molecule}} \times \frac{\text{formula}}{73 \text{ u}} = \frac{2 \text{ formula}}{\text{molecule}} \quad C_6H_4Cl_2 \text{ molecular formula}$$

(b) $40.00 \text{ g C} \dfrac{\text{mol C}}{12.011 \text{ g C}} = 3.330 \text{ mol C} \times \dfrac{1}{3.330 \text{ mol}} = 1 \dfrac{\text{mol C}}{\text{mol C}}$

$$6.71 \text{ g H} \times \frac{\text{mol H}}{1.008 \text{ g H}} = 6.66 \text{ mol H} \times \frac{1}{3.330 \text{ mol C}} = \frac{2 \text{ mol H}}{\text{mol C}}$$

$$53.29 \text{ g O} \times \frac{\text{mol O}}{16.00 \text{ g O}} = 3.331 \text{ mol O} \times \frac{1}{3.330 \text{ mol C}} = \frac{1 \text{ mol O}}{\text{mol C}}$$

CH_2O

formula mass = $12.0 + 2 \times 1.0 + 16.0 = 30.0$

$$\frac{180 \text{ u}}{\text{molecule}} \times \frac{\text{formula}}{30.0 \text{ u}} = \frac{6 \text{ formula}}{\text{molecule}}$$

$C_6H_{12}O_6$ molecular formula

35. (a) $? \text{ mol C} = 72.22 \text{ g C} \times \frac{\text{mol C}}{12.01 \text{ g C}} = 6.013 \text{ mol C} \times \frac{1}{0.334 \text{ mol N}} = \frac{18.00 \text{ mol C}}{\text{mol N}}$

$? \text{ mol H} = 7.07 \text{ g H} \times \frac{\text{mol H}}{1.01 \text{ g H}} = 7.00 \text{ mol H} \times \frac{1}{0.334 \text{ mol N}} = \frac{20.96 \text{ mol H}}{\text{mol N}}$

$? \text{ mol N} = 4.68 \text{ g N} \times \frac{\text{mol N}}{14.01 \text{ g N}} = 0.334 \text{ mol N} \times \frac{1}{0.334 \text{ mol N}} = \frac{1.00 \text{ mol N}}{\text{mol N}}$

$? \text{ mol O} = 16.03 \text{ g O} \times \frac{\text{mol O}}{16.00 \text{ g O}} = 1.002 \text{ mol O} \times \frac{1}{0.334 \text{ mol N}} = \frac{3.00 \text{ mol O}}{\text{mol N}}$

$C_{18}H_{21}NO_3$

(b) $21.9 \text{ g Mg} \times \frac{\text{mol Mg}}{24.31 \text{ g}} = 0.901 \text{ mol Mg} \times \frac{1}{0.898 \text{ mol P}} = \frac{1.00 \text{ mol Mg}}{\text{mol O}}$

$27.8 \text{ g P} \times \frac{\text{mol P}}{38.97 \text{ g}} = 0.898 \text{ mol P} \times \frac{1}{0.898 \text{ mol P}} = \frac{1.00 \text{ mol O}}{\text{mol P}}$

$50.3 \text{ g O} \times \frac{\text{mol O}}{16.00 \text{ g}} = 3.14 \text{ mol O} \times \frac{1}{0.898 \text{ mol P}} = \frac{3.50 \text{ mol O}}{\text{mol P}}$

$Mg_{(1 \times 2)}P_{(1 \times 2)}O_{(3.5 \times 2)} \rightarrow Mg_2P_2O_7$

37. $? \text{ mol C} = 65.44 \text{ g C} \times \frac{\text{mol C}}{12.01 \text{ g C}} = 5.449 \text{ mol C} \times \frac{1}{1.816 \text{ mol O}} = \frac{3.000 \text{ mol C}}{\text{mol O}}$

$? \text{ mol H} = 5.49 \text{ g H} \times \frac{\text{mol H}}{1.008 \text{ g H}} = 5.446 \text{ mol H} \times \frac{1}{1.816 \text{ mol O}} = \frac{2.999 \text{ mol H}}{\text{mol O}}$

$? \text{ mol O} = 29.06 \text{ g O} \times \frac{\text{mol O}}{16.00 \text{ g O}} = 1.816 \text{ mol O} \times \frac{1}{1.816 \text{ mol O}} = \frac{1.000 \text{ O}}{\text{mol O}}$

empirical formula: C_3H_3O

$3 \times 12 + 3 \times 1 + 1 \times 16 = 55$

$\frac{110 \text{ u}}{\text{molecule}} \times \frac{\text{formula}}{55 \text{ u}} = \frac{2 \text{ formula}}{\text{molecule}}$ molecular formula: $C_6H_6O_2$

39. $? \text{ mol Li} = 4.33 \text{ g Li} \times \frac{\text{mol Li}}{6.941 \text{ g Li}} = 0.624 \text{ mol Li} \times \frac{1}{0.623 \text{ mol Cl}} = \frac{1.00 \text{ mol Li}}{\text{mol Cl}}$

$? \text{ mol Cl} = 22.10 \text{ g Cl} \times \frac{\text{mol Cl}}{35.453 \text{ g Cl}} = 0.623 \text{ mol CL} \times \frac{1}{0.623 \text{ mol Cl}}$

$= \frac{1.0 \text{ mol Cl}}{\text{mol Cl}}$

$? \text{ mol O} = 39.89 \text{ g O} \times \frac{\text{mol O}}{15.999 \text{ g O}} = 2.493 \text{ mol O} \times \frac{1}{0.623 \text{ mol Cl}} = \frac{4.00 \text{ mol O}}{\text{mol Cl}}$

$? \text{ mol } H_2O = 33.69 \text{ g } H_2O \times \frac{\text{mol } H_2O}{18.015 \text{ g } H_2O} = 1.870 \text{ mol } H_2O \times \frac{1}{0.623 \text{ mol Cl}}$

$= \frac{3.00 \text{ mol } H_2O}{\text{mol Cl}}$

$LiClO_4 \cdot 3H_2O$

41.

		Actual
$(NH_4)_2SO_4$	$\% N = \dfrac{2\times 14}{28+8+32+64}$	21%
NH_4NO_2	$\% N = \dfrac{2\times 14}{28+4+32}$	43%
NH_4NO_3	$\% N = \dfrac{2\times 14}{28+4+48}$	35%
NH_4Cl	$\% N = \dfrac{14}{14+4+35}$	26%

The first three all have 2 × 14 in the numerator, so the smallest denominator will be the largest percent. If the 28 + 4 is ignored, then it is easy to see that 32 is less than 48 and less than 4 + 32 + 64. The second one is the lesser of the first three. To compare the second with the fourth, notice that although the numerator is double in the second, only part of the denominator is double, so the second has a smaller denominator and larger percent.

43. $1.278 \text{ g C}\dfrac{\text{mol C}}{12.01 \text{ g C}} = 0.1064 \text{ mol C} \times \dfrac{1}{0.1062 \text{ mol S}} = \dfrac{1.00 \text{ mol C}}{\text{mol S}}$

$0.318 \text{ g H}\dfrac{\text{mol H}}{1.008 \text{ g H}} = 0.315 \text{ mol H} \times \dfrac{1}{0.1062 \text{ mol S}} = \dfrac{2.97 \text{ mol H}}{\text{mol S}}$

$3.404 \text{ g S}\dfrac{\text{mol S}}{32.066 \text{ g S}} = 0.1062 \text{ mol S} \times \dfrac{1}{0.1062 \text{ mol S}} = \dfrac{1.00 \text{ mol S}}{\text{mol S}}$

CH_3S empirical formula

$? \dfrac{\text{formula}}{\text{molecule}} = \dfrac{94.19 \text{ U}}{\text{molecule}} \times \dfrac{\text{formula}}{47.09 \text{ U}} = \dfrac{2 \text{ formula}}{\text{molecule}}$

$C_2H_6S_2$ molecular formula

45. $? \text{ mol C} = 1.119 \text{ g } CO_2 \times \dfrac{\text{mol C}}{44.01 \text{ g } CO_2} = 0.02543 \text{ mol C} \times \dfrac{1}{0.00635 \text{ mol S}}$

$= \dfrac{4.00 \text{ mol C}}{\text{mol S}}$

$? \text{ mol H} = 0.229 \text{ g } H_2O \times \dfrac{2 \text{ mol H}}{18.02 \text{ g } H_2O} = 0.02542 \text{ mol H} \times \dfrac{1}{0.00635 \text{ mol S}}$

$= \dfrac{4.03 \text{ mol H}}{\text{mol S}}$

$? \text{ mol S} = 0.407 \text{ g } SO_2 \times \dfrac{\text{mol S}}{64.06 \text{ g } SO_2} = 0.00635 \text{ mol S} \times \dfrac{1}{0.00635 \text{ mol S}}$

$= \dfrac{1.00 \text{ mol S}}{\text{mol S}}$

C_4H_4S

47.

		Actual
(a) CH_3OH	$\dfrac{16}{12+4+16}$	50%

CH_3CH_2OH $\frac{16}{24 + 6 + 16}$ 35%

$CH_3OC(CH_3)_3$ $\frac{16}{60 + 12 + 16}$ 18%

Since the denominator is smaller in CH_3OH, it has the greatest oxygen percent with CH_3CH_2OH next and $CH_3OC(CH_3)_3$ last.

least $CH_3OC(CH_3)_3 < CH_3CH_2OH < CH_3OH$ most

(b) CH_3OH $\frac{16.0 \times 100\%}{12.0 + 4.0 + 16.0} = 50.0\%$

$$\frac{?\ \%\ O}{fuel} = \frac{50.0\%\ O}{CH_3OH} \times \frac{10.5\%\ CH_3OH}{fuel} = \frac{5.25\%\ O}{fuel}$$

Yes, methanol does meet the 2.7% O requirement.

(c) $CH_3OC(CH_3)_3$ $\frac{16.0 \times 100\%}{(5 \times 12.0) + (12 \times 1.01) + 16.0} = 18.2\%$

$$\frac{?\ MTBG}{gasoline} = \frac{\frac{2.7\%\ O}{gasoline} \times 100\%}{\frac{18.2\%\ O}{MTBG}} = \frac{15\%\ MTBG}{gasoline}$$

49. (a) $Cl_2O_5 + H_2O \rightarrow 2\ HClO_3$

(b) $V_2O_5 + 2\ H_2 \rightarrow V_2O_3 + 2\ H_2O$

(c) $4\ Al + 3\ O_2 \rightarrow 2\ Al_2O_3$

(d) $2\ C_4H_{10} + 13\ O_2 \rightarrow 8\ CO_2 + 10\ H_2O$

(e) $Sn + 2\ NaOH \rightarrow Na_2SnO_2 + H_2$

(f) $PCl_5 + 4\ H_2O \rightarrow H_3PO_4 + 5\ HCl$

(g) $2\ CH_3OH + 3\ O_2 \rightarrow 2\ CO_2 + 4\ H_2O$

(h) $3\ Zn(OH)_2 + 2\ H_3PO_4 \rightarrow Zn_3(PO_4)_2 + 6\ H_2O$

51. (a) $2\ CO(g) + 2\ NO(g) \rightarrow 2\ CO_2(g) + N_2(g)$

(b) $C_3H_8(g) + 3\ H_2O(g) \rightarrow 3\ CO(g) + 7\ H_2(g)$

(c) $Mg_3N_2(s) + 6\ H_2O(l) \rightarrow 3\ Mg(OH)_2(s) + 2\ NH_3(g)$

(d) $Pb(s) + PbO_2(s) + 2\ H_2SO_4(aq) \rightarrow 2\ PbSO_4(s) + 2\ H_2O(l)$

53. $? \text{ mol Fe} = 72.3 \text{ g Fe} \times \frac{\text{mol Fe}}{55.85 \text{ g Fe}} = 1.295 \text{ mol Fe} \times \frac{1}{1.295 \text{ mol Fe}} = \frac{1.00 \text{ mol Fe}}{\text{mol Fe}}$

$? \text{ mol O} = 27.7 \text{ g O} \times \frac{\text{mol O}}{16.00 \text{ g O}} = 1.731 \text{ mol O} \times \frac{1}{1.295 \text{ mol Fe}} = \frac{1.34 \text{ mol O}}{\text{mol Fe}}$

$Fe_{(1 \times 3)}O_{(1.34 \times 3)}$ => Fe_3O_4

$3\ Fe_2O_3(s) + H_2(g) \xrightarrow{400\ °C} H_2O(g) + 2\ Fe_3O_4(s)$

55. $4\ NH_3 + 5\ O_2 \rightarrow 4\ NO + 6\ H_2O$

57. (a) $? \text{ mol } CO_2 = 1.8 \times 10^4 \text{ mol } C_8H_{18} \times \frac{16 \text{ mol } CO_2}{2 \text{ mol } C_8H_{18}} = 1.4 \times 10^5 \text{ mol } CO_2$

(b) $? \text{ mol } O_2 = 4.4 \times 10^4 \text{ mol } C_8H_{18} \times \frac{25 \text{ mol } O_2}{2 \text{ mol } C_8H_{18}} = 5.5 \times 10^5 \text{ mol } O_2$

(c) $? \text{ mol } H_2O = 7.6 \times 10^3 \text{ mol } C_8H_{18} \times \frac{18 \text{ mol } H_2O}{2 \text{ mol } C_8H_{18}} = 6.8 \times 10^4 \text{ mol } H_2O$

(d) $? \text{ mol } CO_2 = 2.2 \times 10^4 \text{ mol } O_2 \times \frac{16 \text{ mol } CO_2}{25 \text{ mol } O_2} = 1.4 \times 10^4 \text{ mol } CO_2$

59. $CaCN_2 + 3\ H_2O \rightarrow CaCO_3 + 2\ NH_3$

$Mg_3N_2 + 6\ H_2O \rightarrow 3\ Mg(OH)_2 + 2\ NH_3$

$? \text{ kg } NH_3 = 1 \text{ kg } CaCN_2 \times \frac{\text{mol } CaCN_2}{80 \text{ g } CaCN_2} \times \frac{2 \text{ mol } NH_3}{\text{mol } CaCN_2} \times \frac{17 \text{ g } NH_3}{\text{mol } NH_3}$

$? \text{ kg } NH_3 = 1 \text{ kg } Mg_3N_2 \times \frac{\text{mol } Mg_3N_2}{101 \text{ g } Mg_3N_2} \times \frac{2 \text{ mol } NH_3}{\text{mol } Mg_3N_2} \times \frac{17 \text{ g } NH_3}{\text{mol } NH_3}$

Since the only difference in the two equations is the molar mass of the compounds $CaCN_2$ and Mg_3N_2, the smaller molar mass will produce the most NH_3. $CaCN_2$ produces more than Mg_3N_2. (Actual 0.43 kg to 0.34 kg)

61. $2\ C_{14}H_{30} + 43\ O_2 \rightarrow 28\ CO_2 + 30\ H_2O$

$$? \text{ g } CO_2 = 1.00 \text{ gal } C_{14}H_{30} \times \frac{3.785 \text{ L}}{1.00 \text{ gal}} \times \frac{\text{mL}}{10^{-3} \text{ L}} \times \frac{0.763 \text{ g}}{\text{mL}} \times \frac{\text{mol } C_{14}H_{30}}{198.38 \text{ g } C_{14}H_{30}} \times \frac{28 \text{ mol } CO_2}{2 \text{ mol } C_{14}H_{30}} \times \frac{44.01 \text{ g } CO_2}{\text{mol } CO_2} = 8.97 \times 10^3 \text{ g } CO_2$$

63. (a) $CaCO_3 + 2\ HCl \rightarrow CaCl_2 + CO_2 + H_2O$

$$? \text{ g } CO_2 = 5.05 \text{ g chalk} \times \frac{72.0 \text{ g } CaCO_3}{100.0 \text{ g chalk}} \times \frac{\text{mol } CaCO_3}{100.09 \text{ g } CaCO_3} \times \frac{\text{mol } CO_2}{\text{mol } CaCO_3} \times \frac{44.01 \text{ g } CO_2}{\text{mol } CO_2} = 1.60 \text{ g } CO_2$$

(b) $$? \text{ g } CaCO_3 = 1.31 \text{ g } CO_2 \times \frac{\text{mol } CO_2}{44.01 \text{ g } CO_2} \times \frac{\text{mol } CaCO_3}{\text{mol } CO_2} \times \frac{100.09 \text{ g } CaCO_3}{\text{mol } CaCO_3} = 2.979 \text{ g } CaCO_3$$

$$\frac{2.979 \text{ g } CaCO_3}{4.38 \text{ g chalk}} \times 100\% = 68.0\%\ CaCO_3$$

65. $? \text{ mol } Li_2CO_3 = 0.150 \text{ mol LiOH} \times \frac{\text{mol } Li_2CO_3}{2 \text{ mol LiOH}} = 0.0750 \text{ mol } Li_2CO_3$

$? \text{ mol } Li_2CO_3 = 0.080 \text{ mol } CO_2 \times \frac{\text{mol } Li_2CO_3}{\text{mol } CO_2} = 0.080 \text{ mol } Li_2CO_3$

LiOH is limiting and 0.0750 mol Li_2CO_3 can be produced.

67. $? \text{ mol } O_2 = 4.00\text{g } O_2 \times \dfrac{\text{mol } O_2}{32.00 \text{ g}} = 0.125 \text{ mol } O_2$

0.250 mol O, so Hg is limiting.

? mol O excess = 0.250 mol O – 0.200 mol O = 0.050 mol O excess

$? \text{ g } O_2 = 0.050 \text{ mol O excess} \times \dfrac{\text{mol } O_2}{2 \text{ mol O}} \times \dfrac{32.00 \text{ g}}{\text{mol } O_2} = 0.80 \text{ g } O_2$

(d) is the correct answer.

(a) Some Hg must be used up to make HgO.

(b) The reaction should continue to completion.

(c) The Hg is limiting. The Hg(l) and O_2(g) are *not* in stoichiometric proportions.

69. $? \text{ mol KI} = 481 \text{ g HI} \times \dfrac{1 \text{ mol HI}}{127.9 \text{ g HI}} \times \dfrac{1 \text{ mol KI}}{1 \text{ mol HI}} = 3.76 \text{ mol KI}$

$? \text{ mol KI} = 318 \text{ g } KHCO_3 \times \dfrac{1 \text{ mol } KHCO_3}{100.1 \text{ g } KHCO_3} \times \dfrac{1 \text{ mol KI}}{1 \text{ mol } KHCO_3} = 3.18 \text{ mol KI}$

$KHCO_3$ is the limiting reactant, and the mass of the product is

$? \text{ g KI} = 3.18 \text{ mol KI} \times \dfrac{166.0 \text{ g KI}}{1 \text{ mol KI}} = 528 \text{ g KI.}$

HI is in excess.

$? \text{ g HI}_{\text{ consumed}} = 3.18 \text{ mol KI} \times \dfrac{1 \text{ mol HI}}{1 \text{ mol KI}} \times \dfrac{127.9 \text{ g HI}}{1 \text{ mol HI}} = 407 \text{ g HI}_{\text{ consumed}}$

$481 \text{ g HI}_{\text{initially}} - 407 \text{ g HI}_{\text{consumed}} = 74 \text{ g HI}_{\text{in excess}}$

71. $? \text{ mol ZnS} = 0.488 \text{ g Zn} \times \dfrac{\text{mol Zn}}{65.39 \text{ g Zn}} \times \dfrac{8 \text{ mol ZnS}}{8 \text{ mol Zn}} = 0.007463 \text{ mol ZnS}$

$? \text{ mol ZnS} = 0.503 \text{ g } S_8 \times \dfrac{\text{mol } S_8}{256.53 \text{ g } S_8} \times \dfrac{8 \text{ mol ZnS}}{\text{mol } S_8} = 0.01569 \text{ mol ZnS}$

Zn is limiting.

$? \text{ g ZnS} = 0.007463 \text{ mol ZnS} \times \dfrac{97.46 \text{ g ZnS}}{\text{mol ZnS}} = 0.727 \text{ g ZnS}$ theoretical yield

$? \text{ \% yield} = \dfrac{0.606 \text{ g ZnS actual}}{0.727 \text{ g ZnS theo.}} \times 100\% = 83.4\%$

73. $? \text{ mol } NH_4HCO_3 = 14.8 \text{ g } NH_3 \times \dfrac{\text{mol } NH_3}{17.03 \text{ g } NH_3} \times \dfrac{\text{mol } NH_4HCO_3}{\text{mol } NH_3}$

$= 0.8691 \text{ mol } NH_4HCO_3$

$? \text{ mol } NH_4HCO_3 = 41.3 \text{ g } CO_2 \times \dfrac{\text{mol } CO_2}{44.01 \text{ g } CO_2} \times \dfrac{\text{mol } NH_4HCO_3}{\text{mol } CO_2}$

$= 0.9384 \text{ mol } NH_4HCO_3$

NH_3 is limiting.

$? \text{ g } NH_4HCO_3 = 0.8691 \text{ mol } NH_4HCO_3 \times \dfrac{79.06 \text{ g } NH_4HCO_3}{\text{mol } NH_4HCO_3} \times \dfrac{74.7 \text{ g}}{100.0\text{g}}$

$= 51.3 \text{ g } NH_4HCO_3$

75. (a) $? \text{ M HCl} = \dfrac{6.00 \text{ mol HCl}}{2.50 \text{ L solution}} = 2.40 \text{ M HCl}$

(b) $? \text{ M } Li_2CO_3 = \dfrac{0.000700 \text{ mol } Li_2CO_3}{10.0 \text{ mL solution}} \times \dfrac{\text{mL}}{10^{-3} \text{ L}} = 0.0700 \text{ M } Li_2CO_3$

(c) $? \text{ M } H_2SO_4 = \dfrac{8.905 \text{ g } H_2SO_4}{100.0 \text{ mL solution}} \times \dfrac{\text{mol } H_2SO_4}{98.079 \text{ g } H_2SO_4} \times \dfrac{\text{mL}}{10^{-3} \text{ L}}$

$= 0.9079 \text{ M } H_2SO_4$

(d) $? \text{ M } C_6H_{12}O_6 = \dfrac{439 \text{ g } C_6H_{12}O_6}{1.25 \text{ L solution}} \times \dfrac{\text{mol } C_6H_{12}O_6}{180.2 \text{ g } C_6H_{12}O_6} = 1.95 \text{ M } C_6H_{12}O_6$

(e) $? \text{ M } C_3H_8O_3 = \dfrac{15.50 \text{ mL } C_3H_8O_3}{225.0 \text{ mL}} \times \dfrac{\text{mL}}{10^{-3} \text{ L}} \times \dfrac{1.265 \text{ g}}{\text{mL}}$

$\times \dfrac{\text{mol } C_3H_8O_3}{92.094 \text{ g } C_3H_8O_3} = 0.9463 \text{ M}$

(f) $? \text{ M } C_3H_7OH = \dfrac{35.0 \text{ mL } C_3H_7OH}{250 \text{ mL}} \times \dfrac{\text{mL}}{10^{-3} \text{ L}} \times \dfrac{0.786 \text{ g}}{\text{mL}} \times \dfrac{\text{mol } C_3H_7OH}{60.10 \text{ g } C_3H_7OH}$

$= 1.83 \text{ M } C_3H_7OH$

77. (a) $? \text{ mol NaOH} = 1.25 \text{ L} \times 0.0235 \text{ M NaOH} = 0.0294 \text{ mol NaOH}$

(b) $? \text{ g } C_6H_{12}O_6 = 10.0 \text{ mL} \times \dfrac{10^{-3} \text{ L}}{\text{mL}} \times 4.25 \text{ M} \times \dfrac{180.2 \text{ g } C_6H_{12}O_6}{\text{mol } C_6H_{12}O_6}$

$= 7.66 \text{ g } C_6H_{12}O_6$

(c) $? \text{ mL } C_6H_{15}O_3N = 2.225 \text{ L} \times 0.2500 \text{ M} \times \dfrac{149.19 \text{ g } C_6H_{15}O_3N}{\text{mol } C_6H_{15}O_3N}$

$\times \dfrac{\text{mL}}{1.0985 \text{ g}} = 75.55 \text{ mL } C_6H_{15}O_3N$

(d) $? \text{ mL } C_4H_{10}O = 715 \text{ mL} \times 1.34 \text{ M} \times \dfrac{10^{-3} \text{ L}}{\text{mL}} \times \dfrac{74.12 \text{ g } C_4H_{10}O}{\text{mol } C_4H_{10}O}$

$\times \dfrac{\text{mL}}{0.808 \text{ g}} = 87.9 \text{ mL } C_4H_{10}O$

79. (a) $? \text{ mL} = 0.0867 \text{ mol NaBr} \times \dfrac{1}{0.215 \text{ M}} \times \dfrac{\text{mL}}{10^{-3} \text{ L}} = 403 \text{ mL}$

(b) $? \text{ mL} = 32.1 \text{ g } CO(NH_2)_2 \times \dfrac{\text{mol } CO(NH_2)_2}{60.06 \text{ g } CO(NH_2)_2} \times \dfrac{1}{0.215 \text{ M}} \times \dfrac{\text{mL}}{10^{-3} \text{ L}}$

$= 2.49 \times 10^3 \text{ mL}$

(c) $? \text{ mL} = 715 \text{ mg } CH_3OH \times \dfrac{\text{mol } CH_3OH}{32.04 \text{ g } CH_3OH} \times \dfrac{1}{0.215 \text{ M}} = 104 \text{ mL}$

81. $? \text{ M } HNO_3 = \dfrac{67.0 \text{ g } HNO_3}{100 \text{ g solution}} \times \dfrac{1.40 \text{ g}}{\text{mL}} \times \dfrac{\text{mL}}{10^{-3} \text{ L}} \times \dfrac{\text{mol } HNO_3}{63.02 \text{ g } HNO_3} = 14.9 \text{ M } HNO_3$

83. $? \text{ M} = 25.00 \text{ mL} \times \frac{10^{-3} \text{ L}}{\text{mL}} \times \frac{1.04 \text{ mol}}{\text{L}} \times \frac{1}{0.500 \text{ L}} = 0.0520 \text{ M}$

85. (a) $M_C V_C = M_D V_D$

$$V_C = \frac{M_D V_D}{M_C} = \frac{0.5000 \text{ M HCl x } 2.000 \text{ L}}{6.052 \text{ M HCl}} = 0.1652 \text{ L} \times \frac{\text{mL}}{10^{-3} \text{ L}} = 165.2 \text{ mL}$$

(b) $? \text{ mL} = \frac{7.150 \text{ mg HCl}}{\text{mL}} \times 500.0 \text{ mL} \times \frac{\text{mol HCl}}{36.46 \text{ g HCl}} \times \frac{1}{6.052 \text{ M}} = 16.20 \text{ mL}$

87. (c) 0.17 M NH_3: The concentration must be greater than the average of 0.10 M and 0.20 M, but less than the greater of the two (0.20M).
The actual calculation would be:

$$C = \frac{(0.100 \text{ L} \times 0.100 \text{ mol/L}) + (0.200 \text{ L} \times 0.200 \text{ mol/L})}{0.100 \text{ L} + 0.200 \text{ L}} = 0.167 \text{ M} => 0.17 \text{ M}$$

89. Measure 25.00 mL of 0.0400 M $AgNO_3$ solution 3 times and 10.00 mL once into a flask. Then add 0.8664 g $AgNO_3$.

$$? \text{ mol needed} = (85.0 \text{ ml} \times \frac{10^{-3} \text{ L}}{\text{mL}} \times 0.100 \text{ M}) - (85.0 \text{ mL} \times \frac{10^{-3} \text{ L}}{\text{mL}} \times 0.0400 \text{ M})$$

$$= 5.10 \times 10^{-3} \text{ mol}$$

$$? \text{ g } AgNO_3 \text{ added} = 5.1 \times 10^{-3} \text{ mol} \times \frac{169.88 \text{ g } AgNO_3}{\text{mol } AgNO_3} = 0.866 \text{ g } AgNO_3$$

(An alternate method is listed below.)
Using a volumetric flask, it is necessary to make 100 mL of solution, as that is the smallest volumetric flask available with a volume greater than 80 mL. Measure 100.0 mL of the 0.04000 M $AgNO_3$ solution in the 100.0 mL volumetric flask, remove 20.00 mL with the 10.00 mL pipet, and add 0.815 g of $AgNO_3$.

$$? \text{ mol needed} = 80.0 \text{ mL} \times \frac{10^{-3} \text{ L}}{\text{mL}} \times 0.100 \text{ M} - 80.0 \text{ mL} \times \frac{10^{-3} \text{ L}}{\text{mL}} \times 0.04000 \text{ M}$$

$$= 4.80 \times 10^{-3} \text{ mol needed}$$

$$? \text{ g } AgNO_3 \text{ added} = 4.80 \times 10^{-3} \text{ mol} \times \frac{169.88 \text{ g } AgNO_3}{\text{mol } AgNO_3} = 0.815 \text{ g Ag } NO_3 \text{ added}$$

91. $? \text{ g } BaSO_4 = 635 \text{ mL} \times 0.314 \text{ M } Na_2SO_4 \times \frac{10^{-3} \text{ L}}{\text{mL}} \times \frac{\text{mo lBaSO}_4}{\text{mol } Na_2SO_4}$

$$\times \frac{233.4 \text{ g } BaSO_4}{\text{mol } BaSO_4} = 46.5 \text{ g } BaSO_4$$

93. $? \text{ mol HCl used} = 2.02 \text{ g Al} \times \frac{\text{mol Al}}{26.98 \text{ g Al}} \times \frac{6 \text{ mol HCl}}{2 \text{ mol Al}} = 0.225 \text{ mol HCl used}$

? mol HCl available = 0.400 L × 2.75 M HCl = 1.10 mol HCl

$$? \text{ M} = \frac{1.10 \text{ mol} - 0.225 \text{ mol}}{0.400 \text{ L}} = 2.2 \text{ M}$$

95. $CaCO_3 + 2HCl \rightarrow CaCl_2 + CO_2 + H_2O$

$$? \text{ g } CO_2 = 4.35 \text{ g } CaCO_3 \times \frac{\text{mol } CaCO_3}{100.1 \text{ g } CaCO_3} \times \frac{\text{mol } CO_2}{\text{mol } CaCO_3} \times \frac{44.01 \text{ g } CO_2}{\text{mol } CO_2} = 1.91 \text{ g } CO_2$$

$$? \text{ g } CO_2 = 75.0 \text{ mL} \times 1.50 \text{ M HCl} \times \frac{10^{-3} \text{ L}}{\text{mL}} \times \frac{\text{mol } CO_2}{2 \text{ mol HCl}} \times \frac{44.01 \text{ g } CO_2}{\text{mol } CO_2} = 2.48 \text{ g } CO_2$$

$CaCO_3$ is limiting. 1.91 g CO_2 is produced.

97. (a) $2 \text{ Al} + 6 \text{ HCl} \rightarrow 2 \text{ AlCl}_3 + 3 \text{ H}_2$

$$? \text{ cm}^2 = 0.05 \text{ mL} \times 12.0 \text{ M HCl} \times \frac{10^{-3} \text{ L}}{\text{mL}} \times \frac{2 \text{ mol Al}}{6 \text{ mol HCl}} \times \frac{26.99 \text{ g Al}}{\text{mol Al}} \times \frac{\text{cm}^3}{2.70 \text{ g}} \times \frac{1}{0.10 \text{ mm}} \times \frac{\text{mm}}{10^{-3} \text{ m}} \times \frac{10^{-2} \text{ m}}{\text{cm}} = 0.2 \text{ cm}^2$$

(b)

$$? \text{ drops} = 1.50 \text{ cm}^2 \times 0.065 \text{ mm} \times \frac{10^{-3} \text{ m}}{\text{mm}} \times \frac{\text{cm}}{10^{-2} \text{ m}} \times \frac{8.96 \text{ g}}{\text{cm}^3} \times \frac{\text{mol Cu}}{63.55 \text{ g Cu}} \times \frac{8 \text{ mol } HNO_3}{3 \text{ mol Cu}} \times \frac{1}{6 \text{ M } HNO_3} \times \frac{\text{mL}}{10^{-3} \text{ L}} \times \frac{\text{drop}}{0.05 \text{ mL}} = 12.2 \text{ drops} \Rightarrow 13 \text{ drops}$$

Additional Problems

99. (a)

$$? \text{ g } CaCO_3 = 875 \text{ mg Ca} \times \frac{100.1 \text{ g } CaCO_3}{40.08 \text{ g Ca}} \times \frac{10^{-3} \text{ g}}{\text{mg}} = 2.19 \text{ g } CaCO_3$$

(b)

$$? \text{ g } Ca(C_3H_5O_3)_2 = 875 \text{ mg Ca} \times \frac{218.2 \text{ g } Ca(C_3H_5O_3)_2}{40.08 \text{ g Ca}} \times \frac{10^{-3} \text{ g}}{\text{mg}} = 4.76 \text{ g } Ca(C_3H_5O_3)_2$$

(c)

$$? \text{ g } Ca(C_6H_{11}O_7)_2 = 875 \text{ mg Ca} \times \frac{430.4 \text{ g } Ca(C_6H_{11}O_7)_2}{40.08 \text{ g Ca}} \times \frac{10^{-3} \text{ g}}{\text{mg}} = 9.40 \text{ g } Ca(C_6H_{11}O_7)_2$$

(d)

$$? \text{ g } Ca_3(C_6H_5O_7)_2 = 875 \text{ mg Ca} \times \frac{498.4 \text{ g } Ca_3(C_6H_5O_7)_2}{3 \times 40.08 \text{ g Ca}} \times \frac{10^{-3} \text{ g}}{\text{mg}} = 3.63 \text{ g } Ca_3(C_6H_5O_7)_2$$

101. $? \text{ u/molecule} = \frac{24.31 \text{ u Mg}}{\text{atom Mg}} \times \frac{\text{atom Mg}}{\text{molecule chlorophyll}} \times \frac{100 \text{ u chlorophyll}}{2.72 \text{ u Mg}}$

$= \frac{894 \text{ u}}{\text{molecule}}$

103. $0.210 \text{ g HCl} \times \frac{\text{mol Cl}^-}{36.46 \text{ g HCl}} = 5.76 \times 10^{-3} \text{ mol Cl}^- \times \frac{1}{1.92 \text{ x } 10^{-3} \text{ mol Br}^-}$

$= \frac{3 \text{ mol Cl}^-}{\text{mol Br}}$

$0.155 \text{ g HBr} \times \frac{\text{mol Br}}{80.91 \text{ HBr}} = 1.92 \times 10^{-3} \text{ mol Br}^- \times \frac{1}{1.92 \text{ x } 10^{-3} \text{ mol Br}^-}$

$= \frac{1 \text{ mol Br}^-}{\text{mol Br}^-}$

$BrCl_3$

105. $MCl_2 + 2\, AgNO_3 \rightarrow 2\, AgCl + M(NO_3)_2$

$? \text{ mol } MCl_2 = 1.8431 \text{ g AgCl} \times \frac{\text{mol AgCl}}{143.321 \text{ g AgCl}} \times \frac{\text{mol } MCl_2}{2 \text{ mol AgCl}}$

$= 0.0064300 \text{ mol } MCl_2$

$? \text{ molar mass } MCl_2 = \frac{0.8150 \text{ g } MCl_2}{0.0064300 \text{ mol } MCl_2} = 126.75 \text{ g/mol}$

$? \text{ atomic mass M} = 126.75 \text{ g/mol} - 2 \times 35.45 \text{ g/mol} = 55.85 \text{ g/mol}$ atomic mass
M is iron.

107. $? \text{ g C} = 3.047 \text{ g } CO_2 \times \frac{12.011 \text{ g C}}{44.009 \text{ g } CO_2} = 0.8316 \text{ g C}$

$? \text{ g H} = 1.247 \text{ g } H_2O \times \frac{2.0159 \text{ g H}}{18.015 \text{ g } H_2O} = 0.1395 \text{ g H}$

$? \text{ g O} = 1.525 \text{ g sample} - 0.832 \text{ g C} - 0.140 \text{ g H} = 0.553 \text{ g O}$

$? \text{ mol O} = 0.553 \text{ g O} \times \frac{\text{mol O}}{16.00 \text{ g O}} = 3.456 \times 10^{-2} \text{ mol O} \times \frac{1}{0.03456 \text{ mol O}} = 1.00$

$? \text{ mol C} = 0.832 \text{ g C} \times \frac{\text{mol C}}{12.01 \text{ g C}} = 6.928 \times 10^{-2} \text{ mol C} \times \frac{1}{0.03456 \text{ mol O}} = 2.00$

$? \text{ mol H} = 0.1395 \text{ g H} \times \frac{\text{mol H}}{1.008 \text{ g H}} = 1.384 \times 10^{-1} \text{ mol H} \times \frac{1}{0.03456 \text{ mol O}} = 4.00$

C_2H_4O formula weight 44.0 u

$? \frac{\text{formula}}{\text{molecule}} = \frac{88.1 \text{ u}}{\text{molecule}} \times \frac{\text{formula}}{44.0 \text{ u}} = \frac{2 \text{ formula}}{\text{molecule}}$

$C_4H_8O_2$

There are several structures that could be made of $C_4H_8O_2$. Some of them are:

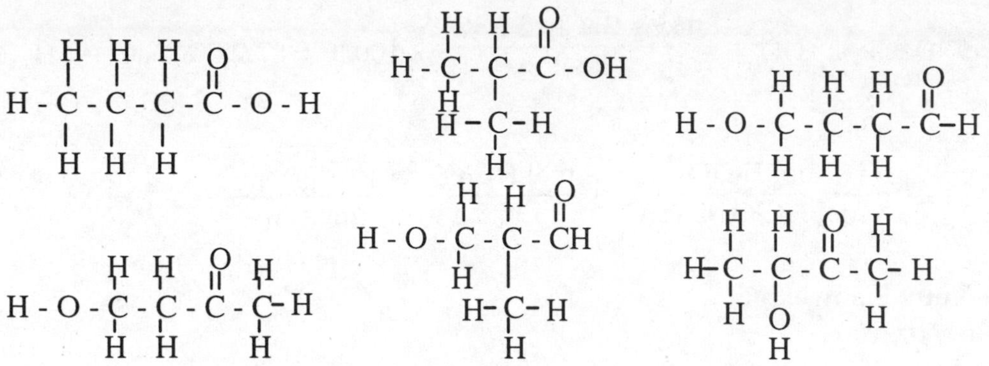

109. $? \text{ mol } C_4H_9Br = 13.0 \text{ g} \times \frac{\text{mol } C_4H_9OH}{74.12 \text{ g } C_4H_9OH} \times \frac{\text{mol } C_4H_9Br}{\text{mol } C_4H_9OH}$

$= 0.1754 \text{ mol } C_4H_9Br$

$? \text{ mol } C_4H_9Br = 21.6 \text{ g} \times \frac{\text{mol NaBr}}{102.9 \text{ g NaBr}} \times \frac{\text{mol } C_4H_9Br}{\text{mol NaBr}} = 0.2099 \text{ mol } C_4H_9Br$

$? \text{ mol } C_4H_9Br = 33.8 \text{ g} \times \frac{\text{mol } H_2SO_4}{98.09 \text{ g } H_2SO_4} \times \frac{\text{mol } C_4H_9Br}{\text{mol } H_2SO_4} = 0.3446 \text{ mol } C_4H_9Br$

C_4H_9OH limiting

$? \text{ g} = 0.1754 \text{ mol } C_4H_9Br \times \frac{137.0 \text{ g } C_4H_9Br}{\text{mol } C_4H_9Br} = 24.0 \text{ g theoretical yield}$

16.8 g actual yield $\qquad ?\% \text{ yield} = \frac{16.8 \text{ g}}{24.0 \text{ g}} \times 100\% = 70.0\% \text{ yield}$

111. $? \text{ g Na} = 250.0 \text{ mL} \times \frac{10^{-3} \text{ L}}{\text{mL}} \times 0.315 \text{ M} \times \frac{2 \text{ mol Na}}{2 \text{ mol NaOH}} \times \frac{22.99 \text{ g Na}}{\text{mol Na}} = 1.81 \text{ g Na}$

113. $37.51 \text{ g C} \times \frac{\text{mol C}}{12.011 \text{ g}} = 3.123 \text{ mol C}$

$3.15 \text{ g H} \times \frac{\text{mol H}}{1.008 \text{ g}} = 3.125 \text{ mol H}$

$59.34 \text{ g F} \times \frac{\text{mol F}}{18.998 \text{ g}} = 3.123 \text{ mol F}$

CHF empirical formula

$? \frac{\text{formula}}{\text{molecule}} = \frac{96.0524\text{u}}{\text{molecule}} \times \frac{\text{formula}}{32.017\text{u}} = \frac{3 \text{ formula}}{\text{molecule}}$

$C_3H_3F_3$ molecular formula

$4\ C_3H_3F_3 + 12\ O_2 \rightarrow 3\ CF_4 + 9\ CO_2 + 6\ H_2O$

115. $? \text{ g } BaCl_2{\bullet}2\ H_2O = (1.6992 \text{ g} - 1.4804 \text{ g}) \times \frac{\text{mol } H_2O}{18.016 \text{ g}} \times \frac{\text{mol } BaCl_2{\bullet}2\ H_2O}{2 \text{ mol } H_2O}$

$\times \frac{244.26}{\text{mol } BaCl_2{\bullet}2\ H_2O} = 1.483 \text{ g } BaCl_2{\bullet}2\ H_2O$

$$? \% \ BaCl_2\bullet 2\ H_2O = \frac{1.483 \text{ g } BaCl_2\bullet 2\ H_2O}{1.6992 \text{ g}} \times 100\ \% = 87.28\ \%\ BaCl_2\bullet 2\ H_2O$$

12.72 % NaCl

117. $$? \ M = \frac{1 \text{ mol } HClO_4}{1 \text{ mol } HClO_4\bullet 2\ H_2O} \times \frac{\text{mol } HClO_4}{136.5 \text{ g}} \times \frac{1.65 \text{ g}}{\text{mL}} \frac{\text{mL}}{10^{-3} \text{ L}} = 12.1 \text{ M}$$

Apply Your Knowledge

119 (a) NH_4NO_3

$$? \ \% \ N = \frac{2 \text{ x } 14.01}{2 \text{ x } 14.01 + 4 \text{ x } 1.01 + 3 \text{ x } 16.00} \times 100\% = 35.00\ \%\ N$$

NPK is 35-0-0

(b) KH_2PO_4

$$?\% \ K_2O = \frac{94 \text{ g } K_2O}{\text{mol } K_2O} \times \frac{1 \text{ mol } K_2O}{2 \text{ mol } KH_2PO_4} \times \frac{\text{mol } KH_2PO_4}{136 \text{ g } KH_2PO_4} \times 100\% = 35\%$$

$$?\% \ P_2O_5 = \frac{142 \text{ g } P_2O_5}{\text{mol } P_2O_5} \times \frac{\text{mol } P_2O_5}{2 \text{ mol } KH_2PO_4} \times \frac{\text{mol } KH_2PO_4}{136 \text{ g } KH_2PO_4} \times 100\% = 52\%$$

K_2HPO_4

$$?\% \ K_2O = \frac{94 \text{ g } K_2O}{\text{mol } K_2O} \times \frac{\text{mol } K_2O}{\text{mol } K_2HPO_4} \times \frac{\text{mol } K_2HPO_4}{174 \text{ g } K_2HPO_4} \times 100\% = 54\%$$

$$?\% \ P_2O_5 = \frac{142 \text{ g } P_2O_5}{\text{mol } P_2O_5} \times \frac{\text{mol } P_2O_5}{2 \text{ mol } K_2HPO_4} \times \frac{\text{mol } K_2HPO_4}{174 \text{ g } K_2HPO_4} \times 100\% = 41\%$$

K_3PO_4

$$?\% \ K_2O = \frac{94 \text{ g } K_2O}{\text{mol } K_2O} \times \frac{3 \text{ mol } K_2O}{2 \text{ mol } K_3PO_4} \times \frac{\text{mol } K_3PO_4}{212 \text{ g } K_3PO_4} \times 100\% = 67\%$$

$$?\% \ P_2O_5 = \frac{142 \text{ g } P_2O_5}{\text{mol } P_2O_5} \times \frac{\text{mol } P_2O_5}{2 \text{ mol } K_3PO_4} \times \frac{\text{mol } K_3PO_4}{212 \text{ g } K_3PO_4} \times 100\% = 33\%$$

0.5 mol of KH_2PO_4 would provide ≈ 18% (0.5 × 35) of K_2O, so one compound is KH_2PO_4. It would also provide ≈ 26% of the P_2O_5.

$NH_4H_2PO_4$

$$?\% \ P_2O_5 = \frac{142 \text{ g } P_2O_5}{\text{mol } P_2O_5} \times \frac{\text{mol } P_2O_5}{2 \text{ mol } NH_4H_2PO_4} \times \frac{\text{mol } NH_4H_2PO_4}{115 \text{ g } NH_4H_2PO_4} \times 100\% = 62\%$$

$$?\% \ N = \frac{14.01 \text{ g N}}{\text{mol N}} \times \frac{\text{mol N}}{\text{mol } NH_4H_2PO_4} \times \frac{\text{mol } NH_4H_2PO_4}{115 \text{ g } NH_4H_2PO_4} \times 100\% = 12\%$$

That is too much P_2O_5 and not enough N.

$(NH_4)_2HPO_4$

$$?\% \ P_2O_5 = \frac{142 \text{ g } P_2O_5}{\text{mol } P_2O_5} \times \frac{\text{mol } P_2O_5}{2 \text{ mol } (NH_4)_2HPO_4} \times \frac{\text{mol } (NH_4)_2HPO_4}{132 \text{ g } (NH_4)_2HPO_4} \times 100 = 54\% \ P_2O_5$$

$$?\%\ N = \frac{14.01\ g\ N}{mol\ N} \times \frac{2\ mol\ N}{mol\ (NH_4)_2HPO_4} \times \frac{mol\ (NH_4)_2HPO_4}{132\ g\ (NH_4)_2HPO_4} \times 100\% = 21\%$$

A 1:1 mole ratio of $(NH4)_2HPO_4$ and KH_2PO_4 would produce a 10-53-18 fertilizer. The 1:1 mole ratio would be a 132:136 $(NH4)_2HPO_4$ to KH_2PO_4 mass ratio, so a 1:1 mass ratio would also provide about the same fertilizer.

(c) $$?\ g\ KNO_3 = 20.0\ g\ K_2O \times \frac{mol\ K_2O}{94.2\ g} \times \frac{2\ mol\ KNO_3}{mol\ K_2O} \times \frac{101.11\ g}{mol\ KNO_3}$$
$$= 42.9\ g\ KNO_3$$

$$?\ g\ N = 42.9\ g\ KNO_3 \frac{mol\ KNO_3}{101.11\ g\ KNO_3} \times \frac{mol\ N}{mol\ KNO_3} \times \frac{14.01\ g}{mol\ N} = 5.94\ g\ N$$

That is 5.95 % N. 14.06 % N is needed from NH_4NO_3.

$$?\ g\ NH_4NO_3 = 14.01\ g\ N \frac{mol\ N}{14.01\ g} \times \frac{mol\ NH_4NO_3}{x\ \ 2\ mol\ N} \times \frac{80.03\ g}{mol\ NH_4NO_3}$$
$$= 40.2\ g\ NH_4NO_3$$

$$?\ NaH_2PO_4 = 20.0\ g\ P_2O_5 \frac{mol\ P_2O_5}{141.94\ g} \times \frac{2\ mol\ P_2O_5}{mol\ P_2O_5} \times \frac{119.98\ g}{mol\ NaH2PO_4}$$
$$= 33.8\ g\ NaH_2PO_4$$

The 20-20-20 fertilizer cannot be made from these three compounds. To make 20 g of N, P_2O_5 and K_2O requires 116.9 g. The percentage of each is less then 20% (17.1%). A mixture could be made that would contain two of the components, but there would not be enough mass left to put in a sufficient amount of the third component.

121. $3\ Mg + N_2 \rightarrow Mg_3N_2$

$$?\ mol\ \ Mg_3N_2 = 35.00\ g\ Mg \times \frac{mol\ Mg}{24.305\ g\ Mg} \times \frac{1\ mol\ Mg_3N_2}{3\ mol\ Mg}$$
$$= 0.4800\ mol\ Mg_3N_2$$

(a) $$?\ mol\ Mg_3N_2 = 15.00\ g\ gas \times \frac{95\ g\ N_2}{100\ g\ gas} \times \frac{1\ mol\ N_2}{28.013\ g\ N_2} \times \frac{mol\ Mg_3N_2}{mol\ N_2}$$
$$= 0.51\ mol\ Mg_3N_2$$

Mg is still limiting

$$?\ g\ Mg_3N_2 = 0.4800\ mol \times \frac{100.93\ g\ Mg_3N_2}{mol\ Mg_3N_2} = 48.45\ g\ Mg_3N_2$$

(b) $$?\ g\ Mg_3N_2 = 15.00\ g\ gas \times \frac{85\ g\ N_2}{100\ g\ gas} \times \frac{1\ mol\ N_2}{28.013\ g\ N_2} \times \frac{mol\ Mg_3N_2}{mol\ N_2}$$
$$= 0.46\ mol\ Mg_3N_2$$

Now N_2 is limiting.

$$?\ g\ Mg_3N_2 = \ 0.46\ mol\ Mg_3N_2 \times \frac{100.93\ g\ Mg_3N_2}{mol\ Mg_3N_2} = 46\ gMg_3N_2$$

(c) $2\ Mg + O_2 \rightarrow 2\ MgO$

$$? \text{ g MgO} = 15.00 \text{ g gas} \times \frac{25\text{ g O}_2}{100\text{ g gas}} \times \frac{\text{mol O}_2}{32.00\text{ g O}_2} \times \frac{2\text{ mol MgO}}{\text{mol O}_2}$$

$$\times \frac{40.304\text{ g MgO}}{\text{mol MgO}} = 9.4\text{ g MgO}$$

$$? \text{ g Mg used} = 15.00 \text{ g gas} \times \frac{25\text{ g O}_2}{100\text{ g gas}} \times \frac{\text{mol O}_2}{32.00\text{ g O}_2} \times \frac{2\text{ mol Mg}}{\text{mol O}_2}$$

$$\times \frac{24.305\text{ g Mg}}{\text{mol Mg}} = 5.7\text{ g Mg used}$$

$? \text{ g Mg left} = 35.00 \text{ g Mg} - 5.7 \text{ g} = 29.3 \text{ g left}$

$$? \text{ mol Mg}_3\text{N}_2 = 29.3 \text{ g Mg} \times \frac{\text{mol Mg}}{24.305\text{ g Mg}} \times \frac{\text{mol Mg}_3\text{N}_2}{3\text{ mol Mg}} = 0.402\text{ mol Mg}_3\text{N}_2$$

$$? \text{ mol Mg}_3\text{N}_2 = 15.00 \text{ g gas} \times \frac{75\text{ g N}_2}{100\text{ g gas}} \times \frac{1\text{ mol N}_2}{28.013\text{ g N}_2} \times \frac{\text{mol Mg}_3\text{N}_2}{\text{mol N}_2}$$

$$= 0.40\text{ mol Mg}_3\text{N}_2$$

Neither is limiting.

$$? \text{ g Mg}_3\text{N}_2 = 0.40 \text{ mol Mg}_3\text{N}_2 \times \frac{100.93\text{ g Mg}_3\text{N}_2}{\text{mol Mg}_3\text{N}_2} = 41\text{ g Mg}_3\text{N}_2$$

Total product = 9.4 g MgO + 41 g Mg_3N_2 = 50 g

123. $40.00 \text{ g C} \times \frac{\text{mol C}}{12.011\text{ g C}} = 3.330 \text{ mol C} \times \frac{1}{3.330\text{ mol C}} = \frac{1\text{ mol C}}{\text{mol C}}$

$$6.71 \text{ g H} \times \frac{\text{mol H}}{1.008\text{ g H}} = 6.66 \text{ mol H} \times \frac{1}{3.330\text{ mol C}} = \frac{2\text{ mol H}}{\text{mol C}}$$

$$53.29 \text{ g O} \times \frac{\text{mol O}}{16.00\text{ g O}} = 3.331 \text{ mol O} \times \frac{1}{3.330\text{ mol C}} = \frac{1\text{ mol O}}{\text{mol C}}$$

CH_2O

$$\frac{?\text{ formula}}{\text{molecule}} = \frac{60\text{ g}}{\text{molecule}} \times \frac{\text{formula}}{30\text{ g}} = \frac{2\text{ formula}}{\text{molecule}}$$

$C_2H_4O_2$

$$22.56 \text{ g P} \times \frac{\text{mol P}}{30.974\text{ g P}} = 0.7283 \text{ mol P} \times \frac{1}{0.7283\text{ mol P}} = \frac{1\text{ mol P}}{\text{mol P}}$$

$$77.44 \text{ g Cl} \times \frac{\text{mol Cl}}{35.453\text{ g Cl}} = 2.1843 \text{ mol Cl} \times \frac{1}{0.7283\text{ mol P}} = \frac{3.00\text{ mol Cl}}{\text{mol P}}$$

PCl_3

$$3.69 \text{ g H} \times \frac{\text{mol H}}{1.008\text{ g H}} = 3.66 \text{ mol H} \times \frac{1}{1.219\text{ mol P}} = \frac{3\text{ mol H}}{\text{mol P}}$$

$$37.77 \text{ g P} \times \frac{\text{mol P}}{30.974\text{ g P}} = 1.219 \text{ mol P} \times \frac{1}{1.219\text{ mol P}} = \frac{1\text{ mol P}}{\text{mol P}}$$

$$58.53 \text{ g O} \times \frac{\text{mol O}}{15.999\text{ g O}} = 3.658 \text{ mol O} \times \frac{1}{1.219\text{ mol O}} = \frac{\text{mol O}}{\text{mol P}}$$

H_3PO_3

$C_2H_4O_2 + PCl_3 \rightarrow H_3PO_3$

$$? \text{ mol } C_2H_4O_2 = 10.000 \text{ g} \times \frac{\text{mol}}{60 \text{ g}} = 0.167 \text{ mol}$$

$$? \text{ mol } PCl_3 = 7.621 \text{ g} \times \frac{\text{mol}}{137.32 \text{ g}} = 0.0555 \text{ mol}$$

$$? \text{ mol } H_3PO_3 = 4.552 \text{ g} \times \frac{\text{mol}}{81.995 \text{ g}} = 0.0555 \text{ mol}$$

$$\frac{0.167 \text{ mol}}{0.0555 \text{ mol}} = 3$$

$3\ C_2H_4O_2 + PCl_3 \rightarrow H_3PO_3 + C_6O_3Cl_3H_9$

leads to 6 C + 3 O + 3 Cl + 9 H

$3\ CH_3COOH + PCl_3 \rightarrow H_3PO_3 + 3\ CH_3C(O)Cl$

Chapter 4

Chemical Reactions in Aqueous Solutions

Exercises

4.1A $[Na^+] = 0.438\ M$

$[Mg^{2+}] = 0.0512M$

$[Cl^-] = 0.438\ M + (2 \times 0.0512\ M) = 0.540\ M$

4.1B $[\text{glucose}] = \dfrac{20.0\ g}{L} \times \dfrac{mol}{180.2\ g} = 0.111\ M\ \text{glucose}$

$$[C_6H_5O_7{}^{3-}] = = \frac{2.9\ g}{L} \times \frac{mol}{258.1\ g} \times \frac{1\ mol\ C_6H_5O_7{}^{3-}}{1\ mol\ Na_3C_6H_5O_7} = 0.0112\ M\ \text{citrate ion}$$

$$[K^+] = \frac{1.5\ g\ KCl}{L} \times \frac{mol\ KCl}{74.55\ g\ KCl} \times \frac{1\ mol\ K^+}{1\ mol\ KCl} = 0.0201\ M\ K^+$$

$$[Na^+] = \frac{3.5\ g\ NaCl}{L} \times \frac{mol\ NaCl}{58.44\ g\ NaCl} \times \frac{1\ mol\ Na^+}{1\ mol\ NaCl} = 0.0599\ M\ Na^+$$

$[Cl^-] = 0.0201\ M + 0.0599\ M = 0.0800\ M$

$[Na^+] = 0.0599\ M + 3 \times 0.0112\ M = 0.0935\ M$

4.2 (a) $Ca(OH)_2(s) + 2\ HCl(aq) \rightarrow CaCl_2(aq) + 2H_2O(l)$

(b) $Ca^{2+}(aq) + 2\ OH^-(aq) + 2\ H^+(aq) + 2\ Cl^-(aq) \rightarrow Ca^{2+}(aq) + 2\ Cl^-(aq) + 2\ H_2O(l)$

(c) $OH^-(aq) + H^+(aq) \rightarrow H_2O(l)$

4.3A
$$?\ mL\ HBr = 25.00\ mL\ Ba(OH)_2 \times \frac{10^{-3}\ L}{mL} \times 0.01580\ M\ Ba(OH)_2 \times \frac{2\ mol\ HBr}{1\ mol\ Ba(OH)_2} \times \frac{L\ HBr}{0.01060\ mol\ HBr} \times \frac{mL}{10^{-3}\ L} = 74.53\ mL\ HBr$$

Notice that the two mL – L conversions can be left out to simplify the equation. Then the unit is millimoles instead of moles.

$$?\ mL\ HBr = 25.00\ mL\ Ba(OH)_2 \times 0.01580\ M\ Ba(OH)_2 \times \frac{2\ mol\ HBr}{1\ mol\ Ba(OH)_2} \times \frac{L\ HBr}{0.01060\ mol\ HBr} = 74.53\ mL\ HBr$$

4.3B
$$?\ mL\ KOH = 2.000\ g\ \text{solution} \times \frac{96.5\ g\ H_2SO_4}{100\ g\ \text{solution}} \times \frac{mol\ H_2SO_4}{98.09\ g\ H_2SO_4} \times \frac{2\ mol\ KOH}{mol\ H_2SO_4} \times \frac{1\ L\ KOH}{0.3580\ mol\ KOH} \times \frac{mL}{10^{-3}\ L} = 110\ mL\ KOH$$

4.4A $? \text{ M } H_2SO_4 = 25.20 \text{ mL NaOH} \times 0.1000 \text{ M NaOH} \times \dfrac{\text{mol } H_2SO_4}{2 \text{ mol NaOH}}$

$\times \dfrac{1}{25.00 \text{ mL } H_2SO_4} = 0.05040 \text{ M } H_2SO_4$

4.4B $? \text{ mL NaOH} = 10.00 \text{ mL} \times \dfrac{4.12 \text{ g } HC_2H_3O_2}{100 \text{ g solution}} \times \dfrac{1.01 \text{ g}}{\text{mL}} \times \dfrac{\text{mL}}{10^{-3} \text{ L}} \times \dfrac{\text{mol } HC_2H_3O_2}{60.05 \text{ g}}$

$\times \dfrac{\text{mol } OH^-}{\text{mol } HC_2H_3O_2} \times \dfrac{1 \text{ L}}{0.550 \text{ mol } OH^-} = 12.6 \text{ mL NaOH}$

4.5 The CH_3NH_2 will cause a dimly lit bulb, as it is a weak base. HNO_3 is a strong acid and will cause a brightly lit bulb. The combination will produce a strong electrolyte and a brightly lit bulb.

$CH_3NH_2 + H_2O \rightleftharpoons CH_3NH_3^+ + OH^-$

$HNO_3 \rightarrow H^+ + NO_3^-$

$H^+ + OH^- \rightarrow H_2O$

$CH_3NH_2 + HNO_3 \rightarrow CH_3NH_3^+ + NO_3^-$

4.6 (a) $MgSO_4(aq) + 2\ KOH(aq) \rightarrow Mg(OH)_2(s) + 2\ K^+(aq) + SO_4^{2-}(aq)$

$Mg^{2+}(aq) + OH^-(aq) \rightarrow Mg(OH)_2(s)$

(b) $2\ FeCl_3(aq) + 3\ Na_2S(aq) \rightarrow Fe_2S_3(s) + 6\ Na^+(aq) + 6\ Cl^-(aq)$

$2Fe^{3+}(aq) + 3\ S^{2-}(aq) \rightarrow Fe_2S_3(s)$

(c) $Sr(NO_3)_2\ (aq) + Na_2SO_4\ (aq) \rightarrow SrSO_4\ (s) + 2\ NaNO_3\ (aq)$

$Sr^{2+}\ (aq) + SO_4^{2-}\ (aq) \rightarrow SrSO_4\ (s)$

4.7 The acid will cause the $Fe(OH)_3$ to dissolve because the H^+ will react with the OH^- to form H_2O.

$$Fe(OH)_3(s) + 3\ H^+(aq) \rightarrow Fe^{3+}(aq) + 3\ H_2O(l)$$

4.8A $? \text{ g NaCl} = 0.9372 \text{ g AgCl} \times \dfrac{\text{mol AgCl}}{143.32 \text{ g AgCl}} \times \dfrac{\text{mol NaCl}}{\text{mol AgCl}} \times \dfrac{58.442 \text{ g NaCl}}{\text{mol NaCl}}$

$= 0.3822 \text{ g NaCl}$

$?\% \text{ NaCl} = \dfrac{0.3822 \text{ g NaCl}}{0.9056 \text{ g sample}} \times 100\% = 42.20\% \text{ NaCl}$

4.8B (a) $? \text{ g AgCl} = 0.540 \text{ M } Cl^- \times 225 \text{ mL} \times \dfrac{\text{mol AgCl}}{\text{mol } Cl^-} \times \dfrac{143.3 \text{ g AgCl}}{\text{mol AgCl}} \times \dfrac{10^{-3} \text{ L}}{\text{mL}}$

$= 17.4 \text{ g AgCl}$

(b) $? \text{ g } Mg(OH)_2 = 0.0512 \text{ M } Mg^{2+} \times 5.00 \text{ L} \times \dfrac{\text{mol } Mg(OH)_2}{\text{mol } Mg^{2+}} \times \dfrac{58.33 \text{ g}}{\text{mol } Mg(OH)_2}$

$= 14.9 \text{ g } Mg(OH)_2$

AgCl and $Mg(OH)_2$ are the precipitates.

4.9

+3 -2	0	+1 +7 -2	+1 -2
(a) Al_2O_3	(b) P_4	(c) $NaMnO_4$	(d) ClO^-

+1+5 -2	+1+5 -1	+1- $\frac{1}{2}$	-2+1 -1
(e) $HAsO_4^{2-}$	(f) $HSbF_6$	(g) CsO_2	(h) CH_3F

+2+1 -1	0+10 -2 -2+1
(i) $CHCl_3$	(j) CH_3COOH

4.10A

$$\overset{+7}{MnO_4^-} + \overset{+3}{C_2O_4^{2-}} + H^+ \rightarrow \overset{+2}{Mn^{2+}} + H_2O + \overset{+4}{CO_2}$$

(-5/Mn; +1/C)

$2\ MnO_4^- + 5\ C_2O_4^{2-} + \_\_\ H^+ \rightarrow 2Mn^{2+} + 10\ CO_2 + \_\_\ H_2O$
balance H and O by inspection
$2\ MnO_4^- + 5\ C_2O_4^{2-} + 16\ H^+ \rightarrow 2Mn^{2+} + 10\ CO_2 + 8\ H_2O$
check charge $2(-) + 5 \times 2(-) + 16(+) = 2 \times 2(+)$
$4(+) = 4(+)$

4.10B

$$\overset{0}{Cl_2} + OH^- \rightarrow \overset{+5}{ClO_3^-} + \overset{-1}{Cl^-} + H_2O$$

(-1; +5)

$3\ Cl_2 + \_\_\ OH^- \rightarrow ClO_3^- + 5\ Cl^- + \_\_\ H_2O$
balance H and O
$3\ Cl_2 + 6\ OH^- \rightarrow ClO_3^- + 5\ Cl^- + 3\ H_2O$
check charge $6(-) = (-) + 5(-)$
$6(-) = 6(-)$

4.11A $Cr_2O_7^{2-}(aq)$ is an oxidizing agent; it will react with a reducing agent. $HNO_3(aq)$ is also an oxidizing agent. There is no reaction between $Cr_2O_7^{2-}(aq)$ and $HNO_3(aq)$. HCl is a reducing agent (with Cl in the oxidation state, –1). It is oxidized by $Cr_2O_7^{2-}(aq)$, probably to $Cl_2(g)$.

4.11B The solution in (a) would be $Zn^{2+}(aq)$ in HCl(aq) instead of unreacted HCl(aq). The solution in (b) would be $Zn^{2+}(aq)$, and $Cu^{2+}(aq)$ in $HNO_3(aq)$ instead of just $Cu^{2+}(aq)$ in $HNO_3(aq)$.

4.12 $$0.2865\text{ g sample} \times \frac{58.01\text{ g Fe}}{100\text{ g sample}} \times \frac{\text{mol } Fe^{2+}}{55.847\text{ g } Fe^{2+}} \times \frac{1\text{ mol } Cr_2O_7^{2-}}{6\text{ mol } Fe^{2+}} \times \frac{1}{0.02250\text{ M } Cr_2O_7^{2-}} \times \frac{\text{mL}}{10^{-3}\text{ L}} = 22.04\text{ mL}$$

Alternatively, because the $K_2Cr_2O_7$ solution in Exercise 4.10 has the same molarity as the $KMnO_4$ of Example 4.10 and the stoichiometric factors are 1 mol $Cr_2O_7^{2-}$/ 6

mol Fe^{2+} and 1 mol MnO_4^- / 5 mol Fe^{2+}, the volume of $K_2Cr_2O_7$(aq) required is 5/6 that of $KMnO_4$(aq).
$5/6 \times 26.45$ mL = 22.04 mL

Review Questions

1. nonelectrolytes: (a), (h)
strong electrolytes: (b), (c), (e), (g)
weak electrolytes: (d), (f)

2. (a) salt (b) strong base (c) salt (d) weak acid (e) strong acid
(f) weak base (g) salt (h) strong base

3. (c) is highest, with $[NO_3^-] = 3 \times 0.040 = 0.12$ M.
(a) and (d) both have $[NO_3^-] = 0.10$ M.
(b) is lowest, with $[NO_3^-] = 0.080$ M.

4. (d) [ions] = 0.075 M

5. 0.10 M NaCl has more ions in solution. It is the only strong electrolyte.

6. 0.10 M H_2SO_4. It produces $[H^+] > 0.10$ M, because ionization is complete in the first ionization step and also occurs to some extent in the second.

13. Nitrates, chlorides (except of Pb^{2+}, Ag^+, Hg_2^{2+}), and sulfates (except Sr^{2+}, Ba^{2+}, Pb^{2+}, and Hg_2^{2+}) are soluble. $PbCrO_4$ must be insoluble.

14. Na_2CO_3. Most carbonates are insoluble.
$CO_3^{2-} + Mg^{2+} \rightarrow MgCO_3(s)$

19. (a) 0 Cr (b) +3 ClO_2^- (c) −2 K_2Se (d) +6 TeF_6 (e) −3 PH_4^+

(f) +4 $CaRuO_3$ (g) +4 $SrTiO_3$ (h) +5 $P_2O_7^{4-}$ (i) +2.5 $S_4O_6^{2-}$ (j) -1 NH_2OH

20. (a) −3 C_2H_6 (b) +2 HCOOH (c) +1 $C_2H_2O_2$
(d) −2 $(CH_3)_2O$ (e) +3 $C_2H_2O_4$

Problems

25. (a) $[Li^+] = 0.0385$ M, $[NO_3^-] = 0.0385$ M
(b) $[Ca^{2+}] = 0.035$ M, $[Cl^-] = 0.070$ M
(c) $[Al^{3+}] = 0.0224$ M, $[SO_4^{2-}] = 0.0336$ M

27. $[Na^+] = 0.0554 + (2 \times 0.0145) = 0.0844$ M,
$[Cl^-] = 0.0554$ M, $[SO_4^{2-}] = 0.0145$ M

29. $$[NO_3^-] = \frac{0.112\text{ g}}{125\text{ mL}} \times \frac{\text{mL}}{10^{-3}\text{ L}} \times \frac{\text{mol Mg(NO}_3)_2\cdot 6\text{ H}_2\text{O}}{256.4\text{ g}} \times \frac{2\text{ mol NO}_3^-}{\text{mol Mg(NO}_3)_2\cdot 6\text{ H}_2\text{O}} = 6.99 \times 10^{-3}\text{ M}$$

31. $[Cl^-] = 0.540$ M
$$?\text{ mg/L} = \frac{0.540\text{ mol}}{\text{L}} \times \frac{35.45\text{ g Cl}^-}{\text{mol Cl}^-} \times \frac{\text{mg}}{10^{-3}\text{ g}} = 1.91 \times 10^4\text{ mg/L}$$

33. $$V = 0.250\text{ L} \times \frac{0.0135\text{ mol Cl}^-}{\text{L}} \times \frac{1\text{ mol MgCl}_2}{2\text{ mol Cl}^-} \times \frac{1\text{ L}}{0.0250\text{ M MgCl}_2} \times \frac{\text{mL}}{10^{-3}\text{ L}} = 67.5\text{ mL}$$

35. (a) < (d) < (c) < (b)
(b) is 3×0.45 M = 1.35 M, (c) is $\frac{1.20\text{ mol}}{2.00\text{ L}} = 0.600 \times 2 = 1.20$ M, d is 0.15 + 0.35 = 0.50 M, and (a) is 0.21 M.

37. (a) $HI(g) \xrightarrow{H_2O} H^+(aq) + I^-(aq)$
(b) $KOH(s) \xrightarrow{H_2O} K^+(aq) + OH^-(aq)$
(c) $HNO_2(aq) \rightleftharpoons H^+(aq) + NO_2^-(aq)$
(d) $H_2PO_4^-(aq) \rightleftharpoons H^+(aq) + HPO_4^{2-}(aq)$
(e) $CH_3NH_2(aq) + H_2O \rightleftharpoons CH_3NH_3^+(aq) + OH^-(aq)$
(f) $CH_3CH_2COOH(aq) \rightleftharpoons CH_3CH_2COO^-(aq) + H^+(aq)$

39. $H^+ + OH^- \rightarrow H_2O(l)$
(a) no reaction
(b) $OH^- + H^+ \rightarrow H_2O(l)$
(c) $NH_3 + H^+ \rightarrow NH_4^+$
(b) is the same.

41. Major reaction: $CaCO_3 + 2\text{ H} \rightarrow Ca^{2+} + H_2CO_3 \rightarrow Ca^{2+} + H_2O + CO_2(g)$
Minor reaction: $CuCO_3 + 2\text{ H} \rightarrow Ca^{2+} + H_2CO_3 \rightarrow Cu^{2+} + H_2O + CO_2(g)$
$CO_3^{2-} + 2\text{ H}^+ \rightarrow H_2O + CO_2(g)$

43. (a) $$?\text{ mL} = 25.00\text{ mL} \times \frac{10^{-3}\text{ L}}{\text{mL}} \times 0.0365\text{ M KOH} \times \frac{\text{mol HCl}}{\text{mol KOH}} \times \frac{1}{0.0195\text{ M}} \times \frac{\text{mL}}{10^{-3}\text{ L}} = 46.8\text{ mL}$$

(b) $? \text{ mL} = 10.00 \text{ mL} \times \frac{10^{-3}\text{ L}}{\text{mL}} \times 0.0116 \text{ M Ca(OH)}_2 \times \frac{2 \text{ mol HCl}}{\text{mol Ca(OH)}_2} \times \frac{1}{0.0195 \text{ M}} \times \frac{\text{mL}}{10^{-3}\text{ L}} = 11.9 \text{ mL}$

(c) $? \text{ mL} = 20.00 \text{ mL} \times \frac{10^{-3}\text{ L}}{\text{mL}} \times 0.0225 \text{ M NH}_3 \times \frac{\text{mol HCl}}{\text{mol NH}_3} \times \frac{1}{0.0195 \text{ M}} \times \frac{\text{mL}}{10^{-3}\text{ L}} = 23.1 \text{ mL}$

45. $CH_3COOH + OH^- \rightarrow H_2O + CH_3COO^-$

$? \text{ M} = 31.45 \text{ mL} \times \frac{10^{-3}\text{ L}}{\text{mL}} \times 0.2560 \text{ M KOH} \times \frac{\text{mol CH}_3\text{COOH}}{\text{mol KOH}} \times \frac{1}{10.00 \text{ mL}} \times \frac{\text{mL}}{10^{-3}\text{ L}} = 0.8051 \text{ M}$

47. (a) $? \text{ mg CaCO}_3 = 38.8 \text{ mL} \times 0.251 \text{ M HCl} \times \frac{\text{mole CaCO}_3}{2 \text{ mole HCl}} \times \frac{100.08 \text{ g CaCO}_3}{\text{mole CaCO}_3} = 487 \text{ mg CaCO}_3$

(b) $? \text{ mg Ca}^{2+} = 38.8 \text{ mL} \times 0.251 \text{ M HCl} \times \frac{\text{mole CaCO}_3}{2 \text{ mole HCl}} \times \frac{40.08 \text{ g Ca}^{2+}}{\text{mole CaCO}_3} = 195 \text{ mg Ca}^{2+}.$

49. $? \text{ mol HCl} = 1 \text{ g NaHCO}_3 \times \frac{\text{mol NaHCO}_3}{84 \text{ g NaHCO}_3} \times \frac{\text{mol HCl}}{\text{mol NaHCO}_3}$

$? \text{ mol HCl} = 1 \text{ g CaCO}_3 \times \frac{\text{mol CaCO}_3}{100 \text{ g CaCO}_3} \times \frac{2 \text{ mol HCl}}{\text{mol CaCO}_3}$

$CaCO_3$ will neutralize more acid. $\left(\frac{2}{100} > \frac{1}{84}\right)$

51. All of the CH_3COOH is neutralized, and there are some excess OH^- ions; thus it is past the equivalence point of the titration by a factor of 10% (1 OH^- for each 10 CH_3COO^-). So (d) 22.00 mL.

53. (a) $2\ I^- + Pb^{2+} \rightarrow PbI_2(s)$
(b) no reaction
(c) $Cr^{3+} + 3\ OH^- \rightarrow Cr(OH)_3(s)$
(d) no reaction
(e) $OH^- + H^+ \rightarrow H_2O(l)$
(f) $HSO_4^- + OH^- \rightarrow H_2O(l) + SO_4^{2-}$

55. (a) $Mg(OH)_2\ (s) + 2\ H^+(aq) \rightarrow Mg^{+2}\ (aq) + 2\ H_2O(l)$
(b) $HCOOH(aq) + NH_3(aq) \rightarrow NH_4^+\ (aq) + HCOO^-(aq)$
(c) no reaction

(d) $Cu^{2+}(aq) + CO_3^{2+}(aq) \rightarrow CuCO_3(s)$

(e) no reaction

57. No, it does not indicate what the powder is because $MgSO_4$ dissolves in aqueous solution and $Mg(OH)_2$ dissolves as a result of an acid-base reaction. Testing the pH ($Mg(OH)_2$ is more basic) or adding Ba^{2+} to precipitate $BaSO_4$ are other tests that would obtain results. The simplest test is to add water to the powder. $MgSO_4$ is soluble; $Mg(OH)_2$ is not.

59. $Cu^{2+}(aq) + CO_3^{2-}(aq) \rightarrow CuCO_3(s)$

61. Add $BaCl_2(aq)$ to one sample. If a precipitate ($BaSO_4$) forms, the solution is $Na_2SO_4(aq)$. If there is no precipitate, add $Na_2SO_4(aq)$ to a second sample. Here, a precipitate ($BaSO_4$) indicates the solution is $Ba(NO_3)_2(aq)$, and no precipitate, that it is $NH_3(aq)$.

63. (a) oxidation. The oxidation state of chromium increases from 2 to 3.
    (b) neither. The oxidation states do not change.
    (c) neither. The oxidation state of nitrogen does not change.

65. (a)

$$\overset{-1}{H}\overset{}{C}l + \overset{0}{O_2} \rightarrow \overset{0}{C}l_2 + \overset{-2}{H_2}O$$

$+ 1/Cl$

$- 2/O$

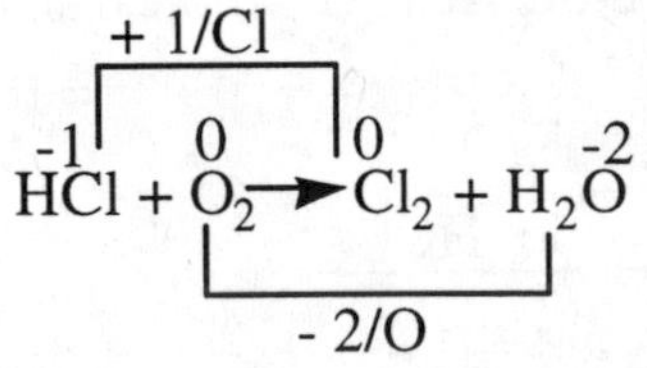

$4\,HCl + O_2 \rightarrow 2\,Cl_2 + 2\,H_2O$

check $0 = 0$

(b)

$$\overset{+2}{N}O + \overset{0}{H_2} \rightarrow \overset{-3}{N}H_3 + \overset{+1}{H_2}O$$

$- 5/N$

$+ 1/H$

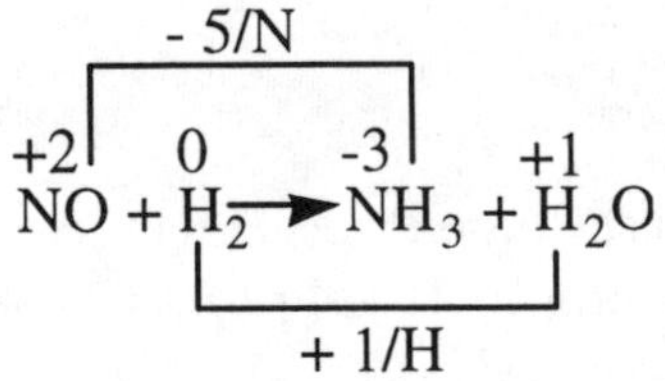

$2\,NO + 5\,H_2 \rightarrow 2\,NH_3 + 2\,H_2O$

(c)

$$\overset{-4}{C}H_4 + \overset{+2}{N}O \rightarrow \overset{0}{N_2} + \overset{+4}{C}O_2 + H_2O$$

$- 2/N$

$+ 8/C$

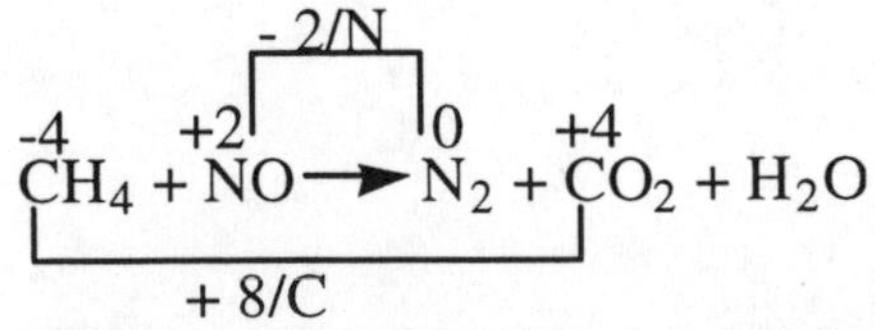

$4\,CH_4 + 16\,NO \rightarrow 8\,N_2 + 4\,CO_2 + 8\,H_2O$

$CH_4 + 4\,NO \rightarrow 2\,N_2 + CO_2 + 2\,H_2O$

(d)

$- 2/O$

$\overset{-3}{N}H_3 + \overset{0}{O_2} \rightarrow \overset{+2}{N}O + H_2\overset{-2}{O}$

$+ 5/N$

$4\ NH_3 + 5\ O_2 \rightarrow 4\ NO + 6\ H_2O$

(e)

$- 4/N$

$\overset{-4}{C}H_4 + \overset{4}{N}O_2 \rightarrow \overset{0}{N_2} + \overset{4}{C}O_2 + H_2O$

$+ 8/C$

$CH_4 + 2\ NO_2 \rightarrow N_2 + CO_2 + 2\ H_2O$

(f)

$+ 1/Cl$

$Ca(\overset{+1}{Cl}O)_2 + H\overset{-1}{Cl} \rightarrow Ca\overset{-1}{Cl_2} + H_2O + \overset{0}{Cl_2}$

$- 2/Cl$

$Ca(ClO)_2 + 4\ HCl \rightarrow CaCl_2 + 2\ H_2O + 2\ Cl_2$

67. (a)

$- 3/Cr$

$Zn + \overset{6}{Cr_2}O_7^{2-} + H^+ \rightarrow Zn^{2+} + Cr^{3+} + H_2O$

$+ 2/Zn$

$3\ Zn + Cr_2O_7^{2-} + 14\ H^+ \rightarrow 3\ Zn^{2+} + 2\ Cr^{3+} + 7\ H_2O$

(b)

$+ 1/I$

$\overset{4}{Se}O_3^{2-} + \overset{-1}{I^-} + H^+ \rightarrow \overset{0}{Se} + \overset{0}{I_2} + H_2O$

$- 4/Se$

$SeO_3^{2-} + 4I^- + 6\ H^+ \rightarrow Se + 2\ I_2 + 3\ H_2O$

(c)

$+ 5/Mn$

$\overset{+2}{Mn^{2+}} + \overset{7}{S_2}O_8^{2-} + H_2O \rightarrow \overset{+7}{Mn}O_4^- + \overset{6}{S}O_4^{2-} + H^+$

$- 1/S$

$2\ Mn^{2+} + 5\ S_2O_8^{2-} + 8\ H_2O \rightarrow 2\ MnO_4^- + 10\ SO_4^{2-} + 16\ H^+$

(d)

$$\overset{3}{CaC_2O_4} + MnO_4^- + H^+ \longrightarrow Ca^{2+} + \overset{4}{Mn^{2+}} + H_2O + CO_2$$

$- 5/Mn$ $+ 1/2$

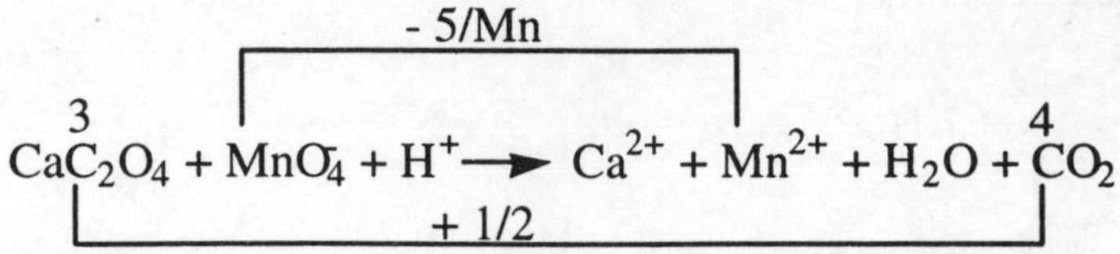

$$5\,CaC_2O_4 + 2\,MnO_4^- + 16\,H^+ \rightarrow 5\,Ca^{2+} + 2\,Mn^{2+} + 8\,H_2O + 10\,CO_2$$

(e)

$$\overset{6}{CrO_4^{2-}} + \overset{-3}{AsH_3} + H_2O \longrightarrow \overset{3}{Cr(OH)_3} + \overset{0}{As} + OH^-$$

$+ 3/As$ $- 3/Cr$

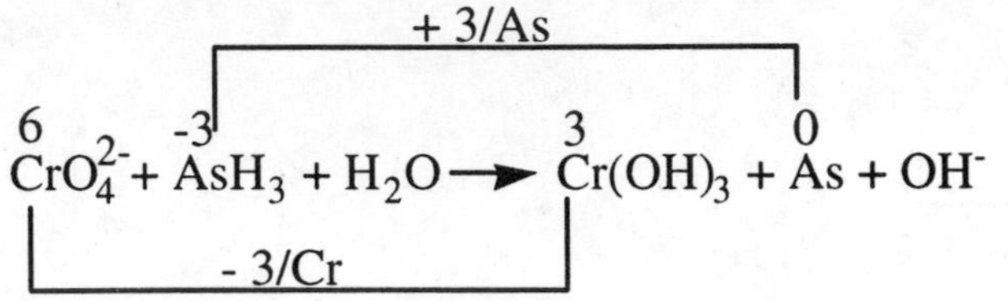

$$CrO_4^{2-} + AsH_3 + H_2O \rightarrow Cr(OH)_3 + As + 2\,OH^-$$

(f)

$$\overset{3}{S_2O_4^{2-}} + \overset{6}{CrO_4^{2-}} + H_2O + OH^- \longrightarrow \overset{3}{Cr(OH)_3} + \overset{4}{SO_3^{2-}}$$

$- 3/Cr$ $+ 1/S$

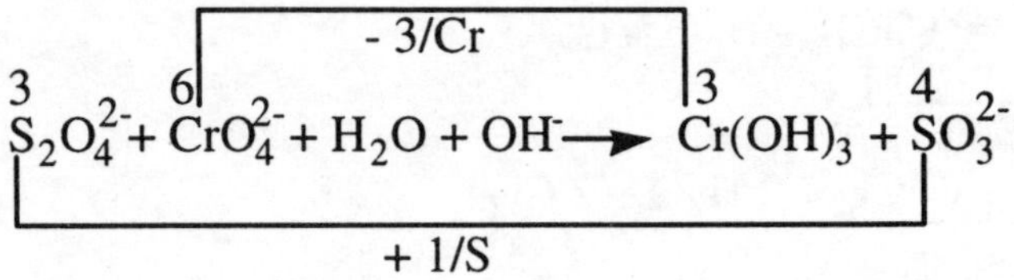

$$3\,S_2O_4^{2-} + 2\,CrO_4^{2-} + 2\,H_2O + 2\,OH^- \rightarrow 2\,Cr(OH)_3 + 6\,SO_3^{2-}$$

(g)

$$\overset{0}{P_4} + H_2O + OH^- \longrightarrow \overset{1}{H_2PO_2^-} + \overset{-3}{PH_3}$$

$+ 1/P$ $- 3/P$

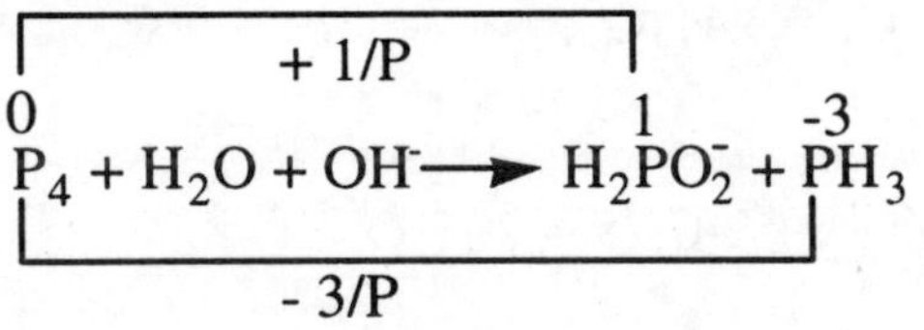

$$P_4 + 3\,H_2O + 3\,OH^- \rightarrow 3\,H_2PO_2^- + PH_3$$

69   (a)

$$\overset{+7}{MnO_4^-} + \overset{3}{C_2H_2O_4} + H^+ \longrightarrow \overset{+2}{Mn^{2+}} + \overset{+4}{CO_2}$$

$+ 1/C$ $- 5/Mn$

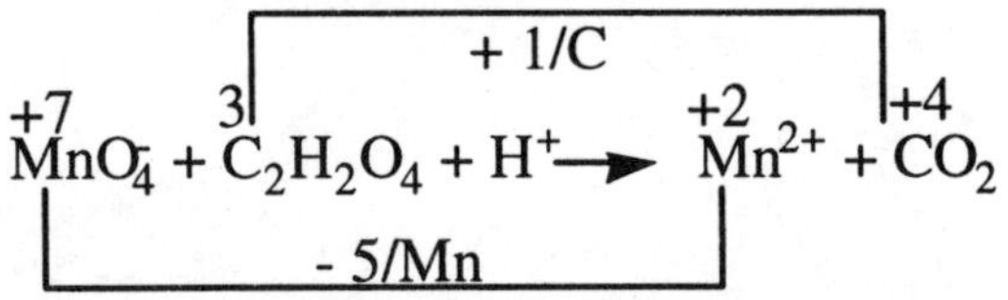

$$2\,MnO_4^- + 5\,C_2H_2O_4 + 6\,H^+ \rightarrow 2\,Mn^{2+} + 10\,CO_2 + 8\,H_2O$$

(b)

$$\overset{+7}{MnO_4^-} + \overset{-1}{C_2H_4O} \longrightarrow \overset{+4}{MnO_2} + \overset{0}{CH_3CO_2^-}$$

$+ 1/C$ $- 3/Mn$

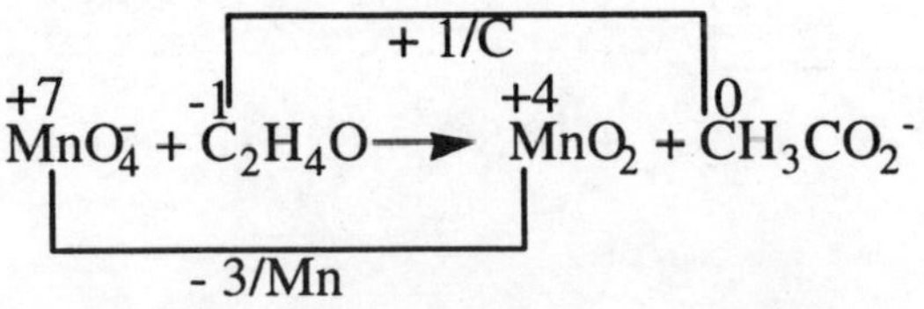

$$2\,MnO_4^- + 3\,C_2H_4O + OH^- \rightarrow 2\,MnO_2 + 3\,CH_3CO_2^- + 2\,H_2O$$

(c)

+ 4/S

$\overset{2}{S_2O_3^{2-}} + \overset{0}{Cl_2} + H^+ \longrightarrow \overset{+6}{HSO_4^-} + \overset{-1}{Cl^-}$

- 1/Cl

$S_2O_3^{2-} + 4\,Cl_2 + 5\,H_2O \rightarrow 2\,HSO_4^- + 8\,Cl^- + 8\,H^+$

71.

	Oxidizing Agent	Reducing Agent
(a)	MnO_4^-	$C_2H_2O_4$
(b)	MnO_4^-	C_2H_4O
(c)	Cl_2	$S_2O_3^{2-}$

73. (a) $Zn + 2H^+ \rightarrow Zn^{2+} + H_2$
(b) $Cu + Zn^{2+} \rightarrow$ no reaction
(c) $Fe + 2\,Ag^+ \rightarrow Fe^{2+} + 2\,Ag$
(d) $Au + H^+ \rightarrow$ no reaction

75. (a) $? \text{ mL } KMnO_4 = 20.00 \text{ mL} \times \dfrac{0.3252 \text{ mol } Fe^{2+}}{\text{L}} \times \dfrac{\text{mol } MnO_4^-}{5 \text{ mol } Fe^{2+}}$

$$\times \frac{\text{L}}{0.1050 \text{ mol } KMnO_4} = 12.39 \text{ mL } KMnO_4$$

(b) $? \text{ mL } KMnO_4 = 1.065 \text{ g } KNO_2 \times \dfrac{\text{mol } KNO_2}{85.103 \text{ g } KNO_2} \times \dfrac{2 \text{ mol } MnO_4^-}{5 \text{ mol } NO_2^-}$

$$\times \frac{\text{L}}{0.1050 \text{ mol } KMnO_4} \times \frac{\text{mL}}{10^{-3} \text{ L}} = 47.67 \text{ mL } KMnO_4$$

77.

- 3/Mn

$\overset{2}{Mn^{2+}} + \overset{7}{MnO_4^-} + OH^- \longrightarrow \overset{4}{MnO_2} + H_2O$

+ 2/Mn

$3\,Mn^{2+} + 2\,MnO_4^- + 4\,OH^- \rightarrow 5\,MnO_2 + 2\,H_2O$

$[Mn^{2+}] = 0.03477 \text{ L} \times 0.05876 \text{ M } MnO_4^- \times \dfrac{3 \text{ mol } Mn^{2+}}{2 \text{ mol } MnO_4^-} \times \dfrac{1}{0.02500 \text{ L}}$

$$= 0.1226 \text{ M } Mn^{2+}$$

79. $6\,Fe^{2+} + Cr_2O_7^{2-} + 14\,H^+ \rightarrow 6\,Fe^{3+} + 2\,Cr^{3+} + 7\,H_2O$

$$? \text{ g Fe} = 29.43 \text{ mL} \times 0.04212 \text{ M } Cr_2O_7^{2-} \times \frac{10^{-3}\text{ L}}{\text{mL}} \times \frac{6 \text{ mol } Fe^{2+}}{\text{mol } Cr_2O_7^{2-}} \times \frac{55.847 \text{ g } Fe^{2+}}{\text{mol } Fe^{2+}} = 0.4154 \text{ g } Fe^{2+}$$

$$?\% \text{ Fe} = \frac{0.4154 \text{ g Fe}}{0.8765 \text{ g sample}} \times 100\% = 47.39\%$$

81. $? \text{ M} = \frac{18.9 \text{ g}}{100 \text{ L}} \times \frac{\text{mol } K^+}{39.10 \text{ g } K^+} = 4.83 \times 10^{-3} \text{ M } K^+$

$? \text{ M} = \frac{365 \text{ g}}{100 \text{ L}} \times \frac{\text{mol } Cl^-}{35.45 \text{ g } Cl^-} = 0.103 \text{ M } Cl^-$

83. $? \text{ mol} = 6.85 \text{ g} \times \frac{98.8 \text{ g NaCl}}{100 \text{ g}} \times \frac{\text{mol NaCl}}{58.44 \text{ g NaCl}} = 0.1158 \text{ mol } Cl^-$

$? \text{ mol} = 6.85 \text{ g} \times \frac{1.2 \text{ g } MgCl_2}{100 \text{ g}} \times \frac{\text{mol } MgCl_2}{95.21 \text{ g } MgCl_2} \times \frac{2 \text{ mol } Cl^-}{\text{mol } MgCl_2} = 0.00173 \text{ mol } Cl^-$

$[Cl^-] = \frac{(0.1158 \text{ mol} + 0.00173 \text{ mol})}{500.0 \text{ mL}} = \frac{0.1175 \text{ mol}}{500.0 \text{ mL}} \times \frac{\text{mL}}{10^{-3} \text{ L}} = 0.235 \text{ M}$

85. The solid is $MgSO_4$. The white precipitate is $BaSO_4$. $BaCl_2$ is water soluble. $MgCO_3$ is not soluble, so it would not form a solution with a high enough $[CO_3^{2-}]$ to form $BaCO_3$.

87. $H_2SO_4 + Na_2CO_3 \rightarrow Na_2SO_4 + H_2CO_3 \rightarrow Na_2SO_4 + H_2O + CO_2$

$$? \text{ kg} = 1.5 \times 10^3 \text{ kg} \times \frac{10^3 \text{ g}}{1 \text{ kg}} \times \frac{93.2 \text{ g}}{100\text{g}} \times \frac{\text{mol } H_2SO_4}{98.08 \text{ g } H_2SO_4} \times \frac{\text{mol } Na_2CO_3}{\text{mol } H_2SO_4} \times \frac{105.99 \text{ g } Na_2CO_3}{\text{mol } Na_2CO_3} \times \frac{1 \text{ kg}}{10^3 \text{ g}} = 1.5 \times 10^3 \text{ kg } Na_2CO_3$$

89. $? \text{ mol} = 220 \text{ mL} \times \frac{1.16 \text{ g}}{\text{mL}} \times \frac{31.4 \text{ g HCl}}{100 \text{ g solution}} \times \frac{\text{mol HCl}}{36.46 \text{ g HCl}} \times \frac{1 \text{ mol } H^+}{1 \text{ mol HCl}} \times \frac{\text{mol } OH^-}{\text{mol } H^+} = 2.198 \text{ mol } OH^-$

$[OH^-] = \frac{2.198 \text{ mol } OH^-}{0.50 \text{ gal}} \times \frac{\text{gal}}{3.785 \text{ L}} = 1.2 \text{ M } OH^-$

91. $? \text{ g } H_2SO_4 = 32.44 \text{ mL} \times \frac{10^{-3} \text{ L}}{\text{mL}} \times 0.00986 \text{ M NaOH} \times \frac{250.0 \text{ mL}}{10.00 \text{ mL}} \times \frac{\text{mol } H_2SO_4}{2 \text{ mol NaOH}} \times \frac{98.09 \text{ g}}{\text{mol}} = 0.3922 \text{ g } H_2SO_4$

$? \text{ \% } H_2SO_4 = \frac{0.3922 \text{ g } H_2SO_4}{1.239 \text{ g sample}} \times 100\% = 31.7\% \ H_2SO_4$

93. $OX_{Nd} + OX_{Ca} + 2 \times OX_{Mn} + 6 \times OX_O = O$
$OX_{Nd} + 2 + 3 + 4 + 6 \times (-2) = O$
$OX_{Nd} = 3$

95. $? \text{ e}^- = 48.97 \text{ mol} \times 0.3000 \text{ M Ce}^{4+} \times \dfrac{1 \text{ mmol e}^-}{\text{mmol Ce}^{4+}} = 14.69 \text{ mmol e}^-$

$? \dfrac{\text{mmol e}^-}{\text{mmol V}^{2+}} = \dfrac{14.69 \text{ mmol e}^-}{25.00 \text{ ml} \times 0.1996 \text{ M}} = \dfrac{2.944 \text{ mmol e}^-}{\text{mmol V}^{2+}}$

V^{2+} goes to V^{5+}

97. (a) $? \text{ g MgCl}_2 = 38.26 \text{ mL} \times \dfrac{10^{-3} \text{ L}}{\text{ml}} \times 0.01000 \text{ M Y}^{4-} \dfrac{1 \text{ mol Mg}^{2+}}{1 \text{ mol Y}^{4-}}$

$\times \dfrac{\text{mol MgCl}_2}{\text{mol Mg}^{2+}} \times \dfrac{95.211 \text{ g}}{\text{mol MgCl}_2} = 0.03643 \text{ g MgCl}_2$

(b) $? \text{ MCl}^- = 38.26 \text{ ml} \times 0.01000 \text{ M Y}^{4-} \times \dfrac{\text{mol Mg}^{2+}}{\text{mol Y}^{4-}} \times \dfrac{2 \text{ mol Cl}^-}{\text{mol Mg}^{2+}}$

$\times \dfrac{1}{40.0 \text{ ml} + 10 \text{ ml} + 38.26 \text{ ml}} = 0.0087 \text{ M Cl}^- \text{ from MgCl}_2$

$? \text{ M Cl}^- = \dfrac{31.5 \text{ g NH}_4\text{Cl}}{1.00 \text{ L}} \times \dfrac{\text{mol NH}_4\text{Cl}}{53.51 \text{ g}} = 0.589 \text{ M Cl}^- \text{ in Stock Buffer}$

$? \text{ M Cl}^- = \dfrac{0.589 \text{ M x } 10 \text{ ml}}{(40.0 + 10 + 38.26)\text{mL}} = 0.0667 \text{ M Cl}^- \text{ from NH}_4\text{Cl}$

$? \text{ M Cl}^- = 0.0667 \text{ M} + 0.0087 \text{ M} = 0.0754 \text{ M Cl}^- \text{ in final solution.}$

99. $Ag^+ + SCN^- \rightarrow AgSCN \text{ (s)}$

$? \text{ g Ag}^+ = 43.56 \text{ mL} \times 0.1005 \text{ M} \times \dfrac{10^{-3} \text{ L}}{\text{mL}} \times \dfrac{\text{mol Ag}^+}{\text{mol SCN}^-} \times \dfrac{107.87 \text{ g}}{\text{mol Ag}^+}$

$= 0.4722 \text{ g Ag}^+$

$? \text{ \% Ag} = \dfrac{0.4722 \text{ g Ag}^+}{0.5039 \text{ g sample}} \times 100\% = 93.72 \text{ \% Ag}$

101. $CH_4O + 6 \text{ ClO}_3^- + 6 \text{ H}^+ \rightarrow CO_2 + 6 \text{ ClO}_2 + 5 \text{ H}_2O$

$? \text{ L CH}_3\text{OH} = 125 \text{ kg ClO}_2 \times \dfrac{10^3 \text{ g}}{\text{kg}} \times \dfrac{\text{mol ClO}_2}{67.45 \text{ g ClO}_2} \times \dfrac{\text{mol CH}_4\text{O}}{6 \text{ mol ClO}_2} \times \dfrac{32.04 \text{ g}}{\text{mol}}$

$\times \dfrac{\text{mL}}{0.791 \text{ g}} = 1.25 \times 10^4 \text{ mL CH}_3\text{OH} \times \dfrac{10^{-3} \text{ L}}{\text{mL}} = 12.5 \text{ L CH}_3\text{OH}$

103. $H_2C_2O_4 + 2 \text{ NaOH} \rightarrow Na_2C_2O_4 + 2 \text{ H}_2O$
$5 \text{ H}_2C_2O_4 + 2 \text{ MnO}_4^- + 6 \text{ H}^+ \rightarrow 10 \text{ CO}_2 + 2 \text{ Mn}^{2+} + 8 \text{ H}_2O$

$? \text{ M KMnO}_4 = 0.03215 \text{ L} \times 0.1050 \text{ M NaOH} \times \dfrac{\text{mol H}_2\text{C}_2\text{O}_4}{2 \text{ mol NaOH}} \times \dfrac{2 \text{ mol MnO}_4^-}{5 \text{ mol H}_2\text{C}_2\text{O}_4}$

$\times \dfrac{1}{0.02812 \text{ L}} = 0.02401 \text{ M KMnO}_4$

Apply Your Knowledge

105. $? \text{ g residue} = ? \text{ g KCl} = 25.00 \text{ mL} \times 1.840 \text{ M} \times \frac{10^{-3} \text{ L}}{\text{mL}} \times \frac{\text{mol KCl}}{\text{mol HCl}}$

$$\times \frac{74.551 \text{ g}}{\text{mol KCl}} = 3.429 \text{ g KCl}$$

All the residue is KCl. All of the HCl became KCl. The calculation can be made using Cl^- as the limiting factor.

$KOH + KCl \rightarrow KCl + H_2O$

$K_2CO_3 + 2\ HCl \rightarrow 2\ KCl + H_2CO_3 \rightarrow H_2O + CO_2(g)$

$HCl + KOH \rightarrow KCl + H_2O$

107. $? \text{ mol } Cl^- = 22.61 \text{ mL} \times 0.1525 \text{ M } AgNO_3 \times \frac{10^{-3} \text{ L}}{\text{mL}} \times \frac{\text{mol } Cl^-}{\text{mol } AgNO_3} =$

$$3.448 \times 10^{-3} \text{ mol } Cl^-$$

Let X be the moles of NaCl.

$$X \times \frac{58.443 \text{ g}}{\text{mol}} + (3.448 \text{ x } 10^{-3} \text{ mol - X}) \times \frac{74.551 \text{ g}}{\text{mol}} = 0.2250 \text{ g}$$

$$\frac{58.443 \text{ X g}}{\text{mol}} + 0.2571 \text{ g} - \frac{74.551 \text{ g}}{\text{mol}} = 0.2250 \text{ g}$$

$$0.0321 \text{ g} = \frac{16.108 \text{ X g}}{\text{mol}}$$

$X = 1.99 \times 10^{-3}$ mol NaCl

1.46×10^{-3} mol KCl

$$? \text{ \% } Na_2O = 1.99 \times 10^{-3} \text{ mol NaCl} \times \frac{\text{mol } Na_2O}{2 \text{ mol NaCl}} \times \frac{61.98 \text{ g}}{\text{mol } Na_2O}$$

$$\times \frac{1}{0.7500 \text{ g sample}} \times 100\% = 8.22 \text{ \% } Na_2O$$

$$? \text{ \% } K_2O = 1.46 \times 10^{-3} \text{ mol KCl} \times \frac{\text{mol } K_2O}{2 \text{ mol HCl}} \times \frac{94.196 \text{ g}}{\text{mol } H_2O}$$

$$\times \frac{1}{0.7500 \text{ g sample}} \times 100\% = 9.17 \text{ \% } K_2O$$

109. $? \text{ g S} = (25.00 \text{ mL} \times 0.00923 \text{ M NaOH} - 13.33 \text{ ml} \times 0.01007 \text{ M HCl})$

$$\times \frac{\text{mol } H_2SO_4}{2 \text{ mol NaOH}} \times \frac{10^{-3} \text{ L}}{\text{mL}} \times \frac{\text{mol S}}{\text{mol } H_2SO_4} \times \frac{32.07 \text{ g S}}{\text{mol S}} = 1.5 \times 10^{-3} \text{ g S}$$

$$? \text{ \% S} = \frac{1.5\text{x}10^{-3} \text{gS}}{4.876 \text{gsample}} \times 100\% = 0.035 \text{ \% S}$$

111. $H_2O_2 + 2\ I^- + 2\ H^+ \rightarrow 2\ H_2O + I_2$

$I_2 + 2\ S_2O_3{}^{2-} \rightarrow 2\ I^- + S_4O_6{}^{2-}$

$$? \text{ g } H_2O_2 = 28.91 \text{ mL} \times \frac{10^{-3}\text{ L}}{\text{mL}} \times 0.1522 \text{ M } Na_2S_2O_3 \times \frac{\text{mol } I_2}{2 \text{ mol } S_2O_3^{2-}}$$

$$\times \frac{\text{mol } H_2O_2}{\text{mol } I_2} \times \frac{34.02 \text{ g } H_2O_2}{\text{mol } H_2O_2} = 7.48 \times 10^{-2} \text{ g } H_2O_2$$

$$? \text{ g} = 10.00 \text{ mL} \times \frac{1.00 \text{ g}}{\text{mL}} = 10.00 \text{ g}$$

$$?\% = \frac{7.48 \times 10^{-2} \text{ g}}{10.00 \text{ g}} \times 100\% = 0.748\%$$

The $H_2O_2(aq)$ is not up to full strength.

Chapter 5

Gases

Exercises

5.1A (a) $? \text{ mmHg} = 0.947 \text{ atm} \times \frac{760 \text{ mmHg}}{\text{atm}} = 720 \text{ mmHg}$

(b) $? \text{ Torr} = 98.2 \text{ kPa} \times \frac{760 \text{ Torr}}{101.325 \text{ kPa}} = 737 \text{ Torr}$

(c) $? \text{ Torr} = 29.95 \text{ in.Hg} \times \frac{2.54 \text{ cm}}{1.000 \text{ in}} \times \frac{10^{-2} \text{ m}}{\text{cm}} \times \frac{\text{mm}}{10^{-3} \text{ m}} \times \frac{\text{Torr}}{\text{mmHg}} = 760.7 \text{ Torr}$

(d) $? \text{ atm} = 768 \text{ Torr} \times \frac{\text{atm}}{760 \text{ Torr}} = 1.01 \text{ atm}$

5.1B $? \text{ kPa} = \frac{1.00 \times 10^2 \text{ N}}{5.00 \text{ cm}^2} \times \left(\frac{\text{cm}}{10^{-2} \text{ m}}\right)^2 \times \frac{\text{Pa}}{\frac{\text{N}}{\text{m}^2}} \times \frac{\text{kPa}}{10^3 \text{ Pa}} = 2.00 \times 10^2 \text{ kPa}$

$? \text{ atm} = 2.00 \times 10^2 \text{ kPa} \times \frac{10^3 \text{ Pa}}{\text{kPa}} \times \frac{\text{atm}}{101325 \text{ Pa}} = 1.97 \text{ atm}$

5.2A $h_{CCl_4} d_{CCl_4} = d_{Hg} h_{Hg}$

$$h_{CCl_4} = \frac{d_{Hg} h_{Hg}}{d_{CCl_4}} = \frac{\frac{13.6 \text{ g}}{\text{cm}^3}}{\frac{1.59 \text{ g}}{\text{cm}^3}} \times 760 \text{ mm} = 6.50 \times 10^3 \text{ mm} = 6.50 \text{ m}$$

5.2B $h_{Hg} = \frac{d_{H_2O} \times h_{H_2O}}{d_{Hg}} = 30.0 \text{ m} \times \frac{\text{mm}}{10^{-3} \text{ m}} \times \frac{1.00 \text{ g/cm}^3}{13.6 \text{ g/cm}^3} = 2.21 \times 10^3 \text{ mmHg}$

$P_{H_2O} = 2.21 \times 10^3 \text{ mmHg} \times \frac{\text{atm}}{760 \text{ mmHg}} = 2.90 \text{ atm}$

The total pressure is 3.90 atm because the atmospheric pressure is 1.00 atm on top of the water.

5.3 (e) $? \text{ mmHg} = 101 \text{ kPa} \times \frac{760 \text{ mmHg}}{101 \text{ kPa}} = 760 \text{ mmHg}$

(f) $? \text{ mmHg} = 103 \text{ kPa} \times \frac{760 \text{ mmHg}}{101 \text{ kPa}} = 775 \text{ mmHg}$

	(d) <	(a) <	(c) <	(e) <	(b) or	(f)
mmHg	<735	745	750	760	>762	775

(f) cannot be placed in relation to (b) as it is not known how much (b) is above 762. (f) is greater than all of the other values.

5.4A $P_1V_1 = P_2V_2$ $\quad P_2 = \frac{P_1V_1}{V_2} = \frac{535 \text{ mL x } 988 \text{ Torr}}{1.05 \text{ L x } \frac{\text{mL}}{10^{-3} \text{ L}}} = 503 \text{ Torr}$

5.4B $P_1V_1 = P_2V_2$ $\quad V_2 = \frac{P_1V_1}{P_2} = \frac{98.7 \text{ kPa x } 73.3 \text{ mL}}{4.02 \text{ atm}} \times \frac{\text{atm}}{101.325 \text{ kPa}} = 17.8 \text{ mL}$

5.5 $P_1V_1 = P_2V_2$

$$P_2 = \frac{P_1V_1}{V_2} = \frac{10.2 \text{ L x } 1208 \text{ Torr}}{30.0 \text{ L}}$$

10.2 L is about $\frac{1}{3}$ of 30, so $\frac{1}{3}$ of 1208 Torr is about 400 Torr, or 400 mmHg.

5.6A $\frac{V_1}{T_1} = \frac{V_2}{T_2}$ $\quad V_2 = \frac{V_1T_2}{T_1} = \frac{692 \text{ L x } (273 + 23)\text{K}}{(273 + 602)\text{K}} = 234 \text{ L}$

5.6B $T_f = T_i \times \frac{V_f}{V_i} = 300 \text{ K} \times \frac{2.25 \text{ L}}{2.00 \text{ L}} = 338 \text{ K}$

$$T(°C) = T(K) - 273.15 = 338 - 273.15 = 65 \text{ °C}$$

5.7 $V_f = V_i \times \frac{T_f}{T_i} = 2.50 \text{ L} \times \frac{180 \text{ °C} + 273}{120 \text{ °C} + 273} = 2.50 \text{ L} \times \frac{453 \text{ K}}{153 \text{ K}}$

T_f is about 3 times the T_i, so the V_f is about 7.5 L. (actual 7.40 L)

5.8A $? \text{ g } C_3H_8 = 50.0 \text{ L } C_3H_8 \times \frac{\text{mol } C_3H_8}{22.4 \text{ L } C_3H_8} \times \frac{44.09 \text{ g } C_3H_8}{\text{mol } C_3H_8} = 98.4 \text{ g } C_3H_8$

5.8B $? \text{ L } CO_2 = 12.0 \text{ in.} \times 12.0 \text{ in.} \times 2.0 \text{ in.} \times \left(\frac{2.54 \text{ cm}}{1.00 \text{ in.}}\right)^3 \times \frac{1.56 \text{ g}}{\text{cm}^3} \times \frac{\text{mol } CO_2}{44.01 \text{ g } CO_2}$

$$\times \frac{22.4 \text{ L } CO_2}{\text{mol } CO_2} = 3.75 \times 10^3 \text{ L } CO_2$$

5.9A $\frac{P_1V_1}{T_1} = \frac{P_2V_2}{T_2}$ $\quad V_1 = V_2$

$$T_2 = \frac{P_2T_1}{P_1} = \frac{8.0 \text{ atm x } (273 + 22)\text{K}}{2.5 \text{ atm}}$$

$$T_2 = 944 \text{ K} - 273 = 671 \text{ °C}$$

5.9B $\frac{P_1V_1}{n_1T_1} = \frac{P_2V_2}{n_2T_2}$, $T_1 = T_2$ and $V_1 = V_2$

$\frac{P_1}{n_1} = \frac{P_2}{n_2}$ or $\frac{P}{n}$ = constant

If the number of molecules increases, the pressure must increase as more molecular collisions with the wall occur. See the graph below.

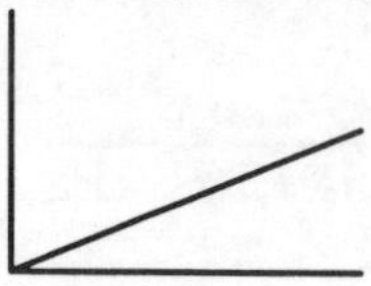

5.10 $n = \frac{PV}{RT} = \frac{3.15\,\text{atm} \times 35.0\,\text{L}}{\frac{0.08206\,\text{L atm}}{\text{K mol}} \times 852\,\text{K}} = 1.58$ moles

5.11A $T = \frac{PV\mathcal{M}}{mR} = \frac{785\text{ Torr x }\frac{\text{atm}}{760\text{ Torr}}\text{ x }5.00\text{ L x }32.00\text{ g}}{15.0\text{ g x }\frac{0.08206\text{ L atm}}{\text{K mol}}} = 134$ K

134 K – 273 = –139 °C where $\mathcal{M}$ is the molar mass and $n = \frac{m}{\mathcal{M}}$

5.11B ? mol O_2 = 25.0 g × $\frac{\text{mol } O_2}{32.00\text{ g } O_2}$ = 0.781 mol O_2

? mol $N_2 = \frac{n_{O_2}RT}{P_{O_2}} \times \frac{P_{N_2}}{RT} = \frac{0.781\,\text{mol } O_2 \times (30+273)\text{K}}{755\,\text{Torr}} \times \frac{734\,\text{Torr}}{(25+273)\text{K}}$

= 0.772 mol N_2

? g N_2 = 0.772 mol N_2 × $\frac{28.02\text{ g N2}}{\text{mol N2}}$ = 21.6 g N_2

5.12 $\mathcal{M} = \frac{mRT}{PV} = \frac{0.440\text{ g x }\frac{0.08206\text{ L atm}}{\text{K mol}}\text{ x }(86 + 273)\text{ K}}{741\text{ mmHg x }179\text{ mL x }\frac{10^{-3}\text{ L}}{\text{mL}}\text{ x }\frac{\text{atm}}{760\text{ mmHg}}}$ $\mathcal{M}$ = 74.3 g/mol

5.13A $\mathcal{M} = \frac{mRT}{PV} = \frac{0.471\text{ g x }\frac{0.08206\text{ L atm}}{\text{K mol}}\text{ x }(98 + 273)\text{ K}}{715\text{ mmHg x }121\text{ mL x }\frac{10^{-3}\text{ L}}{\text{mL}}\text{ x }\frac{\text{atm}}{760\text{ mmHg}}}$ $\mathcal{M}$ = 126 g/mol

5.13B ? mol C = 55.80 g C × $\frac{\text{mol C}}{12.011\text{ g C}}$ = 4.646 mol C × $\frac{1}{2.323\text{ mol O}}$ = 2.00 $\frac{\text{mol C}}{\text{mol O}}$

$$? \text{ mol H} = 7.03 \text{ g H} \times \frac{\text{mol H}}{1.008 \text{ g H}} = 6.974 \text{ mol H} \times \frac{1}{2.323 \text{ mol O}} = 3.00 \frac{\text{mol H}}{\text{mol O}}$$

$$? \text{ mol O} = 37.17 \text{ g O} \times \frac{\text{mol O}}{16.00 \text{ g O}} = 2.323 \text{ mol O} \times \frac{1}{2.323 \text{ mol O}} = 1.00 \frac{\text{mol O}}{\text{mol O}}$$

Empirical formula is C_2H_3O, at 43.04 g/formula unit.

$$\mathcal{M} = \frac{mRT}{PV} = \frac{0.3060 \text{ g} \times \frac{0.08206 \text{ L atm}}{\text{K mol}} \times (100 + 273) \text{ K}}{747 \text{ mmHg} \times 111 \text{ mL} \times \frac{10^{-3} \text{ L}}{\text{mL}} \times \frac{\text{atm}}{760 \text{ mmHg}}}$$

$\mathcal{M} = 85.8$ g/mol

$$\frac{85.8 \text{ g}}{\text{mol}} \times \frac{\text{formula}}{43.04 \text{ g}} = 2 \text{ formula/mol}$$

Molecular formula is $C_4H_6O_2$.

5.14 $PV = nRT$

$$P = \frac{mRT}{V\mathcal{M}} = \frac{dRT}{\mathcal{M}}$$

$$d = \frac{P\mathcal{M}}{RT} = \frac{748 \text{ Torr} \times 30.07 \text{ g/mol} \times \frac{\text{atm}}{760 \text{ Torr}}}{\frac{0.08206 \text{ L atm}}{\text{K mol}} \times (15 + 273) \text{ K}} = 1.25 \text{ g/L}$$

5.15A $$T = \frac{P\mathcal{M}}{Rd} = \frac{785 \text{ Torr} \times 44.09 \text{ g/mol} \times \frac{\text{atm}}{760 \text{ Torr}}}{\frac{0.08206 \text{ L atm}}{\text{K mol}} \times 1.51 \text{ g/L}} = 367 \text{ K}$$

$? \text{ T} = 367 \text{ K} - 273 = 94\ °\text{C}$

5.15B $$? \text{ d} = \frac{P_{NH_3}\mathcal{M}_{NH_3}}{R\,T_{NH_3}} = \frac{1.45 \text{ atm} \times \frac{17.04 \text{ g}}{\text{mol}}}{\frac{0.08206 \text{ L atm}}{\text{mol K}} \times 295.7 \text{ K}} = \frac{1.02 \text{ g}}{\text{L}}$$

$$? \text{ T} = \frac{P_{O_2}\mathcal{M}_{O_2}}{R\,d_{O_2}} = \frac{725 \text{ Torr} \times \frac{32.00 \text{ g}}{\text{mol}}}{\frac{62.36 \text{ L Torr}}{\text{K mol}} \frac{1.02 \text{ g}}{\text{L}}} = 365 \text{ K} - 273 = 92\ °\text{C}$$

5.16A $C_3H_8 + 5\ O_2 \rightarrow 3\ CO_2 + 4\ H_2O$

$$? \text{ L } O_2 = 0.556 \text{ L } C_3H_8 \times \frac{5 \text{ L } O_2}{1 \text{ L } C_3H_8} = 2.78 \text{ L } O_2$$

5.16B $CH_3OCH_3 + 3\ O_2 \rightarrow 2\ CO_2 + 3\ H_2O$

$$? \text{ L } O_2 = 125 \text{ g } C_2H_6O \times \frac{\text{L } C_2H_6O}{1.81 \text{ g } C_2H_6O} \times \frac{3 \text{ L } O_2}{1 \text{ L } C_2H_6O} = 207 \text{ L } O_2$$

5.17A $V = \frac{nRT}{P}$

$$45.8 \text{ kg } CaCO_3 \times \frac{10^3 \text{ g}}{\text{kg}} \times \frac{1 \text{ mol } CaCO_3}{100.09 \text{ g } CaCO_3} \times \frac{\text{mol } CO_2}{\text{mol } CaCO_3} \times \frac{\frac{0.08206 \text{ L atm}}{\text{K mol}} \text{ x } (825 + 273) \text{ K}}{754 \text{ Torr x } \frac{\text{atm}}{760 \text{ Torr}}} = V = 4.16 \times 10^4 \text{ L}$$

5.17B $2\ C_5H_{10} + 15\ O_2 \rightarrow 10\ CO_2 + 10\ H_2O$

$$?\text{ L } C_5H_{10} = \frac{736 \text{ Torr} \times 1.00 \times 10^6 \text{ L}}{\frac{2.36 \text{ L Torr}}{\text{K mol}} \times 298.2 \text{ K}} \times \frac{2 \text{ mol } C_5H_{10}}{10 \text{ mol } CO_2} \times \frac{70.13 \text{ g } C_5H_{10}}{\text{mol } C_5H_{10}} \times \frac{\text{mL}}{0.7445 \text{ g}} \times \frac{10^{-3} \text{ L}}{\text{mL}} = 746 \text{ L } C_5H_{10}$$

5.18A $P_{O_2} = \frac{n_{O_2}RT}{V} = \frac{0.00856 \text{ mol x } \frac{0.08206 \text{ L atm}}{\text{mol K}} \text{ x } 298 \text{ K}}{1.00 \text{ L}}$

$P_{O_2} = 0.209$ atm

$$P_{Ar} = \frac{n_{Ar}RT}{V} = \frac{0.000381 \text{ mol} \times \frac{0.08206 \text{ L atm}}{\text{k mol}} \times 298 \text{ K}}{1.00 \text{ L}} = 0.00932 \text{ atm}$$

$$P_{CO_2} = \frac{n_{CO_2}RT}{V} = \frac{0.00002 \text{ mol} \times \frac{0.08206 \text{ L atm}}{\text{mol K}} \times 298 \text{K}}{1.00 \text{ L}} = 0.0005 \text{ atm}$$

$P_{total} = P_{N_2} + P_{O_2} + P_{Ar} + P_{CO_2} = (0.780 + 0.209 + 0.00932 + 0.0005)$ atm

$P_{total} = 0.999$ atm

5.18B $P_{N_2} = \frac{m_{N_2}RT}{\mathcal{M}V} = \frac{4.05 \text{ g} \times \frac{0.08206 \text{ L atm}}{\text{K mol}} \times (25 + 273) \text{ K}}{28.01 \text{ g/mol} \times 6.10 \text{ L}}$

$P_{N_2} = 0.580$ atm

$$P_{H_2} = \frac{m_{H_2}RT}{\mathcal{M}V} = \frac{3.15 \text{ g} \times \frac{0.08206 \text{ L atm}}{\text{K mol}} \times (25 + 273) \text{ K}}{2.016 \text{ g/mol} \times 6.10 \text{ L}}$$

$P_{H_2} = 6.26$ atm

$$P_{He} = \frac{m_{He}RT}{\mathcal{M}V} = \frac{6.05\,g \times \frac{0.08206\,L\,atm}{K\,mol} \times (25+273)\,K}{4.003\,g/mol \times 6.10\,L}$$

$P_{He} = 6.06$ atm

$P_{total} = P_{N_2} + P_{H_2} + P_{He} = 12.90$ atm

5.19A $P_{N_2} = 0.741 \times 1.000\text{ atm} = 0.741\text{ atm}$

$P_{O_2} = 0.150 \times 1.000\text{ atm} = 0.150\text{ atm}$

$P_{H_2O} = 0.060 \times 1.000\text{ atm} = 0.060\text{ atm}$

$P_{Ar} = 0.009 \times 1.000\text{ atm} = 0.009\text{ atm}$

$P_{CO_2} = 0.040 \times 1.000\text{ atm} = 0.040\text{ atm}$

5.19B $$P = \frac{mRT}{\mathcal{M}V} = \frac{10.5\,g\,C_4H_{10} \times \frac{62.36\,L\,Torr}{K\,mol} \times 296.7\,K}{\frac{58.14\,g}{mol} \times 75.0\,L} = 44.6\text{ Torr (butane)}$$

$P_{total} = P_{CH_4} + P_{C_2H_6} + P_{C_3H_8} + P_{C_4H_{10}}$

$P_{total} = (505 + 201 + 43 + 44.6)\text{ Torr} = 794\text{ Torr}$

$$\chi_{CH_4} = \frac{505\text{ Torr}}{794\text{ Torr}} = 0.636$$

$$\chi_{C_2H_6} = \frac{201\text{ Torr}}{794\text{ Torr}} = 0.253$$

$$\chi_{C_3H_8} = \frac{43\text{ Torr}}{794\text{ Torr}} = 0.054$$

$$\chi_{C_4H_{10}} = \frac{44.6\text{ Torr}}{794\text{ Torr}} = 0.056$$

5.20 The increase to 3.00 atm could be achieved by adding only hydrogen, but the P_{H_2} would be 2.50 atm, not 2.00 atm. To achieve a 3.00 atm pressure with P_{H_2} = 2.00 atm, other gases must supply 1.00 atm pressure. The original He supplies only 0.50 atm. 0.5 mol of any inert gas may be added to supply the partial pressure of 1.00 atm.

5.21A $P_{H_2} = P_{total} - P_{H_2O} = 738\text{ Torr} - 16\text{ Torr} = 722\text{ Torr}$

$$m_{H_2} = \frac{\mathcal{M}PV}{RT} = \frac{2.016 \text{ g/mol x } 722 \text{ Torr x } 246 \text{ mL x } \frac{10^{-3}\text{ L}}{\text{mL}} \text{ x } \frac{\text{atm}}{760 \text{ Torr}}}{\frac{0.08206 \text{ L atm}}{\text{K mol}} \text{ x } (18 + 273) \text{ K}}$$

$= 0.0197$ g H_2

$$m_{H_2O} = \frac{18.02 \text{ g/mol x } 15.5 \text{ Torr x } 246 \text{ mL x } \frac{10^{-3}\text{ L}}{\text{mL}} \text{ x } \frac{\text{atm}}{760 \text{ Torr}}}{\frac{0.08206 \text{ L atm}}{\text{K mol}} \text{ x } (18 + 273)\text{K}} = 0.00379 \text{ g } H_2O$$

$m_{total} = 0.0235$ g

5.21B $2\ KClO_3 \rightarrow 2\ KCl + 3O_2$

$P_{H_2} = P_{total} - P_{H_2O} = 746$ mmHg – 19 mmHg = 727 mmHg

$$? \text{ g } KClO_3 = \frac{727 \text{ mmHg x } 155 \text{ mL x } \frac{10^{-3}\text{ L}}{\text{mL}} \text{ x } \frac{\text{atm}}{760 \text{ mmHg}}}{\frac{0.08206 \text{ L atm}}{\text{K mol}} \text{ x } (21 + 273) \text{ K}} \times \frac{2 \text{ mol } KClO_3}{3 \text{ mol } O_2}$$

$$\times \frac{122.55 \text{ g}}{\text{mol}} = 0.502 \text{ g } KClO_3$$

5.22 The O_2-to-H_2 molar mass ratio is 32/2 = 16. For u_{rms} of O_2 to be greater than 1838 m/s (u_{rms} of H_2 at 0 °C), $T/16$ must be greater than 273 K.
T must be greater than 16 × 273, which simplifies to 15 × 300 = 4500 K. Answer: 5000 K; Actual: 16 × 273 = 4368 K.

5.23A $\frac{r_{N_2}}{r_{Ar}} = \sqrt{\frac{\mathcal{M}_{Ar}}{\mathcal{M}_{N_2}}} = \sqrt{\frac{39.95}{28.01}} = 1.194$

N_2 is 1.19 times faster.

5.23B $\frac{r_{NO}}{r_{(CH_3)_2O}} = \sqrt{\frac{\mathcal{M}_{(CH_3)_2O}}{\mathcal{M}_{NO}}} = \sqrt{\frac{46.0}{30.0}} = 1.24$

NO effuses 1.24 times faster.

$?\% = \frac{0.24}{1.00} \times 100\% = 24\%$ faster

5.24A $\frac{r_{N_2}}{r_{unk}} = \frac{t_{unk}}{t_{N_2}} = \frac{83 \text{ s}}{57 \text{ s}} = \sqrt{\frac{\mathcal{M}_{unk}}{\mathcal{M}_{N_2}}} = 1.46$

$\mathcal{M}_{unk} = (1.46)^2 \times 28.01$ g/mol = 60 g/mol

5.24B $\dfrac{t_{C_4H_{10}}}{t_{O_2}} = \sqrt{\dfrac{\mathcal{M}_{C_4H_{10}}}{\mathcal{M}_{O_2}}}$ $\dfrac{t_{C_4H_{10}}}{123\text{ s}} = \sqrt{\dfrac{58.14}{32.00}}$

$t_{CH_4} = 166\text{ s}$

Review Questions

7. (a) The volume will decrease.
   (b) The volume will decrease.
   (c) The effect is uncertain, as the increase in pressure will cause a volume decrease, and the increase in temperature will cause a volume increase.
   (d) The volume will increase.

8. (a) The pressure will increase.
   (b) The pressure will increase.
   (c) The pressure will increase.
   (d) The effect is uncertain, as the decrease in temperature will cause a pressure decrease while the decrease in volume will cause a pressure increase.

14. (a) A; A higher pressure in container A means more molecules and thus more mass and a higher density.
    (b) They are equal. Equal pressures and temperatures mean the same number of molecules per volume.
    (c) B; A higher temperature means that container A has fewer molecules and thus less mass and a lower density.

Problems

21. (a) $?\text{ mmHg} = 0.985\text{ atm} \times \dfrac{760\text{ mmHg}}{\text{atm}} = 749\text{ mmHg}$

    (b) $?\text{ atm} = 849\text{ Torr} \times \dfrac{\text{atm}}{760\text{ Torr}} = 1.12\text{ atm}$

    (c) $?\text{ kPa} = 642\text{ Torr} \times \dfrac{101.3\text{ kPa}}{760\text{ Torr}} = 85.6\text{ kPa}$

    (d) $?\text{ mmHg} = 15.5\text{ lb/in}^2 \times \dfrac{760\text{ mmHg}}{14.70\text{ lb/in}^2} = 801\text{ mmHg}$

23. $$P_{Hg} = P_{H_2O} \times \frac{d_{H_2O}}{d_{Hg}} = 1250\text{ m }H_2O \times \frac{\dfrac{1.00\text{ g}}{\text{cm}^3}}{\dfrac{13.6\text{ g}}{\text{cm}^3}} = 91.9\text{ m Hg} \times \frac{\text{mm}}{10^{-3}\text{ m}}$$

$$= 9.19 \times 10^4\text{ mm Hg} \times \frac{\text{atm}}{760\text{ mm}} = 121\text{ atm}$$

25. $$h_{CCl_4} = \frac{d_{Hg}h_{Hg}}{d_{CCl_4}} = \frac{\frac{13.6\text{ g}}{\text{cm}^3}}{\frac{1.59\text{ g}}{\text{cm}^3}} \times 25.0\text{ cm} \times \frac{10^{-2}\text{ m}}{\text{cm}} = 2.14\text{ m}$$

27. $$h_{oil} \times \frac{d_{oil}}{d_{Hg}} = h_{Hg}$$

$$h_{Hg} = 44\text{ mm oil} \times \frac{0.789\text{ g/mL}}{13.6\text{ g/mL}} = 2.55\text{ mmHg} = 2.55\text{ Torr}$$

? Torr = 755 Torr + 3 Torr = 758 Torr

29. The height for the closed end of the manometer must be 760 mm for each atmosphere of pressure to be measured. One atmosphere of gas pressure would cause a difference of 760 mm in the heights of the two columns. Two atmospheres would cause a 1520-mm difference.

31. $P_1V_1 = P_2V_2$

(a) $$V_2 = \frac{P_1V_1}{P_2} = 521\text{ mL} \times \frac{1752\text{ Torr}}{752\text{ Torr}} = 1.21 \times 10^3\text{ mL} \times \frac{10^{-3}\text{ L}}{\text{mL}} = 1.21\text{ L}$$

(b) $$V_2 = \frac{P_1V_1}{P_2} = 521\text{ mL} \times \frac{1752\text{ Torr}}{3.55\text{ atm x } \frac{760\text{ Torr}}{\text{atm}}} = 338\text{ mL}$$

(c) $$V_2 = \frac{P_1V_1}{P_2} = 521\text{ mL} \times \frac{1752\text{ Torr}}{125\text{ kPa x } \frac{760\text{ Torr}}{101.3\text{ kPa}}} = 973\text{ mL}$$

33. $$V_2 = \frac{P_1V_1}{P_2} = 10.3\text{ m}^3 \times \frac{4.50\text{ atm}}{1.00\text{ atm}} = 46.4\text{ m}^3$$

35. (a) $$V_2 = \frac{P_1V_1}{P_2} = 60.0\text{ L} \times \frac{150.0\text{ atm}}{750.0\text{ Torr}} \times \frac{760\text{ Torr}}{\text{atm}} = 9.12 \times 10^3\text{ L}$$

(b) $$?\text{ h} = 9.12 \times 10^3\text{ L} \times \frac{\text{min}}{8.00\text{ L}} = 1.14 \times 10^3\text{ min} \times \frac{\text{h}}{60\text{ min}} = 19.0\text{ h}$$

37. $$\frac{V_1}{T_1} = \frac{V_2}{T_2} \qquad V_2 = \frac{V_1T_2}{T_1} = 154\text{ mL} \times \frac{(10.0 + 273.2)\text{K}}{(99.8 + 273.2)\text{K}} = 117\text{ mL}$$

39. $$\frac{V_1}{T_1} = \frac{V_2}{T_2} \qquad \frac{95.0\text{ mL}}{T} = \frac{104.5\text{ mL}}{T + 15}$$

$95.0\,T + 1425 = 104.5\,T$

$T = 150$ K In a 10% increase, the 15 degrees must be 10%.

41. $$?\text{ kg} = 4.55\text{ L} \times 10^3 \times \frac{\text{mol}}{22.4\text{ L}} \times \frac{20.18\text{ g}}{\text{mol}} \times \frac{\text{kg}}{10^3\text{ g}} = 4.10\text{ kg Ne}$$

43. $? \text{ mL} = 125 \text{ mg Ar} \times \frac{\text{mol Ar}}{39.95 \text{ g Ar}} \times \frac{22.4 \text{ L}}{\text{mol}} = 70.1 \text{ mL}$

$? \text{ mL} = 505 \text{ mL} + 70.1 \text{ mL} = 575 \text{ mL}$

45. The answer can be determined by estimation.

(a) 5.0 g $H_2 \approx 2.5$ mol

(b) 50 L SF_6 at STP is a little more than 2.0 mol but less than 2.5 mol.

(c) 1.0×10^{24} is less than 2 mol.

(d) $67 \text{ L} \times \frac{\text{mol}}{22.4 \text{ L}} \times \approx 3 \text{ mol}$

(d) is the largest number of molecules.

<u>Actual calculations:</u>

(a) $? \text{ molecules} = 5.0 \text{ g } H_2 \times \frac{\text{mol } H_2}{2.016 \text{ g } H_2} \times \frac{6.022 \text{ x } 10^{23} \text{ molecules}}{\text{mol}}$

$= 1.5 \times 10^{24}$ molecules

(b) $? \text{ molecules} = 50 \text{ L } SF_6 \times \frac{\text{mol } SF_6}{22.4 \text{ L } SP_6} \times \frac{6.022 \text{ x } 10^{23} \text{ molecules}}{\text{mol}}$

$= 1.3 \times 10^{24}$ molecules

(c) ? molecules = 1.0×10^{24} molecules

(d) $? \text{ molecules} = 67 \text{ L} \times \frac{\text{mol}}{22.4 \text{ L}} \times \frac{6.022 \text{ x } 10^{23} \text{ molecules}}{\text{mol}}$

$= 1.8 \times 10^{24}$ molecules

(d) is the largest number of molecules.

47. $\frac{P_1}{T_1} = \frac{P_2}{T_2}\ P_2 = P_1 \times \frac{T_2}{T_1} = 721 \text{ Torr} \times \frac{(755 + 273)\text{K}}{(25 + 273)\text{K}} = 2.49 \times 10^3 \text{ Torr}$

49. 1 mol at 273 K and 1 atm is 22.4 L.

$\frac{P_1V_1}{T_1} = \frac{P_2V_2}{T_2},\ P_1 = P_2,\ V_2 = V_1 \times \frac{T_2}{T_1} = \frac{22.4 \text{ L x } (25 + 273)\text{K}}{273\text{K}} = 24.5 \text{ L}$

51. $V_2 = V_1 \times \frac{P_1T_2}{P_2T_1} = 2.53 \text{ m}^3 \times \frac{191 \text{ Torr}}{1142 \text{ Torr}} \times \frac{(25 + 273)\text{K}}{(-15 + 273)\text{K}} = 0.489 \text{ m}^3$

53. (a) $V = \frac{nRT}{P} = \frac{1.12 \text{ mol x } \frac{0.08206 \text{ L atm}}{\text{K mol}} \text{ x } (62 + 273)\text{K}}{1.38 \text{ atm}} = 22.3 \text{ L}$

(b) $P = \frac{nRT}{V} = \frac{125 \text{ g CO x } \frac{\text{mol}}{28.01} \text{ x } \frac{0.08206 \text{ L atm}}{\text{K mol}} \text{ x } (29 + 273)\text{K}}{3.96 \text{ L}} = 27.9 \text{ atm}$

(c) $? \text{ mg} = \frac{PV}{RT} = \frac{725 \text{ Torr x } \frac{\text{atm}}{760 \text{ Torr}} \text{ x } 34.5 \text{ mL x } \frac{2.016 \text{ g } H_2}{\text{mol } H_2}}{\frac{0.08206 \text{ L atm}}{\text{K mol}} \text{ x } (273 - 12) \text{ K}} = 3.10 \text{ mg } H_2$

(d) $$P = \frac{mRT}{\mathcal{M}V} = \frac{173 \text{ g N}_2 \text{ x } \frac{\text{mol N}_2}{28.01 \text{ g N}_2} \text{ x } \frac{0.08206 \text{ L atm}}{\text{K mol}} \text{ x } (273 - 0)\text{K}}{8.35 \text{ L}} \times \frac{101.3 \text{ kPa}}{\text{atm}} = 1.68 \times 10^3 \text{ kPa}$$

55. $$V_2 = \frac{nRT_2}{P_2} \qquad n = \frac{P_1V_1}{RT_1}$$

$$V_2 = \frac{P_1V_1}{RT_1} \times \frac{RT_2}{P_2}$$

$$V_2 = \frac{760 \text{ Torr x } 4.65 \text{ L}}{273 \text{ K}} \times \frac{(15 + 273)\text{K}}{756 \text{ Torr}} = 4.93 \text{ L}$$ R will cancel.

57. $$\frac{P_1V_1}{T_1} = \frac{P_2V_2}{T_2}$$

$$V_2 = \frac{V_1P_1T_2}{P_2T_1} = 4.20 \times 10^3 \text{ L} \times \frac{3.00 \text{ atm}}{151 \text{ atm}} \times \frac{(25 + 273)\text{K}}{(17 + 273)\text{K}}$$

$V_2 = 85.7$ L

59. $$\mathcal{M} = \frac{mRT}{PV} = 0.625 \text{ g} \times \frac{\frac{0.08206 \text{ L atm}}{\text{K mol}} \text{ x } (98 + 273)\text{K}}{756 \text{ Torr x } \frac{\text{atm}}{760 \text{ Torr}} \text{ x } 125 \text{ mL x } \frac{10^{-3} \text{ L}}{\text{mL}}} = 153 \text{ g/mol}$$

Molecular mass = 153 u

61. $$? \text{ mol H} = 8.75 \text{ g H} \times \frac{\text{mol H}}{1.008 \text{ g H}} = 8.68 \text{ mol H} \times \frac{1}{7.60 \text{ mol C}} = \frac{1.14 \text{ mol H}}{\text{mol C}}$$

$$? \text{ mol C} = 91.25 \text{ g C} \times \frac{\text{mol C}}{12.01 \text{ g C}} = 7.60 \text{ mol C} \times \frac{1}{7.60 \text{ mol C}} = \frac{1 \text{ mol C}}{\text{mol C}}$$

$C_{(1\times7)}H_{(1.14\times7)} \Rightarrow C_7H_8$ formula mass = 92.1 u/formula

$$n = \frac{PV}{RT} = \frac{761 \text{ Torr x } 435 \text{ mL x } \frac{10^{-3} \text{ L}}{\text{mL}}}{\frac{62.36 \text{ L Torr}}{\text{K mol}} \text{ x } (273 + 115) \text{ K}} = 0.0137 \text{ mol}$$

$$\mathcal{M} = \frac{1.261 \text{ g}}{0.0137 \text{ mol}} = 92.2 \text{ g/mol}$$

$$? = \frac{\text{molar mass}}{\text{formula mass}} = \frac{92.2 \text{ u/molecule}}{92.1 \text{ u/formula unit}} = \frac{1 \text{ formula unit}}{\text{molecule}}$$

Molecular formula is C_7H_8.

63. (a) $$d = \frac{m}{V} = \frac{P\mathcal{M}}{RT} = \frac{1.00 \text{ atm x } \frac{28.01 \text{ g}}{\text{mol}}}{\frac{0.08206 \text{ L atm}}{\text{K mol}} \text{ x } 273 \text{ K}} = 1.25 \text{ g/L} \quad \text{or} \quad \frac{28.01 \text{ g/mol}}{22.4 \text{ L/mol}} = d$$

(b) $$d = \frac{P\mathcal{M}}{RT} = \frac{1.26\text{ atm} \times \frac{39.95\text{ g}}{\text{mol}}}{\frac{0.08206\text{ L atm}}{\text{K mol}} \times (325 + 273)\text{K}} = 1.03\text{ g/L}$$

65. $$T = \frac{P\mathcal{M}}{Rd} = \frac{0.982\text{ atm} \times \frac{32.00\text{ g}}{\text{mol}}}{\frac{0.08206\text{ L atm}}{\text{K mol}} \times \frac{1.05\text{ g}}{\text{L}}} = 365\text{ K}$$

365 K – 273 = 92 °C

67. $$\mathcal{M} = \frac{dRT}{P} = \frac{\frac{4.33\text{ g}}{\text{L}} \times \frac{0.08206\text{ L atm}}{\text{K mol}} \times (445 + 273)\text{K}}{755\text{ mmHg} \times \frac{\text{atm}}{760\text{ mmHg}}} = 256.8\text{ g/mol}$$

$$\frac{256.8\text{ u}}{\text{molecule}} \times \frac{\text{formula unit}}{32.06\text{ u}} = 8.01\,\frac{\text{formula unit}}{\text{molecule}}$$ S_8

69. $$d = \frac{P\mathcal{M}}{RT}$$

	actual
(a) $\dfrac{745\text{ Torr} \times 2\text{ g/mol}}{\frac{62.4\text{ L Torr}}{\text{K mol}} \times (273 - 15)\text{ K}}$	$\dfrac{0.0926\text{ g}}{\text{L}}$
(b) $\dfrac{760\text{ Torr} \times 4\text{ g/mol}}{\frac{62.4\text{ L Torr}}{\text{K mol}} \times 273\text{ K}}$	$\dfrac{0.178\text{ g}}{\text{L}}$
(c) $\dfrac{1.15\text{ atm} \times \frac{760\text{ Torr}}{\text{atm}} \times 16\text{ g/mol}}{\frac{62.4\text{ L Torr}}{\text{K mol}} \times (273 - 10)\text{ K}}$	$\dfrac{0.852\text{ g}}{\text{L}}$
(d) $\dfrac{435\text{ Torr} \times 30\text{ g/mol}}{\frac{62.4\text{ L Torr}}{\text{K mol}} \times (273 + 50)\text{ K}}$	$\dfrac{0.647\text{ g}}{\text{L}}$

By comparing equations, it can be seen that (c) is larger than (b) or (a), because the molar mass in (c) is four times the molar mass in (b) and eight times the molar mass in (a), while temperature and pressure are only slightly changed. Comparing (c) and (d), the molar mass of (c) is more than half that of (d), the pressure is more than twice that of (d), and the temperature is much less than (d). So, (c) has the greatest density.

71. $$?\text{ L } SO_3 = 1.15\text{ L } SO_2 \times \frac{2\text{ mol } SO_3}{2\text{ mol } SO_2} = 1.15\text{ L } SO_3$$

$$?\text{ L } SO_3 = 0.65\text{ L } O_2 \times \frac{2\text{ mol } SO_3}{\text{mol } O_2} = 1.30\text{ L } SO_3$$

SO_2 is limiting. 1.15 L SO_3 is produced.

73. $2\ CH_3CH_2CH_2OH + 9\ O_2 \rightarrow 6\ CO_2 + 8\ H_2O$

$$? \text{ L } CO_2 = 125 \text{ mL } C_3H_8O \times \frac{0.804 \text{ g}}{\text{mL}} \times \frac{\text{mol } C_3H_8O}{60.09 \text{ g } C_3H_8O} \times \frac{6 \text{ mol } CO_2}{2 \text{ mol } C_3H_8O}$$

$$\times \frac{\frac{62.36 \text{ L Torr}}{\text{mol K}} \text{ x } (273 + 26) \text{ K}}{767 \text{ Torr}} = 122 \text{ L } CO_2$$

75. $\frac{PV}{RT} = n$

$$? \text{ mg Mg} = \frac{758 \text{ Torr x } \frac{\text{atm}}{760 \text{ Torr}} \text{ x } 28.50 \text{ mL x } \frac{10^{-3} \text{ L}}{\text{mL}}}{\frac{0.08206 \text{ L atm}}{\text{K mol}} \text{ x } (26 + 273)\text{K}} \times \frac{\text{mol Mg}}{\text{mol } H_2} \times \frac{24.31 \text{ g Mg}}{\text{mol Mg}}$$

$$\times \frac{\text{mg}}{10^{-3} \text{ g}} = 28.2 \text{ mg Mg}$$

77. $P_{N_2} = \chi_{N_2} P_T \qquad \chi_{N_2} = \frac{\text{mol \%}}{100\%}$

$P_{N_2} = 762 \text{ mmHg} \times 0.768 = 585 \text{ mmHg}$

$P_{O_2} = 762 \text{ mmHg} \times 0.201 = 153 \text{ mmHg}$

$P_{CO_2} = 762 \text{ mmHg} \times 0.031 = 24 \text{ mmHg}$

79. (a) $? \text{ mol He} = \frac{1.96 \text{ g He}}{4.003 \text{ g He}} = 0.490 \text{ mol}$

$? \text{ mol } O_2 = \frac{60.8 \text{ g } O_2}{32.00 \text{ g } O_2} = 1.90 \text{ mol}$

$? \text{ mol total} = 0.490 \text{ mol} + 1.90 \text{ mol} = 2.39 \text{ mol}$

$\chi_{He} = \frac{0.490 \text{ mol}}{2.39 \text{ mol}} = 0.205$

$\chi_{O_2} = \frac{1.90 \text{ mol}}{2.39 \text{ mol}} = 0.795$

(b) $$P_{He} = \frac{mRT}{V} = \frac{1.96 \text{ g He x } \frac{0.08206 \text{ L atm}}{\text{K mol}} \text{ x } (25.0 + 273.2)\text{K}}{5.00 \text{ L x } 4.003 \text{ g/mol}} = 2.40 \text{ atm}$$

$$P_{O_2} = \frac{60.8 \text{ g } O_2 \text{ x } \frac{0.08206 \text{ L atm}}{\text{K mol}} \text{ x } (25.0 + 273.2)\text{K}}{5.00 \text{ L x } 32.00 \text{ g/mol}} = 9.30 \text{ atm}$$

(c) $P_{total} = P_{He} + P_{O_2} = 2.40 \text{ atm} + 9.30 \text{ atm} = 11.70 \text{ atm}$

81. $P_{total} = P_{O_2} + P_{H_2O}$

P_{O_2} = 742 Torr – 31.8 Torr = 710 Torr

$$\chi_{O_2} = \frac{P_{O_2}}{P_{total}} = \frac{710 \text{ Torr}}{742 \text{ Torr}} = 0.957$$

83. P_{O_2} = 743 Torr - 19 Torr = 724 Torr

$$n = \frac{PV}{RT} = \frac{724 \text{ Torr x } \frac{\text{atm}}{760 \text{ Torr}} \text{ x } 122 \text{ mL x } \frac{10^{-3} \text{ L}}{\text{mL}}}{\frac{0.08206 \text{ L atm}}{\text{K mol}} \text{ x } (21 + 273)\text{K}} = 4.82 \times 10^{-3} \text{ mol } O_2$$

$$? \text{ g } O_2 = 4.82 \times 10^{-3} \text{ mol } O_2 \times \frac{32.00 \text{ g } O_2}{\text{mol } O_2} = 0.154 \text{ g } O_2$$

$$? \text{ g} = 4.82 \times 10^{-3} \text{ mol } O_2 \times \frac{\text{mol } C_6H_{12}O_6}{6 \text{ mol } O_2} \times \frac{180.16 \text{ g}}{\text{mol}} = 0.145 \text{ g } C_6H_{12}O_6$$

85. $$\frac{r_{H_2}}{r_{He}} = \sqrt{\frac{4.003}{2.016}} = 1.41$$

H_2 will diffuse somewhat faster than He, so b—with a ratio of H_2 to He molecules of 6:4 = 1.5—is the answer.

87. The partial pressures depend on the number of moles of each gas; they can easily be different. However, there can be only one e_k of the gas molecules in the mixture. Because temperature depends on e_k, the two gases must be at the same temperature.

89. (a) $r \propto \frac{1}{t}$

$$\frac{r_{N_2}}{r_X} = \frac{\frac{1}{44 \text{ s}}}{\frac{1}{75 \text{ s}}} = \sqrt{\frac{\mathcal{M}_X}{\mathcal{M}_{N_2}}} = \sqrt{\frac{\mathcal{M}_X}{28.01 \text{ g/mol}}}$$

$$1.70 = \sqrt{\frac{\mathcal{M}_X}{28.01 \text{ g/mol}}}$$

$$2.91 = \frac{\mathcal{M}_X}{28.01 \text{ g/mol}}$$

$\mathcal{M}_X$ = 81.5 g/mol. Molecular weight = 81 u.

(b) $$\frac{r_{N_2}}{r_x} = \frac{\frac{1}{44 \text{ s}}}{\frac{1}{42 \text{ s}}} = \sqrt{\frac{\mathcal{M}_X}{28.01 \text{ g/mol}}}$$

$$0.955 = \sqrt{\frac{\mathcal{M}_X}{28.01\text{ g/mol}}}$$

$\mathcal{M}_X = 0.911 \times 28.01$ g/mol = 26 g/mol. **Molecular weight = 26 u.**

Additional Problems

91. $$? \frac{\text{lb}}{\text{in.}^2} = \frac{32\text{ lb}}{\text{in.}^2} + \frac{15\text{ lb}}{\text{in.}^2} = \frac{47\text{ lb}}{\text{in.}^2}$$

$$?\text{ atm} = \frac{47\text{ lb}}{\text{in.}^2} \times \frac{\text{atm}}{14.7\text{ lb/in}^2} = 3.2\text{ atm}$$

$$PV = nRT = \frac{m}{M}RT$$

$$?\text{ g} = m = \frac{PVM}{Rt} = \frac{3.2\text{ atm} \times 21.1\text{ atm} \times \frac{28.96\text{g}}{\text{mol}}}{0.08206\frac{\text{L atm}}{\text{R mol}} \times 294\text{ K}} = 81\text{ g}$$

$$\frac{P_1}{T_1} = \frac{P_2}{T_2}\; P_2 = \frac{P_1T_2}{T_1} = \frac{\frac{47\text{lb}}{\text{in}^2} \times 317\text{K}}{294\text{K}} = \frac{51\text{ lb}}{\text{in.}^2}$$

$$? \frac{\text{lb}}{\text{in}^2} = \frac{51\text{ lb}}{\text{in.}^2} - \frac{15\text{ lb}}{\text{in.}^2} = \frac{36\text{ lb}}{\text{in.}^2}$$

93. $$V = 5.00\text{ in.}^3 \times \left(\frac{2.54\text{ cm}}{\text{in.}}\right)^3 \times \frac{\text{mL}}{\text{cm}^3} \times \frac{10^{-3}\text{ L}}{\text{mL}} = 0.08194\text{ L}$$

$$P = 195\text{ lb/in.}^2 \times \frac{\text{atm}}{14.696\text{ lb/in.}^2} = 13.27\text{ atm}$$

$$m = \frac{PV\mathcal{M}}{RT} = \frac{13.27\text{ atm x }0.08194\text{ L x }4.003\text{ g/mol}}{\frac{0.08206\text{ L atm}}{\text{K mol}}\text{ x }(20 + 273)\text{K}} = 0.181\text{ g}$$

95. H_2 $\quad 0.0080\text{ g} \times \frac{\text{mol}}{2.016\text{ g}} = 0.0040\text{ mol}$

N_2 $\quad 0.1112\text{ g} \times \frac{\text{mol}}{28.014\text{ g}} = 0.003969\text{ mol}$

O_2 $\quad 0.1281\text{ g} \times \frac{\text{mol}}{31.999\text{ g}} = 0.004003\text{ mol}$

CO_2 $\quad 0.1770\text{ g} \times \frac{\text{mol}}{44.009\text{ g}} = 0.004022\text{ mol}$

C_4H_{10} $\quad 0.2320\text{ g} \times \frac{\text{mol}}{58.124\text{ g}} = 0.003991\text{ mol}$

CCl_2F_2 $\quad 0.4824\text{ g} \times \frac{\text{mol}}{120.91\text{ g}} = 0.003990\text{ mol}$

These data are consistent with Avogadros hypothesis. The amount of each gas is within 1% of 0.00400 mol, strongly suggesting that equal volumes of different gases, compared at the same temperature and pressure, have equal numbers of molecules.

97. Calculate the empirical formula.

$$14.05 \text{ g C} \times \frac{\text{mol}}{12.011 \text{ g}} = 1.1698 \text{ mol C} \times \frac{1}{1.1698 \text{ mol C}} = 1.00$$

$$41.48 \text{ g Cl} \times \frac{\text{mol}}{35.453 \text{ g}} = 1.1700 \text{ mol Cl} \times \frac{1}{1.1698 \text{ mol C}} = 1.00$$

$$44.46 \text{ g F} \times \frac{\text{mol}}{18.998 \text{ g}} = 2.3402 \text{ mol F} \times \frac{1}{1.1698 \text{ mol C}} = 2.00$$

empirical formula $CClF_2$ $\quad 12.01 + 35.45 + 2\,(19.00) = \frac{85.46 \text{ u}}{\text{formula unit}}$

$$\mathcal{M} = \frac{mRT}{PV} = \frac{2.135 \text{ g x } \frac{0.08206 \text{ L atm}}{\text{K mol}} \text{ x } (26.1 + 273.2)\text{K}}{739.2 \text{ mmHg x } \frac{\text{atm}}{760 \text{ mmHg}} \text{ x } 315.5 \text{ mL x } \frac{10^{-3} \text{ L}}{\text{mL}}}$$

$\mathcal{M} = 170.9$ g/mol

$$\frac{170.9 \text{ u}}{\text{molecule}} \times \frac{\text{formula unit}}{85.4 \text{ u}} = \frac{2 \text{ formula unit}}{\text{molecule}} \qquad CClF_2 \times 2 = C_2Cl_2F_4$$

99. $? \text{ L} = \frac{4}{3}\pi r^3 = \frac{4}{3} \times 3.14159 \times (15.3 \text{ m})^3 \times \frac{\text{cm}^3}{10^{-6} \text{ m}^3} \times \frac{\text{L}}{1000 \text{ cm}^3} = 1.49 \times 10^7 \text{ L}$

$$\frac{P_1V_1}{T_1} = \frac{P_2V_2}{T_2}$$

$$V_2 = \frac{P_1V_1T_2}{T_1P_2} = \frac{1.0 \text{ x } 10^{-5} \text{ atm x } 1.49 \text{ x } 10^7 \text{ L x } 298 \text{ K}}{203 \text{ K x } 35 \text{ atm}} = 6.2 \text{ L}$$

101. $? \text{ mol } C_3H_6 = \frac{PV}{RT} = \frac{25.0 \text{ atm} \times 1.50 \text{ L}}{\frac{0.08206 \text{ L atm}}{\text{K mol}} \times 298 \text{ K}} = 1.53 \text{ mol}$

$$? \text{ L } C_3H_6 = \frac{nRT}{P} = \frac{1.53 \text{ mol} \times \frac{0.08206 \text{ L atm}}{\text{K mol}} \times 298 \text{ K}}{755 \text{ mm Hg} \times \frac{1 \text{ atm}}{760 \text{ mm Hg}}} = 37.7 \text{ L}$$

$$? \text{ ft}^3 \; C_3H_6 = 37.7 \text{ L} \times \frac{\text{mL}}{10^{-3} \text{ L}} \times \frac{\text{cm}^3}{\text{mL}} \times \left(\frac{\text{in.}}{2.54 \text{ cm}}\right)^3 \times \left(\frac{\text{ft}}{12 \text{ in.}}\right)^3 = 1.33 \text{ ft}^3$$

$$\% \; C_3H_6 = \frac{1.33 \text{ ft}^3}{72 \text{ ft}^3} \times 100 \,\% = 1.85 \,\% \; C_3H_6$$

No, it is not an explosive mixture.

103. $? \text{ psi} = \dfrac{73 \text{ ton}}{(3.0 \text{ ft})^2} \times \dfrac{2000 \text{ lb}}{\text{ton}} \times \left(\dfrac{\text{ft}}{12 \text{ in.}}\right)^2 = 1.1 \times 10^2 \dfrac{\text{lb}}{\text{in.}^2}$

The air pressure in the bag must exceed 1.1×10^2 psi to lift the 73-ton object.

105. $? \text{ mol } H_2 = n = \dfrac{PV}{RT} = \dfrac{(746\text{-}17)\text{Torr x } 40.71 \text{ mL} \dfrac{10^{-3}}{\text{mL}}}{\dfrac{62.364 \text{ L Torr}}{\text{Kmol}} (273 + 19) \text{ K}} = 1.630 \times 10^{-3} \text{ mol } H_2$

Let X = mass of Lithium

$X \dfrac{\text{mol Li}}{6.941 \text{ g}} \times \dfrac{\text{mol } H_2}{2 \text{ mol Li}} + (0.0297 \text{ g - X}) \times \dfrac{\text{mol Mg}}{24.305 \text{ g}} \times \dfrac{\text{mol } H_2}{\text{mol Mg}} =$

$1.63 \times 10^{-3} \text{ mol } H_2$

$0.07204 \text{ X} + 0.00122 \text{ mol } H_2 - 0.04114 \text{ X} = 1.63 \times 10^{-3} \text{ mol}$

$X = 0.013 \text{ g Li}$

$? \% \text{ Li} = \dfrac{0.013 \text{g}}{0.0297 \text{g}} \times 100\% = 45\% \text{ Li}$

55% Mg

107. $2 \, C_8H_{18} + 25 \, O_2 \rightarrow 16 \, CO_2 + 18 \, H_2O$

$? \text{ L } CO_2 = 265 \text{ mi} \times \dfrac{\text{gal}}{31.2 \text{ mi}} \times \dfrac{3.785 \text{ L}}{\text{gal}} \times \dfrac{\text{mL}}{10^{-3} \text{ L}} \times \dfrac{0.71 \text{ g}}{\text{mL}} \times \dfrac{\text{mol}}{114 \text{ g}}$

$\times \dfrac{16 \text{ mol } CO_2}{2 \text{ mol } C_8H_{18}} \times \dfrac{0.0821 \text{ L atm}}{\text{K mol}} \times \dfrac{(28 + 273)\text{K}}{732 \text{ mmHg x} \dfrac{\text{atm}}{760 \text{ mmHg}}} = 4.1 \times 10^4 \text{ L } CO_2$

109. $? \text{ L } O_2 = \pi r^2 h = 3.14 \times (9 \text{ cm})^2 \times 1.2 \text{ m} \times \dfrac{\text{cm}}{10^{-2} \text{ m}} \times \dfrac{\text{L}}{1000 \text{ cm}^3} = 31 \text{ L } O_2$

$? \text{ mol } O_2 = \dfrac{PV}{RT} = \dfrac{2550 \text{ psi x} \dfrac{1 \text{ atm}}{14.696 \text{ psi}} \text{ x } 31 \text{ L}}{\dfrac{0.0821 \text{ L atm}}{\text{Kmol}} \text{ x } (273 + 19)\text{K}} = 2.2 \times 10^2 \text{ mol } O_2$

$2 \, C_2H_2 + 5 \, O_2 \rightarrow 4 \, CO_2 + 2 \, H_2O$

$? \text{ L } C_2H_2 = \pi r^2 h = 3.14 \times (14 \text{ cm})^2 \times 0.76 \text{ in} \times \dfrac{\text{cm}}{10^{-2}} \times \dfrac{\text{L}}{1000 \text{ cm}^3} = 47 \text{ L } C_2H_2$

$? \text{ mol } C_2H_2 = \dfrac{PV}{RT} = \dfrac{320 \text{ psi x} \dfrac{1 \text{ atm}}{14.696 \text{ psi}} \text{ x } 47 \text{ L}}{\dfrac{0.0821 \text{ L atm}}{\text{Kmol}} \text{ x } (273 + 19)\text{K}} = 43 \text{ mol } C_2H_2$

$? \text{ mol } C_2H_2 = 22 \times 10^2 \text{ mol } O_2 \times \dfrac{2 \text{ mol } C_2H_2}{5 \text{ mol } O_2} = 88 \text{ mol } C_2H_2$

The acetylene tank would empty first.

111. $u_{rms} = \sqrt{\dfrac{3\,RT}{M}}$

$$u_{rms} = \sqrt{\dfrac{3 \times \dfrac{8.3145\ \text{kg·m}^2\text{·s}^{-2}}{\text{mol K}} \times (27 + 273)\text{K}}{0.06407\ \text{kg/mol}}} = 342\ \text{m/s}$$

113. (a) $P = \dfrac{nRT}{V} = \dfrac{1.00\ \text{mol CO}_2 \times \dfrac{0.08206\ \text{L atm}}{\text{K mol}} \times 298\ \text{K}}{2.50\ \text{L}}$

$P = 9.78\ \text{atm}$

(b) $P = \dfrac{nRT}{V\text{-}nb} - \dfrac{n^2a}{V^2}$

$$P = \left(\dfrac{1.00\ \text{mol CO}_2 \times 0.08206\ \text{L atm} \times 298\ \text{K}}{2.50\ \text{L} - 1.00\ \text{mol} \times 0.0427\ \text{L/mol}}\right) - \left(\dfrac{(1.00\ \text{mol})^2\,\dfrac{3.59\ \text{L}^2\ \text{atm}}{\text{mol}^2}}{(2.50\ \text{L})^2}\right)$$

$P = 9.95\ \text{atm} - 0.57\ \text{atm} = 9.38\ \text{atm}$

(c) At high pressures, the molecules are forced closer together. The increased intermolecular interaction of the molecules prevents them from colliding with the wall as often as at lower pressures. Because the molecules exert less force on the wall, the pressure is decreased.

Apply Your Knowledge

115. (a) $?\ \text{mol C} = 4.305\ \text{g CO}_2 \times \dfrac{\text{mol C}}{44.009\ \text{g CO}_2} = 0.09782\ \text{mol C} \times \dfrac{1}{0.09782\ \text{mol C}}$

$= 1.000\ \dfrac{\text{mol C}}{\text{mol C}}$

$?\ \text{mol H} = 2.056\ \text{g H}_2\text{O} \times \dfrac{2\ \text{mol H}}{18.015\ \text{g H}_2\text{O}} = 0.2283\ \text{mol H} \times \dfrac{1}{0.09782\ \text{mol C}}$

$= 2.333\ \dfrac{\text{mol H}}{\text{mol C}}$

$C_{(1\times3)}H_{(2.333\times3)} = C_3H_7 \quad$ formula mass $= 43.09$

$PV = \dfrac{mRT}{M}$

$$M = \dfrac{mRT}{PV} = \dfrac{0.403\ \text{g} \times \dfrac{62.36\ \text{L Torr}}{\text{K mol}} \times (273.2 + 99.8)\ \text{K}}{749\ \text{Torr} \times 145\ \text{mL} \times \dfrac{10^{-3}\ \text{L}}{\text{mL}}} = 86.31\ \text{g/mol}$$

$?\,\dfrac{\text{formula units}}{\text{molecule}} = \dfrac{86.31\,\text{u/molecule}}{43.09\,\text{u/formula units}} = \dfrac{2\ \text{formula units}}{\text{molecule}}$

$C_3H_7 \times 2 = C_6H_{14}$

(b) $CH_3\underset{|}{\overset{}{C}}HCH_2CH_2CH_3 \qquad CH_3CH_2\underset{|}{\overset{}{C}}HCH_2CH_3$
    $\quad\ CH_3 \qquad\qquad\qquad\qquad\quad CH_3$

Two isomers for a 6 C alkane with one methyl group.

117. $$?\frac{ft}{s}\ SO_2 = \sqrt{\frac{3\times\frac{8.3145\,J}{mol\,K}\times 273\,K}{0.06407\frac{kg}{mol}}} \times \frac{cm}{10^{-2}} \times \frac{in.}{2.54cm} \times \frac{ft}{12\ in.} = 1069\frac{ft}{s}$$

$$?\frac{ft}{s}\ CO_2 = \sqrt{\frac{3\times\frac{8.3145\,J}{mol\,K}\times 273\,K}{0.04401\frac{kg}{mol}}} \times \frac{cm}{10^{-2}} \times \frac{in.}{2.54cm} \times \frac{ft}{12\ in.} = 1290\frac{ft}{s}$$

$$?\frac{ft}{s}\ Cl_2 = \sqrt{\frac{3\times\frac{8.3145\,J}{mol\,K}\times 273\,K}{0.07090\frac{kg}{mol}}} \times \frac{cm}{10^{-2}} \times \frac{in.}{2.54cm} \times \frac{ft}{12\ in.} = 1017\frac{ft}{s}$$

$$?\frac{ft}{s}\ C_4H_8 = \sqrt{\frac{3\times\frac{8.3145\,J}{mol\,K}\times 273\,K}{0.05610\frac{kg}{mol}}} \times \frac{cm}{10^{-2}} \times \frac{in.}{2.54cm} \times \frac{ft}{12\ in.} = 1143\frac{ft}{s}$$

C_4H_8 and CO_2, the two higher molecules, have faster than the speed of sound root-mean-square average speeds.

Chapter 6

Thermochemistry

Exercises

6.1A $\Delta U = q + w = (-567\text{ J}) + (+89\text{ J}) = -478\text{ J}$

6.1B $\Delta U = q + w = 41.4\text{ J} = q + (-81.2\text{ J})$
$q = 122.6\text{ J}$
Heat is absorbed by the system.

6.2 (a) Work is done by the gas.
(b) $\Delta U = w$ Since work is done by the system, the internal energy decreases.
(c) Since the internal energy is related to the temperature, the temperature must go down.

6.3A $\Delta H_{rxn} = \frac{1}{2}(285.4\text{ kJ}) = 142.7\text{ kJ}$

6.3B $\Delta H_{rxn} = -\frac{1}{4}(595.5\text{ kJ}) = -148.9\text{ kJ}$

6.4 $N_2O_3(g) \xrightarrow{25\,°C} NO(g) + NO_2(g) \quad \Delta H = 40.5\text{ kJ}$

$$? \text{ kJ/mol} = \frac{0.533\text{ kJ}}{\text{g } N_2O_3} \times \frac{76.02\text{ g}}{\text{mol}} = \frac{40.5\text{ kJ}}{\text{mol}}$$

6.5A $$\Delta H_{rxn} = 12.8\text{ g } H_2 \times \frac{\text{mol } H_2}{2.016\text{ g } H_2} \times \frac{-184.6\text{ kJ}}{\text{mol}} = -1.17 \times 10^3\text{ kJ}$$

6.5B $$? \text{ L} = -1.00 \times 10^6\text{ kJ} \times \frac{\text{mol}}{-890.3\text{ kJ}} \times \frac{0.08206\text{ L atm}}{\text{K mol}} \times \frac{(273 + 25)\text{K}}{745\text{ Torr x } \frac{\text{atm}}{760\text{ Torr}}}$$
$$= 2.80 \times 10^4\text{ L } CH_4$$

6.6A $$C = \frac{q}{\Delta T} = \frac{911\text{ J}}{100°\text{C} - 15°\text{C}} = 11\text{ J/°C}$$

6.6B $$q = \frac{345\text{ J}}{\text{K}} \times (23 - 467)\text{ K} \times \frac{\text{kJ}}{10^3\text{ J}} = -153\text{ kJ}$$

6.7A $q = m$ times specific heat times ΔT

$$q = 814\text{ g} \times \frac{4.182\text{ J}}{\text{g °C}} \times (100.0\text{ °C} - 18.0\text{ °C}) \times \frac{1.00\text{ cal}}{4.184\text{ J}} = 6.67 \times 10^4\text{ cal}$$

$$q = 6.67 \times 10^4\text{ cal} \times \frac{\text{kcal}}{10^3\text{ cal}} = 66.7\text{ kcal}$$

6.7B $$\text{Mass } H_2O = \frac{q}{\text{specific heat} \times \Delta T} = \frac{9.09 \times 10^{10}\text{ J}}{\frac{4.18\text{ J}}{\text{g °C}} \times (55.0 - 5.5)\text{ °C}}$$

$$= 4.39 \times 10^8\text{ g} = 4.39 \times 10^5\text{ kg}$$

6.8A $$T_f - T_i = \frac{q}{m \times \text{specific heat}}$$

$$T_f = 22.5\text{ °C} + \frac{4.22\text{ kJ} \times \frac{10^3\text{ J}}{\text{kJ}}}{454\text{ g} \times 0.128\text{ J/g °C}}$$

$$T_f = 95.1\text{ °C}$$

6.8B $q_{Cu} = q_{H_2O}$

$$m_{Cu} \times \text{specific heat}_{Cu} \times \Delta T = m_{H_2O} \times \text{specific heat} \times \Delta T$$

$$m_{Cu} \times \frac{0.385\text{ J}}{\text{g °C}} \times \Delta T = 145\text{ g } H_2O \times \frac{4.182\text{ J}}{\text{g °C}} \times \Delta T \quad \Delta T \text{ cancels.}$$

$$m_{Cu} = 1.58 \times 10^3\text{ g Cu}$$

6.9A $q_{Ir} = -q_{H_2O}$

$$m_{Ir} \times \text{specific heat} \times \Delta T = -m_{H_2O} \times \text{specific heat}_{H_2O} \times \Delta T$$

$$23.9\text{ g Ir} \times \text{specific heat}_{Ir} \times (22.6\text{ °C} - 89.7\text{ °C}) = -20.0\text{ g } H_2O \times \frac{4.182\text{ J}}{\text{g °C}} \times (22.6\text{ °C} - 20.1\text{ °C})$$

specific heat $_{Ir}$ = 0.13 J/g °C

6.9B $q_{gly} = -q_{Fe}$

$n$sp.ht. $\Delta T = -n$sp.ht. ΔT

$$250.0\text{ mL} \times \frac{1.261\text{ g}}{\text{mL}} \times \text{sp.ht.} \times (44.7\text{ °C} - 23.5\text{ °C}) = -135\text{ g} \times \frac{0.449\text{ J}}{\text{g °C}} \times (44.7\text{ °C} - 225\text{ °C})$$

6683.3 sp.ht. g °C = 10929 J

$$\text{sp.ht.} = \frac{1.64\text{ J}}{\text{g °C}}$$

6.10 $q_{100} = -q_{200}$

$m \times \text{specific heat} \times \Delta T = -m \times \text{specific heat} \times \Delta T$

$100.0\ \text{mL} \times (T_f - 20\ °\text{C}) = -200.0\ \text{mL} \times (T_f - 80\ °\text{C})$

$100.0\ T_f - 2000\ °\text{C} = -200.0\ T_f + 16000\ °\text{C}$

$300.0\ T_f = 18000\ °\text{C}$

$T_f = 60\ °\text{C}$

OR

The heat exchange produces twice the temperature increase in the 100-mL sample, $2X$ as the temperature decreases, X, in the 200-mL sample.

$X + 2X = 60°\quad X = 20°$

The final temperature is $20\ °\text{C} + (2 \times 20°) = 80\ °\text{C} - 20\ °\text{C} = 60\ °\text{C}$

6.11 $?\ \text{mol} = 100.0\ \text{mL} \times 0.500\ \text{M} \times \dfrac{10^{-3}\ \text{L}}{\text{mL}} = 0.0500\ \text{mol product}$

$q_{rxn} = -q_{\text{calorim}} = -m \times \text{specific heat} \times \Delta T$

$q = -200.0\ \text{mL} \times \dfrac{1.00\ \text{g}}{\text{mL}} \times \dfrac{4.182\ \text{J}}{\text{g}\ °\text{C}} \times (23.65\ °\text{C} - 20.29\ °\text{C})$

$q = -2.81 \times 10^3\ \text{J}$

6.12A $\Delta H = -q_{\text{calorim}} = \dfrac{2.81 \times 10^3\ \text{J}}{0.0500\ \text{mol}} \times \dfrac{\text{kJ}}{10^3\ \text{J}} = -56.2\ \text{kJ/mol}$

6.12B $?\ \text{mol}\ H_2O = 125\ \text{mL} \times 1.33\ \text{M HCl} \times \dfrac{10^{-3}\ \text{L}}{\text{mL}} \times \dfrac{\text{mol}\ H_2O}{\text{mol HCl}} = 0.166\ \text{mol}\ H_2O$

$?\ \text{mol}\ H_2O = 225\ \text{mL} \times 0.625\ \text{M NaOH} \times \dfrac{10^{-3}\ \text{L}}{\text{mL}} \times \dfrac{\text{mol}\ H_2O}{\text{mol NaOH}} = 0.141\ \text{mol}\ H_2O$

NaOH is limiting.

$q_{cal} = -q_{rxn}$

$q_{cal} = \dfrac{+57.2\ \text{kJ}}{\text{mol}} \times 0.141\ \text{mol} = 350\ \text{mL} \times \dfrac{1.00\ \text{g}}{\text{mL}} \times \dfrac{\text{kJ}}{10^3\ \text{J}} \times \dfrac{4.18\ \text{J}}{\text{g}\ °\text{C}} \times (T - 24.4\ °\text{C})$

$5.51\ °\text{C} = T - 24.4$

$T = 24.4\ °\text{C} + 5.51\ °\text{C} = 29.9\ °\text{C}$

6.13A $q_{\text{calorim}} = 6.52\ \text{kJ} \cdot °\text{C}^{-1} \times (21.26 - 20.00)\ °\text{C} = 8.22\ \text{kJ}$

$q_{\text{rxn}} = -q_{\text{calorim}} = -8.22\ \text{kJ}$

$\Delta H = \dfrac{12.01\ \text{g}}{\text{mol}} \times \dfrac{-8.22\ \text{kJ}}{0.250\ \text{g}} = \dfrac{-395\ \text{kJ}}{\text{mol}}$

$\text{C(diamond)} + O_2(g) \rightarrow CO_2(g)\qquad \Delta H = -395\ \text{kJ}$

6.13B $?\ \text{kL} = 0.8082\ \text{g} \times \dfrac{\text{mol}}{180.16\ \text{g}} \times \dfrac{-2803\ \text{kJ}}{\text{mol}} = -12.57\ \text{kJ}$

$$\text{heat capacity} = \frac{q_{\text{calorim}}}{\Delta T} = \frac{12.57\,\text{kJ}}{(27.21 - 25.11)°\text{C}} = \frac{5.99\,\text{kJ}}{°\text{C}}$$

6.14A

	ΔH
$2\ CH_4(g) + 4\ O_2(g) \rightarrow 2\ CO_2(g) + 4\ H_2O(l)$	$2 \times (-890.3\ \text{kJ})$
$2\ CO_2(g) \rightarrow 2\ CO(g) + O_2(g)$	$-2 \times (-233.0\ \text{kJ})$
$2\ CH_4(g) + 3\ O_2(g) \rightarrow 2\ CO(g) + 4\ H_2O(l)$	$\Delta H = -1215\ \text{kJ}$

6.14B

$C_2H_4(g) + 3\ O_2(g) \rightarrow 2\ CO_2(g) + 2\ H_2O(l)$	$-1410.9\ \text{kJ}$
$3\ H_2O(l) + 2\ CO_2(g) \rightarrow \frac{7}{2}\ O_2(g) + C_2H_6(g)$	$-\frac{1}{2} \times (-3119.4\ \text{kJ})$
$H_2(g) + \frac{1}{2}\ O_2(g) \rightarrow H_2O(l)$	$\frac{1}{2} \times (-571.6\ \text{kJ})$
$C_2H_4(g) + H_2(g) \rightarrow C_2H_6(g)$	$\Delta H = -137.0\ \text{kJ}$

6.15A $\Delta H° = \Sigma\nu_p \Delta H°_f$ (products) $- \Sigma\nu_r \Delta H°_f$ (reactants)

$\Delta H° = (1\ \text{mol}\ CH_3CH_2OH \times -277.7\ \text{kJ/mol})$

$- (1\ \text{mol}\ H_2O \times -285.8\ \text{kJ/mol} + 1\ \text{mol}\ C_2H_4 \times 52.26\ \text{kJ/mol}) = -44.2\ \text{kJ}$

6.15B $2\ C_4H_{10}(g) + 13\ O_2(g) \rightarrow 8\ CO_2(g) + 10\ H_2O(l)$

$$\Delta H° = 8\ \text{mol}\ CO_2 \times \frac{-393.5\ \text{kJ}}{\text{mol}} + 10\ \text{mol}\ H_2O \times \frac{-285.8\ \text{kJ}}{\text{mol}}$$

$$-2\ \text{mol}\ C_4H_{10} \times \frac{-125.7\ \text{kJ}}{\text{mol}} - 13\ \text{mol} \times 0$$

$\Delta H° = -5755\ \text{kJ}$

6.16A $? \text{kJ} = \left(1\,\text{mol}\,C_2Cl_4 \times \Delta H^{\circ}_f\right) + \left(4\ \text{mol}\ H_2O(l)\ \text{x}\ \frac{-285.8\ \text{kJ}}{\text{mol}}\right)$

$\left(-1\ \text{mol}\ C_2H_4\ \text{x}\ \frac{52.26\ \text{kJ}}{\text{mol}}\right) - \left(4\ \text{mol HCl x}\ \frac{-92.31\ \text{kJ}}{\text{mol}}\right) - (2\ \text{mol}\ O_2\ \text{x}\ 0)$

$= -878.5\ \text{kJ}$

$? \text{kJ} = \left(1\,\text{mol}\,C_2Cl_4 \times \Delta H°_f\right) + -826.2\ \text{kJ} = -878.5\ \text{kJ}$

$$\Delta H°_f = \frac{-878.5\ \text{kJ} + 826.2\ \text{kJ}}{1\ \text{mol}\ C_2Cl_4}$$

$$\Delta H°_f = \frac{-52.3\ \text{kJ}}{\text{mol}\ C_2Cl_4}$$

6.16B $C_4H_4S(l) + 6\ O_2(g) \rightarrow 4\ CO_2(g) + SO_2(g) + 2\ H_2O(l)$

$\Delta H° = \Sigma\nu_p \Delta H°_f$ (products) $- \Sigma\nu_r \Delta H°_f$ (reactants)

$-2523\ \text{kJ} = (4\ \text{mol}\ CO_2 \times -393.5\ \text{kJ/mol}) + (2\ \text{mol}\ H_2O \times -285.8\ \text{kJ/mol})$

$+ (1\ \text{mol}\ SO_2 \times -296.8\ \text{kJ/mol}) - \left(1\,\text{mol}\,C_4H_4S \times \Delta H°_f\right) - (6\ \text{mol}\ O_2 \times 0)$

$-2523\ \text{kJ} = -2442.4\ \text{kJ} - \left(1\,\text{mol}\,C_4H_4S \times \Delta H°_f\right)$

$\Delta H°_f = 81 \text{ kJ /mol}$

6.17 The enthalpy of formation of CH_3OH is 39 kJ/mol greater than that of CH_3CH_2OH, but the enthalpies of the products of the combustion of CH_3CH_2OH are much lower than those of the combustion of CH_3OH [by one mole each of $H_2O(l)$ and $CO_2(g)$]. CH_3CH_2OH has the greater negative heat of combustion.

6.18A $$\Delta H° = \left(1 \text{ mol} \times \Delta H°_f \text{ BaSO}_4\right) - \left(1 \text{ mol Ba}^{2+} \times \frac{-537.6 \text{ kJ}}{\text{mol Ba}^{2+}}\right) - \left(1 \text{ mol SO}_4^{2-} \times \frac{-909.3 \text{ kJ}}{\text{mol SO}_4^{2-}}\right) = -26 \text{ kJ}$$

$$\Delta H°_f = \frac{-1446.9 \text{ kJ} - 26 \text{ kJ}}{\text{mol BaSO}_4} = \frac{-1473 \text{ kJ}}{\text{mol}}$$

6.18B $MgCl_2(aq) + 2\,KOH(aq) \rightarrow Mg(OH)_2(s) + 2\,K^+ + 2\,Cl^-$

$Mg^{2+}(aq) + 2\,Cl^- + 2\,K^+ + 2\,OH^-(aq) \rightarrow Mg(OH)_2(s) + 2\,K^+ + 2\,Cl^-$

The Cl^- and K^+ will cancel to produce a net ionic equation.

$Mg^{2+}(aq) + 2\,OH^-(aq) \rightarrow Mg(OH)_2(s)$

$$\Delta H° = 1 \text{ mol Mg(OH)}_2 \times \frac{-924.5 \text{ kJ}}{\text{mol}} - 2 \text{ mol OH}^- \times \frac{-230.0 \text{ kJ}}{\text{mol}} - 1 \text{ mol Mg}^{2+} \times \frac{-466.9 \text{ kJ}}{\text{mol}} = 2.4 \text{ kJ}$$

Review Questions

14. $2\,Fe(s) + \frac{3}{2}\,O_2(g) \rightarrow Fe_2O_3(s)$

18. carbohydrate $9.0 \text{ g} \times \frac{4.0 \text{ kcal}}{\text{g}} = 36 \text{ kcal}$

protein $2.0 \text{ g} \times \frac{4.0 \text{ kcal}}{\text{g}} = 8.0 \text{ kcal}$

fat $1.0 \text{ g} \times \frac{9.0 \text{ kcal}}{\text{g}} = \underline{9.0 \text{ kcal}}$

53 kcal

? % calories from fat $\frac{9.0 \text{ kcal}}{53 \text{ kcal}} \times 100\% = 17\%$ calories from fat

Problems

19. $\Delta U = 455 \text{ J} - 325 \text{ J} = 130 \text{ J}$

21. 58 cal of work must be done on the system. $58 \text{ cal} \times \frac{4.184 \text{ J}}{\text{cal}} = 2.4 \times 10^2 \text{ J}$

23. Yes, it can do both.

Yes, if the system absorbs as much heat as the work that it does.

25. Endothermic $q = \frac{134\ \text{kJ}}{10.0\ \text{g}} \times \frac{18.02\ \text{g}}{\text{mol}} = 241\ \text{kJ /mol}$

$q_p = \Delta H$ At constant pressure there is pressure–volume work that is included in q_p.

27. $CaCO_3(s) \rightarrow CaO(s) + CO_2(g)$ $\Delta H = \frac{8.90\ \text{kJ}}{0.0500\ \text{mol}} = \frac{178\ \text{kJ}}{\text{mol}}$

29. (a) $\Delta H^\circ_{rxn} = \frac{1}{3}(-46\ \text{kJ}) = -15\ \text{kJ}$

(b) $\Delta H^\circ_{rxn} = -\frac{1}{2}(-46\ \text{kJ}) = 23\ \text{kJ}$

31. $\frac{1}{4}P_4(s) + \frac{3}{2}Cl_2(g) \rightarrow PCl_3(g)$ $\Delta H^\circ = -287\ \text{kJ}$

$? \text{kJ} = \frac{9.27\ \text{kJ}}{1.00\ \text{g P}_4} \times \frac{123.88\ \text{g}}{\text{mol}} \times 0.25\ \text{mol} = -287\ \text{kJ}$

33. $? \text{kJ} = 500\ \text{g CaO} \times \frac{1\ \text{mol CaO}}{56.08\ \text{g CaO}} \times \frac{-65.2\ \text{kJ}}{\text{mol}} = -581\ \text{kJ}$

35. $? \text{mol O}_2 = 150.0\ \text{g Na}_2\text{O}_2 \times \frac{\text{mol Na}_2\text{O}_2}{77.978\ \text{g Na}_2\text{O}_2} \times \frac{\text{mol O}_2}{2\ \text{mol Na}_2\text{O}_2} = 0.9618\ \text{mol O}_2$

$? \text{mol O}_2 = 50.0\ \text{mL H}_2\text{O} \times \frac{1.00\ \text{g}}{\text{mL}} \times \frac{\text{mol H}_2\text{O}}{18.02\ \text{g}} \times \frac{\text{mol O}_2}{2\ \text{mol H}_2\text{O}} = 1.39\ \text{mol O}_2$

Na_2O_2 is limiting.

$\Delta H = \frac{287\ \text{kJ}}{2\ \text{mol Na}_2\text{O}_2} \times 150.0\ \text{g Na}_2\text{O}_2 \times \frac{\text{mol Na}_2\text{O}_2}{77.98\ \text{g Na}_2\text{O}_2} = -276\ \text{kJ}$

37. $? \text{L C}_2\text{H}_6 = -1.00 \times 10^6\ \text{kJ} \times \frac{2\ \text{mol C}_2\text{H}_6}{3.12 \times 10^3\ \text{kJ}} \times \frac{62.36\ \text{L Torr}}{\text{K mol}} \times \frac{(273 + 23)\ \text{K}}{751\ \text{Torr}}$

$= 1.58 \times 10^4\ \text{L}$

39. $? \text{g CaO} = \frac{1.00\ \text{L CH}_4 \times 772\text{Torr}}{\frac{62.36\ \text{L Torr}}{\text{K mol}} \times (273.2 + 21.5)\ \text{K}} \times \frac{-890.3\ \text{kJ}}{\text{mol CH}_4} \times \frac{\text{mol CaO}}{-65.2\ \text{kJ}}$

$\times \frac{56.08\ \text{g CaO}}{\text{mol CaO}} = 32.2\ \text{g CaO}$

41. $\Delta H = \frac{3.12 \times 10^3\ \text{kJ}}{2\ \text{mol C}_2\text{H}_6} \times \frac{\text{mol C}_2\text{H}_6}{30.07\ \text{g}}$

$$\Delta H = \frac{5.76\times10^3\ \text{kJ}}{2\ \text{mol}\ C_4H_{10}} \times \frac{\text{mol}\ C_4H_{10}}{58.12\ \text{g}}$$

$\frac{3.12}{30} > \frac{5.76}{58}$, so ethane gives off more heat per gram.

43. $q = C\Delta T$

$$C = \frac{112\ \text{J}}{45\ °\text{C} - 18\ °\text{C}} = 4.1\ \text{J/°C}$$

45. (a) $q = 20.0\ \text{g} \times \frac{4.182\ \text{J}}{\text{g}\ °\text{C}} \times (96.0\ °\text{C} - 20.0\ °\text{C}) \times \frac{\text{kJ}}{10^3\ \text{J}} = 6.36\ \text{kJ}$

(b) $q = 120.0\ \text{g} \times \frac{2.46\ \text{J}}{\text{g}\ °\text{C}} \times (44.5\ °\text{C} - (-\ 10.5\ °\text{C})) \times \frac{\text{kJ}}{10^3\ \text{J}} = 16.2\ \text{kJ}$

47. $$\Delta T = \frac{q}{m \text{ x specific heat}} = \frac{93.5\ \text{J}}{48.7\ \text{g} \times \frac{0.128\ \text{J}}{\text{g}\ °\text{C}}} = 15.0\ °\text{C}$$

$t_f = \Delta T + t_i = 15.0\ °\text{C} + 27.0\ °\text{C} = 42.0\ °\text{C}$

49. $-q_{\text{metal}} = q_{\text{water}}$

$q_{\text{metal}} = 10.25\ \text{g} \times \text{specific heat} \times (22.03\ °\text{C} - 99.10\ °\text{C})$

$q_{\text{water}} = 20.0\ \text{g} \times \frac{4.182\ \text{J}}{\text{g}\ °\text{C}} \times (22.03\ °\text{C} - 18.51\ °\text{C})$

$q_{\text{water}} = 294.4\ \text{J} = -q_{\text{metal}} = -\ (-\ 790.0\ \text{g°C} \times \text{specific heat})$

specific heat = 0.373 J/g°C

51. $q_{Fe} = -q_{H_2O}$

$m_{Fe}\ \text{sp. ht.}_{(Fe)}\ \Delta T = -m_{H_2O}\ \text{sp. ht.}_{(H_2O)}\ \Delta T$

$$1.35\ \text{kg} \times \frac{10^3\ \text{g}}{\text{kg}} \times \frac{0.449\ \text{J}}{\text{g}\ °\text{C}} \times (39.6 - t_i)°\text{C} = -0.817\ \text{kg} \times \frac{10^3\ \text{g}}{\text{kg}} \times \frac{4.182\ \text{J}}{\text{g}\ °\text{C}} \times (39.6 - 23.3)°\text{C}$$

$$2.400 \times 10^4\ \text{J} - \frac{606.2\ \text{J}}{°\text{C}} t_i = -5.569 \times 10^4\ \text{J}$$

$$\frac{606.2\ \text{J}\ t_i}{°\text{C}} = 5.569 \times 10^4\ \text{J} + 2.400 \times 10^4\ \text{J}$$

$$t_i = \frac{7.969\times10^4\ \text{J}}{606.2\ \text{J/°C}} = 131\ °\text{C}$$

53. $q_{Cu} = 2.25 \text{ g Cu} \times \dfrac{0.385 \text{ J}}{\text{g °C}} \times (100 - 25)\text{ °C}$

$q_{Pb} = 4.50 \text{ g Pb} \times \dfrac{0.128 \text{ J}}{\text{g °C}} \times (100 - 25)\text{ °C}$

$q_{Ag} = 1.87 \text{ g Ag} \times \dfrac{0.235 \text{ J}}{\text{g °C}} \times (100 - 25)\text{ °C}$

By estimating	actual
$q_{Cu} = 2 \times 0.4 \times 75 = 0.8 \times 75$	65 J
$q_{Pb} = 5 \times 0.1 \times 75 = 0.5 \times 75$	43 J
$q_{Ag} = 2 \times 0.2 \times 75 = 0.4 \times 75$	33 J

Thus, Cu must absorb the most heat, with Pb next.

55. $q_{\text{ calorim}} = \text{mass} \times \text{specific heat} \times \Delta T$

$q_{\text{ calorim}} = 1000 \text{ mL} \times \dfrac{1.00 \text{ g}}{\text{mL}} \times \dfrac{4.182 \text{ J}}{\text{g °C}} \times (23.21 - 20.00)\text{°C}$

$q_{\text{ calorim}} = 1.342 \times 10^4 \text{ J} \times \dfrac{\text{kJ}}{10^3 \text{ J}} = 13.42 \text{ kJ}$

$? \text{ mol} = 500.0 \text{ mL soln} \times 0.500 \text{ M NaOH} \times \dfrac{\text{mol } H_2O}{1 \text{ mol NaOH}} \times \dfrac{10^{-3} \text{ L}}{\text{mL}} = 0.2500 \text{ mol}$

$\Delta H = q_{rxn} = -q_{\text{ calorim}} = \dfrac{-13.42 \text{ kJ}}{0.2500 \text{ mol}} = -\dfrac{53.7 \text{ kJ}}{\text{mol}}$

57. $q_{rxn} = -q_{calorimeter} = -n\text{sp.ht.}\Delta T$

$q_{rxn} = -35.0 \text{ g } H_2O \times \dfrac{4.182 \text{ J}}{\text{g °C}} \times (19.4 \text{ °C} - 22.7 \text{ °C})$

$q_{rxn} = 483 \text{ J}$

$\Delta H = \dfrac{483 \text{ J}}{1.50 \text{ g}} \times \dfrac{\text{kJ}}{10^3 \text{ J}} \times \dfrac{80.05 \text{ g}}{\text{mol } NH_4NO_3} = \dfrac{25.8 \text{ kJ}}{\text{mol } NH_4NO_3}$

59. $q_{calorim} = \text{heat capacity} \times \Delta T$

$q_{calorim} = \dfrac{4.62 \text{ kJ}}{\text{°C}} \times (22.28 - 20.45)\text{°C} = 8.455 \text{ kJ}$

$q_{rxn} = -q_{calorim} = -\dfrac{8.455 \text{ kJ}}{0.309 \text{ g}} = -27.4 \text{ kJ/g coal}$

61. $q_{\text{ calorim}} = \text{heat capacity} \times \Delta T = -q_{rxn}$

$$\text{heat capacity} = \dfrac{\dfrac{(16.5\text{kJ})}{\text{g}} \times 2.00\text{g}}{(25.67 - 22.83)\text{°C}} = \dfrac{11.6 \text{ kJ}}{\text{°C}}$$

63. $q_{naphthalene} = -q_{calorimeter} = -\text{heat capacity} \times \Delta T$

$$\text{heat capacity} = \frac{\frac{5153.5\,\text{kJ}}{\text{mol}} \times \frac{\text{mol}}{128.16\,\text{g}} \times 1.108\,\text{g}}{5.92\,°\text{C}}$$

heat capacity = 7.526 kJ /°C

$$q_{\text{thymol}} = -q_{\text{calorimeter}} = -\text{heat capacity} \times \Delta T = -\frac{7.526\,\text{kJ}}{°\text{C}} \times 6.74\,°\text{C}$$

$q_{\text{thymol}} = -50.73$ kJ

$$\Delta H_{\text{comb}}(\text{thymol}) = \frac{q}{\text{mol}} = \frac{-50.73\ \text{kJ}}{1.351\ \text{g} \times \frac{\text{mol}}{150.2\ \text{g}}} = -5.64 \times \frac{10^3\ \text{kJ}}{\text{mol}}$$

65. $2\,NO_2 \rightarrow 2\,O_2 + N_2$ —(33.2 kJ)

$N_2 + 2\,O_2 \rightarrow N_2O_4$ 9.2 kJ

$2\,NO_2 \rightarrow N_2O_4$ $\Delta H° = -24.0$ kJ

67. $C_3H_8 + 5\,O_2 \rightarrow 3\,CO_2 + 4\,H_2O$ −2219.9 kJ

$3\,CO_2 \rightarrow \frac{3}{2}\,O_2 + 3\,CO$ 3 × (−283.0 kJ)

$C_3H_8 + \frac{7}{2}\,O_2 \rightarrow 3\,CO + 4\,H_2O$ $\Delta H° = -1370.9$ kJ

69.

	ΔH (kJ)
$4\,NH_3(g) \rightarrow 6\,H_2(g) + 2\,N_2(g)$	-2 x -92.22
$2\,N_2(g) + 2\,O_2(g) \rightarrow 4NO(g)$	2 x 180.5
$6\,H_2(g) + 3\,O_2(g) \rightarrow 6\,H_2O\,(l)$	3 x -571.6

$4\,NH_3(g) + 5\,O_2(g) \rightarrow 4\,NO(g) + 6\,H_2O\,(l)$ $\Delta H = -1170$ kJ

71. $\Delta H°_{rxn} = \Sigma\nu_p\Delta H°_f\,(\text{products}) - \Sigma\nu_r\Delta H°_f\,(\text{reactants})$

(a) $\Delta H°_{rxn} = \Delta H°_f[NH_4Cl(s)] - \Delta H°_f[NH_3(g)] - \Delta H°_f[HCl(g)]$

$\Delta H°_{rxn} = -314.4\ \text{kJ} - (-46.11\ \text{kJ} - 92.31\ \text{kJ}) = -176.0\ \text{kJ}$

(b) $\Delta H°_{rxn} = \Delta H°_f[NH_4NO_3\,(s)] - \Delta H°_f[NH_3\,(g)] - \Delta H°_f[HNO_3\,(l)]$

$\Delta H°_{rxn} = -365.6\ \text{kJ} - (-46.11\ \text{kJ} - 173.2\ \text{kJ}) = -146.3\ \text{kJ}$

(c) $\Delta H°_{rxn} = \Delta H°_f[CaCl_2(s)] + \Delta H°_f[Mg(s)] - \Delta H°_f[MgCl_2(s)] - \Delta H°_f[Ca(s)]$

$\Delta H°_{rxn} = -795.8\ \text{kJ} + 0 - (-641.3\ \text{kJ} - 0) = -154.5\ \text{kJ}$

(d) $\Delta H°_{rxn} = \Delta H°_f[CO_2(g)] + \Delta H°_f[Fe(s)] - \Delta H°_f[FeO(s)] - \Delta H°_f[CO(g)]$

$\Delta H°_{rxn} = -393.5\ \text{kJ} + 0\ \text{kJ} - (-272\ \text{kJ} - 110.5\ \text{kJ}) = -11\ \text{kJ}$

73. $\Delta H°_{rxn} = \Sigma\nu_p\Delta H°_f\,(\text{products}) - \Sigma\nu_r\Delta H°_f(\text{reactants})$

$\Delta H°_{rxn} = 2\Delta H°_f[SO_2(g)] + 2\Delta H°_f[Zn\,(s)] - 2\Delta H°_f[ZnS(s)] - 3\Delta H°_f[O_2(g)]$

$- 2\Delta H°_f[ZnS(s)] = 878.2\ \text{kJ} - 2\ \text{mol} \times -296.8\ \text{kJ/mol} - 2\ \text{mol} \times -348.3\ \text{kJ/mol}$

$+ 3\ \text{mol} \times 0\ \text{kJ/mol} = 412.0\ \text{kJ}$

$$\Delta H^\circ_f[\text{ZnS (s)}] = \frac{412.0\text{ kJ}}{-2\text{ mol}} = -206.0\text{ kJ /mol}$$

75. $\frac{-23.50\text{ kJ}}{1.050\text{ g}} \times \frac{106.12\text{ g}}{\text{mol}} = \frac{-2.3751 \times 10^3\text{ kJ}}{\text{mol}}$

$C_4H_{10}O_3 + 5\ O_2(g) \rightarrow 4\ CO_2(g) + 5\ H_2O(l)$

$\Delta H^\circ_{rxn} = \Sigma \nu_p \Delta H^\circ_f(\text{products}) - \Sigma \nu_r \Delta H^\circ_f(\text{reactants})$

$-2375.1\text{ kJ} = 4\text{ mol } CO_2 \times -393.5\text{ kJ/mol} + 5\text{ mol } H_2O \times -285.8\text{ kJ/mol}$

$-5\text{ mol } O_2 \times 0 - 1\text{ mol } C_4H_{10}O_6\ \Delta H^\circ_f$

$\Delta H^\circ_f = -628\text{ kJ/mol}$

77. $\Delta H^\circ = \Delta H^\circ_f(NH_3) + \Delta H^\circ_f(H_2O) - \Delta H^\circ_f(NH_4) - \Delta H^\circ_f(OH^-)$

$\Delta H^\circ = 46.11\text{ kJ} + (-285.8\text{ kJ}) - (-132.5\text{ kJ}) - (-230.0\text{ kJ})$

$\Delta H^\circ = 30.6\text{ kJ}$

79. CH_4 –890 kJ

$C_2H_6\ \frac{-1560\text{ kJ}}{2} = -780\text{ kJ}$ $C_6H_{14}\ \frac{-4163\text{ kJ}}{6} = -693\text{ kJ}$

$C_3H_8\ \frac{-2220\text{ kJ}}{3} = -740\text{ kJ}$ $C_7H_{16}\ \frac{-4811\text{ kJ}}{7} = -687\text{ kJ}$

$C_4H_{10}\ \frac{-2879\text{ kJ}}{4} = -720\text{ kJ}$ $C_8H_{18}\ \frac{-5450\text{ kJ}}{8} = -681\text{ kJ}$

$C_5H_{12}\ \frac{-3536\text{ kJ}}{5} = -707\text{ kJ}$

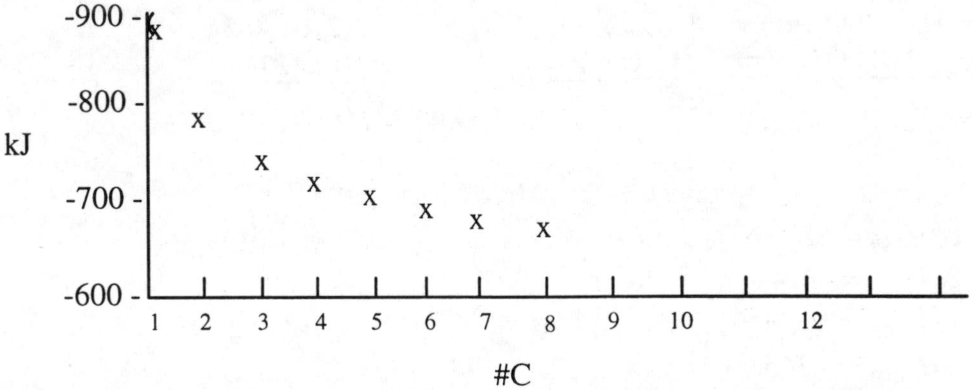

The curve is almost linear for 6 through 8 C. Extending that line produces –687 –(–693) = 6; –681 –(–687) = 6; –681 +12 = –669; –669 + 12 = –657

$$\Delta H\ (C_{10}H_{22}) = \frac{-669\text{ kJ}}{\text{C}} \times 10\text{C} \approx -6700\text{ kJ}$$

$$\Delta H\ (C_{12}H_{26}) = \frac{-657\text{ kJ}}{\text{C}} \times 12\text{C} \approx -7900\text{ kJ}$$

81. $?\ \% = 15.0\text{ g} \times \frac{50.1\text{ g fat}}{100\text{ g}} \times \frac{9.0\text{ Cal}}{\text{g}} \times \frac{100\%}{0.20 \times 2500\text{ Cal}} = 14\%$

83. (a) $KE = \frac{1}{2}\ m\ \underline{u}^2 = \frac{1}{2}\frac{M}{N_A}\left(\frac{3RT}{M}\right) = \frac{3}{2}\frac{RT}{N_A}$

$$KE = \frac{3}{2} \times \frac{8.3145 \frac{J}{mol\ K} \times 273\ K}{6.022 \times 10^{23}\ atoms/mol} = 5.65 \times 10^{-21}\ J$$

The KE of a single molecule, no matter how fast it is moving, is not likely to match the KE of the much more massive "BB," thus the calculation in (a) could be skipped.

(b) $KE = \frac{1}{2}\ m\ \underline{u}^2 = \frac{1}{2} \times 1.0\ g \times \frac{kg}{10^3\ g} \times \left(100\ \frac{m}{s}\right)^2$

$KE = 5.00\ J$

(c) $q = m \times \text{specific heat} \times \Delta T$

$$q = 10\ mL \times \frac{1.0\ g}{mL} \times \frac{4.18\ J}{g\ °C} \times (21\ °C - 20\ °C)$$

$q = 42\ J$

(c) is the greatest

85. $q_{Fe} = -q_{H_2O}$

$m_{Fe}\ sp.ht.(Fe)\ \Delta T = -m_{H_2O}\ sp.ht.(H_2O)\Delta T$

$$100\ g\ \frac{0.449\ J}{g\ °C}(t_f\ 100)\ °C = -100\ g\ \frac{4.182\ J}{g\ °C}(t_f\ 20)\ °C$$

$$-4.49 \times 10^3\ J + \frac{44.9\ J}{°C}t_f = +8.364 \times 10^3\ J - \frac{418.2\ J}{°C}t_f$$

$$\frac{463.1\ J}{°C}t_f - 12.85 \times 10^3\ J$$

$t_f = 27.8\ °C$

87. $q_{H_2O} = -q_{H_2O}$

$m_{H_2O}\ sp.ht.\Delta T = -m_{H_2O}\ sp.ht.\Delta T$

$$225\ mL\ \frac{1.00\ g}{mL}\ sp.ht.(t_f - 20.7\ °C) = -334\ mL\ \frac{1.00\ g}{mL}\ sp.ht.(t_f - 35.4\ °C)$$

$225t_f - 4658\ °C = -334t_f + 11824\ °C$

$559t_f = 16482\ °C$

$t_f = 29.5\ °C$

89. (a)

Al	$0.902 \times 26.98 = 24.3$
Cu	$0.385 \times 63.55 = 24.5$
Fe	$0.449 \times 55.85 = 25.1$
Pb	$0.128 \times 207.2 = 26.5$
Hg	$0.139 \times 200.6 = 27.4$

Ag $0.235 \times 107.9 = 25.4$
S $0.706 \times 32.07 = 22.6$ Average 25.2

(b) y = mx + b is the straight line equation. To make Dulong and Petit's law fit a straight line, plot either $\frac{1}{\mathcal{M}}$ versus specific heat or $\frac{1}{\text{specific heat}}$ versus molar mass.

$C = \mathcal{M} \times \text{specific heat}$

$\text{specific heat} = \frac{C}{\mathcal{M}}$ where $X = \frac{1}{\mathcal{M}}$

(c) $? \text{ g/mol} = \dfrac{25.2}{\frac{0.421\text{ J}}{\text{g °C}}} = 59.9 \text{ g/mol}$ actual 58.9 g/mol

(d) $? \text{ J/g °C} = \dfrac{25.2}{47.88 \text{ g/mol}} = \dfrac{0.526\text{ J}}{\text{g °C}}$

$-q_{Ti} = q_{H_2O}$

$$-25.5 \text{ g} \times \frac{0.526\text{ J}}{\text{g °C}} \times (27.4 - 99.7)\text{ °C} = V_{H_2O} \times \frac{1.00\text{ g}}{\text{mL}} \times \frac{4.182\text{ J}}{\text{g °C}} \times (27.4 - 24.6)\text{ °C}$$

$$969.8 \text{ J} = V_{H_2O} \times \frac{11.71\text{J}}{\text{mL}}$$

$V_{H_2O} = 83 \text{ mL}$

91. $? \text{ mol} = 15.5 \text{ g } H_2 \times \dfrac{\text{mol}}{2.016 \text{ g}} \times \dfrac{\text{mol } H_2O}{\text{mol } H_2} = 7.688 \text{ mol}$

$? \text{ mol} = 84.5 \text{ g } O_2 \times \dfrac{\text{mol}}{32.00 \text{ g}} \times \dfrac{\text{mol } H_2O}{0.500 \text{ mol } O_2} = 5.281 \text{ mol}$

Oxygen is limiting.

$? \text{ kJ} = 84.5 \text{ g } O_2 \times \dfrac{\text{mol}}{32.00 \text{ g}} \times \dfrac{241.8 \text{ kJ}}{0.500 \text{ mol } O_2} = 1.28 \times 10^3 \text{ kJ}$

93. $$n = \frac{PV}{RT} = \frac{744 \text{ mmHg} \times \frac{\text{atm}}{760 \text{ mmHg}} \times 215 \text{ L}}{\frac{0.08206 \text{ L atm}}{\text{K mol}} \times (273.2 + 24.5) \text{ K}} = 8.616 \text{ moles}$$

The following $\Delta H°_{comb}$ were taken from Problem 85, but the values can be calculated using $\Delta H°_f$ values from the Appendix for the complete combustion reactions.

$$q_{CH_4} = 8.616 \text{ mol} \times \frac{0.830 \text{ moles } CH_4}{\text{mol gas}} \times \frac{890 \text{ kJ}}{\text{mol}} = 6.36 \times 10^3 \text{ kJ}$$

$$q_{C_2H_6} = 8.616 \text{ mol} \times \frac{0.112 \text{ mol } C_2H_6}{\text{mol gas}} \times \frac{1560 \text{ kJ}}{\text{mol}} = 1.51 \times 10^3 \text{ kJ}$$

$$q_{C_3H_8} = 8.616 \text{ mol} \times \frac{0.058 \text{ mol } C_3H_8}{\text{mol gas}} \times \frac{2220 \text{ kJ}}{\text{mol}} = \underline{1.11 \times 10^3 \text{ kJ}}$$

$$8.98 \times 10^3 \text{ kJ}$$

95. It is necessary to compare only the 15% ethanol with 15% gasoline. If 100 g is assumed, then the 15% of each is 15 g of each.

$CH_3CH_2OH + 3O_2 \rightarrow 2\ CO_2 + 3\ H_2O$

$$\Delta H^\circ_{rxn} = \frac{-1273.7 \text{ kJ}}{\text{mol } CH_3CH_2OH}$$

ethanol

$$? \text{ kJ} = 15 \text{ g } C_2H_5OH \times \frac{\text{mol } C_2H_5OH}{46.1 \text{ g } C_2H_5OH} \times \frac{-1273.7 \text{ kJ}}{\text{mol } C_2H_5OH} = -4.1 \times 10^2 \text{ kJ}$$

octane

$$? \text{ kJ} = 15 \text{ g } C_8H_{18} \times \frac{\text{mol } C_8H_{18}}{114 \text{ g } C_8H_{18}} \times \frac{-5450 \text{ kJ}}{\text{mol } C_8H_{18}} = -7.2 \times 10^2 \text{ kJ}$$

Gasoline (octane) produces more heat on combustion.

97. $2\ C_{12}H_{26}(l) + 37\ O_2(g) \rightarrow 24\ CO_2(g) + 26\ H_2O(g)$

$2\ C_{12}H_{26}(l) + 37\ O_2(g) \rightarrow 24\ CO_2(g) + 26\ H_2O(l)$

Because ΔH_f° $[H_2O(l)]$ is more negative than ΔH_f° $[H_2O(g)]$ the second reaction should liberate the greater quantity of heat, by -572 kJ/mol $C_{12}H_{26}$.

$$\frac{? \text{ kJ}}{\text{mol } C_{12}H_{26}} - \left(\frac{-285.8 \text{ kJ}}{\text{mol}} \quad \frac{-241.8 \text{ kJ}}{\text{mol}}\right) \times \frac{26 \text{ mol } H_2O}{2 \text{ mol } C_{12}H_{26}} - \frac{-572 \text{ kJ}}{\text{mol } C_{12}H_{26}}$$

99. (a) Because ΔH_f is a state function, ΔH_f values of $H_2SO_4(aq)$ depend on concentration.

(b) When concentrated solutions are diluted, there is a temperature increase.

$\Delta H_{rxn} = \Sigma\Delta H_f(\text{prod}) - \Sigma\Delta H_f(\text{react})$

$\Delta H_{rxn} = \Delta H_f$ dilute $- \Delta H_f$ concentrated

The dilute solution has a more negative value of ΔH, making the ΔH_{rxn} negative and the dilution an exothermic process; heat is given off.

Apply Your Knowledge

101. $$3.294 \text{ g } CO_2 \times \frac{\text{mol C}}{44.010 \text{ g } CO_2} = 0.07484 \text{ mol C} \times \frac{1}{0.07484 \text{ mol C}} = \frac{1 \text{ mol C}}{\text{mol C}}$$

$$1.573 \text{ g } H_2O \times \frac{2 \text{ mol H}}{18.01 \text{ g } H_2O} = 0.1746 \text{ mol H} \times \frac{1}{0.07984 \text{ mol C}} = \frac{2.333 \text{ mol H}}{\text{mol C}}$$

$C_{(1 \times 3)}H_{(2.333 \times 3)} \Rightarrow C_3H_7$

Molecular formula is twice the empirical formula.

Molecular formula = C_6H_{14}

(a) $2\ C_6H_{14} + 19\ O_2 \rightarrow 12\ CO_2\ (g) + 14\ H_2O\ (l)$

$\Delta H_{rxn} = \Sigma\Delta H_f^\circ\ (prod) - \Sigma\Delta H_f^\circ\ (react)$

$$\Delta H_{rxn} = 12\ mol \times -\frac{393.5\ kJ}{mol} + 14\ mol\ x\ -\frac{285.8\ kJ}{mol} - 2\ mol \times -\frac{204.6\ kJ}{mol}$$

$= -8314.0$ kJ

(b) $CH_3CH(CH_3)CH_2CH_2CH_3$ or $CH_3CH_2CH(CH_3)CH_2C$

(c) $$?\ mL\ H_2O = 25.0\ m^3 \times \left(\frac{cm}{10^{-2}\ m}\right)^3 \times \frac{mL}{cm^3} \times\ = 2.50 \times 10^7\ mL$$

$$?\ gal\ C_6H_{14} = 2.50 \times 10^7\ mL \times \frac{1.00\ g}{mL} \times (33.0 - 19.2)^\circ C \times \frac{4.184\ J}{g^\circ C} \times \frac{kJ}{10^3\ J}$$

$$\times \frac{mol}{4157\ kJ} \times \frac{86.178\ g}{mol} \times \frac{mL}{0.6532\ g} \times \frac{10^{-3}\ L}{mL} \times \frac{1.059\ qt}{1.000\ L} \times \frac{gal}{4\ qt} = 12.1\ gal$$

103. $W = F \times d = m \times gravity \times d$

$$?\ kJ = 4 \times 58.0\ kg \times 9.807\ \frac{m}{s^2} \times 1450\ m \times \frac{kJ}{10^3\ J} = 3.30 \times 10^3\ kJ \text{ to climb mountain}$$

$C_6H_{12}O_6 + 6\ O_2 \rightarrow 6\ CO_2\ (g) + 6\ H_2O\ (l)$

$\Delta H = 6\ mol\ CO_2 \times -393.5\ kJ/mol + 6\ mol\ H_2O \times -285.8\ kJ/mol$

$- 1\ mol\ C_6H_{12}O_6 \times -1273.3\ kJ/mol = -2802.5\ kJ$

$$?\ g = 3.30 \times 10^3\ kJ \times \frac{mol}{2802.5\ kJ} \times \frac{100\ kJ\ total}{70\ kJ\ work} \times \frac{180.2\ g}{mol} = 3.0 \times 10^2\ glucose$$

105. $PV = \frac{m}{M}\ RT \qquad M = \frac{mRT}{PV}$

$$M = \frac{11.103\ g \times \frac{62.364\ L\ Torr}{K\ mol} \times 298.15\ K}{769.9\ Torr \times 582\ mL \times \frac{10^{-3}\ L}{mL}} = 463\ \frac{g}{mol}$$

$$?\ g\ C = 2.108\ g\ CO_2 \times \frac{12.01\ g\ L}{44.01\ g\ CO_2} = 0.5752\ g\ C$$

$$?\ g\ H = 1.294\ g\ H_2O \times \frac{2.016\ g\ H}{18.016\ g\ H_2O} = 0.1448\ g\ H$$

$?\ g\ O = 1.103\ g - 0.5752\ g\ C - 0.1448\ g\ H = 0.383\ g\ O$

$$0.5752\ g\ C \times \frac{mol\ C}{12.01\ g\ C} = 0.04790\ mol\ C \times \frac{1}{0.0239\ mol\ O} = \frac{2.00\ mol\ C}{mol\ O}$$

$$0.1448\ g\ H \times \frac{mol\ H}{1.008\ g\ H} = 0.1437\ mol\ H \times \frac{1}{0.0239\ mol\ O} = \frac{6.01\ mol\ H}{mol\ O}$$

$$0.383\ g\ O \times \frac{mol\ O}{16.00\ g\ O} = 0.0239\ mol\ O \times \frac{1}{0.0239\ mol\ O} = 1\ \frac{mol\ O}{mol\ O}$$

C_2H_6O

$C_2H_6O + 3\ O_2 \rightarrow 2\ CO_2 + 3\ H_2O$

$$\Delta H° = \frac{5.015\ \text{kJ}}{°\text{C}} \times (31.94 - 25.00)°\text{C} \times \frac{1}{1.103\ \text{g}} \times \frac{46.0\ \text{g}}{\text{mol}} = 1.45 \times 10^3\ \text{kJ}$$

Chapter 7

Atomic Structure

Exercises

7.1A $\nu = \frac{c}{\lambda} = \frac{3.00 \text{ x } 10^8 \text{ m/s}}{1.07 \text{ mm}} \times \frac{\text{mm}}{10^{-3} \text{ m}} = 2.80 \times 10^{11} \text{ Hz}$

7.1B $\lambda = \frac{c}{\nu} = \frac{3.00 \text{ x } 10^8 \text{ m/s}}{9.76 \text{ x } 10^{13} \text{ s}^{-1}} \times \frac{\text{nm}}{10^{-9}} = 3.07 \times 10^3 \text{ nm}$

7.2 The microwave has a longer wavelength than the visible light of the television set.

7.3A $E = h\nu = \frac{6.626\times10^{-34} \text{ J}\cdot\text{s}}{\text{photon}} \times \frac{2.89\times10^{10}}{\text{s}} = \frac{1.91\times10^{-23} \text{ J}}{\text{photon}}$

7.3B $E = \frac{hc}{\lambda} = \frac{\frac{6.626\times10^{-34} \text{ J}\cdot\text{s}}{\text{photon}} \times \frac{3.00\times10^{8} \text{ m}}{\text{s}}}{235 \text{ nm} \times \frac{10^{9} \text{ m}}{\text{nm}}} = \frac{8.46\times10^{-19} \text{ J}}{\text{photon}}$

7.4A $E = \frac{hc}{\lambda} = \frac{\frac{6.626\times10^{-34} \text{ J}\cdot\text{s}}{\text{photon}} \times \frac{3.00\times10^{8} \text{ m}}{\text{s}}}{400 \text{ nm} \times \frac{10^{9} \text{ m}}{\text{nm}}} \times \frac{6.022\times10^{23} \text{ photon}}{\text{mole}} \times \frac{\text{kJ}}{10^3 \text{ J}}$

$= \frac{2.99\times10^{2} \text{ kJ}}{\text{mol}}$

7.4B $\lambda = \frac{\frac{6.022\times10^{23} \text{ photon}}{\text{mol}} \times 6.626\times10^{-34} \text{ J}\cdot\text{s} \times 3.00\times10^{8} \text{ m/s}}{\frac{100 \text{ kJ}}{\text{mol}} \times \frac{10^3 \text{ J}}{\text{kJ}}} \times \frac{\text{nm}}{10^{9} \text{ m}}$

$= 1.20 \times 10^3 \text{ nm}$

7.5 $E_6 = \frac{-B}{n^2} = \frac{-2.179 \text{ x } 10^{-18} \text{ J}}{6^2} = -6.053 \times 10^{-20} \text{ J}$

7.6 $\Delta E = B \times \left(\frac{1}{n_i^2} - \frac{1}{n_f^2}\right) = 2.179 \times 10^{-18} \times \left(\frac{1}{2^2} - \frac{1}{4^2}\right)$

$= 2.179 \times 10^{-18} \times (0.1875) = 4.086 \times 10^{-19}\ \text{J}$

7.7A $E_i = \frac{-B}{n^2} = \frac{-2.179 \text{ x } 10^{-18}\ \text{J}}{4^2} = -1.362 \times 10^{-19}\ \text{J}$

$E_f = \frac{-B}{n^2} = \frac{-2.179 \text{ x } 10^{-18}\ \text{J}}{1^2} = -2.179 \times 10^{-18}\ \text{J}$

$\Delta E = E_f - E_i = -2.043 \times 10^{-18}\ \text{J}$

$\nu = \frac{?E}{h} = \frac{2.043 \times 10^{-18}\ \text{J}}{6.626 \times 10^{-34}\ \text{J} \cdot \text{s}} = 3.083 \times 10^{15}\ \text{s}^{-1}$

7.7B $\Delta E = B \times \left(\frac{1}{n_i^2} - \frac{1}{n_f^2}\right) = 2.179 \times 10^{-18} \times \left(\frac{1}{5^2} - \frac{1}{2^2}\right)$

$\Delta E = 2.179 \times 10^{-18} \times (-0.21) = -4.576 \times 10^{-19}\ \text{J}$

$\nu = \frac{\Delta E}{h} = \frac{4.576 \times 10^{-19}\ \text{J}}{6.626 \times 10^{-34}\ \text{J} \cdot \text{s}} = 6.906 \times 10^{14}\ \text{s}^{-1}$

$\lambda = \frac{c}{\nu} = \frac{2.998 \text{ x } 10^8\ \text{m/s}}{6.906 \text{ x } 10^{14}\ \text{s}^{-1}} = 4.341 \times 10^{-7}\ \text{m} \times \frac{\text{nm}}{10^{-9}\ \text{m}} = 434.1\ \text{nm}$ visible region

7.8 (c) requires energy to be emitted. Since the energy difference between two successive energy levels is smaller the higher the energy levels, (a) is larger than (d). Comparing (a) and (b) produces $\left(\frac{1}{1^2} - \frac{1}{2^2}\right)$ versus $\left(\frac{1}{3^2} - \frac{1}{8^2}\right)$ or 0.75 versus 0.11, so (a) represents the greatest amount of energy absorbed.

7.9 $\lambda = \frac{h}{mu} = \frac{\frac{6.626 \text{ x } 10^{-34}\ \text{kg m}^2\ \text{s}}{\text{s}^2}}{1.67 \text{ x } 10^{-27}\ \text{kg x } 3.79 \text{ x } 10^3\ \text{m/s}}$

$\lambda = 1.05 \times 10^{-10}\ \text{m} \times \frac{\text{nm}}{10^{-9}\ \text{m}} = 0.105\ \text{nm}$

7.10 (a) not possible. For $l = 1$, m_l must be between +1 and –1.
(b) All values are possible.
(c) All values are possible.
(d) not possible. For $l = 2$, m_l must be between +2 and –2.

7.11 (a) 5 p $l = 1$ $m_l = -1, 0, 1$ 3 orbitals
(b) $l = 0$ $m_l = 0$ 1 orbital
$l = 1$ $m_l = -1, 0, 1$ 3 orbitals
$l = 2$ $m_l = -2, -1, 0, 1, 2$ 5 orbitals

$l = 3$ $m_l = -3, -2, -1, 0, 1, 2, 3$ 7 orbitals

$n^2 = 4^2 = 16$ orbitals 16 orbitals

(c) $l = 3$ $m_l = -3, -2, -1, 0, 1, 2, 3$ f subshell consists of seven orbitals

Review Questions

4. When using atomic numbers to order elements by atomic mass, there are some discrepancies because the atomic masses don't always increase from one element to the next. An example is tellurium and iodine. In ordering by atomic mass, iodine would come first, but tellurium has one less proton.

10. At the microscopic level, changes are quantized—that is, occurring only in discrete steps or intervals. At the macroscopic level, those available intervals are so close together that they appear to be continuous.

13. The negative value is an indication that this energy is due to a force of attraction. The closer to the nucleus, the more attraction, so the more negative the energy. If the equation were $E_n = -B \times n^2$, the value would become more negative the farther from the nucleus. The equation $E_n = -B/n^2$ gives a smaller negative value the farther from the nucleus, becoming zero for an infinite value of n.

16. Bohr predicted precise orbitals for electrons, and this conflicts with the Heisenberg uncertainty principle of not knowing both the precise location and momentum of an electron. Schrödinger's wave equations gave probabilities instead of precise orbits. Bohr predicted a precise orbit at 52.9 pm, and Schrödinger found the most likely electron probability at 52.9 pm.

19. (a) 3d (b) 2s (c) 4p (d) 4f

20. (a) $n = 3$ $l = 0,\ l = 1,\ l = 2$ 3 subshells

(b) $n = 2$ $l = 0,\ l = 1$ 2 subshells

(c) $n = 4$ $l = 0,\ l = 1,\ l = 2,\ l = 3$ 4 subshells

Problems

23. Charge = 1.602×10^{-19} C

1.044×10^{-8} kg/C $\times 1.602 \times 10^{-19}$ C $= 1.672 \times 10^{-27}$ kg

25. 6.4×10^{-19} C $= 4 \times 1.6 \times 10^{-19}$ C

3.2×10^{-19} C $= 2 \times 1.6 \times 10^{-19}$ C

4.8×10^{-19} C $= 3 \times 1.6 \times 10^{-19}$ C

8.0×10^{-19} C $= 5 \times 1.6 \times 10^{-19}$ C

All values are integral multiples of 1.6×10^{-19} C, the charge on an electron or on a proton.

27. Values are atomic units of mass and charge followed by SI units.

(a) $^{80}Br^-$ –80:1 -8.29×10^{-7} kg/C

(b) $^{18}O^{2-}$ $-9:1$ -9.33×10^{-8} kg/C
(c) $^{40}Ar^{+}$ $+40:1$ 4.15×10^{-7} kg/C

Values are approximate because mass numbers were used instead of actual atomic masses.

29. $69.9243 \times 0.205 = 14.33$
$71.9217 \times 0.274 = 19.71$
$72.9234 \times 0.078 = 5.69$
$73.9219 \times 0.365 = 26.98$
$75.9214 \times 0.076 = \underline{5.77}$
$72.48 \Rightarrow 72.5$

31. $? \text{ s} = 4.4 \times 10^{9} \text{ km} \times \frac{10^{3} \text{ m}}{\text{km}} \times \frac{\text{s}}{3.00 \text{ x } 10^{8} \text{ m}} = 1.5 \times 10^{4} \text{ s}$

33. (a) $\lambda = \frac{c}{\nu} = \frac{2.998 \times 10^{8} \text{ m/s}}{992 \text{ kHz} \times \frac{10^{3} \text{ Hz}}{\text{kHz}}} = 302 \text{ m}$ radio

(b) $\lambda = \frac{c}{\nu} = \frac{2.998 \times 10^{8} \text{ m/s}}{400 \text{ MHz} \times \frac{10^{6} \text{ Hz}}{\text{MHz}}} = 0.75 \text{ m}$ radio

(c) $\lambda = \frac{c}{\nu} = \frac{3.00 \text{ x } 10^{8} \text{ m/s}}{\frac{5.2 \text{ x } 10^{14}}{\text{s}}} \times \frac{\text{nm}}{10^{-9} \text{ m}} = 5.8 \times 10^{2} \text{ nm}$ visible

35. $\frac{1}{4} \lambda = 175 \text{ nm}$
$\lambda = 700 \text{ nm}$ red
$\nu = \frac{3.00 \text{ x } 10^{8} \text{ m/s}}{700 \text{ nm x } \frac{10^{-9} \text{ m}}{\text{nm}}} = 4.29 \times 10^{14} \text{ s}^{-1}$

37. $E = h\nu = 6.626 \times 10^{-34} \text{ J}\cdot\text{s} \times 7.42 \times 10^{14} \text{ s}^{-1} = 4.92 \times 10^{-19} \text{ J}$

$E = \frac{hc}{\nu} = \frac{6.626 \text{ x } 10^{-34} \text{ J s x } \frac{3.00 \text{ x } 10^{8} \text{ m}}{\text{s}}}{655 \text{ nm x } \frac{10^{-9} \text{ m}}{\text{nm}}} = 3.03 \times 10^{-19} \text{ J}$

The violet light has greater energy than the red 655-nm light. Energy increases as wavelength decreases and frequency increases.

39. $\nu = \frac{c}{\lambda} = \frac{2.998 \text{ x } 10^8 \text{ m/s}}{780 \text{ nm x } \frac{10^{-9}\text{ m}}{\text{nm}}} = 3.844 \times 10^{14} \text{ s}^{-1}$

$E = h\nu = \frac{6.626 \text{ x } 10^{-34} \text{ J s}}{\text{photon}} \times 3.844 \times 10^{14} \text{ s}^{-1} = \frac{2.55 \text{ x } 10^{-19} \text{ J}}{\text{photon}}$

$E = \frac{2.55 \text{ x } 10^{-10} \text{ J}}{\text{photon}} \times \frac{6.022 \text{ x } 10^{23} \text{ photons}}{\text{mol}} \times \frac{1 \text{ kJ}}{1000 \text{ J}} = 154 \text{ kJ/mol}$

41. $\nu = \frac{E}{h} = \frac{7.21 \text{ x } 10^{-19} \text{ J}}{6.626 \text{ x } 10^{-34} \text{ J s}} = 1.088 \times 10^{15} \text{ s}^{-1}$

$\lambda = \frac{c}{\nu} = \frac{2.998 \text{ x } 10^8 \text{ m/s}}{1.088 \text{ x } 10^{15} \text{ s}^{-1}} = 2.76 \times 10^{-7} \text{ m} \times \frac{\text{nm}}{10^{-9} \text{ m}} = 276 \text{ nm}$

43. $\lambda = \frac{hc}{E} = \frac{6.626\times10^{-34} \text{ J}\cdot\text{s}\times\frac{3.00\times10^8 \text{ m}}{\text{s}}}{\frac{191 \text{ kJ}}{\text{mol photon}}\times\frac{10^3 \text{ J}}{\text{kJ}}\times\frac{\text{mol photon}}{6.022\times10^{23}\text{ photon}}} = 6.27 \times 10^{-7} \text{ m} \times \frac{\text{nm}}{10^{-9} \text{ m}}$

$= 627 \text{ nm}$ orange

45. (a) $E_i = \frac{-B}{n^2} = \frac{-2.179 \text{ x } 10^{-18} \text{ J}}{5^2} = -8.716 \times 10^{-20} \text{ J}$

$E_f = \frac{-B}{n^2} = \frac{-2.179 \text{ x } 10^{-18} \text{ J}}{1^2} = -2.179 \times 10^{-18} \text{ J}$

$\Delta E = E_f - E_i = -2.179 \times 10^{-18} \text{ J} - (-8.716 \times 10^{-20} \text{ J})$

$\Delta E = -2.092 \times 10^{-18} \text{ J}$

(b) $E_i = \frac{-B}{n^2} = \frac{-2.179 \text{ x } 10^{-18} \text{ J}}{5^2} = -8.716 \times 10^{-20} \text{ J}$

$E_f = \frac{-B}{n^2} = \frac{-2.179 \text{ x } 10^{-18} \text{ J}}{4^2} = -1.362 \times 10^{-19} \text{ J}$

$\Delta E = E_f - E_i = -1.362 \times 10^{-19} \text{ J} - (-8.716 \times 10^{-20} \text{ J}) = -4.90 \times 10^{-20} \text{ J}$

47. (a) $E_i = \frac{-B}{n^2} = \frac{-2.179 \text{ x } 10^{-18} \text{ J}}{3^2} = -2.421 \times 10^{-19} \text{ J}$

$E_f = \frac{-B}{n^2} = \frac{-2.179 \text{ x } 10^{-18} \text{ J}}{2^2} = -5.448 \times 10^{-19} \text{ J}$

$\Delta E = E_f - E_i = -3.027 \times 10^{-19} \text{ J}$

$\nu = \frac{?E}{h} = \frac{3.027 \text{ x } 10^{-19} \text{ J}}{6.626 \text{ x } 10^{-34} \text{ J s}} = 4.568 \times 10^{14} \text{ s}^{-1}$

(b) $E_i = \frac{-B}{n^2} = \frac{-2.179 \text{ x } 10^{-18} \text{ J}}{4^2} = -1.362 \times 10^{-19} \text{ J}$

$E_f = \frac{-B}{n^2} = \frac{-2.179 \text{ x } 10^{-18} \text{ J}}{1^2} = -2.179 \times 10^{-18} \text{ J}$

$\Delta E = E_f - E_i = -2.043 \times 10^{-18} \text{ J}$

$$\nu = \frac{?E}{h} = \frac{2.043 \text{ x } 10^{-18} \text{ J}}{6.626 \text{ x } 10^{-34} \text{ J s}} = 3.083 \times 10^{15} \text{ s}^{-1}$$

49. $E_n = -2.179 \times 10^{-21} \text{ J} = -B \times \frac{1}{n^2} = -2.179 \times 10^{-18} \text{ J} \times \frac{1}{n^2}$

That would make $\frac{1}{n^2} = 10^{-3}$; $n^2 = 10^3$ or n = 31.6, which is not an integer number.

51. $\Delta E = B \times \left(\frac{1}{1^2} - \frac{1}{\infty^2}\right)$

$= 2.179 \times 10^{-18} \text{ J} \times (1 - 0)$

$= 2.179 \times 10^{-18} \text{ J} \times \frac{6.022 \text{ x } 10^{23} \text{ electron}}{\text{mol electrons}} \times \frac{\text{kJ}}{10^3 \text{ J}}$

$\Delta E = 1.312 \times \frac{10^3 \text{ kJ}}{\text{mol}}$

53. $\lambda = \frac{h}{mu} = \frac{6.626 \text{ x } 10^{-34} \text{ kg m}^2 \text{ s}^{-1}}{1.67 \text{ x } 10^{\ 27} \text{ kg x } 2.55 \text{ x } 10^6 \text{ m s}^{-1}} = 1.56 \times 10^{-13} \text{ m} \times \frac{\text{nm}}{10^{-9} \text{ m}}$

$= 1.56 \times 10^{-4}$ nm.

55. $u = \frac{h}{m\lambda} = \frac{6.626 \text{ x } 10^{-34} \text{ kg m}^2 \text{ s}^{-1}}{9.109 \text{ x } 10^{-31} \text{ kg x } 84.4 \text{ nm x } \frac{10^{\ 9} \text{ m}}{\text{nm}}} = 8.62 \times 10^3$ m/s

57. (a) For p orbitals, $l = 1$ n must be at least 2 for $l = 1$.
(b) For f orbitals, $l = 3$ n must be at least 4 for $l = 3$.

59. If $n = 5$, $l = 0, 1, 2, 3, 4$.
If $l = 3$, $m_l = -3, -2, -1, 0, 1, 2, 3$.

61. (a) permissible
(b) If $n = 3$, l cannot be 3; l cannot be greater than $n - 1$.
(c) permissible
(d) $n = 0$ is not permissible.

63. (a) For 3s, $n = 3$, $l = 0$, $m_l = 0$
(b) For 5f, $n = 5$, $l = 3$, $m_l = -3, -2, -1, 0, 1, 2, 3$
(c) 3p subshell

65. (a) $n = 3$ $\quad m_s = \pm\frac{1}{2}$ $\quad$ posssible orbitals: 3d, 4d, 5d, etc.
(b) $l = 1$ $\quad$ 2p orbital
(c) $l = 3$ or 2 $\quad m_s = \pm\frac{1}{2}$ $\quad$ possible orbitals 4f and 4d
(d) $n = 1, 2, 3, \ldots$ $m_l = 0$ $\quad$ possible orbitals: 1s, 2s, 3s, …

67. $n = 3, l = 2, m_l = 0, m_s = -\frac{1}{2}$

$n = , l = 2, m_l = \pm 1, \pm 2, ms = \pm \frac{1}{2}$

69. $196 \times 0.00146 = 0.286$
$198 \times 0.1002 = 19.8$
$199 \times 0.1684 = 33.5$
$200 \times 0.2313 = 46.3$
$201 \times 0.1322 = 26.6$
$202 \times 0.2980 = 60.2$
$204 \times 0.0685 = \underline{14.0}$
200.7

It is approximate because mass numbers are used instead of the isotopic masses.

71. $T = \frac{1}{\nu}$ Period and frequency are inversely related.

$\frac{1\text{s}}{60 \text{ cycles}} = 1.7 \times 10^{-2}$ s/cycle

73. The visible radiation given off by Fronkensteen is caused by excited electrons falling back to lower-energy orbitals in helium atoms. The energies of the emitted radiation are the same as the energy differences between levels in the atmospheric helium atoms. The radiation from Fronkensteen would be absorbed by the helium atoms in the atmosphere of the planet. Radiation from other stars emitted by atoms other than helium is not absorbed.

75. To a scientist the term "quantum jump" means from one energy level to the next. In the microscopic world, those levels are distinguishable, and one level may be rather widely separated from the next allowable level, thus representing a significant energy change. In the macroscopic world, successive energy levels are very closely spaced, and the energy difference between two adjacent levels is essentially undetectable. Thus, quantum jumps in the microscopic world are often very significant changes, whereas in the macroscopic world they would not be.

77. $\Delta E = E_8 - E_1 = 0 - \frac{-2.179 \text{ x } 10^{-18} \text{ J}}{1^2} = 2.179 \times 10^{-18}$ J

$\lambda = \frac{hc}{E} = \frac{6.626 \text{ x } 10^{-34} \text{ J s x } 2.998 \text{ x } 10^{8} \text{ m/s}}{2.179 \text{ x } 10^{-18} \text{ J}} \times \frac{\text{nm}}{10^{-9} \text{ m}} = 91.16$ nm minimum

$?E = E_2 - E_1 = \frac{-2.179 \text{ x } 10^{-18} \text{ J}}{2^2} - \frac{-2.179 \text{ x } 10^{-18} \text{ J}}{1^2} = 1.634 \times 10^{-18}$ J

$\lambda = \frac{hc}{E} = \frac{6.626 \text{ x } 10^{-34} \text{ J s x } 2.998 \text{ x } 10^{8} \text{ m/s}}{1.634 \text{ x } 10^{-18} \text{ J}} \times \frac{\text{nm}}{10^{-9} \text{ m}} = 121.6$ nm maximum

79. $E = \frac{hc}{\lambda} = \frac{6.626 \text{ x } 10^{-34} \text{ J s x } 2.998 \text{ x } 10^{8} \text{ m/s}}{486.1 \text{ nm x } 10^{-9} \text{ m/nm}} = 4.087 \times 10^{-19} \text{ J}$

$$\Delta E = \frac{-B}{n^2} - \frac{-B}{m^2} = -B\left(\frac{1}{n^2} - \frac{1}{m^2}\right)$$

$$4.087 \times 10^{-19} \text{ J} = 2.179 \times 10^{-18} \text{ J} \times \left(\frac{1}{n^2} - \frac{1}{m^2}\right)$$

$$\frac{1}{n^2} - \frac{1}{m^2} = 0.1876$$

$$\frac{1}{1^2} = 1 \quad \frac{1}{2^2} = 0.2500 \quad \frac{1}{3^2} = 0.1111 \quad \frac{1}{4^2} = 0.0625 \quad \frac{1}{5^2} = 0.0400$$

$\frac{1}{2^2} - \frac{1}{4^2} = 0.1875$ Thus, the transition is $n = 4 \longrightarrow n = 2$

81. $\Delta E = B \times \left(\frac{1}{n_i^2} - \frac{1}{n_f^2}\right)$

If level 1 –> 2 $\Delta E = B \times \left(\frac{1}{1} - \frac{1}{4}\right) = \frac{3}{4} B$,

then level 1 –> 3 (c) $\Delta E = B \times \left(1 - \frac{1}{9}\right) = \frac{8}{9} \times B = \frac{32}{27} \times \frac{3}{4} \times B$

83. $E_1 = \frac{2^2 \text{ x } 2.179 \text{ x } 10^{-18} \text{ J}}{1^2} = 8.716 \times 10^{-18} \text{ J}$

85. Planck $E = h\nu$ Einstein

$E = \frac{hc}{\lambda}$ $E = mc^2$

$$\frac{hc}{\lambda} = mc^2$$

$$\lambda = \frac{hc}{mc^2} = \frac{h}{mc}$$

In the deBroglie equation, the velocity of the particles u corresponds to the c in this derivation.

87. $\Delta E = \text{B} \times \left(\frac{1}{1^2} - \frac{1}{5^2}\right) = 2.179 \times 10^{-18} \text{ J} \times \left(1 - \frac{1}{25}\right) = 2.092 \times 10^{-18} \text{ J}$

$$\lambda = \frac{hc}{E} = \frac{6.626 \text{ x } 10^{-34} \text{ J s x } \frac{2.998 \text{ x } 10^{8} \text{ m}}{\text{s}}}{2.092 \text{ x } 10^{-18} \text{ J}} = 9.496 \times 10^{-8} \text{ m}$$

$$u = \frac{h}{m\lambda} = \frac{\frac{6.626 \text{ x } 10^{-34} \text{ kg m}^2 \text{ s}}{\text{s}^2}}{9.11 \text{ x } 10^{-31} \text{ kg x } 9.496 \text{ x } 10^{-8} \text{ m}} = 7.66 \times 10^{3} \text{ m/s}$$

89. $E = h\nu = 6.626 \times 10^{-34} \text{ J s} \times 8.4 \times 10^{9} \text{ s}^{-1} = 5.566 \times 10^{-24} \text{ J/photon}$

$$\frac{4 \text{ x } 10^{-21} \text{ watt}}{5.566 \text{ x } 10^{-24} \text{ J /photon}} \times \frac{\text{J}}{\text{watt s}} = 7 \times 10^{2} \text{ photon/s}$$

91. The volume of a shell 52.9 pm from the nucleus is so much larger than a small volume element at the nucleus that even though the probability of finding the electron in any one volume element 52.9 pm from the nucleus is less; the total probability of all the volume elements at that distance is greater than in a small volume element at the nucleus.

93. $$? \frac{\text{kg}}{\text{molecule}} = \frac{720 \text{ g}}{\text{mol}} \times \frac{\text{kg}}{10^{3} \text{ g}} \times \frac{\text{mole}}{6.022 \times 10^{23} \text{ molecules}} = 1.20 \times 10^{-24} \frac{\text{kg}}{\text{molecule}}$$

$$\lambda = \frac{h}{mv} = \frac{6.626 \times 10^{-34} \frac{\text{kg m}^2}{\text{s}}}{1.20 \times 10^{-24} \text{kg} \times 220 \frac{\text{m}}{\text{s}}} = 2.51 \times 10^{-12} \text{m}$$

$$? \text{ nm} = 2.51 \times 10^{-12} \text{ m} \times \frac{\text{nm}}{10^{-9}} = 2.51 \times 10^{-3} = 0.00251 \text{ nm}$$

The calculated value is a factor of 10 less than the observed value.

95. The level of energies are:

$$E_1 = \frac{-13}{n^2} = \frac{-2.179 \times 10^{-18} \text{ J}}{1^2} = -2.179 \times 10^{-18} \text{ J}$$

$E_2 = -5.448 \times 10^{-19}$ J

$E_3 = -2.421 \times 10^{-19}$ J

$E_4 = -1.362 \times 10^{-19}$ J

$E_5 = -8.716 \times 10^{-20}$ J

$E_6 = -6.053 \times 10^{-20}$ J

$E_7 = -4.447 \times 10^{-20}$ J

(a) $$E = \frac{hc}{\lambda} = \frac{6.626 \times 10^{-34} \text{ Js} \times 2.998 \times 10^{8} \frac{\text{m}}{\text{s}}}{750 \text{ nm} \times \frac{10^{-9} \text{m}}{\text{nm}}} = 2.649 \times 10^{-19}$$

The transition from 3 $\rightarrow$ 2 has the energy of 3.027 x 10^{-19} J.

$$\lambda = \frac{hc}{\Delta E} = \frac{6.626 \times 10^{-34} \text{ Js} \times 2.998 \times 10^{8} \frac{\text{m}}{\text{s}}}{3.027 \times 10^{-19} \text{ J}} = 6.56 \times 10^{-7} \text{m}$$

This is too short; there is no such line in the hydrogen spectrum.

(b) $$E = \frac{hc}{\lambda} = \frac{6.626 \times 10^{-34} \text{ Js} \times 2.998 \times 10^{8} \frac{\text{m}}{\text{s}}}{1800 \text{ nm} \times \frac{10^{-9} \text{m}}{\text{nm}}} = 1.103 \times 10^{-19} \text{ J}$$

$$E = \frac{hc}{\lambda} = \frac{6.626\times10^{-34}\,\text{Js}\times2.998\times10^{8}\,\frac{\text{m}}{\text{s}}}{1900\,\text{nm}\times\frac{10^{-9}\,\text{m}}{\text{nm}}} = 1.046\times10^{-19}\,\text{J}$$

The electron transition from $n = 4$ to $n = 3$ has the energy 1.059 x 10^{-19}. That is in the energy range, so a line does exist.

Apply Your Knowledge

97. (a) $\nu = \frac{E}{h} = \frac{3.42\text{ x }10^{-19}\text{ J}}{6.626\text{ x }10^{-34}\text{ Js}} = 5.16\times10^{14}\text{ s}^{-1}$

(b) $\lambda = \frac{c}{\nu} = \frac{2.998\times10^{8}\text{ m/s}}{5.16\times10^{14}\text{ s}^{-1}} = 5.81\times10^{-7}\text{ m}\times\frac{\text{nm}}{10^{-9}\text{ m}} = 581\text{ nm}$

The longer wavelength (1000 nm) has less energy and would not have enough energy to expel an electron.

$$E = \frac{hc}{\lambda} = \frac{6.626\text{ x }10^{-34}\text{ J x }\frac{3.00\text{ x }10^{8}\text{ m}}{\text{s}}}{1000\text{ nm x }\frac{10^{-9}\text{ m}}{\text{nm}}} = 1.99\times10^{-19}\text{ J}$$

which is less than the work function for cesium.

(c) $E = \frac{hc}{\lambda} = \frac{6.626\text{ x }10^{-34}\text{ J x }\frac{2.99\text{ x }10^{8}\text{ m}}{\text{s}}}{425\text{ nm x }\frac{10^{-9}\text{ m}}{\text{nm}}} = 4.67\times10^{-19}\text{ J}$

$E_{\text{excess}} = 4.67\times10^{-19}\text{ J} - 3.42\times10^{-19}\text{ J} = 1.25\times10^{-19}\text{ J}$

$E = \frac{1}{2}mu^2$

$u = \sqrt{\frac{2E}{m}} = \sqrt{\frac{2\text{ x }1.25\text{ x }10^{-19}\text{ J}}{9.109\text{ x }10^{-31}\text{kg}}} = 5.24\times10^{5}\text{ m/s}$

99. (a) $q = m \times \text{sp.ht.} \times \Delta T$

$q = 345\text{ g}\times\frac{4.18\text{ J}}{\text{g }^\circ\text{C}}\times(99.8 - 26.5)^\circ\text{C} = 1.057\times10^{5}\text{ J}$

$$E = \frac{hc}{\lambda} = \frac{6.626\text{ x }10^{-34}\text{ J x }\frac{2.998\text{ x }10^{8}\text{ cm}}{\text{s}}}{12.2\text{ cm x }\frac{10^{-2}\text{ m}}{\text{cm}}} = 1.628\times10^{-24}\text{ J/photon}$$

$$\frac{1.057\text{ x }10^{5}\text{ J}}{1.628\text{ x }10^{-24}\text{ J/photon}}\times\frac{1\text{ mol photons}}{6.022\text{ x }10^{23}\text{ photon}} = 1.08\times10^{5}\text{ mol photons}$$

(b) $?\text{ sec} = 1.057\times10^{5}\text{ J}\times\frac{1}{700\text{ w}}\times\frac{\text{ws}}{\text{J}} = 151\text{ sec}$

$?\text{ min} = 151\text{ s}\times\frac{\text{min}}{60\text{ s}} = 2.52\text{ min}$

(c) $\Delta E = \dfrac{hc}{\lambda} = \dfrac{6.626\times10^{-34}\ \text{Js}\times2.998\times10^{8}\ \frac{\text{m}}{\text{s}}}{12.2\ \text{cm}\times\frac{10^{-2}\ \text{m}}{\text{cm}}} = 1.63 \times 10^{-24}$ g

$$\Delta E = -B \times \left(\frac{1}{(n+1)^2} - \frac{1}{n^2}\right)$$

$$\frac{\Delta E}{B} = \frac{1.63\times10^{-24}\ \text{J}}{2.179\times10^{-18}} = \left(\frac{1}{(n+1)^2} - \frac{1}{n^2}\right) = 7.47 \times 10^{-7}$$

This is a very small energy difference and is probably impossible for hydrogen to produce. A 12.1-cm wavelength would be produced by a transition from n = 139 to n = 138 where the electron is almost a free electron.

101. $r_n = n^2a_o = 3^2 \times 53$ pm = 477 pm

(a) $\dfrac{nh}{2\pi} = \text{mvr}$

$$\text{v} = \frac{nh}{2\pi mr}$$

$$\text{v} = \frac{3\times6.626\times10^{-34}\ \text{kg m}^2\ \text{s}^{-1}}{2\times3.1416\times477\ \text{pm}\times\frac{10^{-12}\ \text{m}}{\text{pm}}\times9.109\times10^{-31}\ \text{kg}} = 7.3 \times 10^{5}\ \frac{\text{m}}{\text{s}}$$

(b) $C = \pi d = 3.142 \times 2 \times 477\ \text{pm} \times \dfrac{10^{-12}\ \text{m}}{\text{pm}} = 3.00 \times 10^{-9}$ m

$$?\ \text{revolutions} = 1\text{s} \times 7.3 \times 10^{5}\ \frac{\text{m}}{\text{s}} \times \frac{\text{revolution}}{3.00\times10^{-9}\ \text{m}}$$

$$= 2.43 \times 10^{14}\ \text{revolutions}$$

Chapter 8

Electron Configurations, Atomic Properties, and the Periodic Table

Exercises

8.1 (a) P $1s^22s^22p^63s^23p^3$

[Ne] $3s^23p^3$

	3s	3p		
[Ne]	↑↓	↑	↑	↑

(b) Cl $1s^22s^22p^63s^23p^5$

[Ne] $3s^23p^5$

	3s	3p		
[Ne]	↑↓	↑↓	↑↓	↑

8.2A (a) Mo $1s^22s^22p^63s^23p^63d^{10}4s^24p^64d^45s^2$

is the answer from the periodic table. $4d^55s^1$ is actual.

$[Kr]4d^45s^1$

(b) Bi $1s^22s^22p^63s^23p^63d^{10}4s^24p^64d^{10}4f^{14}5s^25p^65d^{10}6s^26p^3$

$[Xe]4f^{14}5d^{10}6s^26p^3$

8.2B (a) Sn $[Kr]4d^{10}5s^25p^2$

(b)

	4d					5s
Zr [Kr]	↑	↑				↑↓

8.3A Se^{2-} $[Ar]3d^{10}4s^24p^6$

Pb^{2+} $[Xe]4f^{14}5d^{10}6s^26p^0$

8.3B I^- $[Kr]4d^{10}5s^25p^6$

Cr^{3+} $[Ar]3d^34s^0$

8.4A (a) 2 (b) 4 (c) 0 (d) 0 (e) 0 (f) 4

8.4B

(a) K	$[Ar]4s^1$	paramagnetic	one odd e^-
(b) Hg	$[Xe]4f^{14}5d^{10}6s^2$	diamagnetic	all paired
(c) Ba^{2+}	[Xe]	diamagnetic	all paired
(d) N	$[He]2s^22p^3$	paramagnetic	3 unpaired e^-
(e) F^-	[Ne]	diamagnetic	all paired
(f) Ti^{2+}	$[Ar]3d^2$	paramagnetic	2 unpaired e
(g) Cu^{2+}	$[Ar]3d^9$	paramagnetic	one unpaired e^-

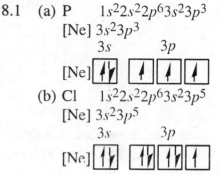

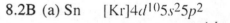

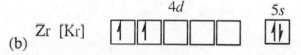

8.5 (a) F < N < Be
(b) Be < Ca < Ba
(c) F < Cl < S
(d) Mg < Ca < K

8.6A All are isoelectronic, so the more protons, the smaller the size.
$Y^{3+} < Sr^{2+} < Rb^{+} < Br^{-} < Se^{2-}$

8.6B

	Ca^{2+}	Cr^{2+}	Cs^{+}	Cl^{-}	Cr^{3+}	K^{+}
p^{+}	20	24	55	17	24	19
e^{-}	18	22	54	18	22	18

smallest $Cr^{3+} < Cr^{2+} < Ca^{2+} < K^{+} < Cs^{+} < Cl^{-}$ largest
Cr^{3+} has one less electron than Cr^{2+}.
Cr^{2+} has more protons than Ca^{2+}.
Ca^{2+} has more protons than K^{+}.
K^{+} has more protons than Cl^{-}.
One might expect Cs^{+} to be largest because it is an outer shell more than the others (actually two shells more than the others), but actual radii values are Cs^{+} as 169 pm and Cl^{-} as 181 pm. Cl^{-} is larger than the others because it has more electrons than protons, an anion. It has electrons in the same shell as the others but with fewer protons.

8.7 (a) Be < N < F
(b) Ba < Ca < Be
(c) S < P< F
(d) K < Ca < Mg

8.8 Because the Se atom is larger than the S atom, the value must be less than 450 kJ/mol but a positive number. The value must be 400 kJ/mol.

8.9 (a) O is more nonmetallic than P.
(b) S is more nonmetallic than As.
(c) F is more nonmetallic than P.

8.10

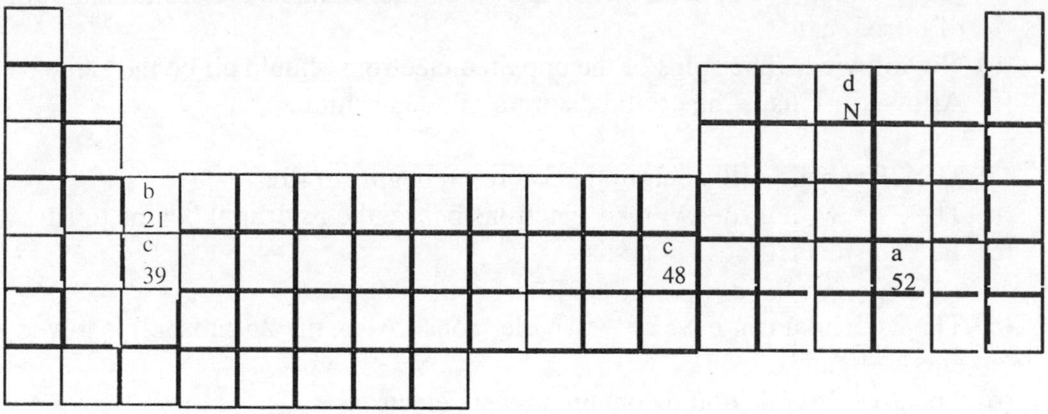

(c) Cadmium is the highest; yttrium is the lowest.

Review Questions

2. The orbital energies in (b) and (c) are identical for the H atom; those in (c) are also identical for other atoms.

4. (a) There are two electrons in the first shell, two electrons in the first subshell (s), and five electrons in the second subshell (p) of the second shell. Fluorine.
(b) There are two electrons in the first shell, two electrons in the first subshell (s), and six electrons in the second subshell (p) of the second shell. The third shell has two electrons in the first subshell (s), six in the second (p), and 10 in the third (d). The fourth shell has one electron in the first subshell (s). Copper.
(c) There are two electrons in the first shell, two electrons in the s subshell of the second shell, and one electron in each of the p orbitals (p_x, p_y, p_z) of the second shell. Nitrogen.
(d) There is one electron in the third level, which is the level that follows a filled neon configuration. Sodium.
(e) The argon configuration is filled, and there are two electrons in the first $3d$ orbital and one each in the other four $3d$ orbitals. The fourth shell has two electrons in the first subshell (s). Iron.

5. (a) ns; (b) np (c) $(n-1)d$ (d) $(n-2)f$.

7. (a) 4 (b) 8 (c) 7 (d) 3 (e) 2

13. Yes, there has to be at least one unpaired electron.
No, an even atomic-number element may have one electron in each of several orbitals in a subshell and thus be paramagnetic C– $1s^2 2s^2 2p_x^1 2p_y^1$.

Problems

19. (a) Not allowed. One pair of electrons in a $2p$ orbital has parallel spins.
(b) Not allowed. There are three electrons in one orbital.
(c) Allowed. This is the orbital diagram of nitrogen.

(d) Allowed, but the spins are often drawn up first. This is the orbital diagram of nitrogen.
(e) Not allowed. The spins of the unpaired electrons should all be the same.
(f) Allowed. This is the orbital diagram of magnesium.

21. (a) The $2p$ subshell fills before the $3s$ orbital begins to fill.
(b) The $2p$ subshell fills with six electrons before the $3s$ orbital begins to fill.
(c) The $2d$ subshell does not exist.

23. (a) The $2s$ orbital can have only two electrons. Also, the $2p$ subshell is lower energy than $3s$.
(b) The $2p$ subshell should contain only six electrons.
(c) There is no $2d$ subshell.

25. (a) Al $1s^22s^22p^63s^23p^1$
(b) Cl $1s^22s^22p^63s^23p^5$
(c) Na $1s^22s^22p^63s^1$
(d) B $1s^22s^22p^1$
(e) He $1s^2$
(f) O $1s^22s^22p^4$
(g) C $1s^22s^22p^2$
(h) Li $1s^22s^1$
(i) Si $1s^22s^22p^63s^23p^2$

27. (a) Ba [Xe] $6s^2$
(b) Rb [Kr] $5s^1$
(c) As [Ar] $3d^{10}4s^24p^3$
(d) F [He] $2s^22p^5$
(e) Se [Ar] $3d^{10}4s^24p^4$
(f) Sn [Kr] $4d^{10}5s^25p^2$

29. (a) C 

(b) O 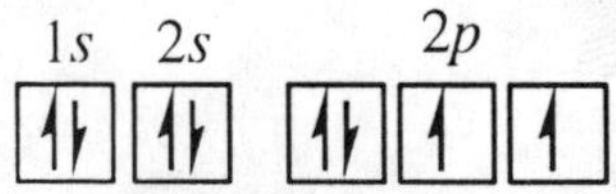

(c) K 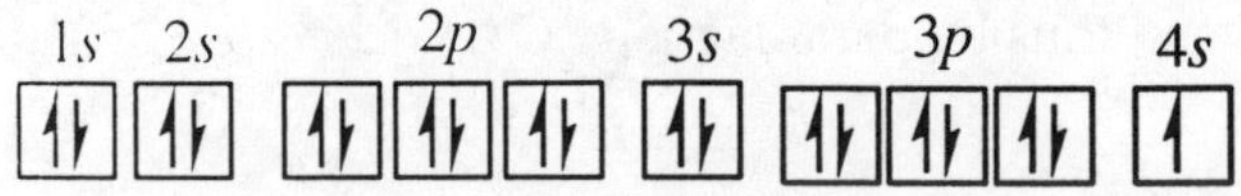

(d) Al 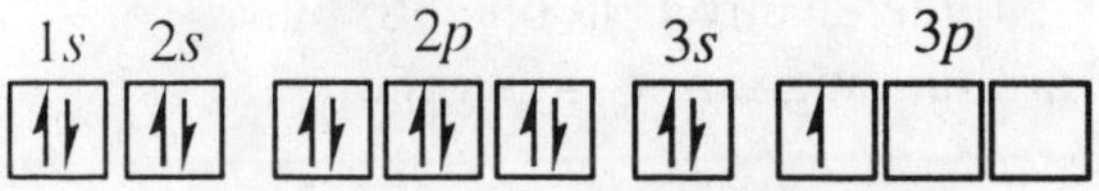

(e) S 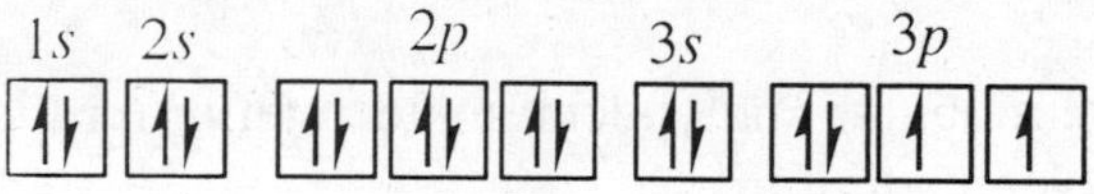

(f) Mg 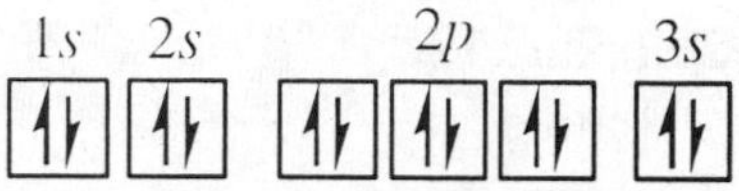

31. Hf [Xe] $4f^{14}5d^{2}6s^{2}$ Hf is in the 6th period and Group 4B, and it is the second element of the $5d$ transition elements.

[Xe] 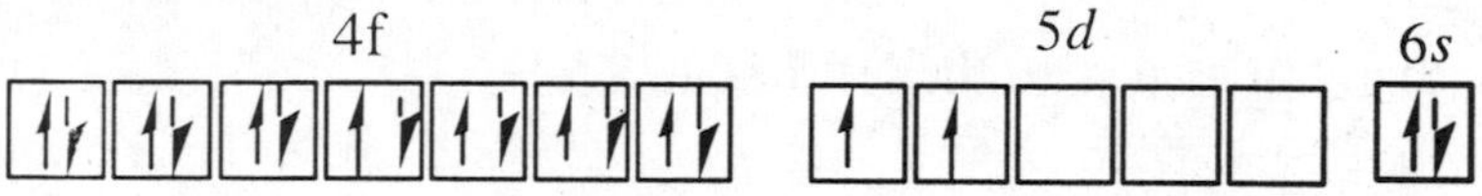

33.

(a) Br^{-} [Ar] 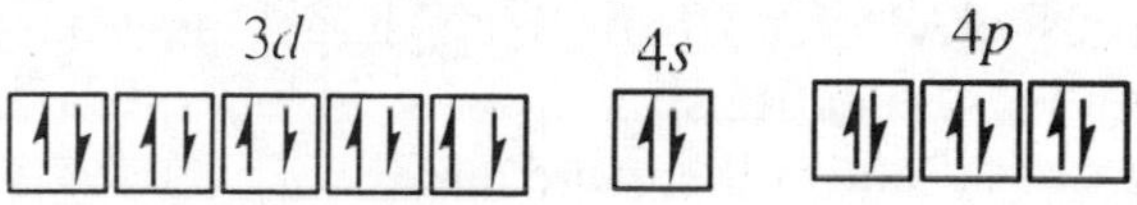

or [Kr]

(b) Ni^{2+} [Ar] 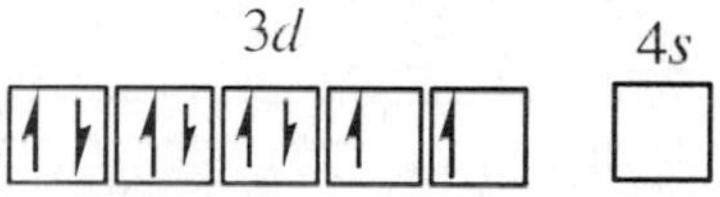

(c) Sb^{3+} [Kr] 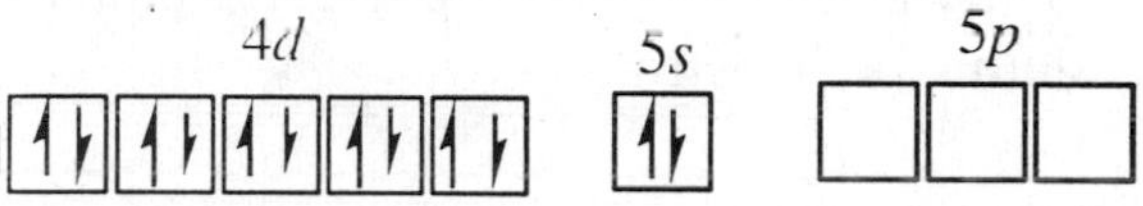

(d) Te^{2-} [Kr] 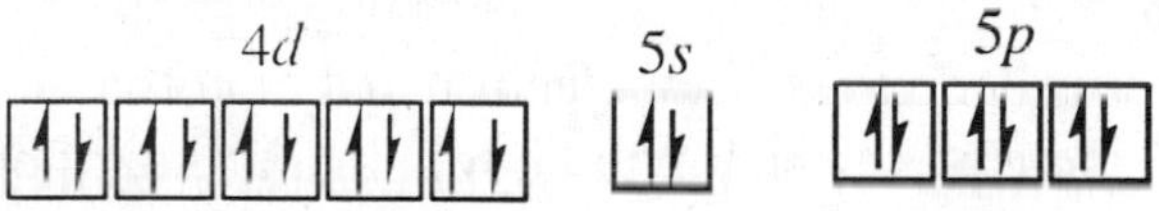

or [Xe]

35. $1s^{2}2s^{2}2p^{6}3s^{2}3p^{6}3d^{7}$
(a) Mn- $1s^{2}2s^{2}2p^{6}3s^{2}3p^{6}4s^{2}3d^{5}$
(b) Co^{3+}-$1s^{2}2s^{2}2p^{6}3s^{2}3p^{6}3d^{6}$
(c) Ni^{2+}-$1s^{2}2s^{2}2p^{6}3s^{2}3p^{6}3d^{8}$
(d) Cu^{2+}-$1s^{2}2s^{2}2p^{6}3s^{2}3p^{6}3d^{9}$
None are correct.

37.

(a)	period 2nd	group 8A
(b)	3rd	4A
(c)	3rd	1A
(d)	2nd	2A
(e)	2nd	5A
(f)	3rd	3A

39. (a) 5
(b) 32 ($4s^{2}4p^{6}4d^{10}4f^{14}$)

(c) 5 (nitrogen, phosphorus, arsenic, antimony, bismuth)
(d) 2
(e) 10 (yttrium, zirconium, niobium, molybdenium, technetium, ruthenium, rhodium, palladium, silver, cadmium)

41.

(a) S	$[Ne]3s^23p_x^23p_y3p_z$	2 unpaired e^-	paramagnetic
(b) Ba	$[Xe]6s^2$	all paired	diamagnetic
(c) V^{2+}	$[Ar]3d^34s^0$	3 unpaired e^-	paramagnetic
(d) O^{2-}	[Ne]	all paired	diamagnetic
(e) Ag		odd atomic number	paramagnetic

43. (a) S is larger than Cl, as it is to the left of Cl.
(b)

	Cl^-	S^{2-}
p^+	17	16
e^-	18	18

Cl^- and S^{2-} are isoelectronic. Thus S^{2-} is larger, as it has fewer protons to pull in the same number of electrons.
(c) Mg is to the left of Al and is therefore larger.
(d)

	Mg^{2+}	F^-
p^+	12	9
e^-	10	10

Mg^{2+} and F^- are isoelectronic. Thus F^- is larger, as there are fewer protons to pull in the same number of electrons.

45. (a) Al < Mg < Na
The atomic radius decreases as an electron and a proton are added to an atom. Atomic radius increases from top to bottom in a group of elements and from right to left in a period of elements.
(b) Mg < Ca < Sr
Strontium has one more shell of electrons than calcium, which has one more shell than magnesium. Atomic radius increases from top to bottom in a group of elements and from right to left in a period of elements.

47. The lower left-hand corner of the periodic table will have the largest atoms. Size increases to the left and down the periodic table.

49. (a) Ca is larger than Mg, which in turn is larger than Cl, so Ca is larger than Cl.
(b) K^+ is smaller than Cl^- because it has more protons. F^- is also smaller than Cl^-, so it is impossible to predict which is larger.

51. (a) Ionization energies decrease with increasing size or down the periodic table, so Ba < Ca < Mg.
(b) Ionization energies decrease with increasing size or to the right on the periodic table. Al < P < Cl

(c) Ne is a noble gas with a high I (ionization energy). Cl is below (larger than) F, so $I_F < I_{Cl}$. Na is to the left of Cl, so $I_{Na} < I_{Cl}$. The transition elements are about the same as Ca, which has I larger than Na but less than Cl.
Na < Fe < Cl < F < Ne.

53. The ionization energy increases for each electron removed, as positive charge builds up on the ions formed. Removing the three outer-shell electrons leaves a stable noble-gas configuration. The 4th electron is very difficult to remove because it must come from the shell $n = 2$, which is at a much lower energy than the shell $n = 3$.

55. The 7A atoms are the most negative in their periods. An electron to be gained can get closer to the positively charged nucleus of a 7A atom than to other atoms in the period, and so it is gained most readily.

57. An extra electron is more readily accepted by a Si atom, where it can enter an empty 3p orbital, than by a P atom, where the electron would have to pair up with an existing 3p electron, because each 3p orbital in P is already half filled.

59. The atomic mass divided by density gives a molar volume, which should be related to the atomic size. Na - 23.7, Mg - 14.0, Al - 9.99, Si - 12.1, P - 14.1, S - 15.5, Cl - 17.5, Ar - 24.1,K - 45.4, Ca - 25.9, Sc - 15.0, Cr - 7.23, Co - 6.62, Zn - 9.17, Ga - 11.8, As - 15.9, Br - 19.7, Kr - 29.5, Rb - 55.9, Sr - 34.5.

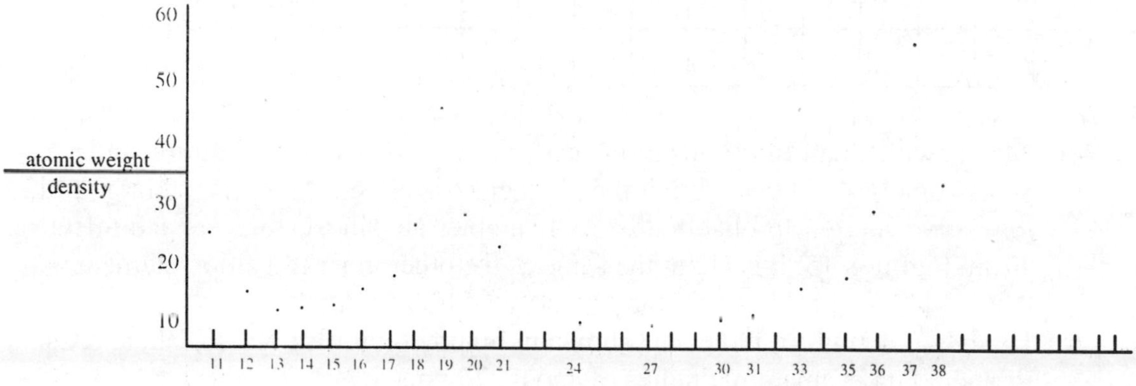

atomic number

The overall shape of the curve is similar to that for atomic radius versus atomic number. The alkali metals (Group 1A) are at the peaks, and there is less variation for the elements in between. The values for the elements 13–17 and 30–36 increase with atomic number rather than decrease.

61. Metallic character increases as electron removal becomes easier—that is, larger radius, lower ionization potential, and location in the periodic table (including designation as a metal, nonmetal, or metalloid). K is to the left of Ca, which is to the left and below Al. Rb is below K. $I_{Al} < I_{Ca}$. Bi is to the right of Al. Ge is a metalloid. P is a nonmetal but is also to the right and above the other elements.

Least P < Ge < Bi < Al < Ca < K < Rb most metallic.

63. (a) Ge (b) S (c) Tl (d) Ne (e) Se
Tl < Ge < Se < S < Ne greatest

65. (a) $Cl_2(g) + 2\ Br^-(aq) \rightarrow 2\ Cl^-(aq) + Br_2(l)$
(b) $I_2(s) + F^-(aq) \rightarrow$ no reaction
(c) $Br_2(l) + 2\ I^-(aq) \rightarrow 2\ Br^-(aq) + I_2(s)$

67. (a) $N_2O_5(s) + H_2O(l) \rightarrow 2\ HNO_3(aq)$
(b) $MgO(s) + 2\ CH_3COOH(aq) \rightarrow Mg(CH_3COO)_2(aq) + H_2O(l)$
(c) $Li_2O(s) + H_2O(l) \rightarrow 2\ LiOH(aq)$

69.

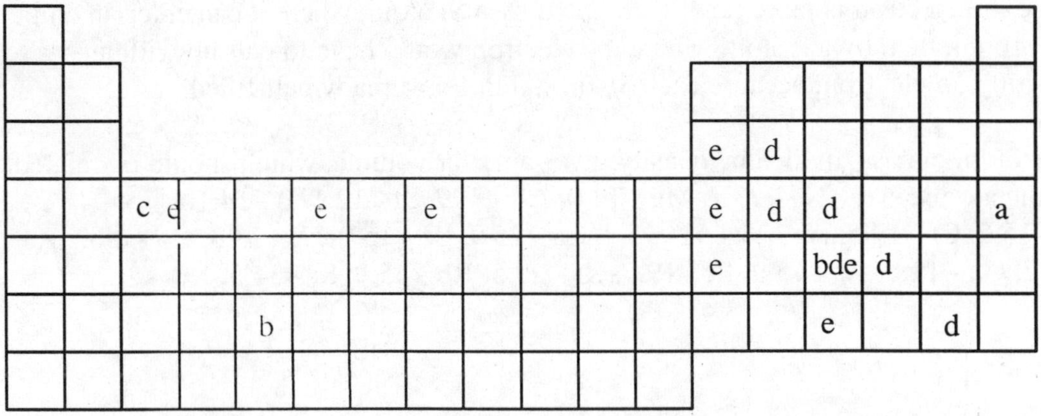

71. (a) Metals with small first-ionization energies, such as cesium, rubidium, and potassium, probably do exhibit the photoelectric effect. Metals with large first-ionization energies probably do not. Examples are zinc, cadmium, and mercury.
(b) From Figure 8.15, it is about the same as the other inner transition elements—that is, the other elements in the *f* block, about 600 kJ/mol.
(c) By extrapolating from the data shown in Figure 8.9 for Li, Na, K, Rb, and Cs, Fr should have an atomic radius of about 275 pm.

73 (a) Adding one electron to silicon would produce a half-filled *p* subshell. Adding one electron to Al results in only two electrons in the subshell.
(b) Adding one electron to phosphorus would be adding to a half-filled subshell, but adding one electron to silicon would produce a half-filled subshell.

75. (a) 37 electrons

$n = 1$	$l = 0$	$m_l = 0$	$m_s = -1/2,\ 0,\ +1/2$	$3\ e^-$	
$n = 2$	$l = 0$	$m_l = 0$	$m_s = -1/2,\ 0,\ +1/2$	$3\ e^-$	$3\ e^-$
	$l = 1$	$m_l = -1$	$m_s = -1/2,\ 0,\ +1/2$	$3\ e^-$	
		$m_l = 0$	$m_s = -1/2,\ 0,\ +1/2$	$3\ e^-$	

		$m_l = 1$	$m_s = -1/2,\ 0,\ +1/2$	3 e⁻	
n = 3	l = 0	$m_l = 0$	$m_s = -1/2,\ 0,\ +1/2$	3 e⁻	15 e⁻
	l = 1	$m_l = -1$	$m_s = -1/2,\ 0,\ +1/2$	3 e⁻	
		$m_l = 0$	$m_s = -1/2,\ 0,\ +1/2$	3 e⁻	
		$m_l = 1$	$m_s = -1/2,\ 0,\ +1/2$	3 e⁻	
	l = 2	$m_l = -2$	$m_s = -1/2,\ 0,\ +1/2$	3 e⁻	
		$m_l = -1$	$m_s = -1/2,\ 0,\ +1/2$	3 e⁻	
		$m_l = 0$	$m_s = -1/2,\ 0,\ +1/2$	3 e⁻	
		$m_l = 1$	$m_s = -1/2,\ 0,\ +1/2$	3 e⁻	36 e⁻
		$m_l = 2$	$m_s = -1/2,\ 0,\ +1/2$	3 e⁻	37–39 e⁻
n = 4	l = 0	$m_l = 0$	$m_s = -1/2,\ 0,\ +1/2$	3 e⁻	

Rb = $1s^3 2s^3 2p^9 3s^3 3p^9 3d^7 4s^3$

(b)

n = 1	l = 0	$m_l = 0$	$m_s = \pm 1/2$	2 e⁻	
	l = 1	$m_l = -1$	$m_s = \pm 1/2$	2 e⁻	2e⁻
		$m_l = 0$	$m_s = \pm 1/2$	2 e⁻	
		$m_l = 1$	$m_s = \pm 1/2$	2 e⁻	
n = 2	l = 0	$m_l = 0$	$m_s = \pm 1/2$	2 e⁻	8e⁻
	l = 1	$m_l = -1$	$m_s = \pm 1/2$	2 e⁻	10e⁻
		$m_l = 0$	$m_s = \pm 1/2$	2 e⁻	
		$m_l = 1$	$m_s = \pm 1/2$	2 e⁻	
	l = 2	$m_l = 2$	$m_s = \pm 1/2$	2 e⁻	16 e⁻
		$m_l = -1$	$m_s = \pm 1/2$	2 e⁻	
		$m_l = 0$	$m_s = \pm 1/2$	2 e⁻	
		$m_l = 1$	$m_s = \pm 1/2$	2 e⁻	
		$m_l = 2$	$m_s = \pm 1/2$	2 e⁻	
				26e⁻	

etc., etc., etc.

Rb = $1s^2 1p^6 2s^2 2p^6 2d^{10} 3s^2 3p^6 3d^1 4s^2$

Na 11e⁻ for Case a, Na $1s^3 2s^3 2p^5$

Sodium would not be in the same group. Sodium would be a main-group element, Rb a transition element.

For Case b, Na $1s^2 1p^6 2s^2 2p^1$

Na and Rb would not be in the same group. Na would be a main-group element, Rb a transition element.

77. (a) $E_1 = \dfrac{-2.179 \text{ x } 10^{-18}\text{ J}}{1^2} \times \dfrac{6.022 \text{ x } 1023 \text{ e}-}{\text{mol e}-} \times \dfrac{\text{kJ}}{10^3\text{ J}} = -1.312 \times 10^3 \text{ kJ/mol}$

$I_1 = E_\infty - E_1 = 1.312 \times 10^3$ kJ/mol

(b) The shortest wavelength is the largest E.

The shortest wavelength in Balmer series is $\lambda \quad - \lambda_2$. The longest wavelength in the Lyman series is $\lambda_2 - \lambda_1$.

In energy terms, $I_1 = (E_\infty \quad - E_2) + (E_2 - E_1)$. $I_1 = E_\infty \quad - E_1$, which is the same value calculated above, because $E_\infty \; = 0$.

$$I_1 = -\text{B} \times \left(\frac{1}{\infty} - \frac{1}{2^2}\right) + -\text{B} \times \left(\frac{1}{2^2} - \frac{1}{1^2}\right)$$

$$I_1 = -2.179 \times 10^{-18}\ \text{J} \times \left(0 - \frac{1}{4}\right) \ - 2.179 \times 10^{-18}\ \text{J} \times \left(\frac{1}{4} - 1\right)$$

$$I_1 = 0.545 \times 10^{-18}\ \text{J} \ + 1.634 \times 10^{-18}\ \text{J}$$

$$I_1 = 2.179 \times 10^{-18}\ \text{J} \times \frac{6.022 \text{ x } 10^{23} \text{ e}}{\text{mol e}} \times \frac{\text{kJ}}{10^3 \text{ J}}$$

$$I_1 = 1.312 \times 10^3 \text{ kJ/mol}$$

79. $Cl(g) + e^- \rightarrow Cl^-(g)$ ΔH = electron affinity = –349 kJ is the second step.

For the overall process, $\Delta H = \left(\frac{1}{2} \times 242.8 \text{ kJ}\right) - 349 \text{ kJ} = -228$ kJ./mol Cl^-. The overall process is exothermic.

81. (c) The first and third are about the same size with the second one double that size.

Apply Your Knowledge

83. $$\frac{22.99 \text{ g}}{\text{mol}} \times \frac{\text{cm}^3}{0.971 \text{ g}} \times \frac{\text{mol}}{6.022 \text{ x } 10^{23} \text{ atoms}} = \frac{3.93 \text{ x } 10^{-23} \text{ cm}^3}{\text{atom}}$$

$$V = \frac{4}{3}\pi r^3$$

$$r = \sqrt[3]{\frac{3 \times 3.93 \times 10^{-23} \text{cm}^3 \text{/atom}}{4\pi}} \quad 2.10 \times 10^{-8} \text{ cm} \times \frac{10^{-2} \text{ m}}{\text{cm}} \times \frac{\text{pm}}{10^{-12} \text{ m}} = 210 \text{ pm}$$

Actual 180 pm.
The estimate is not exact, because not all of the space is taken up by the atoms. Because the estimated atomic volume is too large, the estimate of the atomic radius is too high.

85 (a) In would have been put where Sr is.
With the atomic mass of 75.4 g/mol, In would have been between As and Se, but the 2+ charge is not consistent with that placement.

$$\text{Atomic weight} = \frac{6.4}{0.055} = 116$$

$$17.5 \text{ g O} \times \frac{\text{mole}}{16.00 \text{ g O}} = 1.09 \text{ mol O} \times \frac{\text{mol O}}{0.711 \text{ mol In}} = \frac{1.54 \text{ mol O}}{\text{mol In}}$$

$$82.5 \text{ g In} \times \frac{\text{mol In}}{116 \text{ g In}} = 0.711 \text{ mol} \times \frac{\text{mol In}}{0.711 \text{ mol In}} = \frac{1 \text{ mol In}}{\text{mol In}}$$

$In_{1 \times 2}O_{1.5 \times 2} \Rightarrow In_2O_3$
A 3+ ion of atomic mass 116 goes where In is now.
Mendelev would have put it where it is now.

(b) $$\text{Atomic mass} = \frac{6.4}{0.0276} = 232 \text{ g/mol}$$

$$62.66 \text{ g U} \times \frac{\text{mol}}{240 \text{ g}} = \frac{0.261}{0.261} = 1$$

$$37.34 \text{ g Cl} \times \frac{\text{mol}}{35.45\text{g}} = \frac{1.05}{0.261} = 4.04$$

UCl_4

The Dulong and Petit molar mass is 232 g/mol, and the number of Cl atoms per U atom is very close to an integer.

87. (a)

Element	z	ν s^{-1}	$\sqrt{\nu}$
Ca	20	8.902×10^{17}	9.438×10^{8}
Ti	22	1.088×10^{18}	1.043×10^{9}
V	23	1.191×10^{18}	1.091×10^{9}
Cr	24	1.303×10^{18}	1.142×10^{9}
Mn	25	1.421×10^{18}	1.192×10^{9}
Fe	26	1.542×10^{18}	1.242×10^{9}
Co	27	1.669×10^{18}	1.292×10^{9}
Ni	28	1.805×10^{18}	1.344×10^{9}
Cu	29	1.937×10^{18}	1.392×10^{9}
Zn	30	2.076×10^{18}	1.441×10^{9}

$$\nu = \frac{C}{\lambda} = \frac{3.00 \times 10^{8} \text{ m/s}}{3.368 \times 10^{-8} \text{ cm}} \times \frac{\text{cm}}{10^{-2} \text{ m}} = 8.907 \times 10^{-17} \text{ s}^{-1}$$

$$\sqrt{\nu} = \sqrt{A}\ z - \sqrt{A}\ b$$

plot z on the horizontal axis and $\sqrt{\nu}$ on the vertical axis.

slope = $5.0 \times 10^{7} = \sqrt{A}$

$A = 2.5 \times 10^{15}$

Intercept = -3.4×10^{7}

b = 1.18

(b) $$\sqrt{A} = \frac{1.242 \times 10^{9} - 1.142 \times 10^{9}}{26 - 24}$$

$\sqrt{A} = 5.00 \times 10^{7}$

$A = 2.5 \times 10^{15}$

$\sqrt{\nu} = \sqrt{A}\ z - \sqrt{A}\ b$ thus $\sqrt{A}\ b = \sqrt{A}\ z - \sqrt{\nu}$

$$b = \frac{1.20 \times 10^{9} - 1.142 \times 10^{9}}{5.00 \times 10^{7}} = 1.16$$

The A values are the same, but the b values differ slightly. The long distance on the horizontal axis probably limited the accuracy of the graphical value.

(c) $\nu = A(z - b)^2 = 2.50 \times 10^{15} \times (43.00 - 1.16)^2 = 4.38 \times 10^{18}$ s^{-1}

$$\lambda = \frac{C}{\nu} = \frac{3.00 \times 10^{8} \text{ m/s}}{4.38 \times 10^{18} \text{ s}^{-1}} \times \frac{\text{cm}}{10^{-2} \text{ m}} = 6.85 \times 10^{-9} \text{ cm}$$

Chapter 9

Chemical Bonds

Exercises

9.1 (a) $:\overset{\cdot\cdot}{\underset{\cdot\cdot}{Ar}}:$ (b) $\cdot Ca \cdot$ (c) $\cdot \overset{\cdot}{\underset{\cdot\cdot}{Br}}:$ (d) $:\overset{\cdot}{As}\cdot$ (e) $K \cdot$ (f) $\cdot \overset{\cdot}{\underset{\cdot\cdot}{Se}} \cdot$

9.2A

$$\cdot Ba \cdot + 2 :\overset{\cdot}{I}: \longrightarrow :\overset{\cdot\cdot}{\underset{\cdot\cdot}{I}}: Ba^{2+} :\overset{\cdot\cdot}{\underset{\cdot\cdot}{I}}:^-$$

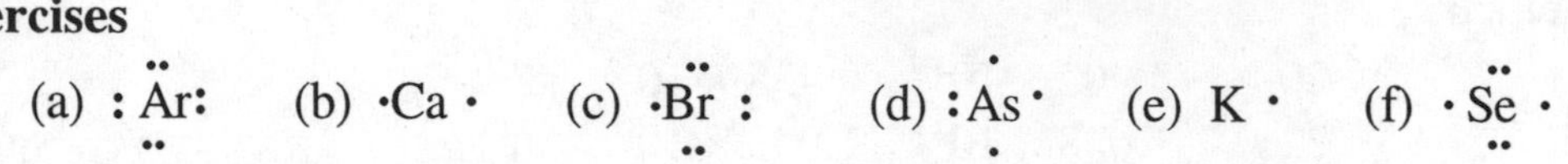

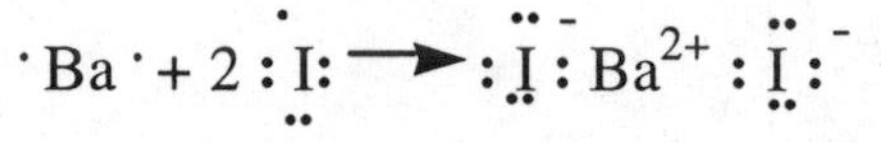

barium iodide BaI_2

9.2B

$$\cdot \overset{\cdot}{Al}\cdot \quad \cdot \overset{\cdot\cdot}{\underset{\cdot\cdot}{O}}\cdot \qquad \qquad [Al]^{3+} \qquad \left[:\overset{\cdot\cdot}{\underset{\cdot\cdot}{O}}:\right]^{2-}$$
$$\cdot \overset{\cdot\cdot}{\underset{}{O}}\cdot \longrightarrow \qquad \qquad \left[:\overset{\cdot\cdot}{\underset{\cdot\cdot}{O}}:\right]^{2-}$$
$$\cdot \overset{\cdot}{Al}\cdot \quad \cdot \overset{\cdot\cdot}{\underset{\cdot\cdot}{O}}\cdot \qquad \qquad [Al]^{3+} \qquad \left[:\overset{\cdot\cdot}{\underset{\cdot\cdot}{O}}:\right]^{2-}$$

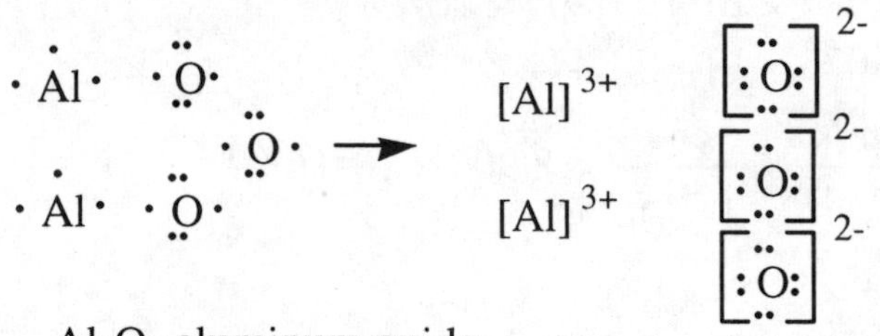

Al_2O_3 aluminum oxide

9.3A $Li\ (s) + \frac{1}{2}\ F_2\ (g) \rightarrow LiF\ (s)$ $\Delta H°_f = ?$

This is the sum of the steps 1–5 in the Born-Haber cycle.

$Li\ (s) \rightarrow Li\ (g)$	$\Delta H_1 = 159\ kJ$
$\frac{1}{2}\ F_2\ (g) \rightarrow F\ (g)$	$\Delta H_2 = \frac{1}{2}(159\ kJ)$
$Li\ (g) \rightarrow Li^+\ (g) + e^-$	$\Delta H_3 = 520\ kJ$
$F\ (g) + e^- \rightarrow F^-\ (g)$	$\Delta H_4 = -328\ kJ$
$\underline{Li^+\ (g) + F^-\ (g) \rightarrow LiF\ (s)}$	$\underline{\Delta H_5 = -1047}$
$Li\ (s) + \frac{1}{2}\ F_2\ (g) \rightarrow LiF\ (s)$	$\Delta H°_f =$

$\Delta H°_f = (159 + 80 + 520 + (-328) - 1047)\ kJ = -617\ kJ$
$\Delta H°_f = -617\ kJ/mol\ LiF\ (s)$

9.3B $Li^+\ (g) + Cl^-\ (g) \rightarrow LiCl\ (s)$ $\Delta H_{LE} = ?$

$Li\ (s) \rightarrow Li\ (g)$	$\Delta H_1 = 159\ kJ$
$\frac{1}{2}\ Cl_2\ (g) \rightarrow Cl\ (g)$	$\Delta H_2 = \frac{1}{2}(243\ kJ)$
$Li\ (g) \rightarrow Li^+\ (g) + e^-$	$\Delta H_3 = 520\ kJ$
$Cl\ (g) + e^- \rightarrow Cl^-\ (g)$	$\Delta H_4 = -349\ kJ$
$\underline{Li^+\ (g) + Cl^-\ (g) \rightarrow LiCl\ (s)}$	$\underline{\Delta H_5 = lattice\ energy}$

$$Li\,(s) + \frac{1}{2}\,Cl_2\,(g) \rightarrow LiCl\,(s) \qquad\qquad \Delta H°_f = -409\ kJ$$

$\Delta H°_f = -409\ kJ = (159 + 122 + 520 + (-349))\ kJ + \text{lattice energy}$
$\text{lattice energy} = (-409 - 159 - 122 - 520 + 349)\ kJ = -861\ kJ$
The lattice energy is - 861 kJ/mol LiCl (s)

9.4   (a) Ba < Ca < Be   (b) Ga < Ge < Se   (c) Te < S < Cl   (d) Bi < P < S

9.5   C - Cl   C - H   C - Mg   C - O   C - S
      0.5   0.4   1.3   1.0   -
    C - S < C - H < C - Cl < C - O < C - Mg

9.6A   H-N̈-N̈-H
          | |
          H H

9.6B
         II   II
         | |
     H - C - C - C̈l :
         | | ··
         H H

9.7A   S̈ = C = Ö

9.7B   :N - Ö - F̈ :
        ||
        :O:

9.8A [: C = N:]⁻

9.8B
$$\left[\begin{array}{c} H \\ | \\ H-P-H \\ | \\ H \end{array}\right]^{+}$$

9.9A   : Ö = N̈ = C̈l :

9.9B
       H Ö: H
       | || |
    H-C-C-Ö-C-H
       | ·· |
       H   H

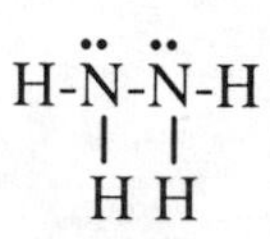

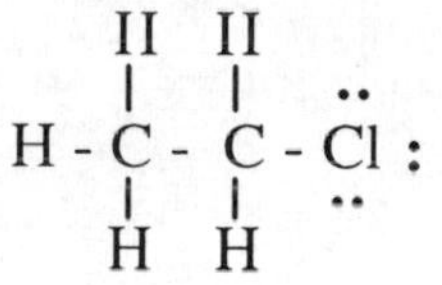

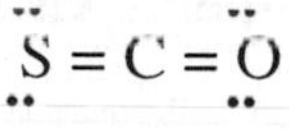

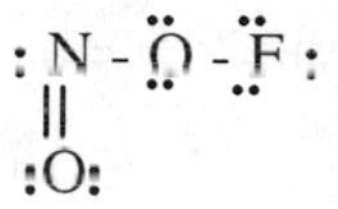

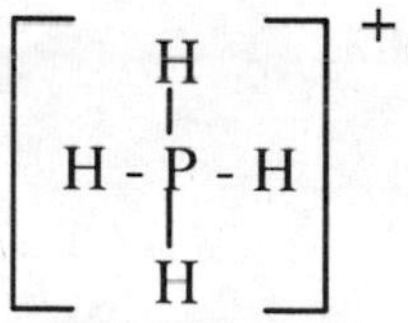

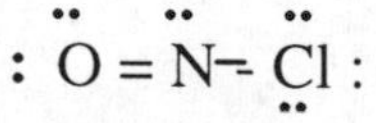

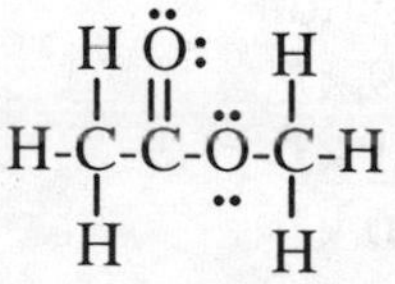

9.10

$$\left[\ddot{:O} - \overset{\displaystyle N}{\underset{\displaystyle :\ddot{O}:}{|}} = \ddot{O} \right]^{-} \longleftrightarrow \left[:\ddot{O} - \overset{\displaystyle N}{\underset{\displaystyle :\ddot{O}:}{\parallel}} - \ddot{O}: \right]^{-} \longleftrightarrow \left[\ddot{O} = \overset{\displaystyle N}{\underset{\displaystyle :\ddot{O}:}{|}} - \ddot{O}: \right]^{-}$$

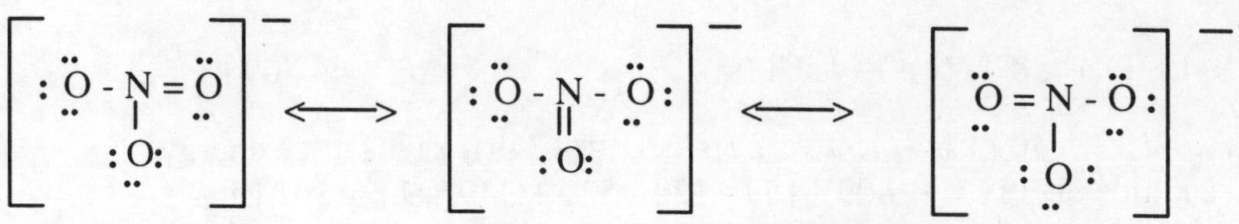

This resonance hybrid, which involves equal contributions from these three equivalent resonance structures, has N-to-O bonds with bond lengths and bond energies intermediate betweeen single and double bonds.

9.11A $:\ddot{C}l - \overset{\displaystyle P}{\underset{\displaystyle :\ddot{C}l:}{|}} - \ddot{C}l:$

9.11B (a)   (b)

$.\ :\ddot{F} - \overset{\displaystyle \ddot{C}l}{\underset{\displaystyle :\ddot{F}:}{|}} - \ddot{F}:$ $:\ddot{F} - \overset{\displaystyle :\ddot{F}:}{\underset{\displaystyle :\ddot{F}:}{\overset{\displaystyle |}{S}}} - \ddot{F}:$

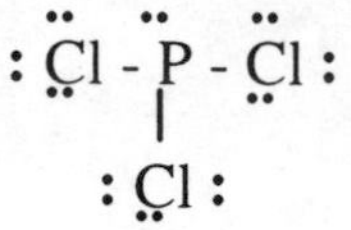

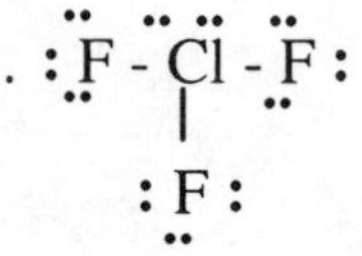

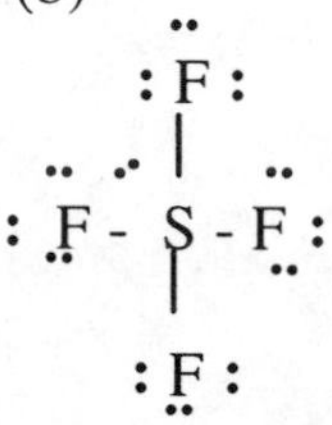

9.12 (a) Incorrect. The molecule ClO_2 has 19 valence electrons, but the Lewis structure shows 20, one too many.

(b) Correct.

(c) Incorrect. The structure shown has 26 electrons rather than the 24 available. The correct structure is

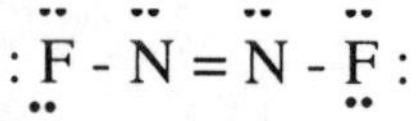

$:\ddot{F} - \ddot{N} = \ddot{N} - \ddot{F}:$

9.13A

$:\ddot{F} : \ddot{O} : \ddot{F}:$ B.L. $= \frac{1}{2}$ (145 pm) $+ \frac{1}{2}$ (143 pm) = 144 pm

9.13B

a. $H - \overset{\displaystyle \ddot{N}}{\underset{\displaystyle H}{|}} - \overset{\displaystyle N}{\underset{\displaystyle :O:}{|}} = \ddot{O}:$ N-to-N bond length = 145 pm

9.14 ΔH bond breakage ΔH bond formation

1 mol CH	414 kJ		1 mol C - Cl	-339 kJ
1 mol Cl_2	243 kJ		1 mol H - Cl	-431 kJ
sum	657 kJ		sum	-770 kJ

$\Delta H = \Delta H_{\text{breakage}} + \Delta H_{\text{formation}}$

$\Delta H = 657$ kJ $+ (-770$ kJ$) = -113$ kJ

Review Questions

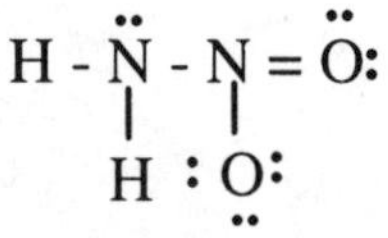

3.

    (a) Na·   (b) :Ö: (c) ·Si·   (d) :Br·   (e) ·Ca· (f) ·As·

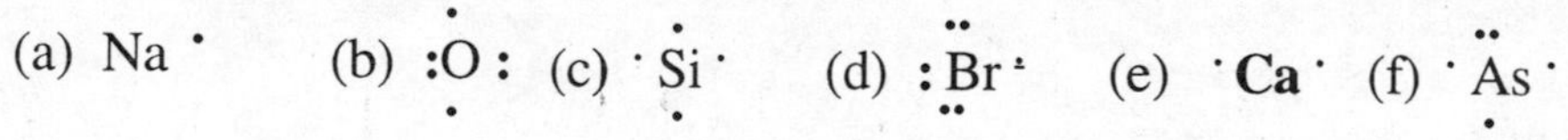

5.

    (a) ·Ca· + 2 ·Br: ⟶ $\left[:\ddot{Br}:\right]^{-}$ Ca²⁺ $\left[:\ddot{Br}:\right]^{-}$

    (b) ·Ba· + ·Ö· ⟶ Ba²⁺ $\left[:\ddot{O}:\right]^{2-}$

    (c) 2 ·Al· + 3 :Ṡ: ⟶ 2 [Al]³⁺ + 3 $\left[:\ddot{S}:\right]^{2-}$

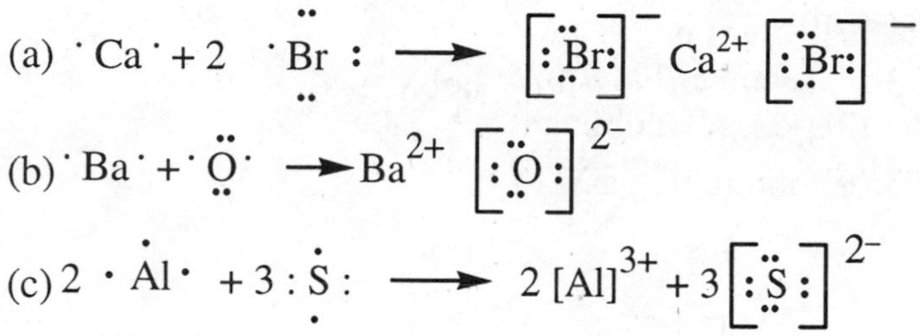

6.   The only such compound is lithium hydride: Li⁺ :H⁻

7.

    — lone pairs

    — bonding pair   or :Ï—Ï: where the dash line

    represents the bonding pair

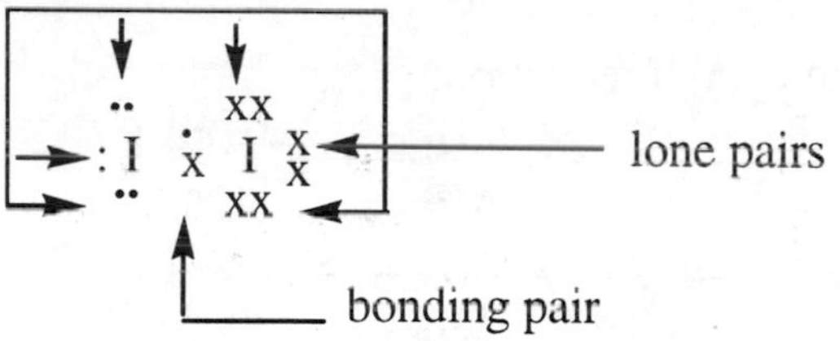

8.   δ+ H $\overset{x}{F}$:δ- or δ+ H–F̈:δ–

10.   (a) N is more electronegative than S.
     (b) Cl is more electronegative than B.
     (c) F is more electronegative than As.
     (d) O is more electronegative than S.

11.   (a) F is more electronegative than Br.
     (b) Br is more electronegative than Se.
     (c) Cl is more electronegative than As.
     (d) N is more electronegative than H.

12.   (a) ionic   (b) polar covalent
     (c) ionic   (d) polar covalent
     (e) ionic   (f) ionic
     (g) nonpolar covalent (h) nonpolar covalent
     (i) polar covalent

13.   (a) 1   (b) 4 (c) 2 (d) 1 (e) 3 (f) 1

18.   Unsaturated hydrocarbons are carbon-hydrogen compounds, which have double or triple bonds.

(a), (b), and (c) are unsaturated.

Problems

21. (a) Cr^{3+} - $1s^22s^22p^63s^23p^63d^3$ not noble gas
(b) Sc^{3+} - $1s^22s^22p^63s^23p^6$ noble gas
(c) Zn^{2+} - $1s^22s^22p^63s^23p^63d^{10}$ not noble gas
(d) Te^{2-} - $1s^22s^22p^63s^23p^63d^{10}4s^24p^64d^{10}5s^25p^6$ noble gas
(e) Zr^{4+} - $1s^22s^22p^63s^23p^63d^{10}4s^24p^6$ noble gas
(f) Cu^+ - $1s^22s^22p^63s^23p^63d^{10}$ not noble gas

23.

(a) $K^+ + [:\ddot{\underset{..}{I}}:]^-$ (b) $2[:\ddot{\underset{..}{F}}:]^- + Ba^{2+}$

(c) $2\ Rb^+ + [:\ddot{\underset{..}{S}}:]^{2-}$ (d) $2\ Al^{3+} + 3[:\ddot{\underset{..}{O}}:]^{2-}$

25. Equations must add together to produce the ΔH_f equation.

$Na^+ (g) + F^- (g) \rightarrow NaF (s)$	$\Delta H_{LE} =$ -914 kJ
$Na (g) \rightarrow Na^+ (g) + e^-$	$\Delta H_{IE} =$ 496 kJ
$Na (s) \rightarrow Na (g)$	$\Delta H_{sub} =$ 107 kJ
$F (g) + e^- \rightarrow F^- (g)$	$\Delta H_{EA} =$ -328 kJ
$\frac{1}{2}F_2 (g) \rightarrow F (g)$	$\Delta H_{dis} = \frac{1}{2}(159\ kJ)$
$Na (s) + \frac{1}{2}F_2 (g) \rightarrow NaF (s)$	ΔH_f = -559 kJ

Appendix C ΔH_f = - 573.7 kJ

27. Equations must add together to produce the lattice energy equation.
$Cs^+ (g) + Br^- (g) \rightarrow CaBr (s)$
(See next page for calculation.)

$Br^- (g) \rightarrow Br (g) + e^-$	$-\Delta H_{EA}$ = - (- 325 kJ/mol)
$Br (g) \rightarrow \frac{1}{2} Br_2 (g)$	$-\Delta H_{dis}$ = - (96.4 kJ/mol Br)
$\frac{1}{2} Br_2 (g) \rightarrow \frac{1}{2} Br_2 (l)$	$-\frac{1}{2}\Delta H_{sub}$ = $-\frac{1}{2}$(30.9 kJ/mol)
$Cs (s) + \frac{1}{2} Br_2 (l) \rightarrow CsBr(s)$	ΔH°_f = - 405.8 kJ/mol
$Cs(g) \rightarrow Cs(s)$	ΔH_{sub} = - (76.1 kJ/mol)
$Cs^+ (g) + e^- \rightarrow Cs (g)$	$-\Delta H_{IE}$ = - (376 kJ/mol)
$Br^- (g) + Cs^+ (g) \rightarrow CsBr (s)$	$-\Delta H_{LE}$ = - 645 kJ

$[\Delta H_{dis} = \Delta H_f\ Br(g) - \Delta H_f \frac{1}{2}\ Br_2(g)]$

29. Equations must add together to produce the lattice energy equation.
$2Li^+ (g) + O^{2-} (g) \rightarrow Li_2O (s)$

$2\,Li^+(g) + 2\,e^- \rightarrow 2\,Li(g)$	$-2\Delta H_{IE}$	=	-2 (520 kJ/mol)
$2\,Li(g) \rightarrow 2\,Li(s)$	$-2\Delta H_{sub}$	=	-2(159.4 kJ/mol)
$2\,Li(s) + \frac{1}{2}\,O_2(g) \rightarrow Li_2O(s)$	ΔH_f°	=	-597.94 kJ/mol
$O(g) \rightarrow \frac{1}{2}\,O_2(g)$	ΔH_{dics}	=	$-\frac{1}{2}$ (498 kJ/mol) O_2
$O^-(g) \rightarrow O(g) + e^-$	ΔH_{EA1}	=	-(-141 kJ/mol)
$O^{2-}(g) \rightarrow O(g) + e^-$	$-\Delta H_{EA2}$	=	- (744 kJ/mol)
$2Li^+(g) + O^{2-}(g) \rightarrow Li_2O(s)$	ΔH_{LE}	=	-2809 kJ

31.

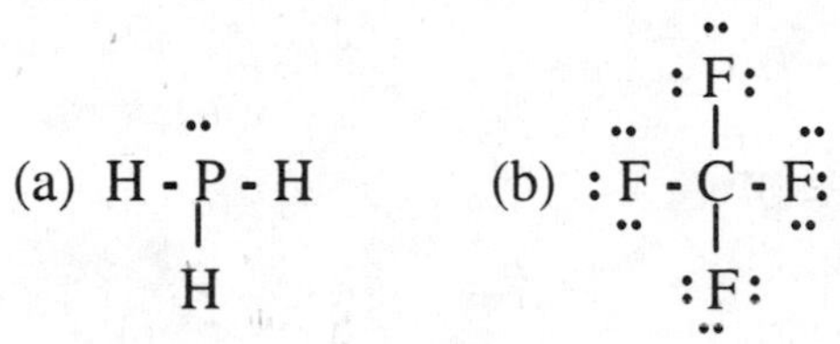

33.

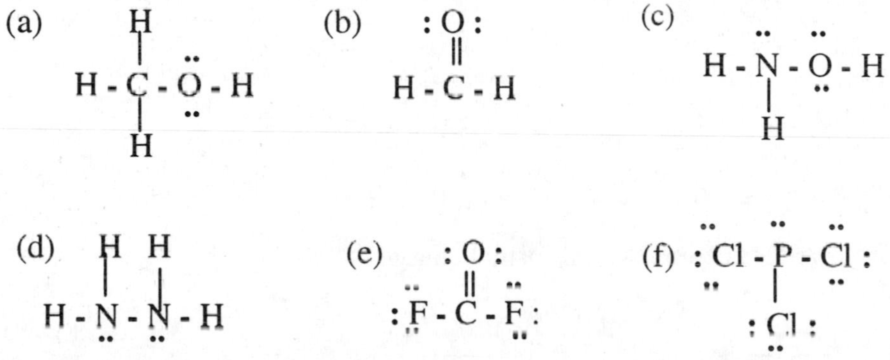

35. (a) The cyanate ion should have a triple bond between the C and N.
(b) The acetylide ion has 10 electrons, with a lone pair on both C atoms.
(c) is correct.
(d) Nitrogen monoxide has only one lone pair and one unpaired electron on N.

37. (a) B < N < F (b) Ca < As < Br (c) Ga < C < O

39.

(a) F — F < $\overset{\delta+}{\text{Cl}}$ — $\overset{\delta-}{\text{F}}$ < $\overset{\delta+}{\text{Br}}$ — $\overset{\delta-}{\text{F}}$ < $\overset{\delta+}{\text{I}}$ — $\overset{\delta-}{\text{F}}$ < $\overset{\delta+}{\text{H}}$ — $\overset{\delta-}{\text{F}}$

4.0 3.0 4.0 2.8 4.0 2.5 4.0 2.1 4.0

(b) H — H < $\overset{\delta+}{\text{H}}$ — $\overset{\delta-}{\text{I}}$ < $\overset{\delta+}{\text{H}}$ — $\overset{\delta-}{\text{Br}}$ < $\overset{\delta+}{\text{H}}$ — $\overset{\delta-}{\text{Cl}}$ < $\overset{\delta+}{\text{H}}$ — $\overset{\delta-}{\text{F}}$

2.1 2.1 2.5 2.1 2.8 2.1 3.0 2.1 4.0

41.

(a) $:\overset{..}{O} = \overset{..}{S} - \overset{..}{O}:$

 O $6 - \frac{1}{2}(4) - 4 = 0$

 S $6 - \frac{1}{2}(6) - 2 = 1$

 O $6 - \frac{1}{2}(2) - 6 = -1$

(b) $: C \equiv O:$

 C $4 - \frac{1}{2}(6) - 2 = -1$

 O $6 - \frac{1}{2}(6) - 2 = 1$

(c) $[H - \overset{..}{\underset{..}{O}} - \overset{..}{\underset{..}{O}}:]^-$

 H $1 - \frac{1}{2}(2) - 0 = 0$

 O $6 - \frac{1}{2}(4) - 4 = 0$

 O $6 - \frac{1}{2}(2) - 6 = -1$

(d) $H - C \equiv N:$

 H $1 - \frac{1}{2}(2) - 0 = 0$

 C $4 - \frac{1}{2}(8) - 0 = 0$

 N $5 - \frac{1}{2}(6) - 2 = 0$

43.

(a) $[: \overset{}{\underset{.}{C}} = N = \overset{..}{\underset{..}{O}}:]^-$

 C $4 - \frac{1}{2}(4) - 4 = -2$

 N $5 - \frac{1}{2}(8) - 0 = 1$

 O $6 - \frac{1}{2}(4) - 4 = 0$

(b) $[: N \equiv C - \overset{..}{\underset{..}{O}}:]^-$

 N $5 - \frac{1}{2}(6) - 2 = 0$

 C $4 - \frac{1}{2}(8) - 0 = 0$

 O $6 - \frac{1}{2}(2) - 6 = -1$

(b) is preferred since the overall formal charge is less.

45.

$$\left[H - \overset{..}{\underset{..}{O}} - \overset{}{\underset{|}{C}} = \overset{..}{\underset{..}{O}} \right]^- \longleftrightarrow \left[H - \overset{..}{\underset{..}{O}} - \overset{}{\underset{\|}{C}} - \overset{..}{\underset{..}{O}}: \right]^-$$

$$:\overset{}{\underset{..}{O}}: :\overset{}{\underset{..}{O}}:$$

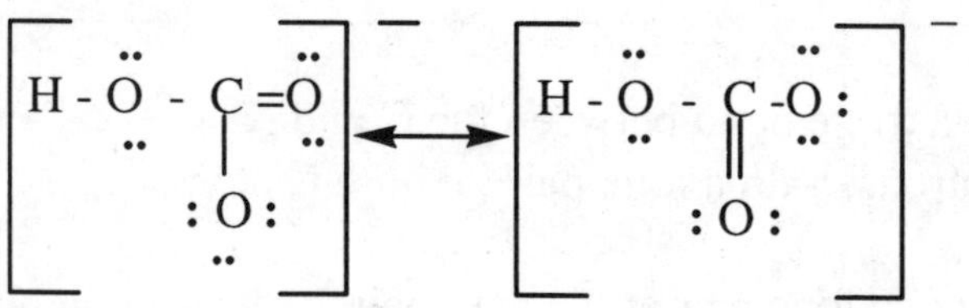

The resonance hybrid is a combination of these two structures, where the bonds of the C atom to the terminal O atoms are intermediate between single and double bonds, and the C atom and the terminal O atoms share a pair of delocalized electrons.

47.

$$H - \overset{..}{\underset{..}{O}} - N = \overset{..}{\underset{..}{O}} H - \overset{..}{\underset{..}{O}} - \overset{}{\underset{|}{N}} = \overset{..}{\underset{..}{O}} \longleftrightarrow H - \overset{..}{\underset{..}{O}} - \overset{}{\underset{\|}{N}} - \overset{..}{\underset{..}{O}}:$$

$$:\overset{}{\underset{..}{O}}: :\overset{}{\underset{..}{O}}:$$

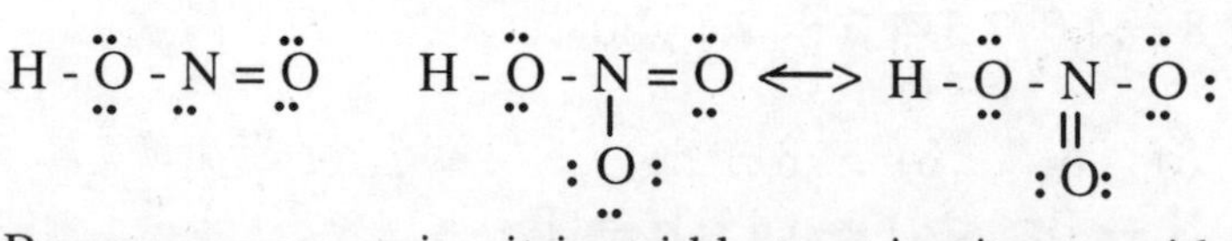

Resonance occurs in nitric acid but not in nitrous acid. In nitric acid two oxygens are equivalent. The O-N bonds are between a single and a double bond. A pair of electrons is shared by both oxygens and the nitrogen.

47.

$$H - \ddot{O} - N = \ddot{O} \qquad H - \ddot{O} - \underset{\underset{\displaystyle :\underset{\cdot\cdot}{O}:}{|}}{N} = \ddot{O} \longleftrightarrow H - \ddot{O} - \underset{\underset{\displaystyle :\underset{\cdot\cdot}{O}:}{\|}}{N} - \ddot{O}:$$

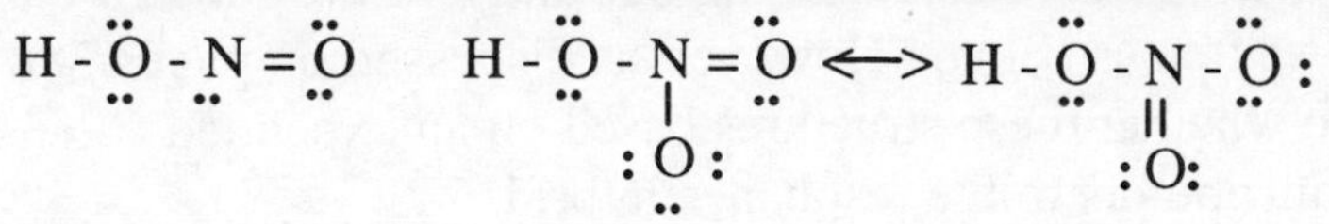

Resonance occurs in nitric acid but not in nitrous acid. In nitric acid, the two oxygen atoms are equivalent. The O-N bonds are between a single and a double bond. A pair of electrons is shared by both oxygen atoms and the nitrogen atom.

49.

(a) $\cdot \ddot{N} = \ddot{O}$

(b) $:\ddot{F} - \underset{\underset{\displaystyle :\ddot{F}:}{|}}{\ddot{Cl}} - \ddot{F}:$

(c) $:\ddot{Cl} - \underset{\underset{\displaystyle :\ddot{Cl}:}{|}}{B} - \ddot{Cl}:$

(d) $:\ddot{F} - \underset{\underset{\displaystyle :\ddot{F}:}{|}}{\overset{\overset{\displaystyle :\ddot{F}:}{|}}{\ddot{Se}}} - \ddot{F}:$

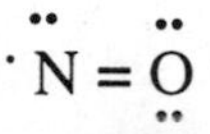

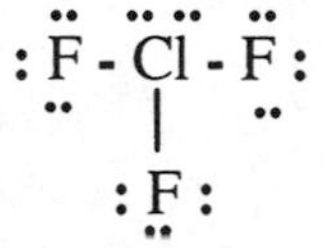

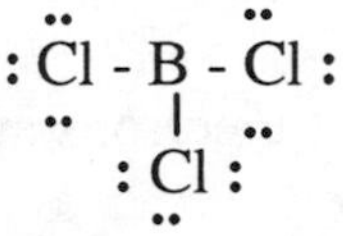

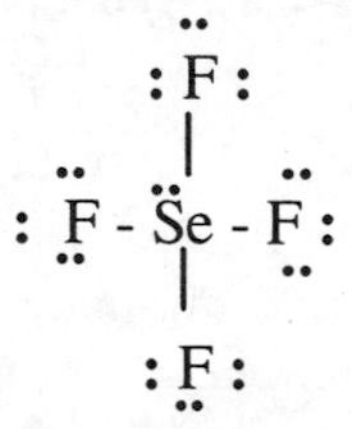

NO is an odd-electron molecule; BCl_3 has an incomplete octet. ClF_3 and SeF_4 have expanded shells.

51.

(a)
$:\ddot{S} - \underset{\underset{\displaystyle :\ddot{F}:}{|}}{\ddot{S}} - \ddot{F}:$

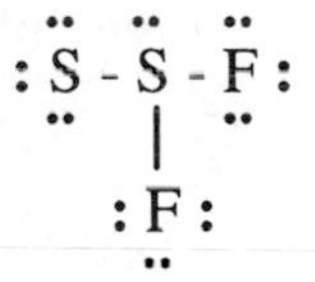

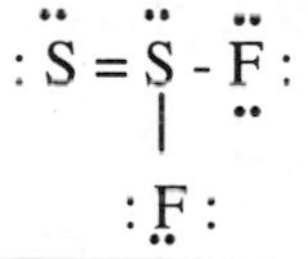

F $7 - 7 = 0$	S $6 - 6 = 0$
F $7 - 7 = 0$	S $6 - 6 = 0$
S $6 - 7 = -1$	F $7 - 7 = 0$
S $6 - 5 = 1$	F $7 - 7 = 0$

$:\ddot{S} = \underset{\underset{\displaystyle :\ddot{F}:}{|}}{\ddot{S}} - \ddot{F}:$

The first structure does not have an expanded valence shell. The second has an expanded valence shell but no formal charges. Both structures are likely to contribute to a resonance hybrid. A molecule bonded F-S-S-F would be an isomer, not a resonance structure.

(b) $\left[:\ddot{I} - \ddot{I} - \ddot{I}: \right]^-$

(c) $H - \ddot{O} - \underset{\underset{\displaystyle :\ddot{O}:}{\|}}{C} - \ddot{O} - H$

(d) $[:C \equiv N:]^-$

(e) $\left[\underset{\underset{\displaystyle :\ddot{F}:}{|}}{:\ddot{F} - \overset{\overset{\displaystyle :\ddot{F}: \quad \ddot{F}:}{\diagdown \diagup}}{S}} - \ddot{F}: \right]^-$

(f) $\left[:\ddot{O} - \underset{\underset{\displaystyle :\ddot{O}:}{|}}{Br} - \ddot{O}: \right]^- \longleftrightarrow \left[:\ddot{O} = \underset{\underset{\displaystyle :\ddot{O}:}{|}}{Br} - \ddot{O}: \right]^- \longleftrightarrow \left[:\ddot{O} = \underset{\underset{\displaystyle :\ddot{O}:}{|}}{Br} = \ddot{O}: \right]^-$

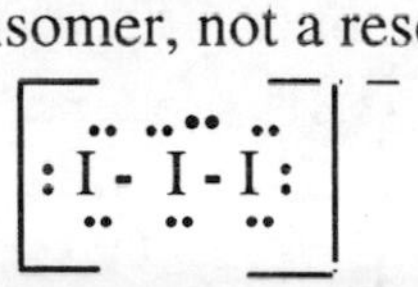

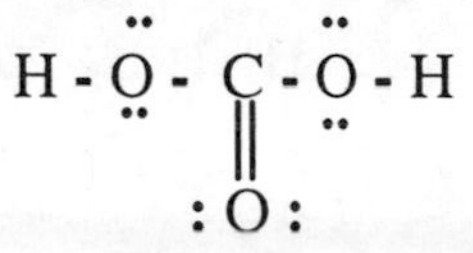

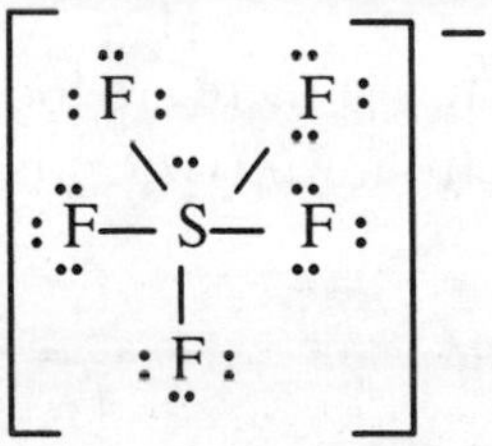

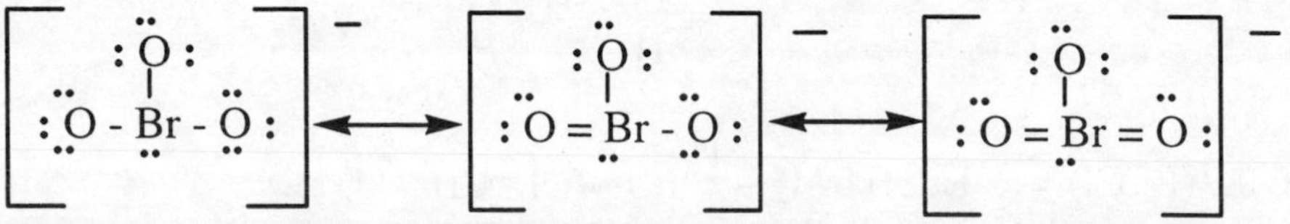

(3 equivalent structures) (3 equivalent structures)
The distribution of formal charges improves from left to right.

Several possible resonance structures involving one and two bromine-to-oxygen double bonds can be written as above. However, as discussed on pages 388-389, it is questionable whether these structures based on an expanded valence shell are important contributors to the resonance hybrid.

53.

$$\text{(a)} \quad \begin{array}{c} H \quad H \\ | \quad\; | \\ H-C-C-\ddot{O}-H \\ | \quad\; \\ H \\ | \\ H-C-H \\ | \\ H \end{array} \qquad \text{(b)} \quad \begin{array}{c} \;\; :O: \\ \;\; \| \\ H-C-\ddot{O}-H \end{array} \qquad \text{(c)} \quad \begin{array}{c} H \qquad\;\; H \\ | \qquad\;\; | \\ H-C-\ddot{O}-C-H \\ | \qquad\;\; | \\ H \qquad\;\; H \end{array}$$

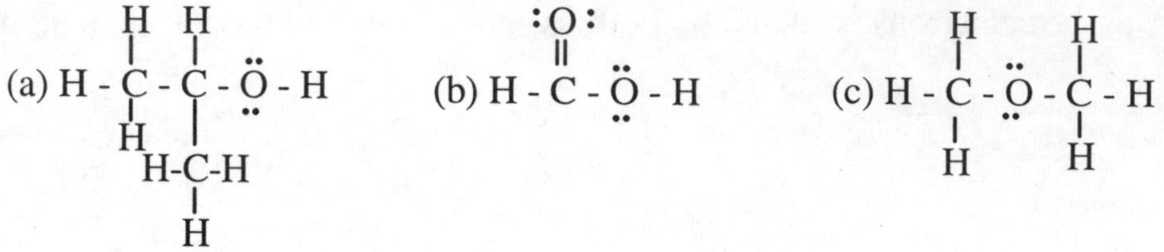

55. (a) There are only six electrons around C.

$$\left[\ddot{\underset{..}{S}} = C = \ddot{\underset{..}{N}} \right]$$

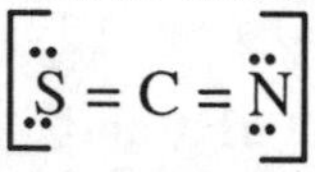

(b) There are only 16 electrons, instead of the required 17.

$$:\ddot{O} = \dot{N} - \ddot{\underset{..}{O}}: \longleftrightarrow :\ddot{\underset{..}{O}} - \dot{N} = \ddot{O}:$$

(c) Nitrogen cannot have an expanded shell. All the N-to-Cl bonds must be single bonds.

$$:\ddot{\underset{..}{Cl}} - \ddot{N} - \ddot{\underset{..}{Cl}}:$$
$$\qquad\; | $$
$$\qquad :\underset{..}{Cl}:$$

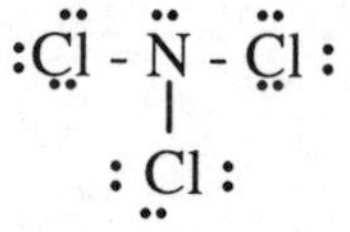

57. $? \text{ mol C} = 53.31 \text{ g C} \times \dfrac{\text{mol C}}{12.011 \text{ g C}} = 4.438 \text{ mol C} \times \dfrac{1}{2.220 \text{ mol O}} = \dfrac{1.999 \text{ mol C}}{\text{mol O}}$

$? \text{ mol H} = 11.18 \text{ g H} \times \dfrac{\text{mol H}}{1.0079 \text{ g H}} = 11.09 \text{ mol H} \times \dfrac{1}{2.220 \text{ mol O}} = \dfrac{4.995 \text{ mol H}}{\text{mol O}}$

$? \text{ mol O} = 35.51 \text{ g O} = \dfrac{\text{mol O}}{15.999 \text{ g O}} = 2.220 \text{ mol O} \times \dfrac{1}{2.220 \text{ mol O}} = \dfrac{1.000 \text{ mol O}}{\text{mol O}}$

$$C_2H_5O \qquad \begin{array}{c} H \quad H \\ | \quad\; | \\ H-C=C-O-H \\ | \\ H \end{array} \qquad \begin{array}{c} H \quad H \\ | \quad\; | \\ H-C-C=O-H \\ | \\ H \end{array} \qquad \begin{array}{c} \;\; H \quad H \\ \;\; | \quad\; | \\ H-\ddot{O}-C-C^{\textbf{.}} \\ \;\; | \quad\; | \\ \;\; H \quad H \end{array}$$

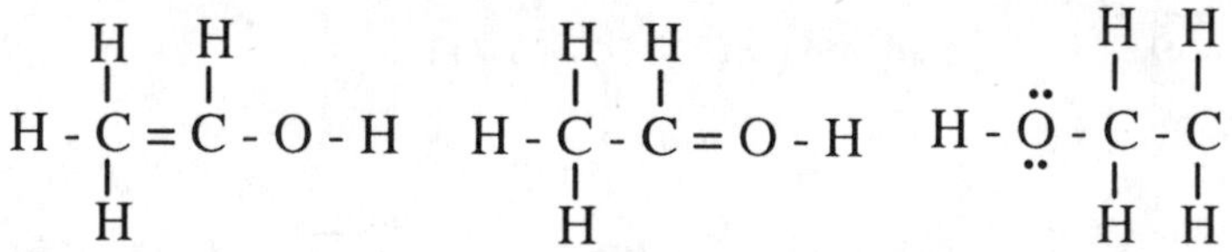

These are not good structures; there are too many bonds on the C or the O, or the structure has an odd electron. $C_4H_{10}O_2$ is better.

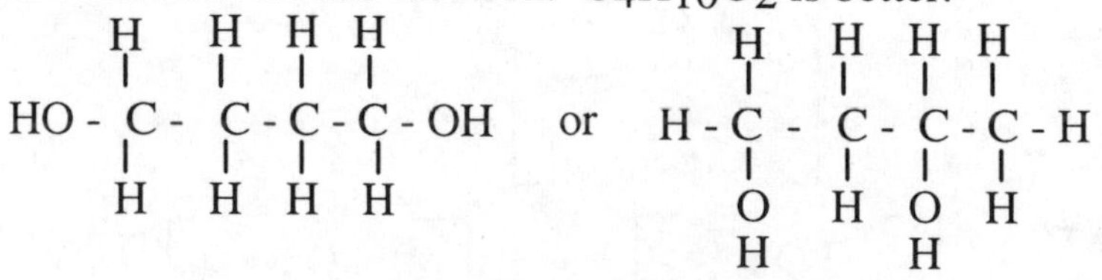

$$\begin{array}{c} H \quad H \;\; H \;\; H \\ | \quad\; | \quad | \quad | \\ HO-C-C-C-C-OH \\ | \quad\; | \quad | \quad | \\ H \quad H \;\; H \;\; H \end{array} \qquad \text{or} \qquad \begin{array}{c} H \quad H \;\; H \;\; H \\ | \quad\; | \quad | \quad | \\ H-C-C-C-C-H \\ | \quad\; | \quad | \quad | \\ O \quad H \;\; O \;\; H \\ | \qquad\quad | \\ H \qquad\quad H \end{array}$$

59. (a) $(B)L. (I - Cl) = \dfrac{1}{2} B.L. (I - I) + \dfrac{1}{2} B.L. (Cl - Cl)$

$B.L. = \dfrac{1}{2} (266 \text{ pm}) + \dfrac{1}{2} (199 \text{ pm}) = 233 \text{ pm}$

(b) B.L. (C - F) = $\frac{1}{2}$ B.L. (C - C) + $\frac{1}{2}$ B.L. (F - F)

B.L. = $\frac{1}{2}$(154 pm) + $\frac{1}{2}$(143 pm)

B.L. = 149 pm

Estimates are likely to be too high because the bonds are polar. In general, the more polar the bond, the more the bond length is shortened from calculated value.

61. The smallest percent difference should be between those atoms that are closest in electronegativity. H 2.1 F 4.0 Cl 3.5 Br 2.8 I 2.5
The smallest percent difference should be in HI.

63. NN double bond NF single bonds
$:\ddot{\underset{..}{F}} - \ddot{N} = \ddot{N} - \ddot{\underset{..}{F}}:$

65. ΔH = BE (H_2) + BE (F_2) - 2BE (HF)
ΔH = 436 kJ + 159 kJ - 2 mol × 565 kJ/mol = - 535 kJ

67. It is convenient to consider only the bonds that change instead of all of the bonds.

broken bonds		formed bonds	
C - H	414 kJ	C - Cl	- 339 kJ
Cl-Cl	243 kJ	H - Cl	- 431 kJ
	657 kJ		-770 kJ

exothermic ΔH = 657 kJ - 770 kJ = -113 kJ

69.

$:\ddot{O} = \ddot{O} - \ddot{\underset{..}{O}}: \quad + :\ddot{\underset{..}{O}}: \longrightarrow 2\ \ddot{O} = \underset{..}{\ddot{O}}$

$\Delta H = \Delta H_{\text{broken bonds}} + \Delta H_{\text{formed bonds}}$

-391.9 kJ = $2\Delta H_{O_3}$ - 2 ΔH_{O_2}

-391.9 kJ = $2\Delta H_{O_3}$ + 2 mol (-498 kJ/mol)

ΔH_{O_3} = 302 kJ/O_3 bond

71. ΔH_{rxn} = bonds formed + bonds broken
C(graphite) + 2 H_2(g) → CH_4(g)
-74.81 kJ = -4 mol ΔH_{CH} + 2 mol $\Delta H_{H\text{-}H}$ + 1 mol $\Delta H_{C\text{-}C}$
-74.81 kJ = -4 mol × ΔH_{CH} + 2 mol × 436 kJ/mol + 1 mol × 717 kJ
-1664 kJ = -4 mol × ΔH_{CH}
$\Delta H_{CH} = \frac{-1664\text{ kJ}}{-4\text{ mol}}$ = 416 kJ/mol
The tabulated value of 414 kJ/mol compares very well.

73. (a) $H_2C = CH_2$ (b) $H_2C = CHCH_2CH_3$

(c) $HC \equiv CH$ (d) $H_3CC \equiv CCH_2CH_3$

75. (a) $H_2C = CH_2 + H_2 \rightarrow H_3C - CH_3$
(b) $HC \equiv CH + 2H_2 \rightarrow H_3C - CH_3$

77. (a) $\Delta H = \Delta H_{\text{broken bonds}} + \Delta H_{\text{formed bonds}}$

$\Delta H = 1 \text{ mol} \times \Delta H_{C=C} + 4 \text{ mol} \times \Delta H_{C\text{-}H} + 1 \text{ mol} \times \Delta H_{H\text{-}H} + 1 \text{ mol } \Delta H_{C\text{-}C} + 6 \text{ mol} \Delta H_{C\text{-}H}$

$\Delta H = 1 \text{ mol} \times 611 \text{ kJ/mol} + 4 \text{ mol} \times 414 \text{ kJ/mol} + 1 \text{ mol} \times 436 \text{ kJ/mol} + 1 \text{ mol} \times (-347 \text{ kJ/mol}) + 6 \text{ mol} \times (-414 \text{ kJ/mol})$

$\Delta H = -128 \text{ kJ}$

(b) $\Delta H = \Delta H_f(C_2H_6) - \Delta H_f(C_2H_4) - \Delta H_{f_{H_2}}$

$\Delta H = 1 \text{ mol } C_2H_6 \times (-84.68 \text{ kJ/mol}) - 1 \text{ mol } C_2H_4 \times 52.26 \text{ kJ/mol} - 1 \text{ mol } H_2 \times 0$

$\Delta H = -136.94 \text{ kJ}$

The ΔH_f value is more negative.

79. (a) $-[-CH_2CH_2CH_2CH_2CH_2CH_2CH_2CH_2-]-$

(b) 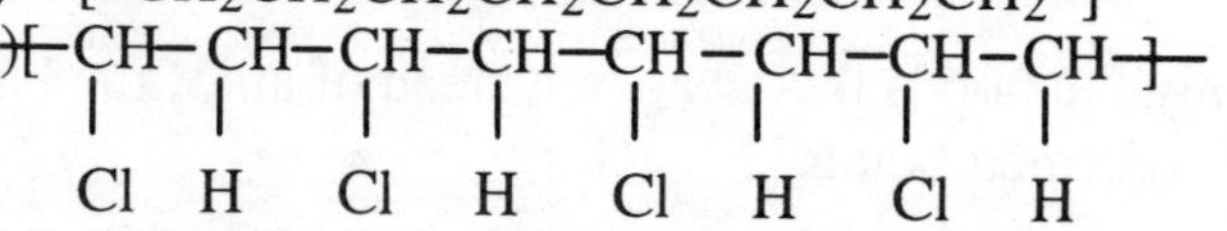

(c) 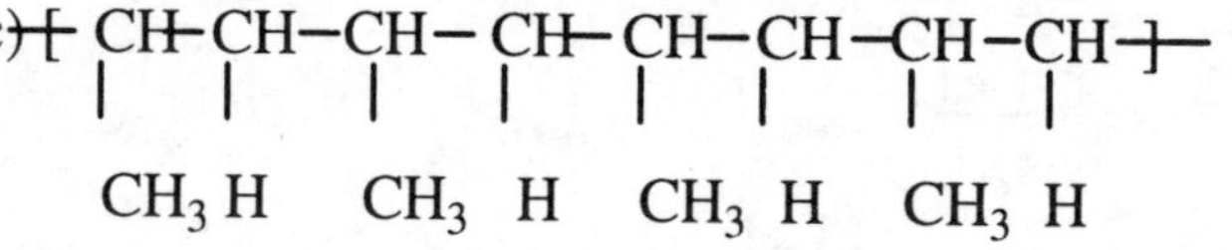

Additional Problems

81. In alkanes, the Lewis structures and structural formulas are identical because all electron pairs are bonding pairs (represented by a single dash line). There are no lone-pair electrons. In other organic compounds there may be lone-pair electrons to be shown as dots. For example, in organic compounds containing N or O atoms, there will be lone-pair electrons to be shown as dots on the O or N atoms.

83.

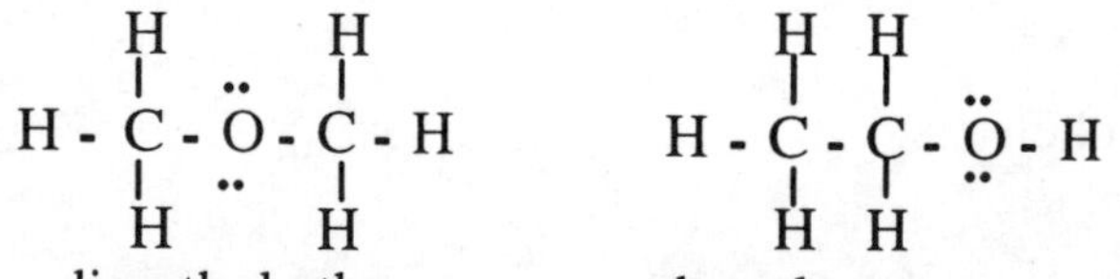

dimethyl ether ethanol

85.

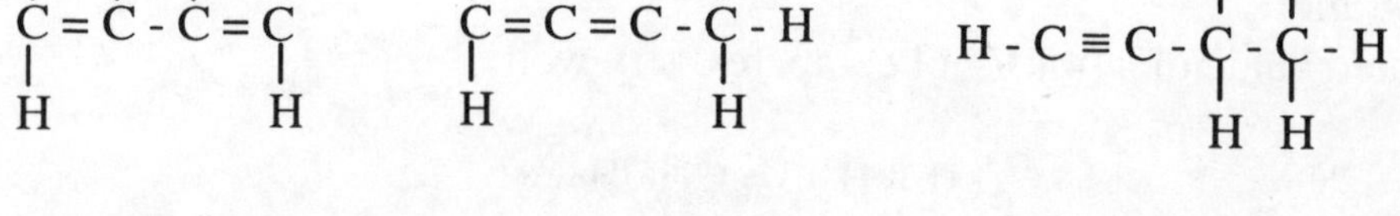

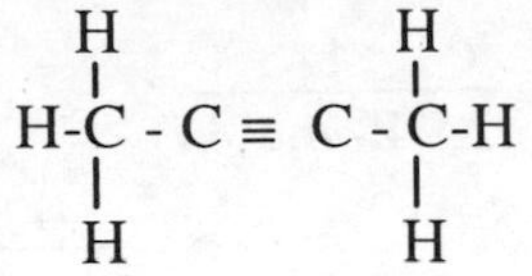 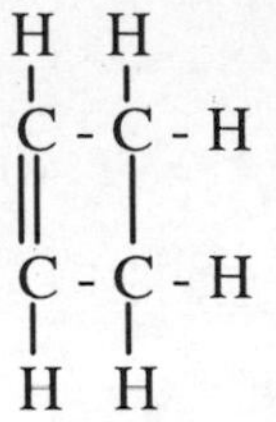

87.

$H - \ddot{N} - N \equiv N: \longleftrightarrow H - \ddot{N} = N = \ddot{N}:$

89. $N_2(g) + O_2(g) \rightarrow 2\ NO(g)$ $\qquad \Delta H_f = 90.25$ kJ/mol

$\Delta H = \Delta H_{\text{broken bonds}} + \Delta H_{\text{formed bonds}}$

$2 \text{ mol } \Delta H_f = \Delta H_{N_2} + \Delta H_{O_2} + 2 \text{ mol } \Delta H_{NO}$

$2 \text{ mol} \times 90.25 \text{ kJ/mol} = \frac{946 \text{ kJ}}{\text{mol}} + \frac{498 \text{ kJ}}{\text{mol}} + 2 \text{ mol } \Delta H_{NO}$

$\Delta H_{NO} = -632 \text{ kJ/mol} = -\Delta H_{diss}$

$\Delta H_{diss} = 632$ kJ/mol

Table value 590 kJ/mol

91. $Mg(s) + \frac{1}{2} Cl_2(g) \rightarrow MgCl(s)$ $\qquad \Delta H_f$

$Mg(s) \rightarrow Mg(g)$	150 kJ
$Mg(g) \rightarrow Mg^+(g) + e^-$	738 kJ
$\frac{1}{2} Cl_2(g) \rightarrow Cl(g)$	$\frac{1}{2}$ mol × 243 kJ/mol
$Cl(g) + e^- \rightarrow Cl^-(g)$	- 349 kJ
$Mg^+(g) + Cl^-(g) \rightarrow MgCl(s)$	- 676 kJ
$Mg(s) + \frac{1}{2} Cl_2(g) \rightarrow MgCl(s)$	$\Delta H_f =$ - 15 kJ

93. $H(g) + e^- \rightarrow H^-(g)$ $\qquad \Delta H_{EA} = ?$

The equations have to be added together to generate the above equation.

$Na(s) + \frac{1}{2} H_2(g) \rightarrow NaH(s)$	$\Delta H_f =$ -56.27 kJ/mol
$Na(g) \rightarrow Na(s)$	$-\Delta H_{sub} =$ -(107) kJ/mol
$Na^+(g) + e^- \rightarrow Na(g)$	$-\Delta H_{IE} =$ -(496) kJ/mol
$NaH(s) \rightarrow Na^+(g) + H^-(g)$	$-\Delta H_{LE} =$ -(812) kJ/mol
$H(g) \rightarrow \frac{1}{2} H_2(g)$	$-\Delta H_{diss} =$ -(218.0) kJ/mol
$H(g) + e^- \rightarrow H^-(g)$	$\Delta H_{EA} =$ -65 kJ/mol

95. The formal charge on HClO are H 0, Cl 1, and O −1. The formal charges on HOCl are H 0, Cl 0, and O 0. HOCl is best.

Apply Your Knowledge

97. (a) $? \text{ mol O} = 47.04 \text{ g O} \times \dfrac{\text{mol O}}{15.999 \text{ g O}} = 2.940 \text{ mol O} \times \dfrac{1}{2.948 \text{ mol O}}$

$$= \dfrac{1.000 \text{ mol O}}{\text{mol O}}$$

$? \text{ mol C} = 52.96 \text{ g C} \times \dfrac{\text{mol C}}{12.011 \text{ g C}} = 4.409 \text{ mol C} \times \dfrac{1}{2.940 \text{ mol O}}$

$$= \dfrac{1.500 \text{ mol C}}{\text{mol O}}$$

$C_{1.500 \times 2} O_{1 \times 2} = C_3O_2$ formula mass = 68 g/formula unit

$$M = \dfrac{mRT}{PV} = \dfrac{0.507 \text{ g} \times \dfrac{62.36 \text{ L mm Hg}}{\text{K mol}} \times 298 \text{ K}}{752 \text{ mm Hg} \times 184 \text{ mL} \times \dfrac{10^{-3} \text{ L}}{\text{mL}}} = 68.1 \text{ g mol}$$

Empirical formula is the molecular formula of the oxide.
Because the sole product in the reaction of the oxide with water has a molar mass of 104 g/mol, the reaction product must have two water added to the oxide.

$\text{mol}(H_2O) \text{ added} = (104 \text{ g/mol} - 68 \text{ g/mol}) \times \dfrac{\text{mol}}{18 \text{ g}} = 2 \text{ mol added}$

$C_3O_2 + 2 H_2O \rightarrow H_4C_3O_4$

(b) $\dfrac{? \text{ mol H}}{\text{mol compound}} = \dfrac{29.52 \text{ mol} \times 0.5050 \text{ M} \times \dfrac{10^{-3} \text{ L}}{\text{mol}}}{0.507 \text{ g} \times \dfrac{\text{mol}}{68 \text{ g}}} \times \dfrac{1 \text{ mol H}^+}{1 \text{ mol NaOH}}$

$$= \dfrac{1.49 \times 10^{-2} \text{ mol H}^+}{7.46 \times 10^{-3} \text{ mol}} = \dfrac{2.000 \text{ mol H}^+}{\text{mol compound}}$$

From the titration data, we find 2 mol ionizable H atoms per mole of compound. Two of the four H atoms in $H_4C_3O_4$ are bonded in a different way than the other two. Usually H needs to be bonded to O to be acidic.

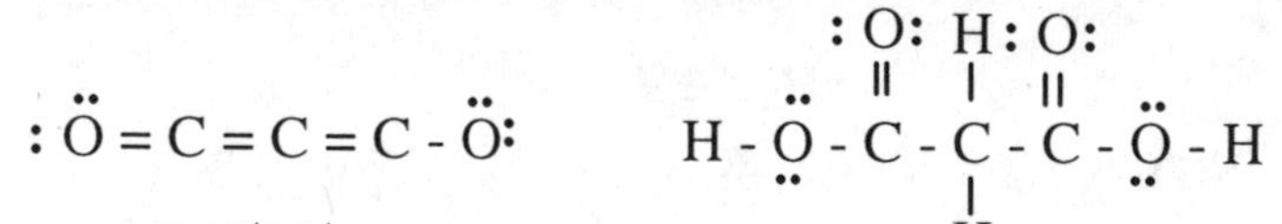

$$:\overset{..}{O} = C = C = C - \overset{..}{O}:$$
reactant

$$H - \overset{..}{\underset{..}{O}} - \overset{:O:}{\underset{\underset{H}{|}}{\overset{\parallel}{C}}} - \overset{H}{\underset{|}{C}} - \overset{:O:}{\overset{\parallel}{C}} - \overset{..}{\underset{..}{O}} - H$$
product

99. (a)

MgO (s) → Mg²⁺ + O²⁻	−lattice energy	−(−3795 kJ)
Mg (s) + ½ O₂ (g) → MgO (s)	$\Delta H°_f$	−601.7 kJ
Mg (g) → Mg (s)	$-\Delta H°_{sub}$	−(146 kJ)
Mg⁺ + e⁻ → Mg (g)	$-IP_1$	−(738 kJ)
Mg²⁺ + e⁻ → Mg⁺ (g)	$-IP_2$	−(1451 kJ)
O⁻ → e⁻ + O	$-EA_1$	−(−141 kJ)
O → ½ O₂	−Bond Energy	−1/2(498 kJ)
O⁻ + e⁻ → O²⁻	EA_2	750 kJ

(b) 750 kJ

Chapter 10

Bonding Theory and Molecular Structure

Exercises

10.1A

(a) $SiCl_4$ 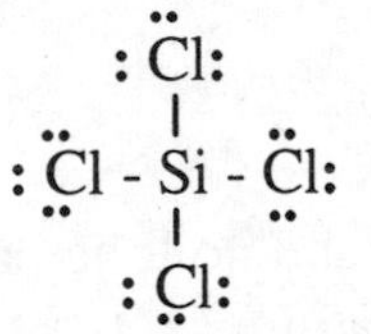

The four bonding groups of electrons (AX_4) produce a tetrahedral electron-group geometry and molecular geometry.

(b) $SbCl_5$ 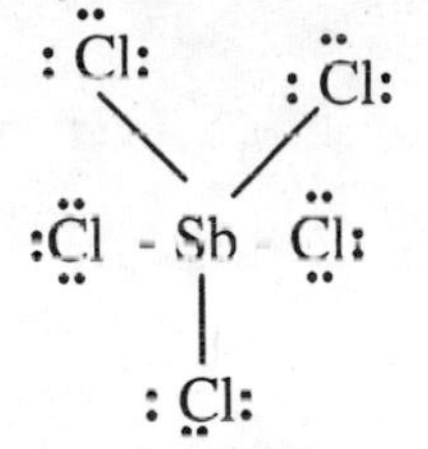

The five bonding groups of electrons (AX_5) produce both a trigonal bipyramidal electron-group geometry and molecular geometry.

10.1B

(a) 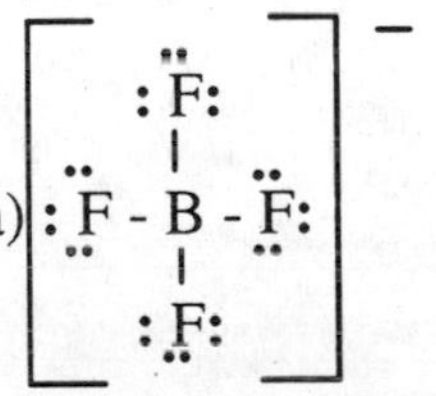

The four bonding groups of electrons (AX_4) produce a tetrahedral electron-group geometry and molecular geometry.

(b) 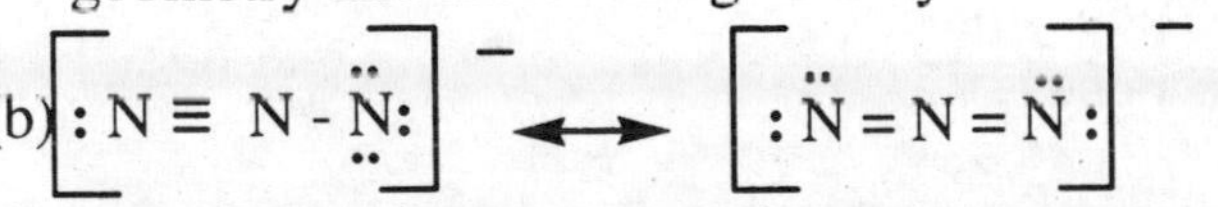

The two bonding groups of electrons (AX_2) produce a linear electron-group geometry and molecular geometry.

10.2A SF_4: There are five groups of electrons, but one pair is a lone pair of electrons VSEPR notation. AX_4E. Electron-group geometry is trigonal bipyramidal. In the seesaw structure (Table 10.1); two LP-BP repulsions at 90° and two at 120°; in the pyramidal structure given, three LP-BP at 90° and one at 180°. Because 90° LP-BP interactions are especially unfavorable, the seesaw structure is adopted so that the lone pair is on the equatorial plane of the structure.

10.2B There are five groups of electrons, but two pairs are lone pairs of electrons: VSEPR notation (AX_3E_2). Those two pairs will be on the equatorial plane of the structure, producing a T-shape molecular shape from the trigonal bipyramidal electron-group geometry. In the T-shaped structure (Table 10.1), four LP-BP repulsions at 90°; in the trigonal planar structure, six LP-BP repulsions at 90°. The T-shaped structure is observed.

10.3A

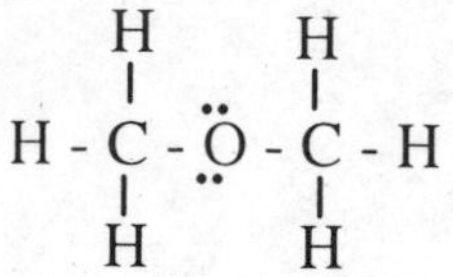

$$\begin{array}{ccc} H & & H \\ | & \ddot{} & | \\ H-C-\ddot{O}-C-H \\ | & \ddot{} & | \\ H & & H \end{array}$$

For both carbon atoms there are four bonding groups and no lone pairs of electrons (AX_4). So, the molecular shape around each carbon atom is tetrahedral. Around the O atom are four electron-groups but two of the electron groups are lone pairs (AX_2E_2), so the C-O-C molecular shape is bent.

10.3B

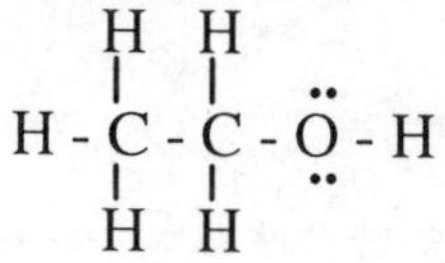

$$\begin{array}{ccc} H & H & \\ | & | & \ddot{} \\ H-C-C-\ddot{O}-H \\ | & | & \ddot{} \\ H & H & \end{array}$$

For both carbon atoms there are four bonding groups and no lone pairs of electrons, (AX_4), so the shape around each carbon atom is tetrahedral. Around the O atom there are four electron groups, but two of them are lone pairs, (AX_2E_2), so the COH bond is "bent."

10.4A

(a)
$$\begin{array}{c} F \\ | \\ F-B-F \end{array}$$
trigonal planar
symmetric
nonpolar

(b) $\ddot{O} = \ddot{S} - \ddot{O}:$
bent
nonsymmetric
polar

(c) $:\ddot{Br} - \ddot{Cl}:$
linear
different electronegativities
polar

(d) $:N \equiv N:$
linear
same electronegativity
nonpolar

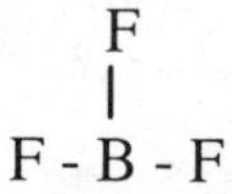

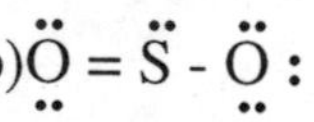

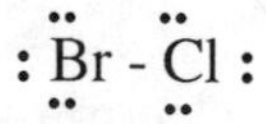

10.4B

(a)

$$
\begin{array}{c}
: \ddot{O}: \\
| \\
\ddot{O} = S - \ddot{O}: \\
\end{array}
$$

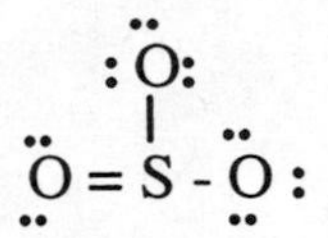

3 resonance structures
trigonal planar
symmetric
nonpolar

(b)

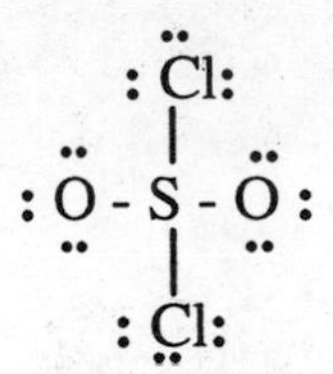

tetrahedral
symmetric
different electronegativities
polar

(c)

$$
\begin{array}{c}
: \ddot{F}: \\
| \\
: \ddot{F} - \ddot{Cl}: \\
| \\
: \ddot{F}: \\
\end{array}
$$

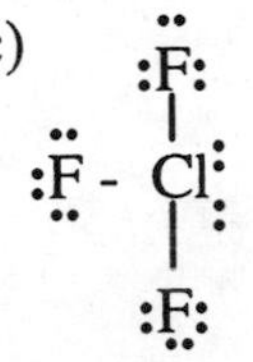

T-shape
nonsymmetric
polar

(d)

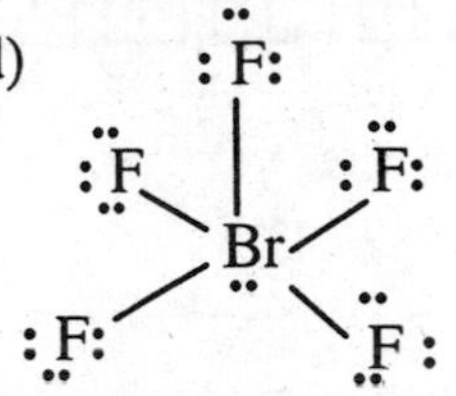

square pyramidal
nonsymmetric
polar

10.5 NOF is probably 110°, and NO₂F is 118°. The lone pair of electrons on N in NOF is more repulsive than are the bonding pairs in the N-to-O bonds in NO₂F.

10.6A

$SiCl_4$

$$
\begin{array}{c}
: \ddot{Cl}: \\
| \\
: \ddot{Cl} - Si - \ddot{Cl}: \\
| \\
: \ddot{Cl}: \\
\end{array}
$$

VSEPR notation AX₄
electron group-geometry: tetrahedral
tetrahedral molecular geometry
hybridization scheme sp^3

Si [Ne]sp^3

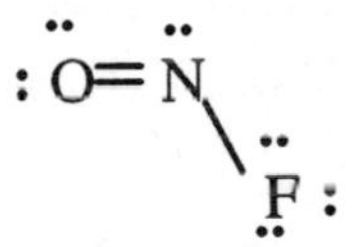

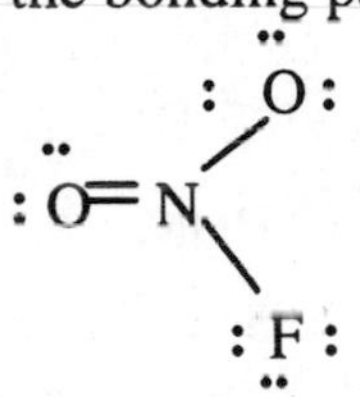

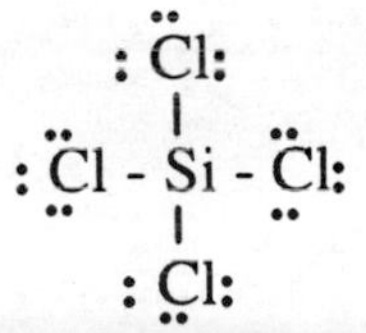

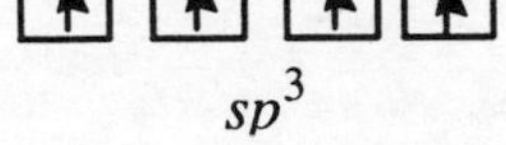

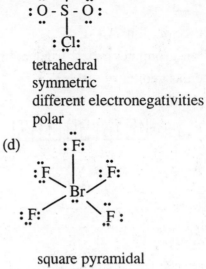

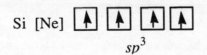

10.6B

$$\left[\ddot{:I} - \ddot{I} - \ddot{I}: \right]^{-}$$

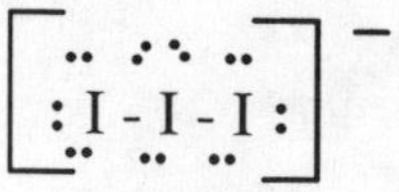

VSEPR notation: AX_2E_3

electron-group geometry: trigonal bipyramidal

molecular geometry linear

hybridization scheme: sp^3d

I [Kr] $4d^{10}$ ⥮ ⥮ ⥮ ↑ ↑

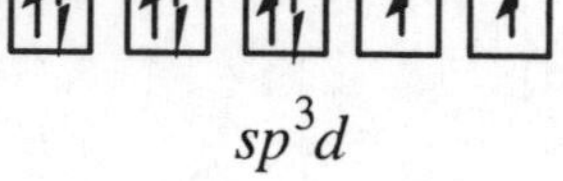

sp^3d

10.7A

$$\begin{array}{c} H \\ | \\ H - C - \ddot{O} - H \\ | \\ H \end{array}$$

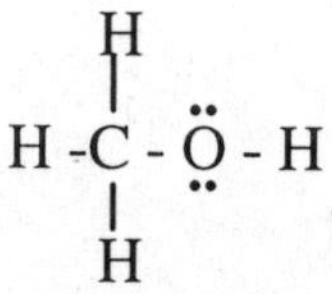

(a) Around the C is tetrahedral. AX_4

Around the O is bent. AX_2E_2

(b) C is sp^3, O is sp^3

(c)

σ O(sp^3) - C(sp^3)

σ O(sp^3) - H(1s)

σ C(sp^3) - H(1s)

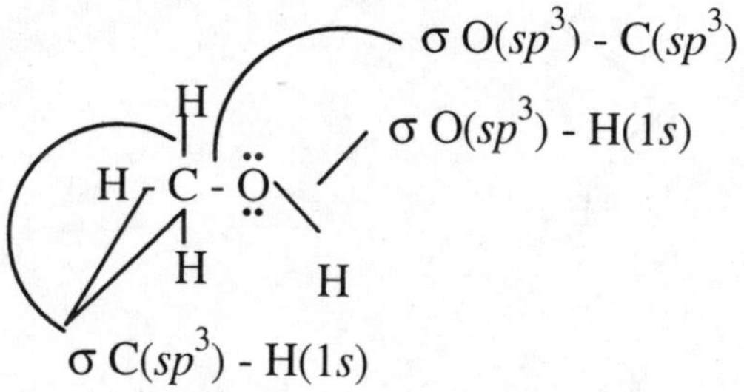

10.7B : N ≡ C - C ≡ N:

(a) Around each C is linear. AX_2

(b) C is sp, N is $2p$

π C ($2p$) - N ($2p$)

π C ($2p$) - N ($2p$)

π C ($2p$) - N 10.8

π C ($2p$) -

: N ≡ C - C ≡ N:

σ C (sp) - N ($2p$)

σ C (sp) - N

σ C (sp) - C ($2p$)

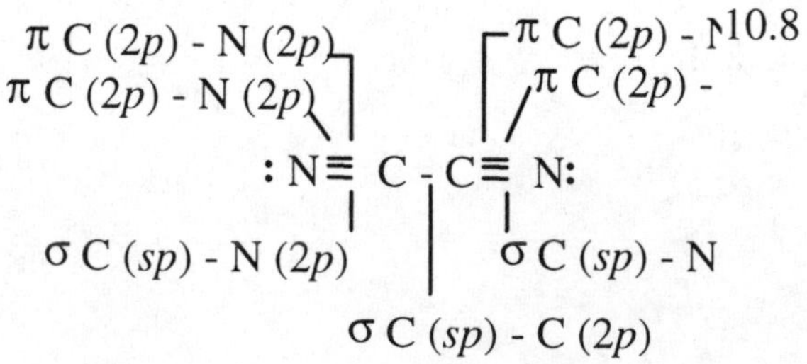

(a) 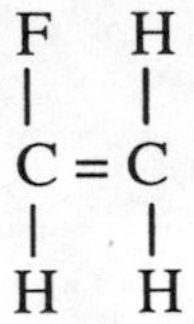

Fluorine has so much greater electronegativity that the molecule is polar.

(b) 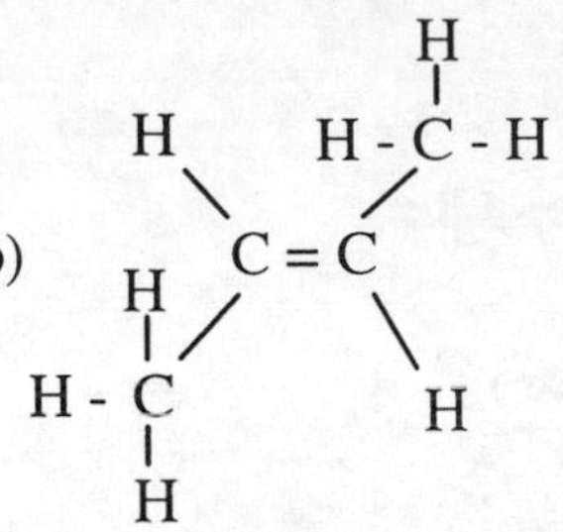

There is little electronegativity difference between H and C, and the CH_3 groups balance each other. So the molecule is nonpolar.

(c) H - C ≡ C - H

It is all symmetric, so it is nonpolar.

(d) 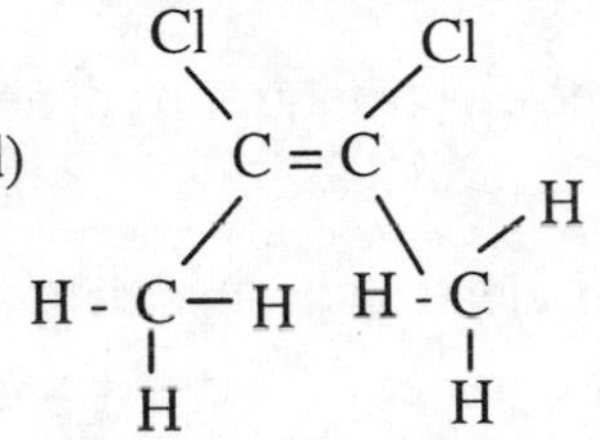

Both Cl are on one side, so it is polar.

10.9 H_2^- would have three electrons—two in bonding molecular orbitals and one in an antibonding molecular orbital. The bond order would be $\frac{1}{2}$. The molecular ion would exist but would be only slightly stable.

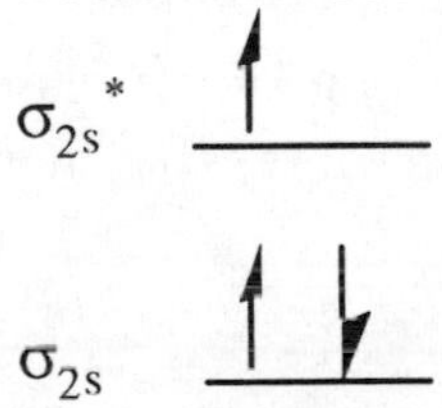

10.10 $BO = \frac{8\text{-}3}{2} = \frac{5}{2}$

Either the N_2 or O_2 molecular orbital diagram is useful, and both produce the same result because the σ sp^b and π $2p^b$ are completely filled in both diagrams, with one electron in the π sp^* orbital in each case.

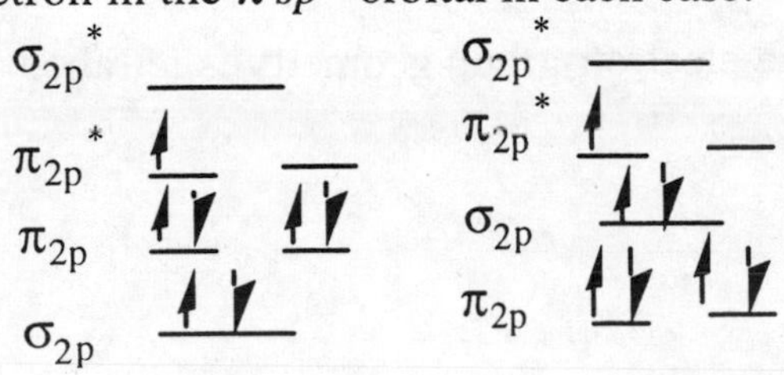

NO | NO

N_2 diagram | O_2 diagram

Review Questions

3.

    (a) 180°   :Cl̤ - Be - C̤l :

    (b) 120°   Ö = S̈ - Ö :

    (c) 109.5°   H - Ö - H

13.

    :Ö : N :: Ö :   The three electron groups denote a *sp²* hybridization on N.

    :C̤l : N :: Ö :   Three electron groups denote a *sp²* hybridization on N.

    H :Ö : N : H
           Ḧ   Four electron groups denote *sp³* hybridization on N.

Problems

21.

    (a) H — Ö — H   (b) [:Ö — Cl :]⁻   (c) [:I — I — I :]⁻

        AX_2E_2 AXE_3 AX_2E_3

23. (a) :S̈ :: C :: S̈ : Bent is not probable because the electron-group geometry is linear.

    (b)

        [:Ö : Cl : Ö :]⁻
             :O :

    Pyramidal is the probable shape. The electron-group geometry is tetrahedral, and there is one lone pair.

    (c) H — P̈ — H
            H

    Planar is not probable, because the electron-group geometry is tetrahedral.

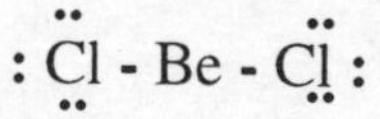

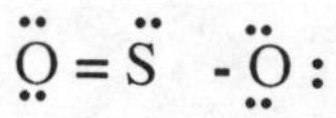

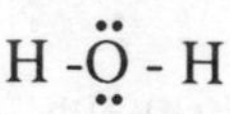

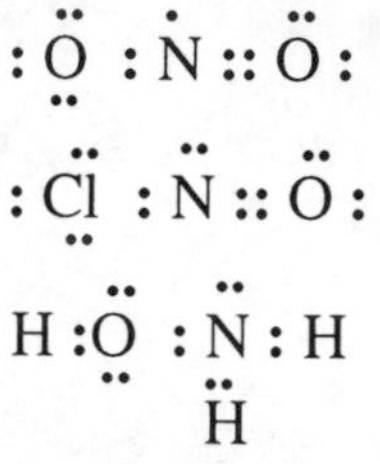

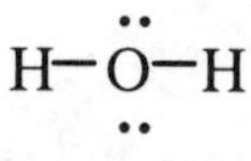

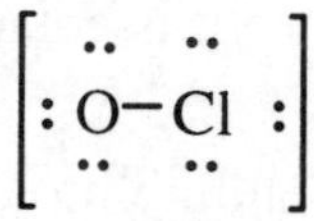

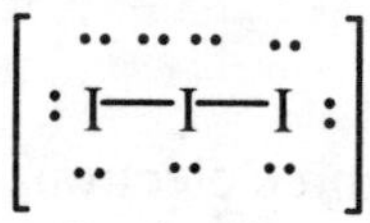

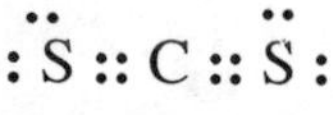

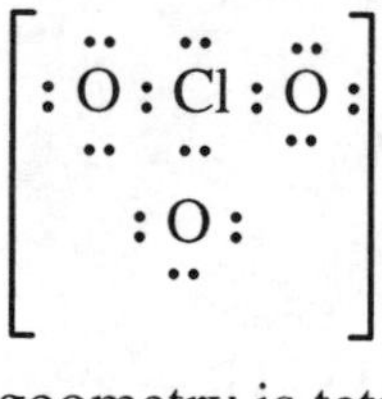

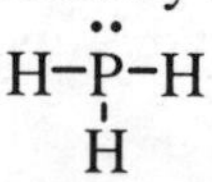

25.

    :F̈ - B - F̈ :   :F̈ - C̤l - F̈:
        :F̈ :   :F̈ :

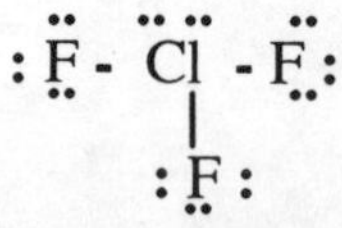

BF₃ has three electron groups, all bonding pairs, and so BF₃ is trigonal planar. ClF₃ has three bonding pairs and two lone pairs. The electron-group geometry is trigonal bipyramidal, and the molecular geometry is T-shaped.

27.

(a) 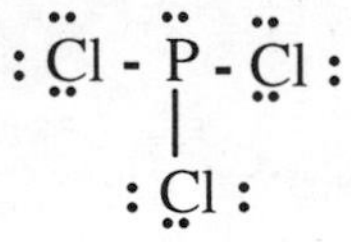 4 electron groups
3 bond pairs
1 lone pair
trigonal pyramidal

(b) 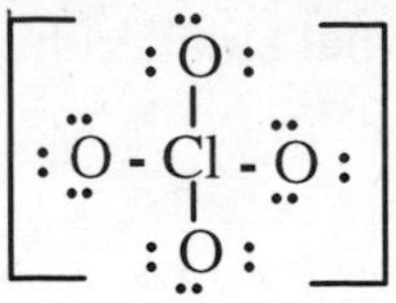 4 electron groups
4 bond pairs
tetrahedral

(c) 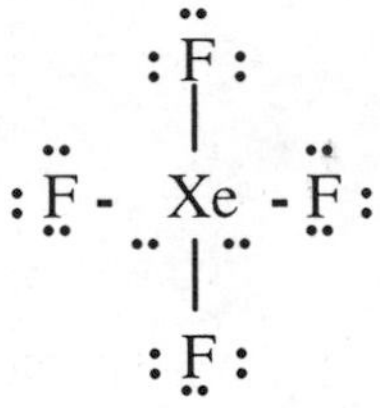 6 electron groups
4 bond pairs
2 lone pairs
square planar

(d) 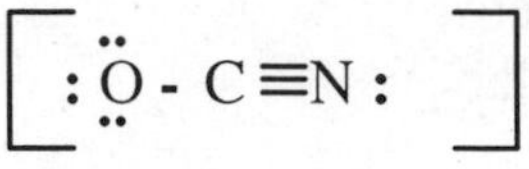 2 electron groups
both bonding
linear

(e) 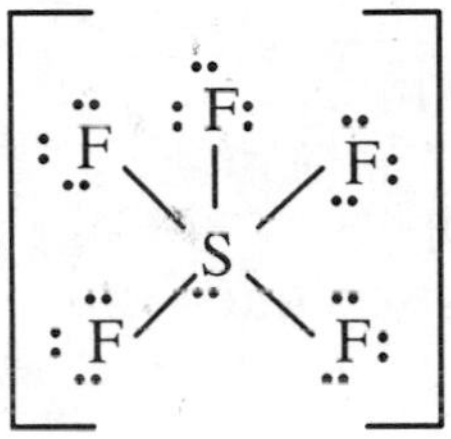 6 electron groups
5 bond pairs
1 lone pair
square pyramidal

29. (a)

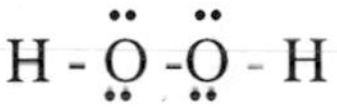

The structure is "bent" at the O-O-H bonds. The overall shape is a nonplanar "zigzag."

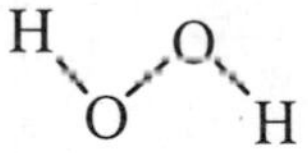

(b)

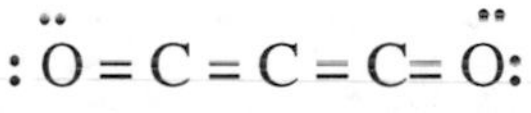

The shape about each C is linear.

(c) 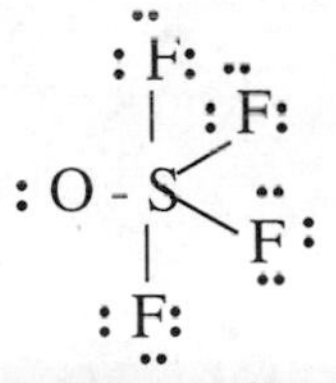 A trigonal bipyramidal shape with S as the only central atom. The O is in an equatorial position. That is, the 4 fluorine atoms will be in the seesaw shape.

(d) 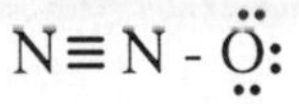 2 electron groups
2 bond pairs
linear

(e) 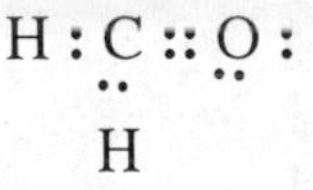 3 electron groups
3 bond pairs
trigonal planar

31.

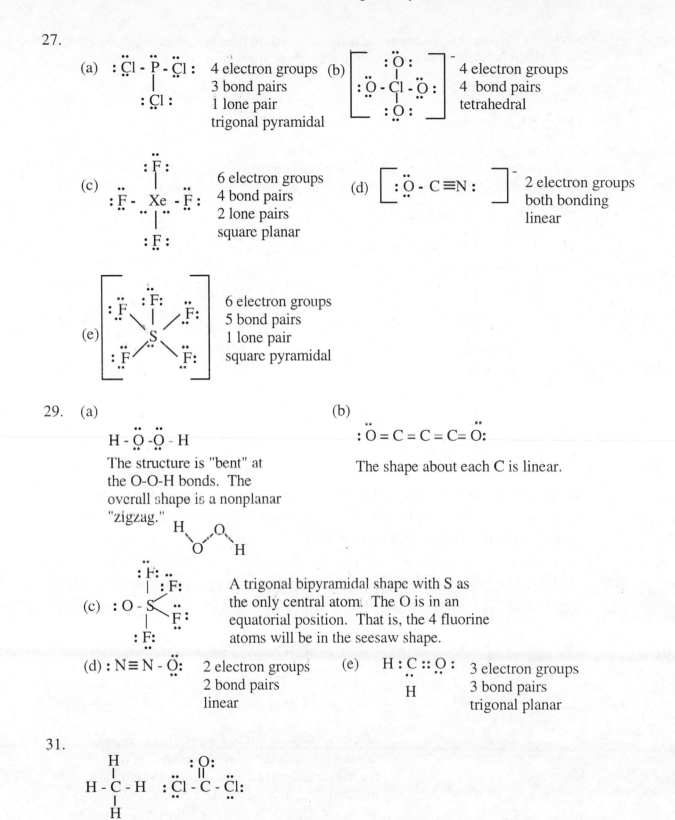

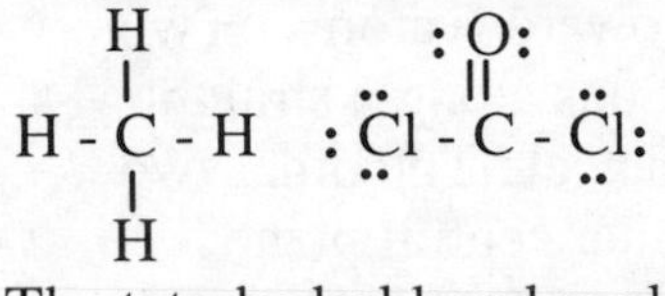

The tetrahedral bond angles are closer to the predicted angles in CH_4, as all of the bonding pairs have the same repulsion. In $COCl_2$, electron-group repulsions are strongest between the two lone pairs of electrons on the O atom, forcing the Cl-O-Cl angle to be somewhat smaller than a tetrahedral angle.

33.

(a) S = C = S

linear
symmetric
nonpolar

(b) O = N = O

bent
nonsymmetric
polar

(c)

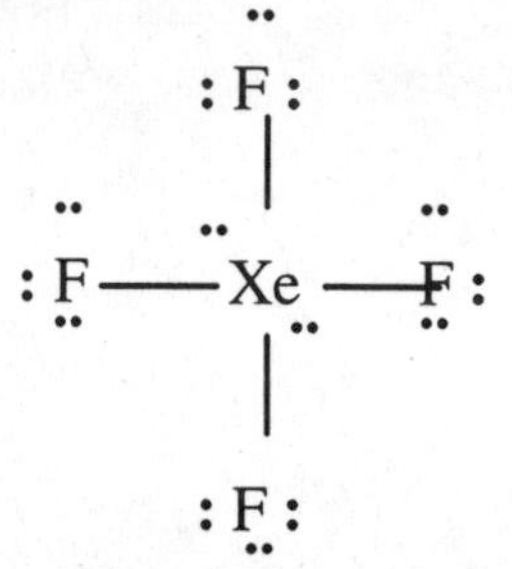

square planar
symmetric
nonpolar

(d)

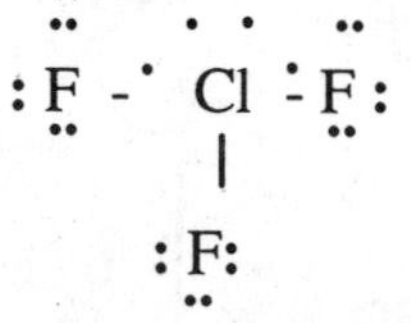

T-shape
nonsymmetric
polar

35.

H - O - H   F - O - F

Both H_2O and OF_2 are bent molecules with a similar bond angle (AXE_e); both are polar because of EN differences. H_2O should have the greater dipole moment because ΔEN is greater in H_2O than in OF_2.

37.

(a)

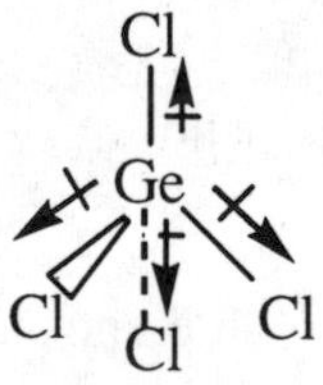

no resultant
dipole moment

(b)

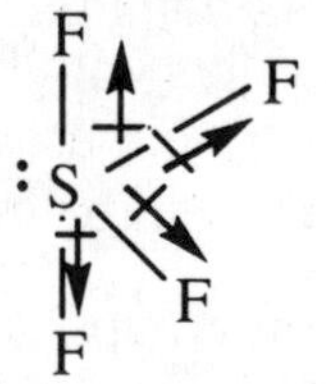

SF_4 has a seesaw shape (Table 10.1). Bond dipoles are directed toward F atoms. Two are in opposite directions along a straight line and effectively cancel. The other two are in the trigonal planar central plane (equatorial plane). The molecule is polar, but LP electrons on the S atom tend to counteract the two S-F bond dipoles in the central plane to some extent.

39. Li$_2$ is formed by an overlap of the 2*s* orbitals. F$_2$ is formed by the overlap of 2*p* orbitals along the axis between the two nuclei. The F$_2$ has a greater bond energy because the *p* orbitals overlap more.

41.

(a) :F̈ - Ö - F̈ :

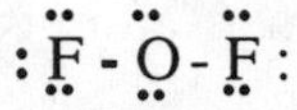

molecular shape bent

electron-group geometry

tetrahedral *sp*^{3}

AX$_2$E$_2$

(b) $$\left[\begin{array}{c} H \\ | \\ H - N - H \\ | \\ H \end{array} \right]^+$$

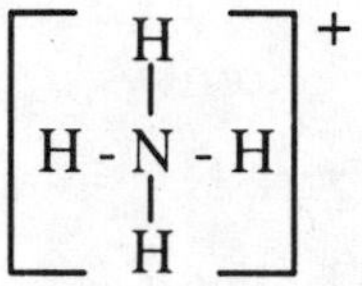

molecular shape tetrahedral

electron-group geometry

tetrahedral *sp*^{3}

AX$_4$

(c) Ö = C = Ö

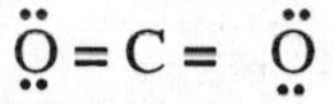

molecular shape linear

electron-group geometry

linear *sp*

AX$_2$

(d) :C̈l - C - C̈l :   (with :O: double-bonded above C)

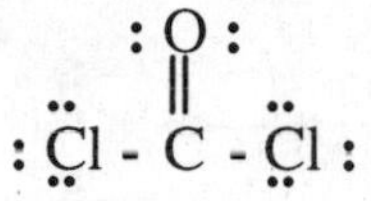

molecular shape trigonal planar

electron-group geometry

trigonal planar *sp*^{2}

AX$_3$

(e) : N ≡ C – C ≡ N :

A linear molecule; both C atoms are *sp* hybridized with linear electron geometry.

(f) H - N̈ = C = Ö

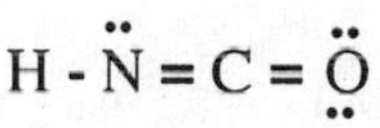

The N-C-O portion is linear shape and electron geometry; the C atom is *sp* hybridized. The H-N-C portion is trigonal planar electron geometry with a bent shape; the N is *sp*^{2} hybridized.

(g) 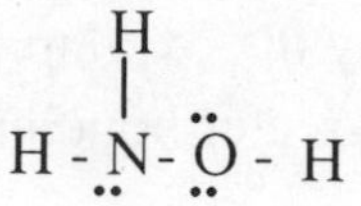

(h) 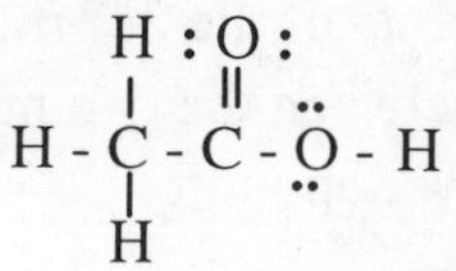

The NH_2-O portion is pyramidal with the electron geometry tetrahedral; the N atom is *sp*^{3}hybridized. The N-O-H portion is bent with tetrahedral electron geometry; the O atom is *sp*^{3} hybridized.

The CH_3-C portion is tetrahedral shape and electron geometry; the first C atom is *sp*^{3} hybridized. The C-CO-O portion is trigonal planar electron geometry and shape; the second C atom is *sp*^{2} hybridized. The C-O-H portion is bent shape, tetrahedral electron geometry, and *sp*^{3} hybridized.

43.

(a) 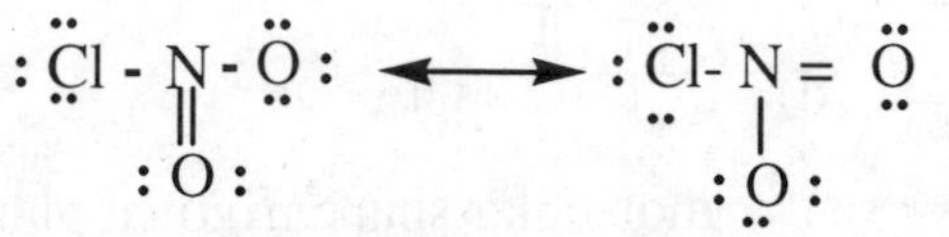 trigonal planar

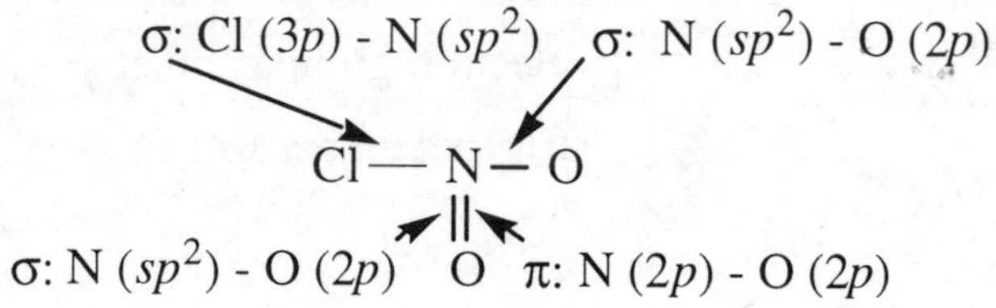
σ: Cl (3p) - N (sp²)
σ: N (sp²) - O (2p)
Cl — N — O
σ: N (sp²) - O (2p)
O
π: N (2p) - O (2p)

(b): 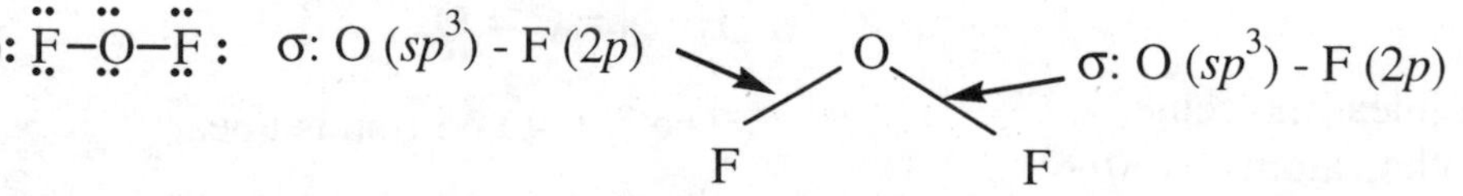
σ: O (sp³) - F (2p)
O
σ: O (sp³) - F (2p)
F
F

(c) 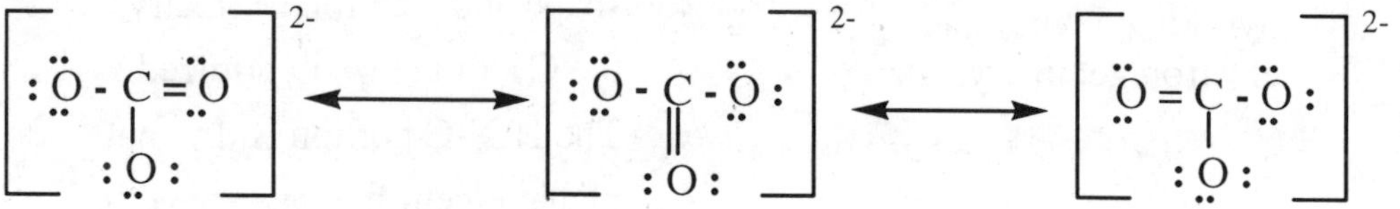

trigonal planar

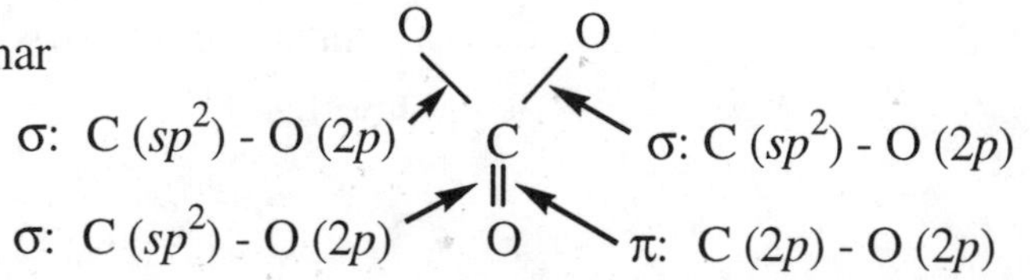
O
O
σ: C (sp²) - O (2p)
C
σ: C (sp²) - O (2p)
σ: C (sp²) - O (2p)
O
π: C (2p) - O (2p)

45.

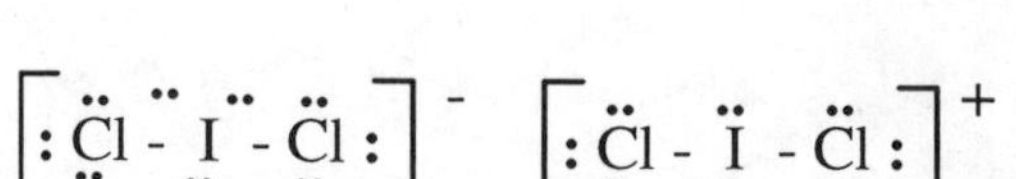

The electron-group geometry in the ICl_2^- ion is that of AX_2E_3 trigonal bipyramidal (the ion is linear). The hybridization scheme is sp^3d. The electron-group geometry in the ICl_2^+ ion is that of AX_2E_2, tetrahedral, (the ion is bent). The hybridization scheme is sp^3.

47.

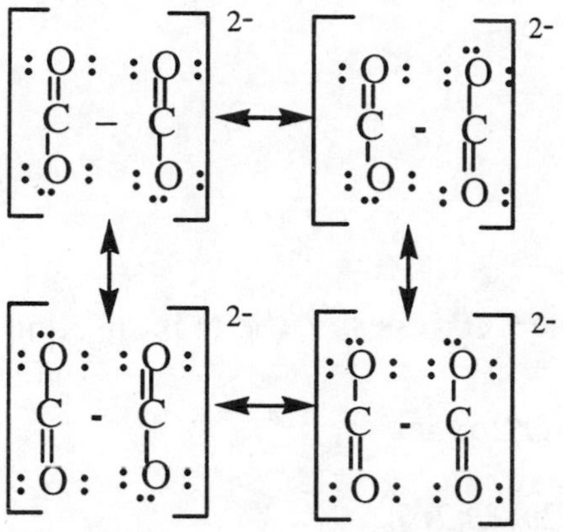

The 157 pm is the length for a single bond between the carbon atoms. The 125 pm is not short enough for a double bond between C and O, but too short for a C-O single bond. It would be correct for a delocalized pair of electrons shared by two O and C. Repulsions between lone-pair electrons on the O atoms bonded to the same C atom cause the O-C-O bonds to open up to a larger angle (126°) than the 120° angles corresponding to sp^2 hybridization. The C atoms are sp^2 hybridized.

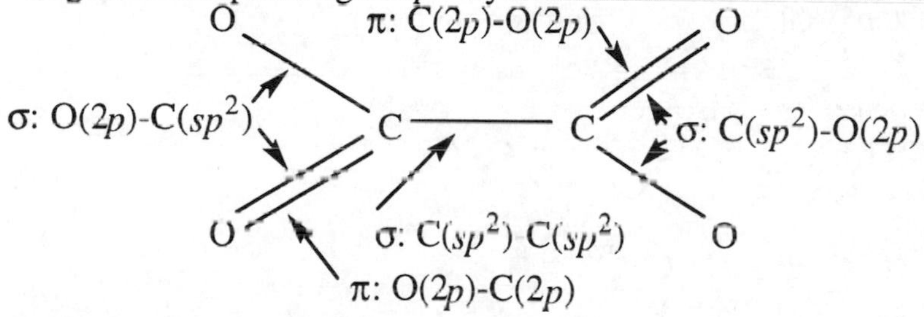

49. (a) NH_2OH

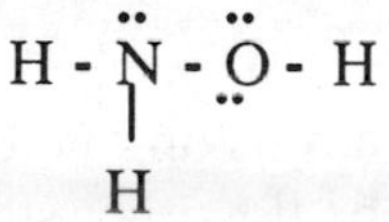

$$H - \overset{..}{\underset{|}{N}} - \overset{..}{\underset{..}{O}} - H$$
$$H$$

The hybridization for N is sp^3 and for O is sp^3.

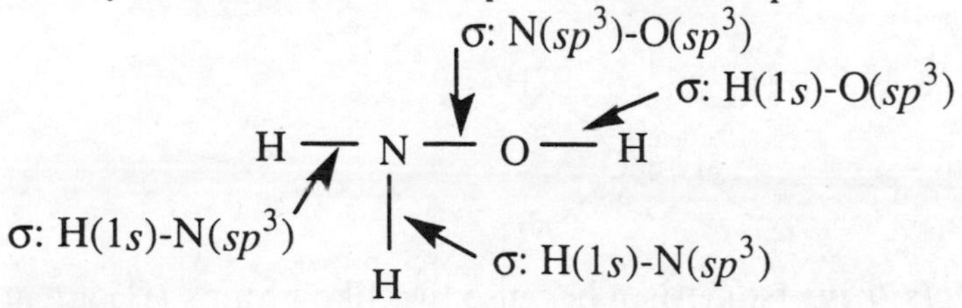

σ: N(sp^3)-O(sp^3)
σ: H($1s$)-O(sp^3)
σ: H($1s$)-N(sp^3)
σ: H($1s$)-N(sp^3)

(b) HNCO

$$H - \overset{..}{N} = C = \overset{..}{\underset{..}{O}}$$

The hybridization for N is sp^2, for C is sp, and O is unhybridized.

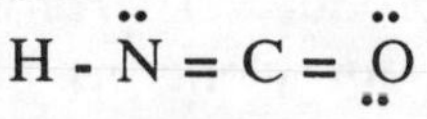

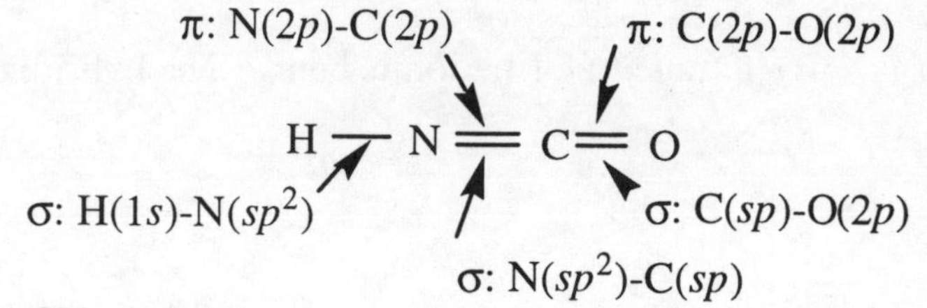

(c) CH_3CONH_2

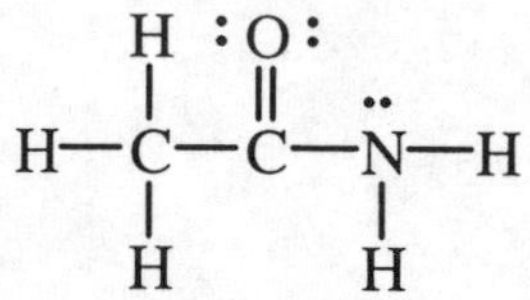

The hybridization on the CH_3 C is sp^3, the CO C is sp^2, the N is sp^3, and O is unhybridized.

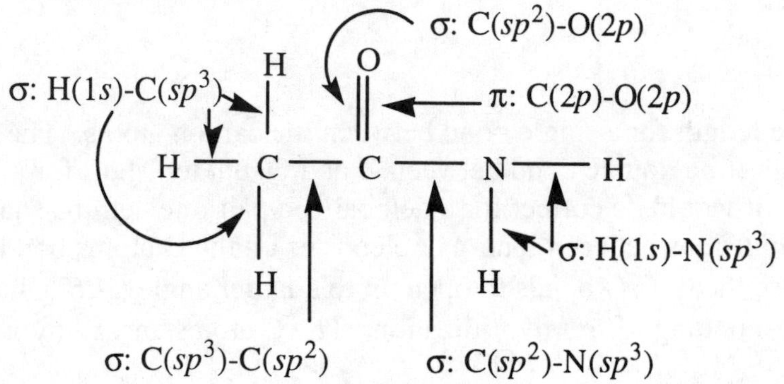

51. (a) and (c) can be *cis-trans* isomers.

(a)

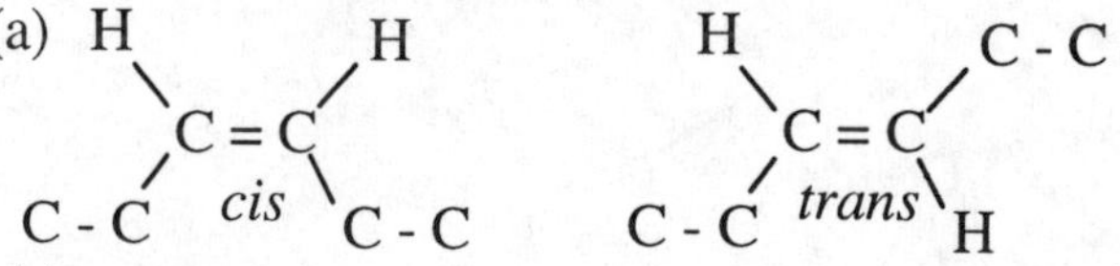

(c)

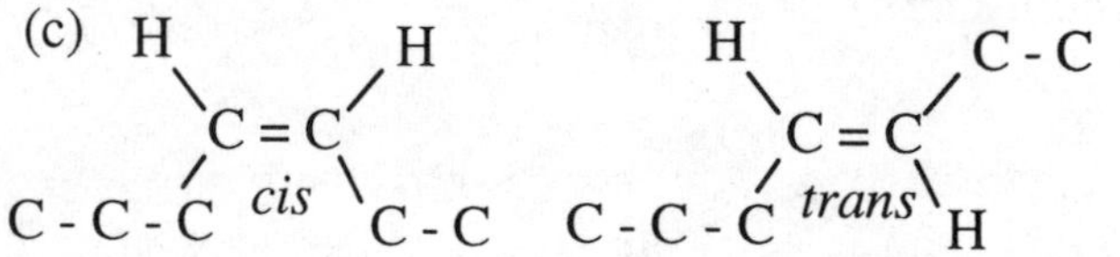

There is no *cis-trans* isomerism for (b) because two H atoms on one C atom at the double bond allows one structure to be flipped over to be the other structure.

53.

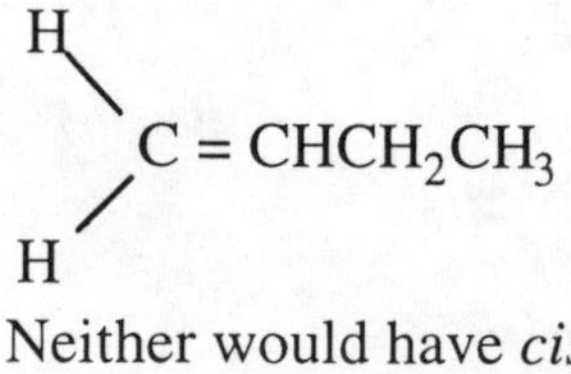

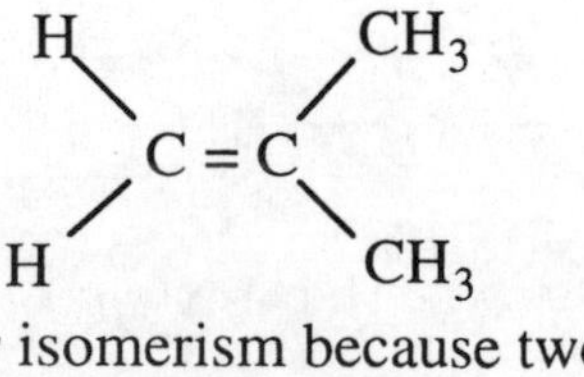

Neither would have *cis-trans* isomerism because two like groups (H) are attached to the carbons. Replacing one of the H with Cl would produce *cis-trans* isomers of 1-butene but would not produce *cis-trans* isomers for isobutylene because the second carbon has two identical groups (CH_3) bonded to it.

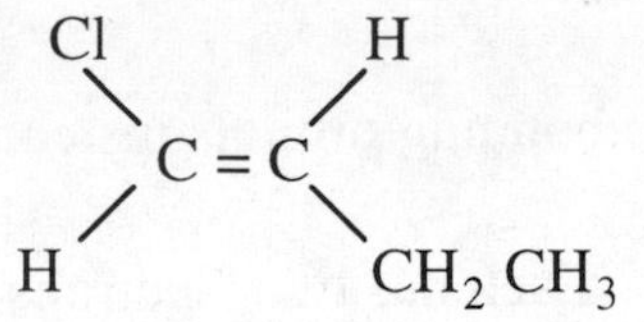

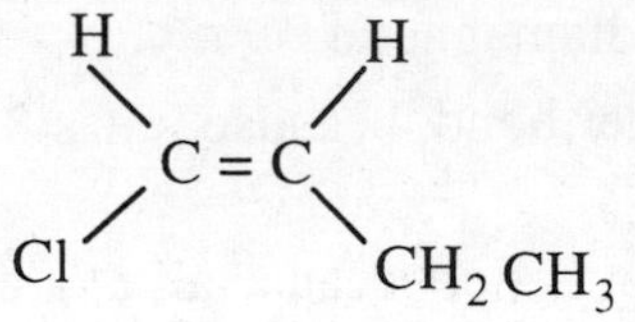

55. F_2^+ has a greater bond energy because it has two fewer antibonding electrons. The electron lost by F_2 to become F_2^+ comes from an antibonding MO, therefore strengthening the F–F bond. The electron gained by F_2 to become F_2^+ goes into an antibonding MO, therefore weakening the F–F bond in F_2^-.

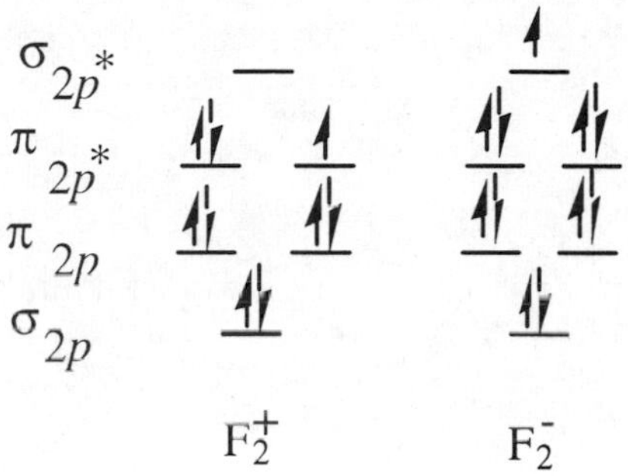

57.

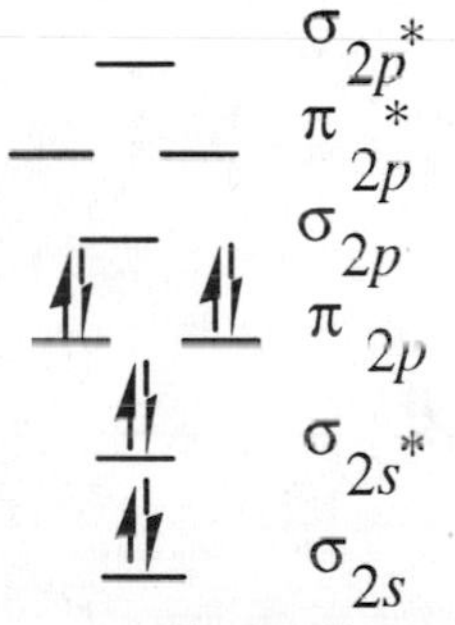

$$BO = \frac{6-2}{2} = 2$$

C :::: C   BO = 4

The Lewis structure needs as much overlap as possible to provide as many electrons as possible to produce octets. The MO theory would also indicate an σ and 2 π bonds, but it would also indicate a σ*, so the bond order would be less.

59.

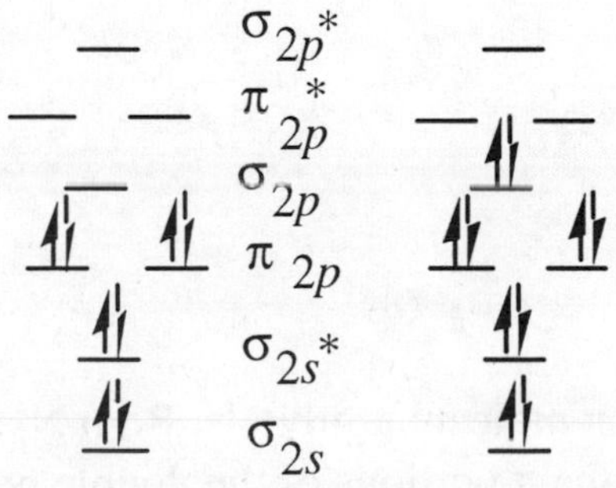

	CN^+	CN^-
BO	2	3

diamagnetic   diamagnetic

(a) CN^- is the stronger bond, because it has 8 e^- in the bonding orbitals instead of six.

(b) Neither is paramagnetic, because all e^- are paired. (Each has two electrons in antibonding MOs.)

61.

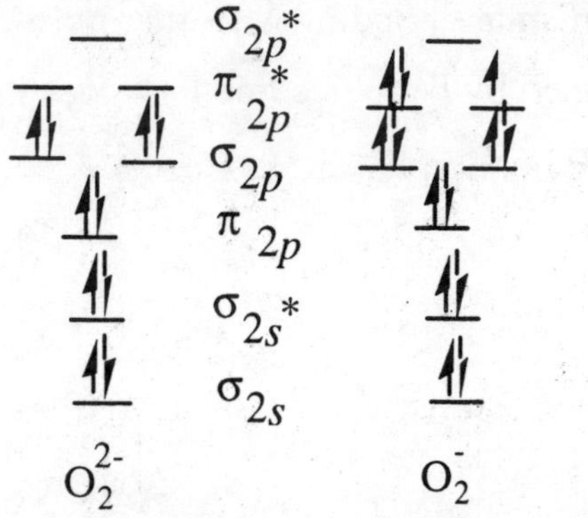

diamagnetic	paramagnetic
BO $\frac{8-6}{2} = 1$	$\frac{8-5}{2} = 1.5$
peroxide ion	superoxide ion

63.

(a) 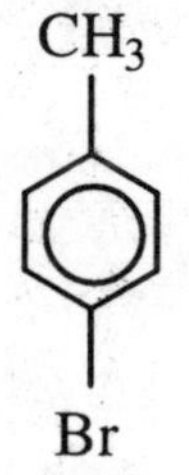

(b) 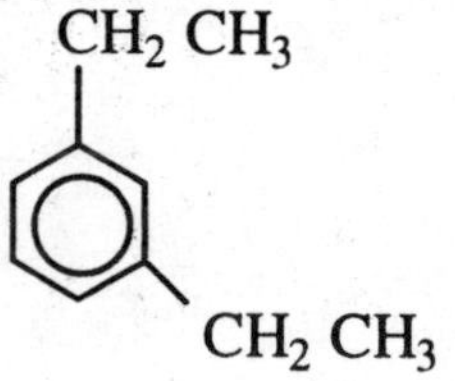

(c) 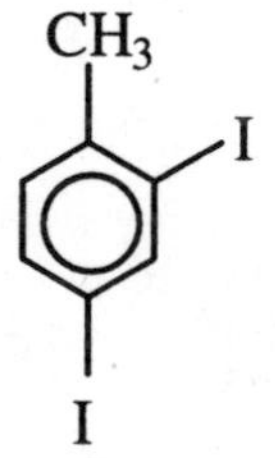

65. (a) 2,6 - dichlorophenol
(b) 1,3,5 - trinitrobenzene
(c) 2,5 - dibromotoluene

67.

(a) 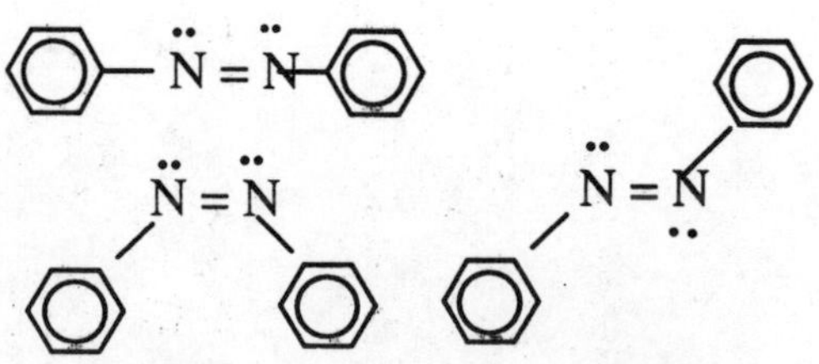

cis   *trans*

(b) All C are sp^2. The delocalized e^- are in π molecular orbitals Both N are sp^2. There has to be one p orbital for each of the π portions of the double bond.

(c) C - C - C bond angles are 120°.
C - C - N bond angles are 120°.

H-C-C- bond angles are also 120°.
C - N - N bond angles are a little less than 120°, because the lone pairs are extra repulsive.

Additional Problems

69. No. Although thc statcment is correct for diatomic molecules, for molecules with more than two atoms, bond dipoles of different bonds may cancel each other, resulting in a nonpolar molecule.

71. The fluorine atom has a greater electronegativity than the oxygen atoms have, so the fluorine end of the molecule acquires a slight negative charge, and the oxygen end a slight positive charge.

73. (a)

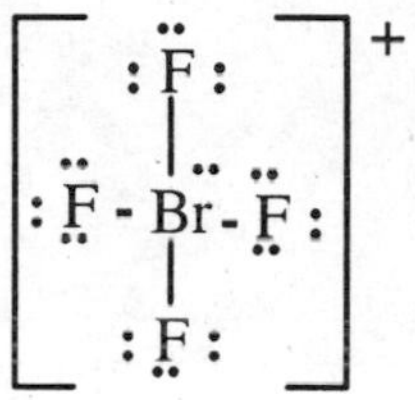

The ion has (5x7) - 1 = 34 valence electrons. The central Br atom is surrounded by five electron groups--4 bonding pairs and one lone pair: AX_4E. This corresponds to a "seesaw" shape.

(b)

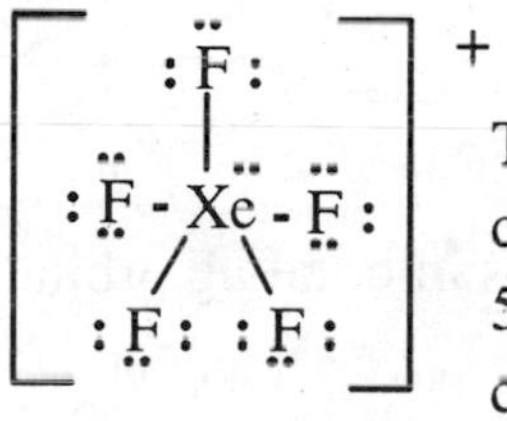

The ion has 8 + (5 x 7) - 1 = 42 valence electrons. The central Xe atom is surrounded by six electron groups--5 bonding pairs and one lone pair: AX_5E. This corresponds to a square pyramidal shape.

75. ethane < methylamine < methanol < fluoromethane
C_2II_6 CII_3NH_2 CH_3OH CH_3F
Ethane has a zero dipole moment. The other three increase as the electronegativity of N to O to F increase.

77.

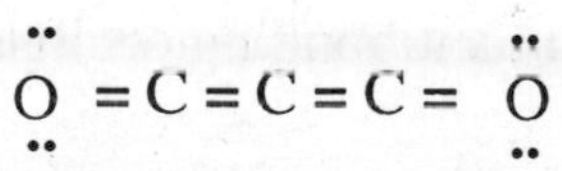

According to VSEPR theory, the molecule should be linear, and this corresponds to *sp* hybridization of all three C atoms.

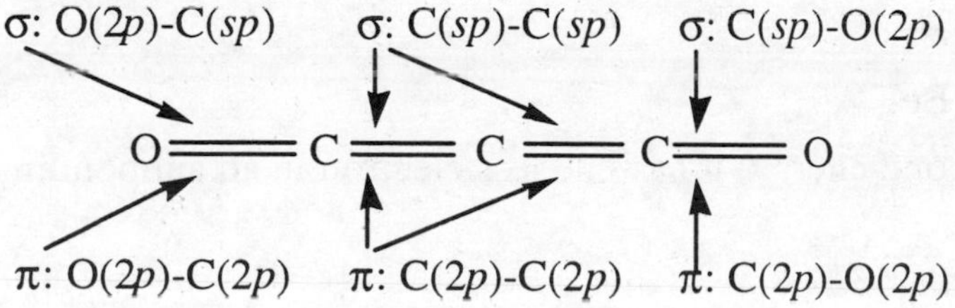

79.

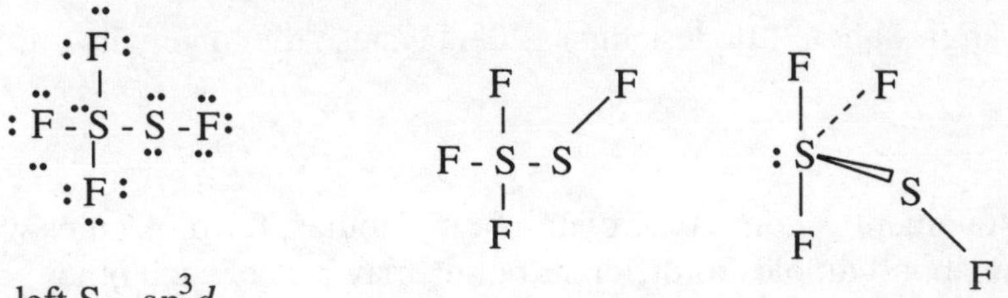

left S sp^3d

right S sp^3

The shape around the left S is seesaw.

The shape around the right S is bent.

81. (a)

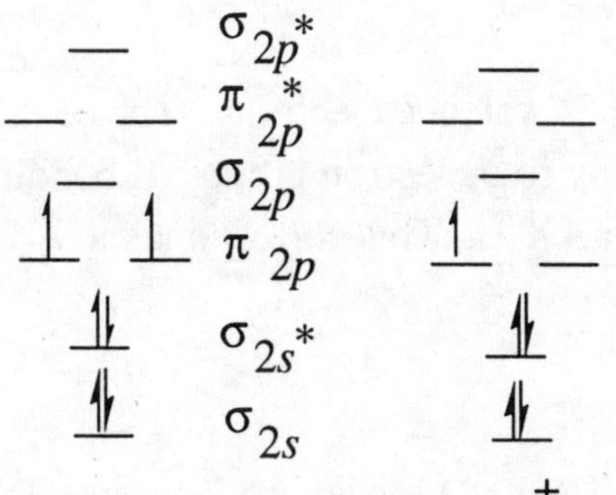

B_2 B_2^+

B_2 has the greater bond energy; it has more electrons in bonding orbitals

(b)

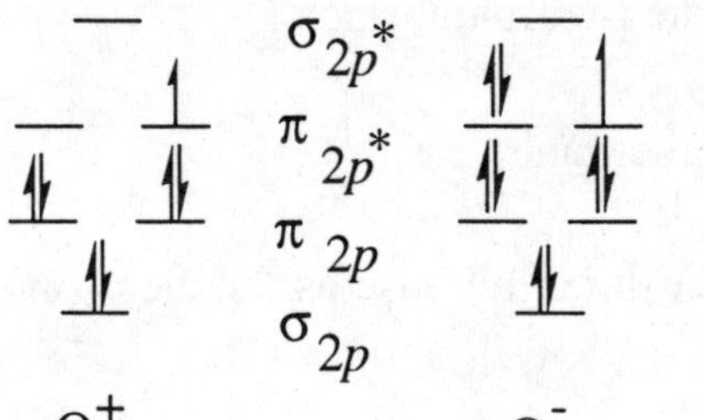

O_2^+ O_2^-

O_2^+ has the greater bond energy; it has less electrons in antibonding orbitals.

(c)

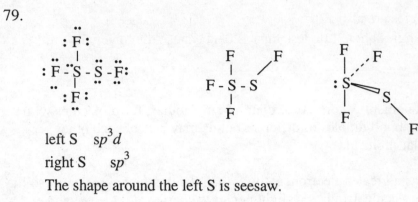

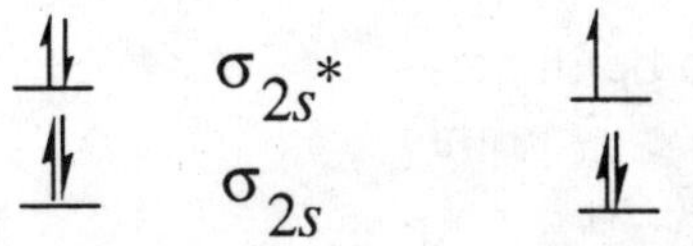

Be_2 Be_2^+

Be_2^+ has the greater bond energy; it has one less electron in an antibonding orbital.

(d)

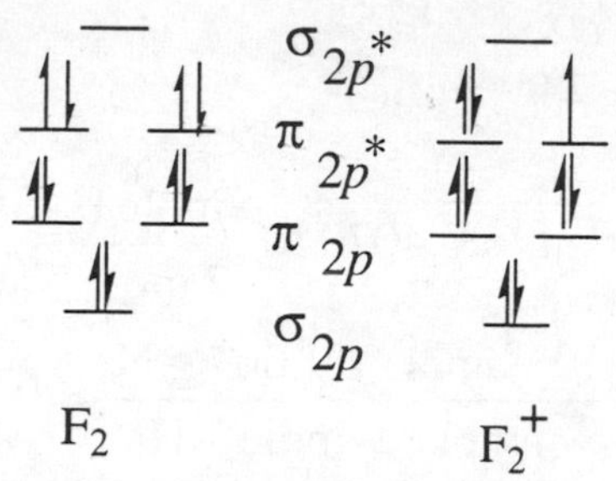

F_2^+ has the greater bond energy; it has one less electron in the antibonding orbital.

(e)

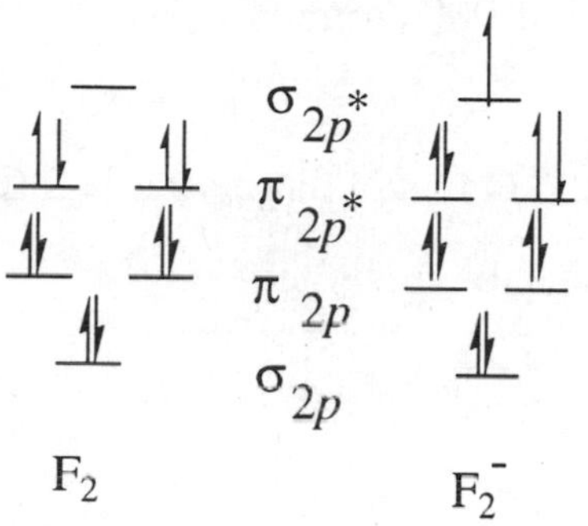

F_2 has the greater bond energy; it has one less electron in antibonding orbitals.

83. electron charge = 1.602×10^{-19} C
HCl bond length = Cl^- ionic radii = 127 pm (from Table 9.1)

$$\mu = \delta d = 1.602 \times 10^{-19}\ \text{C} \times 127\ \text{pm} \times \frac{10^{-12}\ \text{m}}{\text{pm}} \times \frac{\text{D}}{3.34 \times 10^{-30}\ \text{C m}} = 6.09\ \text{D}$$

$$\%\ \text{ionic character} = \frac{1.07}{6.09} \times 100\% = 17.6\%$$

85. HF

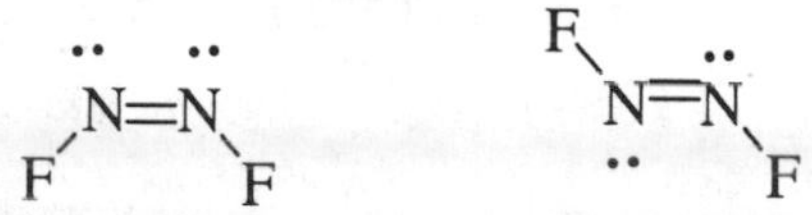

$\mu = 1.820$ $\mu = 0.160$ $\mu = 0$

The HF molecule has the entire charge difference along one line so the dipole moment is large. The *cis* – N_2F_2 has F atoms on one side of the double bond and lone pairs of electrons on the opposite side, resulting in a dipole moment. In *trans*-N_2F_2 the effects of the F atoms and the lone pair of electrons both cancel out.

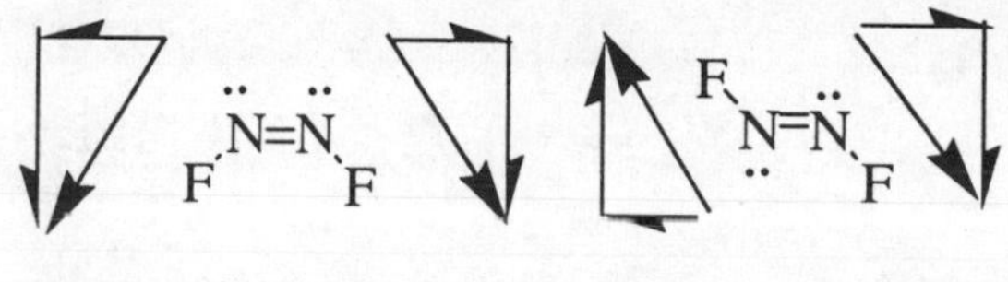

Apply Your Knowledge

87. $? \# H = 57.05\,u \times \dfrac{5.29\,u}{100.0\,u\,\text{sample}} \times \dfrac{\text{atom H}}{1.008\,u\,H} = 2.99\,\text{atom}$

$? \text{ g N} = (50.00\,\text{mL} \times 0.2800\,\text{M }H_2SO_4 - 36.49\,\text{mL} \times 0.4070\,\text{M }\dfrac{\text{mole }H_2SO_4}{2\,\text{mole NaOH}})$

$\times \dfrac{2\,\text{mole }NH_3}{\text{mol }H_2SO_4} \times \dfrac{10^{-3}\,\text{mol}}{\text{mmol}} \times \dfrac{14.007\,\text{g N}}{\text{mole }NH_3} = 0.1842\,\text{g N}$

$? \% \text{ N} = \dfrac{0.1842\,\text{g N}}{0.7500\,\text{g sample}} \times 100\,\% = 24.56\,\% \text{ N}$

$? \# N = 57.05\,u \times \dfrac{24.56\,u\,N}{100.0\,u\,\text{sample}} \times \dfrac{\text{atom N}}{14.007\,\text{g N}} = 1.000\,\text{atom}$

$57.05 - 3 \times 1.008 - 1 \times 14.007 = 40.02$

The only combination of atomic masses of C and O that adds to 40 u is 2 C and 1 O.

C_2H_3ON

(a)

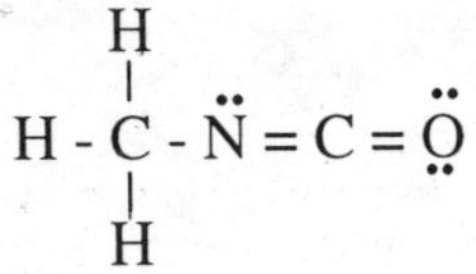

$$H - \overset{\overset{\displaystyle H}{|}}{\underset{\underset{\displaystyle H}{|}}{C}} - \ddot{N} = C = \ddot{O}\!\!:$$

(b) CH_3 C sp^3

$\quad\quad$ N sp^2

$\quad\quad$ C sp

$\sigma:H(1s)\text{-}C(sp^3)$

$\sigma:C(sp^3)\text{-}N(sp^2)$

$\sigma:N(sp^2)\text{-}C(sp)$

$\sigma:C(sp)\text{-}O(2p)$

$\pi:C(2p)\text{-}O(2p)$

$\pi:N(2p)\text{-}C(2p)$

$$H - \overset{\overset{\displaystyle H}{|}}{\underset{\underset{\displaystyle H}{|}}{C}} - N = C = O$$

(c) H - C - H $109.5°$

$\quad\;$ H - C - N $109.5°$

$\quad\;$ C - N - C   slightly less than $120°$

$\quad\;$ N = C = O $180°$

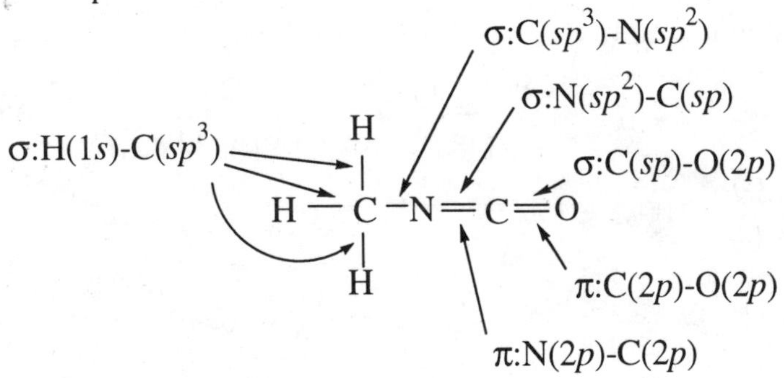

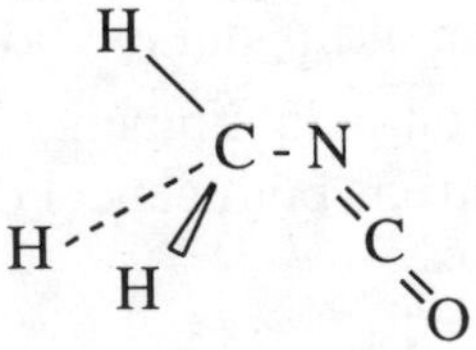

89. (a) $Li_2 \rightarrow Li_2^+ + e^-$

$\quad\quad$ $Be_2 \rightarrow Be_2^+ + e^-$

$\quad\quad$ $B_2 \rightarrow B_2^+ + e^-$

$C_2 \rightarrow C_2^{+} + e^-$
$N_2 \rightarrow N_2^{+} + e^-$
$O_2 \rightarrow O_2^{+} + e^-$
$F_2 \rightarrow F_2^{+} + e^-$
$Ne_2 \rightarrow Ne_2^{+} + e^-$

(b) Yes, Ne_2^+ has more bonding electrons than antibonding electrons. BO = ½.

(c) The antibonding orbital in F_2 that the electron comes from is at a higher energy level than the *p* orbital in F.

(d) The antibonding orbital in O_2 that the electron comes from is at a higher energy level than the *p* orbital in O, so O_2 has a lower first-ionization energy than O.

(e) The bonding orbital in N_2 that the electron comes from is at a lower level than the *p* orbital in N, so N_2 has a higher first-ionization energy than N.

(f) The first ionization potential increases from $O_2 < O < N < N_2$. Thus molecular nitrogen has a higher ionization energy than molecular oxygen. Helium also has a higher ionization potential because of its small size.

91. (a) Top to bottom: *trans* unsaturated, *cis* unsaturated, saturated fatty acid.

(b) The *trans* isomers, like the saturated fats, can pack together more tightly than can the *cis* isomer.

(c) Unsaturated fatty acids will react with bromine or permanganate.

Chapter 11

States of Matter and Intermolecular Forces

Exercises

11.1A $\Delta H_{vap} = \frac{652 \text{ J}}{1.50 \text{ g } C_6H_6} \times \frac{78.11 \text{ g } C_6H_6}{\text{mol } C_6H_6} \times \frac{\text{kJ}}{10^3 \text{ J}} = 34.0 \text{ kJ/mol}$

11.1B raise temperature

$$\Delta H = 0.750 \text{ L} \times \frac{0.789 \text{ g}}{\text{mL}} \times \frac{\text{mL}}{10^{-3} \text{ L}} \times \frac{2.46 \text{ J}}{\text{g °C}} \times \frac{\text{kJ}}{10^3 \text{ J}} \times (25.0 - 0.0) \text{ °C} = 36.4 \text{ kJ}$$

vaporize

$$\Delta H = 0.10 \times 0.750 \text{ L ethanol} \times \frac{\text{mL}}{10^{-3} \text{ L}} \times \frac{0.789 \text{ g}}{\text{mL}} \times \frac{\text{mol } C_2H_5OH}{46.07 \text{ g } C_2H_5OH} \times \frac{43.3 \text{ kJ}}{\text{mol}} = 55.6 \text{ kJ}$$

$\Delta H_{total} = 36.4 \text{ kJ} + 55.6 \text{ kJ} = 92.0 \text{ kJ}$

11.2 CS_2 $1.00 \text{ kg} \times \frac{10^3 \text{ g}}{\text{kg}} \times \frac{27.4 \text{ kJ}}{\text{mol}} \times \frac{\text{mol}}{76.13 \text{ g}} = 360 \text{ kJ}$

CCl_4 $1.00 \text{ kg} \times \frac{10^3 \text{ g}}{\text{kg}} \times \frac{37.0 \text{ kJ}}{\text{mol}} \times \frac{\text{mol}}{153.81 \text{ g}} = 241 \text{ kJ}$

CH_3OH $1.00 \text{ kg} \times \frac{10^3 \text{ g}}{\text{kg}} \times \frac{38.0 \text{ kJ}}{\text{mol}} \times \frac{\text{mol}}{32.04 \text{ g}} = 1.19 \times 10^3 \text{ kJ}$

C_8H_{18} $1.00 \text{ kg} \times \frac{10^3 \text{ g}}{\text{kg}} \times \frac{41.5 \text{ kJ}}{\text{mol}} \times \frac{\text{mol}}{114.24 \text{ g}} = 363 \text{ kJ}$

C_2H_5OH $1.00 \text{ kg} \times \frac{10^3 \text{ g}}{\text{kg}} \times \frac{43.3 \text{ kJ}}{\text{mol}} \times \frac{\text{mol}}{46.07 \text{ g}} = 940 \text{ kJ}$

H_2O $1.00 \text{ kg} \times \frac{10^3 \text{ g}}{\text{kg}} \times \frac{44.0 \text{ kJ}}{\text{mol}} \times \frac{\text{mol}}{18.02 \text{ g}} = 2.44 \times 10^3 \text{ kJ}$

$C_6H_5NH_2$ $1.00 \text{ kg} \times \frac{10^3 \text{ g}}{\text{kg}} \times \frac{52.3 \text{ kJ}}{\text{mol}} \times \frac{\text{mol}}{93.13 \text{ g}} = 562 \text{ kJ}$

CCl_4, with a relatively low ΔH value and a large molar mass, requires the smallest quantity of heat for its vaporization.

$$11.3\text{A } P = \frac{mRT}{VM} = \frac{1.100 \text{ g x } \frac{0.08206 \text{ L atm}}{\text{K mol}} \text{ x } (273 + 39)\text{K x } \frac{760\text{mmHg}}{\text{atm}}}{335 \text{ mL x } \frac{10^{-3} \text{ L}}{\text{mL}} \text{ x } 159.80 \text{ g/mol}} = 4.00 \times 10^2 \text{ mmHg}$$

11.3B $m = \frac{PVM}{RT} = \frac{19.8 \text{ mm } H_2O \times 275 \text{ mL} \times \frac{10^{-3} \text{ L}}{\text{mL}} \times 18.02 \text{ g/mol}}{\frac{62.36 \text{ L mmHg}}{\text{K mol}} \times (273 + 22)\text{K}} = 5.33 \times 10^{-3} \text{ g } H_2O$

11.4 Methane has a $T_c = -82.4$ °C at $P_c = 45$ atm. Room temperature is above T_c, so methane stays a gas at all pressures. The pressure gauge would work fine.

11.5 $P = \frac{nRT}{V} = \frac{1.05 \text{ mol} \times \frac{0.08206 \text{ L atm}}{\text{K mol}} \times (273.2 + 30.0)\text{K}}{2.61 \text{ L}}$

$P = 10$ atm

This pressure greatly exceeds the vapor pressure of H_2O at 30.0 °C, so some of the vapor condenses to liquid. The sample cannot be all liquid, however, because 1.05 mol $H_2O(l)$ occupies a volume of less than 20 mL. The final condition reached is one of liquid and vapor in equilibrium.

11.6A IBr has the greater molecular mass and therefore the greater London forces. Both molecules are somewhat polar, but the electronegativity differences are small. IBr is a solid, and BrCl is a gas.

11.6B (a)Toluene is less polar because CH_3 is less electronegative than NH_2. It requires less energy to separate molecules of toluene, so it has a lower boiling point.

(b) The *trans* isomer is nonpolar because the chlorine atoms are balanced on the molecules. *Trans*-1,2-dichloroethene would have the lower boiling point.

(c) The para isomer is nonpolar because the chlorine atoms are symmetrical on the molecule. On the ortho isomer, the chlorine atoms are on one side, so the molecule is polar. Para-dichlorobenzene has the lower boiling point.

11.7 NH_3, yes. H is bonded to N, a small electronegative atom.

CH_4, no. C is not electronegative enough for hydrogen bonding.

C_6H_5OH, yes. H is bonded to O, a small electronegative atom. Hydrogen bonding is somewhat overshadowed by dispersion forces associated with the large organic part of the molecule.

$CH_3\overset{\overset{\displaystyle O}{\|}}{C}OH$, yes. H is bonded to O, a small electronegative atom

H_2S, no. There is a little hydrogen bonding because although S is electronegative, S is too large to permit much hydrogen bonding.

H_2O_2, yes. H is bonded to O, a small electronegative atom.

11.8 $CsBr < KI < KCl < MgF_2$

A cesium ion is larger than a potassium ion. An iodide ion is larger than a chloride ion, which is larger than a fluoride ion. A magnesium ion is a small cation with a 2^+ charge and exerts stronger interionic attractions than do either K^+ or Cs^+ ions.

11.9 (a) In a bcc cell the distance from one corner of the cell to the opposite corner is 4r. That distance is the hypotenuse of a right triangle whose other sides are the length (ℓ) of the cell (down one corner) and the diagonal of the bottom face of the cell. The diagonal distance from the Pythogorian theorem is $d_d = \sqrt{\ell^2 + \ell^2} = \sqrt{2\ell^2}$. The distance from one corner of the cell to the other is $d = (2\ell^2 + \ell^2)^{\frac{1}{2}}$.

Since d = 4r, then

$4r = (2\ell^2 + \ell^2)^{\frac{1}{2}} = (3\ell^2)^{\frac{1}{2}}$

$4 \times 124.1 \text{ pm} = \sqrt{3}\ \ell$

$\ell = 286.6 \text{ pm}$

(b) $V = \ell^3 = (286.6 \text{ pm})^3 \times \left(\frac{10^{-12} \text{ m}}{\text{pm}}\right)^3 \times \left(\frac{\text{cm}}{10^{-2} \text{ m}}\right)^3 = 2.354 \times 10^{-23} \text{ cm}^3$

11.10 In a bcc cell there are two atoms per cell—one in the center and $\frac{1}{8}$ at each of the eight corners.

$m_{\text{cell}} = \frac{55.847 \text{ g}}{\text{mol}} \times \frac{\text{mol}}{6.022 \text{ x } 10^{23} \text{ atoms}} \times \frac{2 \text{ atoms}}{\text{cell}} = 1.855 \times 10^{-22} \text{ g/cell}$

$d = \frac{m}{V} = \frac{1.855 \text{ x } 10^{-22} \text{ g/cell}}{2.354 \text{ x } 10^{-23} \text{ cm}^3\text{/cell}} = 7.880 \text{ g/cm}^3$

Review Questions

3. (a) Increases in temperature increase the vapor pressure.
   (b), (c), and (d) have no effect. The vapor pressure will be the same whatever the volume of liquid and vapor and the area of contact between them, as long as both phases exist together.

16. N_2 has a slightly lower molar mass than O_2, so it would have weaker intermolecular forces and a lower boiling point. -196°C is a good and correct choice.

Problems

17. $\Delta H = 1.00 \text{ kg } CS_2 \times \frac{10^3 \text{ g}}{\text{kg}} \times \frac{\text{mol } CS_2}{76.15 \text{ g } CS_2} \times \frac{27.4 \text{ kJ}}{\text{mol}} = 360 \text{ kJ}$

19. $\Delta H_T = \Delta H_1 + \Delta H_2$

18.0 °C $\rightarrow$ 25.0 °C $\Delta H_1 = \text{mass} \times \text{sp. ht.} \times \Delta T$

$\Delta H_1 = 25.0 \text{ g} \times \frac{4.18 \text{ J}}{\text{g °C}} \times (25.0 - 18.0) \text{ °C} \times \frac{\text{kJ}}{1000 \text{ J}}$

$\Delta H_1 = 0.73 \text{ kJ}$

liquid $\rightarrow$ vapor $\Delta H_2 = n\Delta H_{\text{vap}}$

$\Delta H_2 = 25.0 \text{ g} \times \frac{\text{mol}}{18.02 \text{ g}} \times \frac{44.0 \text{ kJ}}{\text{mol}} = 61.0 \text{ kJ}$

$\Delta H_T = \Delta H_1 + \Delta H_2 = 61.7\ \text{kJ}$

21. $? \text{ g } C_3H_8 = 0.750 \text{ L } H_2O \times \frac{\text{mL}}{10^{-3} \text{ L}} \times \frac{1 \text{ g}}{\text{mL}} \times \frac{\text{mol } H_2O}{18.02 \text{ g } H_2O} \times \frac{44.0 \text{ kJ}}{\text{mol}} \times \frac{\text{mol } C_3H_8}{2.22 \text{ x } 10^3 \text{ kJ}}$

$\times \frac{44.09 \text{ g } C_3H_8}{\text{mol } C_3H_8} = 36.4 \text{ g } C_3H_8$

23. In an oven the heat is a dry heat. In steam, there is the heat transfer associated with the temperature difference plus a very large quantity of heat from the condensation of steam. The transfer of heat from air in the oven to the hand occurs slowly, and so the hand warms slowly. Above the boiling water there is an almost instantaneous transfer of a large quantity of heat, from condensed steam (the heat of condensation) directly into the hand.

25. (a) about 425 mmHg
(b) about 77 °C

27. $n = \frac{PV}{RT} = \frac{1.00 \text{ mmHg} \times \frac{\text{atm}}{760 \text{ mmHg}} \times 486 \text{ mL} \times \frac{10^{-3} \text{ L}}{\text{mL}}}{\frac{0.08206 \text{ L atm}}{\text{K mole}} \times (1360 + 273) \text{ K}} = 4.77 \times 10^{-6} \text{ mol}$

$? \text{ atoms Ag} = 4.77 \times 10^{-6} \text{ mol} \times \frac{6.022 \text{ x } 10^{23} \text{ atoms}}{\text{mol}} = 2.87 \times 10^{18} \text{ atoms Ag}$

29. $P = \frac{mRT}{MV} = \frac{0.480 \text{ g x } \frac{0.08206 \text{ L atm}}{\text{K mol}} \text{ x } (40.0 + 273.2)\text{K}}{\frac{153.81 \text{ g}}{\text{mol}} \text{ x } 285 \text{ mL x } \frac{10^{-3} \text{ L}}{\text{mL}}} = 0.281 \text{ atm}$

$P = 0.281 \text{ atm} \times \frac{760 \text{ mmHg}}{\text{atm}} = 214 \text{ mmHg}$

31. $P = \frac{mRT}{MV} = \frac{1.82 \text{ g x } \frac{0.08206 \text{ L atm}}{\text{K mol}} \text{ x } (30.0 + 273.2)\text{K}}{\frac{18.02 \text{ g}}{\text{mol}} \text{ x } 2.55 \text{ L}} \times \frac{760 \text{ mmHg}}{\text{atm}} = 749 \text{ mmHg}$

Because the calculated vapor of $H_2O(g)$ (749 mmHg) greatly exceeds the vapor pressure of water at 30.0 °C, some of the water vapor condenses. The final mixture is one of liquid and vapor. The final condition cannot be liquid exclusively because 1.82 g of $H_2O(l)$ only occupies a volume of less than 2 mL. The rest of the 2.55-L flask must be filled with $H_2O(g)$.

33. The high heat capacity of water means that the heat goes into the water instead of heating the cup above the water temperature. When the water does boil, the heat is used for the large heat of vaporization of water, and the temperature of the cup stays at the

temperature of the water as long as liquid water remains. The temperature of the boiling water is below the ignition temperature of the paper cup. The boiling point is below 100 °C because the barometric pressure is less than 1 atm.

35. A gas cannot be liquefied above T_c, regardless of the pressure applied. A gas can be either liquefied or solidified by a sufficient lowering of the temperature, regardless of its pressure. A gas can always be liquefied by an appropriate combination of pressure and temperature changes.

37. $\Delta H = n\Delta H_{fus} = 3.5 \text{ cm} \times 2.6 \text{ cm} \times 2.4 \text{ cm} \times \frac{0.92 \text{ g}}{\text{cm}^3} \times \frac{\text{mol}}{18.02 \text{ g}} \times \frac{6.01 \text{ kJ}}{\text{mol}} = 6.7 \text{ kJ}$

39. $\Delta H_{ice} = n\Delta H_{fus} = 0.506 \text{ kg} \times \frac{10^3 \text{ g}}{\text{kg}} \times \frac{\text{mol}}{18.02 \text{ g}} \times \frac{6.01 \text{ kJ}}{\text{mol}} = 169 \text{ kJ}$

$$\Delta H_{H_2O} = \text{mass} \times \text{sp.ht.} \times \Delta T = 315 \text{ mL} \times \frac{1.0 \text{ g}}{\text{mL}} \times \frac{4.18 \text{ J}}{\text{g °C}} \times (0.0 \text{ °C} - 20.2\text{° C}) \times \frac{\text{kJ}}{10^3 \text{ J}}$$
$$= -26.6 \text{ kJ}$$

Not enough heat is liberated in cooling the liquid water to 0 °C to melt all the ice. Ice will remain after thermal equilibrium is reached.

41.

	actual (kJ)
(a) $q = 3.0 \text{ mol ice} \times \frac{6.01 \text{ kJ}}{\text{mol}}$	18
(b) $q = 10.0 \text{ g} \times \frac{\text{mol}}{18.01 \text{ g}} \times \frac{44.0 \text{ kJ}}{\text{mol}}$	24
(c) $q = 2 \text{ mol ice} \times 6.01 \text{ kJ/mol} + 2 \text{ mol} \times \frac{0.075 \text{ kJ}}{\text{mol °C}} \times (10\text{-}0) \text{ °C}$	14
(d) $q = 1.0 \text{ mol} \times 50 \text{ kJ/mol}$	50

(d) requires the most heat. It is the largest ΔH value. Only (b) is close, but it is far less than (d).

43.

(a) and (b)

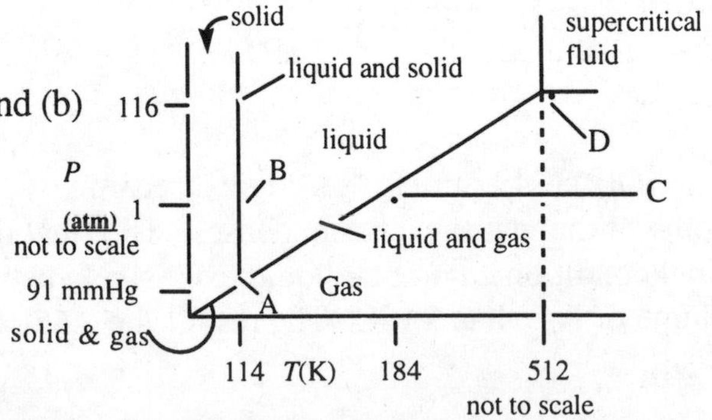

(Note: The computer program used to draw this graph is limited in its ability to draw curved lines accurately. Thus the curved line for the vapor-pressure curve looks straight in this drawing. Also, know that the lines for the vapor-pressure

curve and the sublimation curve are not the same or the extension of each other; they are separate lines. In the vicinity of the triple point, the sublimation curve has a greater slope than the vapor-pressure curve.)

(c) The solid rises in temperature until it reaches the melting point (which is slightly above normal melting point). The solid changes to liquid. The liquid temperature rises until the boiling point, which is above the normal boiling point. The liquid vaporizes and then increases in temperature.

45.

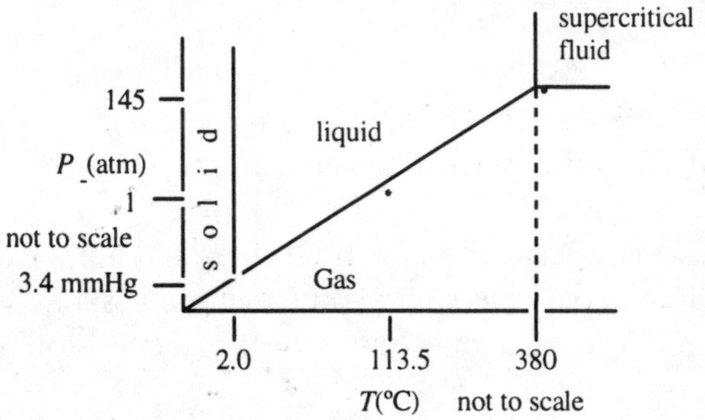

(See note in Problem 43.)

47. (a) liquid (The vapor pressure of water at 88.15 °C is about 500 mmHg.)
(b) liquid and vapor (0.0313 atm is V.P. at 25 °C)
(c) liquid (This pressure is above the normal melting point pressure, and the fusion curve slopes left.)
(d) solid (A pressure of 0.100 atm is well above the sublimation pressure of ice at -10 °C.)

49. Both substances are nonpolar, but because of its lower molecular mass, CS_2 should have weaker dispersion forces than CCl_4. As a result, it should have the lower boiling point.

51. Diethyl ether is polar, so it has dipole-dipole forces; pentane is nonpolar. Both have about the same molecular mass. Diethyl ether has the higher melting point.

53. C_6H_5COOH (122 u) has a high molecular mass, dipole-dipole forces, and hydrogen bonds; it should be a solid. $CH_3(CH_2)_8CH_3$ (142 u) has only dispersion forces as intermolecular forces and is likely to be a liquid. C_6H_{14} (86 u) has only dispersion forces, but with a lower molar mass than $CH_3(CH_2)_8CH_3$, its forces are weaker producing a lower boiling point than $CH_3(CH_2)_8CH_3$. C_6H_{14} is also likely to be a liquid. $(CH_3CH_2)_2O$ (74 u) has dipole-dipole and dispersion forces; it is also likely to be a liquid.

55.

molecular mass, u	28	30	46	86	74
	N_2	< NO <	$(CH_3)_2O$ <	C_6H_{14} <	C_4H_9OH
boiling point	−195	−192	polar −25	69	H bond 118

The higher molecular mass of C_6H_{14} is more important than the polarity of $(CH_3)_2O$. C_4H_9OH clearly has the highest boiling point, because of its molecular mass and hydrogen bonding.

57. Cl_2 < CCl_4 < CsCl < $MgCl_2$
-103 -23 ionic 646 ionic 708
Cl_2 and CCl_4 are both nonpolar, with Cl_2 having the lower molecular mass. $MgCl_2$ has a smaller, more highly charged cation than does CsCl.

59. 1-octanol has H bonds, and octane has only dispersion forces, so 1-octanol should have the stronger intermolecular forces and the higher surface tension.

61. Water "wets" glass, because adhesive forces between water molecules and glass exceed cohesive forces in H_2O (l). The situation is the reverse for some materials, such as Teflon, and water does not "wet" them. Wetting agents lower the surface tension of water and improve its wetting ability.

63. (a) The larger, upper unit is a unit cell, but not the smaller, lower one. The larger cell can be shifted left, right, up, and down with no gaps. That is, the bottom edge of one cell coincides with the top edge of the identical cell below it. In contrast, the bottom edge of the small cell does not coincide with the top edge of the identical cell below it; there is a gap between them.

(b) $(4 \times \frac{1}{4}) + 2 = 3$ hearts $(4 \times \frac{1}{2}) + 1 = 3$ diamonds $(4 \times \frac{1}{2}) + 1 = 3$ clubs

65.

(a)

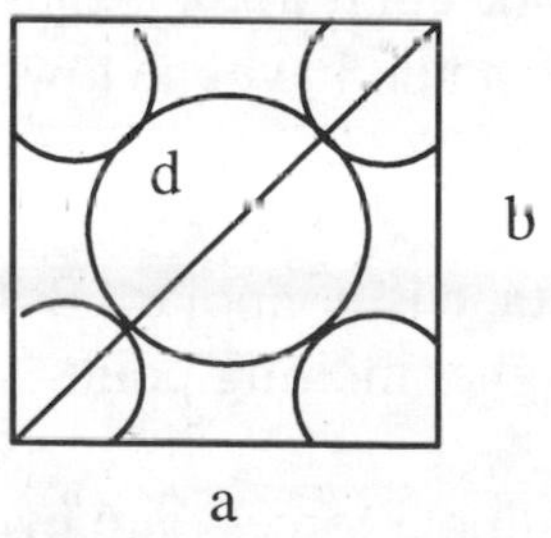

Three atoms are in contact along the diagonal. The length of the diagonal = $4r$

Pythagorean formula $a^2 + b^2 = d^2$
For unit cell $a = b = l$ length of cell

$$l^2 + l^2 = 2l^2 = d^2 = (4r)^2$$
$$2l^2 = 16r^2$$
$$l = \frac{4r}{\sqrt{2}} = \frac{4 \text{ x } 144.4 \text{ pm}}{\sqrt{2}} = 408.4 \text{ pm}$$

(b) $V = l^3 = (408.4 \text{ pm})^3 \times \frac{(10^{-10}\text{cm})^3}{\text{pm}^3} = 6.812 \times 10^{-23} \text{ cm}^3$

(c) $d = \dfrac{m}{V} = \dfrac{\dfrac{107.87\text{ g}}{\text{mol}} \times \dfrac{\text{mol}}{6.022 \times 10^{23}\text{ atom}} \times \dfrac{4\text{ atom}}{\text{unit cell}}}{6.812 \times 10^{-23}\text{ cm}^3} = 10.52\text{ g/cm}^3$

$6 \times \frac{1}{2}$ atoms on face + $8 \times \frac{1}{8}$ atoms on corner = 4 atoms/unit cell

67. (a) For a Cs^+ in the center of Cl^- ions, the distance from one corner to the opposite corner is 2×169 pm + 2×181 pm = 700 pm. That line is the hypotenuse for a right triangle that has one side the length of the cell (down the corner) and one side as the diagonal of the bottom of the cell ($\sqrt{2}\ \ell$). Since $C^2 = a^2 + b^2$, the distance from one corner to the other is

$\sqrt{(\sqrt{2}\ell)^2 + \ell^2} = 700\text{ pm}$

$\ell = 404\text{ pm}$

(b) $V = \ell^3 = (404\text{ pm})^3 \times \left(\dfrac{10^{-10}\text{ cm}}{\text{pm}}\right)^3 = 6.59 \times 10^{-23}\text{ cm}^3$

for each cell there are one Cs and $8 \times \frac{1}{8}$ Cl^-, so $\dfrac{\text{molar mass}}{\text{cell}} = \dfrac{168.4\text{ g}}{\text{mole}}$

mass of unit cell = $6.59 \times 10^{-23}\text{cm}^3 \times 3.9889\text{ g/cm}^3 = 2.63 \times 10^{-22}\text{g}$

$N_A = \dfrac{168.4\text{g CsCl/mol}}{2.63 \times 10^{-22}\text{g CsCl/formula unit}} = 6.4 \times 10^{23}\ \dfrac{\text{formula unit}}{\text{mol}}$

69. Ca^{2+} $\quad 6 \times \frac{1}{2} + 8 \times \frac{1}{8} = 4$

F 8

Ca_4F_8 => CaF_2

Coordination number for F^- is 4.

Coordination number for Ca^{2+} is 8.

No. Since there are twice as many F^- as Ca^{2+}, it would be expected that Ca^{2+} would have twice the coordination number.

Additional Problems

71. vapor pressure = 23.8 mmHg

$n = \dfrac{PV}{RT} = \dfrac{23.8\text{ mmHg} \times 1.00\text{ L}}{\dfrac{62.4\text{ mmHg L}}{\text{K mol}} \times 298\text{ K}}$

estimate $n \approx \dfrac{20}{60 \times 300} = \dfrac{1}{900} = 1.1 \times 10^{-3}$

so choose 1.2×10^{-3}

actual 1.28×10^{-3} moles

73. $\dfrac{P_1V_1}{T_1} = \dfrac{P_2V_2}{T_2}$, $P_2 = \dfrac{P_1V_1}{V_2} = \dfrac{750.0\text{ mmHg x }150.0\text{mL}}{172\text{ mL}}$

For N_2 gas $P_2 = 654$ mmHg

$$P_{\text{total}} = P_{N_2} + P_{C_6H_6}$$

$$P_{C_6H_6} = 750.0 \text{ mmHg} - 654 \text{ mmHg} = 96 \text{ mmHg}$$

75. Hexanoic acid has a higher molecular mass, but more important reasons are the dipole-dipole forces and hydrogen bonds that are present in hexanoic acid and absent in 2-methylbutane.

77. (a) CH_3CH_2OH Hydrogen bonding is most important. There are also dipole-dipole and dispersion forces.
(b) CH_3CH_2Cl Dipole-dipole is the most important force. There are also dispersion forces.
(c) HCOONa Ionic bonding is the most important force.

79. The density of liquid water rises from its value at the melting point to a maximum at 3.98 °C, because at this temperature hydrogen bonding between water molecules occurs to the greatest extent. Above this temperature its density falls with temperature in a customary fashion. Thus, for every density in a range from 0 °C to 3.98 °C, there is another temperature in a range extending a few degrees above 3.98 °C at which the same density is observed.

81. (a) $\log p = 7.5547 - \dfrac{1002.7}{-75.0 + 247.89} = 1.755$

$10^{\log p} = p = 10^{1.755} = 56.9 \text{ mmHg}$

(b) The normal boiling point is the boiling point at atm pressure.

$$\log(760) = 7.5547 - \frac{1002.7}{t + 247.89}$$

$$2.8808 = 7.5547 - \frac{1002.7}{t + 247.89}$$

$$4.6739 = \frac{1002.7}{t + 247.89}$$

$$t + 247.89 = \frac{1002.7}{4.6739} = 214.53$$

$$t = 214.53 - 247.89 = -33.36 \text{ °C}$$

(c) The critical point is the last point on the vapor-pressure curve.

405.6 K - 273.2 = 132.4 °C

$$\log p = 7.5547 - \frac{1002.7}{132.4 + 247.89} = 4.918$$

$$10^{\log p} = p = P_c = 10^{4.918} = 8.279 \times 10^4 \text{ mmHg} = 108.9 \text{ atm}$$

83. $\Delta H_{fus} < \Delta H_{vap} = \Delta H_{cond} < \Delta H_{sub}$

This must be the case because $\Delta H_{vap} > \Delta H_{fus}$; $\Delta H_{cond} = -\Delta H_{vap}$; and $\Delta H_{sub} = \Delta H_{fus} + \Delta H_{vap}$.

Often enthalpy values of chemical reactions are greater than these values because chemical reactions break chemical bonds instead of simply overcoming intermolecular forces. Reactions that have about the same energy for bond breaking as bond forming have a small ΔH and would be exceptions.

85. The energy to change ice to water at 100 °C -

$$\Delta H_{\text{total}} = n\Delta H_{\text{fus}} + n \times \text{sp.ht.} \times \Delta T$$

$$\Delta H_{\text{total}} = 5.00\ \text{mol} \times \frac{6.01\ \text{kJ}}{\text{mol}} + 5.00\ \text{mol} \times \frac{18.02\ \text{g}}{\text{mol}} \times \frac{4.18\ \text{J}}{\text{g}\ {}^\circ\text{C}} \times \frac{\text{kJ}}{10^3\ \text{J}} \times (100.0\ {}^\circ\text{C} - 0.0\ {}^\circ\text{C}) = 67.7\ \text{kJ}$$

How much steam must condense to liberate the 67.7 kJ required to melt the ice and raise the temperature of the liquid water to 100 °C?

$$\text{no. moles of steam} = -67.7\ \text{kJ} \times \frac{1\ \text{mol}}{-40.6\ \text{kJ}} = 1.67\ \text{mol}$$

$$\text{total mass of liquid water} = 5.00\ \text{mol} + 1.67\ \text{mol} = 6.67\ \text{mol} \times \frac{18.02\ \text{g}}{\text{mol}} = 120\ \text{g liquid}$$

$$\text{total mass of steam} = (50.0\ \text{moles} - 1.7\ \text{moles}) \times \frac{18.02\ \text{g}}{\text{mole}} = 870\ \text{g steam}$$

87.

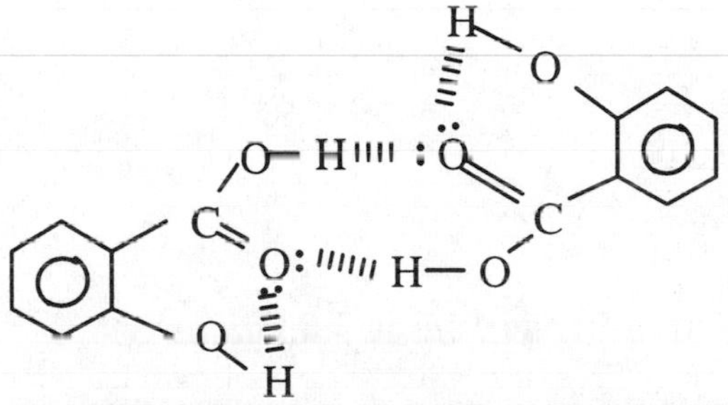

89. Sodium chloride is FCC for the Cl^-. In Figure 11.33, replace Na^+ by Mg^{2+} and Cl^- by O^{2-}.

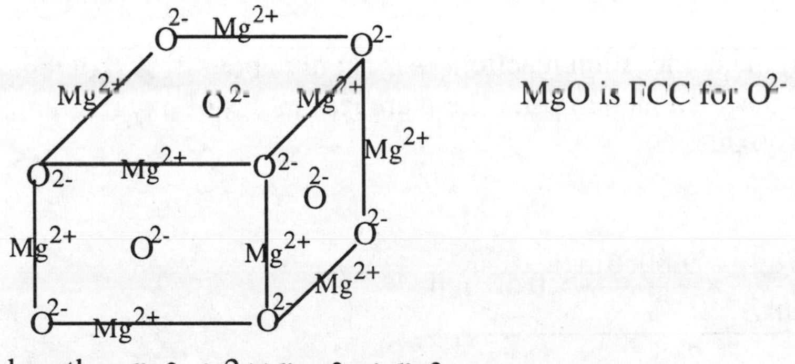

$$\text{length} = r_{O^{2-}} + 2 \times r_{Mg^{2+}} + r_{O^{2-}}$$

$$\text{length} = 140\ \text{pm} + 2 \times 65\ \text{pm} + 140\ \text{pm} = 410\ \text{pm}$$

MgO will be like NaCl. The ratio of radii of Mg^{2+} to O^{2-} is 0.46, more similar to the ratio of radii of Na^+ to Cl^- (0.55) than to the ratio of radii of Cs^+ to Cl^- (0.93). The 0.46 ratio falls in the range for filling octahedral holes. MgO and NaCl are fcc.

	O^{2-}	Mg^{2+}	Cl^-	Na^+	Cs^+
radius (pm)	140	65	181	99	169

91. diagonal of unit cell face

$4x = b$

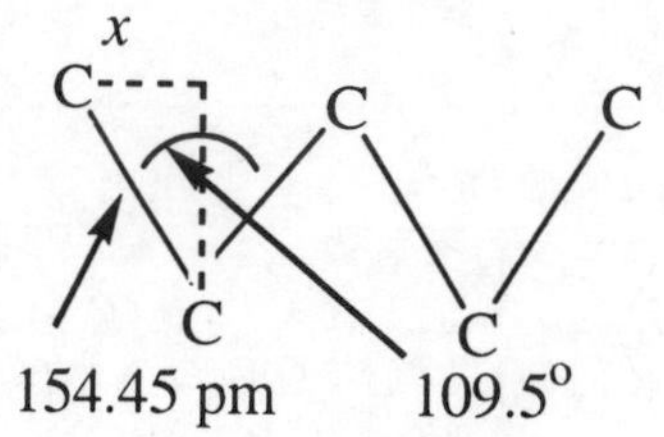

$$\sin\left(\frac{109.5}{2}\right) = \frac{x}{154.45}$$

$x = 126.1$ pm

$4x = 504.4$ pm $= b$

$2l^2 = b^2$

$l = b/\sqrt{2}$

$l = 356.7$ pm

$V = l^3 = (356.7 \times 10^{10}\text{cm})^3 = 4.538 \times 10^{-23}\ \text{cm}^3$

$$\frac{\#\ \text{atoms}}{\text{cell}} = 8\ \text{corners} \times \frac{1}{8} + 6\ \text{face} \times \frac{1}{2} + 4 = 8\ \text{atoms/cell}$$

$$d = \frac{m}{V} = \frac{8\ \text{atoms} \times \dfrac{12.01\,\text{g}}{\text{mole}} \times \dfrac{\text{mole}}{6.022 \times 10^{23}\ \text{atoms}}}{4.538 \times 10^{-23}\ \text{cm}^3}$$

$d = 3.516\ \text{g/cm}^3$

The tabulated value in CRC is 3.51 g/cm^3 and in Lange 14/e is 3.53 g/cm^3.

93. Gasoline should have a higher proportion of the more volatile isooctane for cold weather (November in Minnesota) and a lower proportion for warm weather (May in North Carolina).

95. It is oxygen that reacts in the combustion reaction. Air is only about 21% oxygen. Thus the pressure of oxygen is about 525 psi. There are less moles of oxygen than there are moles of nitrous oxide.

Apply Your Knowledge

97. $$\gamma = \frac{hdgr}{2} = \frac{1.10\,\text{cm} \times \dfrac{0.789\,\text{g}}{\text{mL}} \times \dfrac{9.8066\,\text{m}}{\text{s}^2} \times 0.50\,\text{mm}}{2}$$

$$\gamma = \frac{1.10\,\text{cm}}{2} \times \frac{10^{-2}\text{ m}}{\text{cm}} \times \frac{0.789\text{ g}}{\text{mL}} \times \frac{\text{kg}}{10^3\text{ g}} \times \frac{\text{mL}}{\text{cm}^3} \times \frac{\text{cm}^3}{(10^{-2})^3\text{ m}} \times \frac{9.8066\text{ m}}{\text{s}^2}$$

$$\times 0.50\text{ mm} \times \frac{10^{-3}\text{ m}}{\text{mm}} = \frac{0.021\text{ J}}{\text{m}^2}$$

99. (a) $? \text{ m} = 900\text{ ft} \times \frac{12\text{ in}}{\text{ft}} \times \frac{2.54\text{ cm}}{\text{in}} \times \frac{10^{-2}\text{ m}}{\text{cm}} = 274\text{ m}$

$1917\text{ m} \times \frac{1}{274\text{ m}} = 7.00$

$1.00\text{ atm} - \frac{1}{30} \times 1.00\text{ atm} = 0.967\text{ atm}$

$0.967\text{ atm} - \frac{1}{30} \times 9.67\text{ atm} = 0.934\text{ atm}$

$0.934\text{ atm} - \frac{1}{30} \times 0.934\text{ atm} = 0.903\text{ atm}$

$0.903\text{ atm} - \frac{1}{30} \times 0.903\text{ atm} = 0.873\text{ atm}$

$0.873\text{ atm} - \frac{1}{30} \times 0.873\text{ atm} = 0.844\text{ atm}$

$0.844\text{ atm} - \frac{1}{30} \times 0.844\text{ atm} = 0.816\text{ atm}$

$0.816\text{ atm} - \frac{1}{30} \times 0.816\text{ atm} = 0.789\text{ atm}$

$? \text{ mmHg} = 0.789\text{ atm} \times \frac{760\text{ mmHg}}{\text{atm}} = 599\text{ mmHg}$

(b) from table — 588.6 mmHg is 93.0 °C
610.9 mmHg is 94.0 °C

$599 - 589 = 10$
$611 - 599 = 12$
The temperature is closer to 93 °C than to 94 °C.

101. At the triple point, the pressure should be equal.

$$6.9379 - \frac{861.34}{t + 246.33} = 9.7051 - \frac{1444.2}{t + 267.13}$$

$$-\frac{861.34}{t + 246.33} = 2.7672 - \frac{1444.2}{t + 267.13}$$

$$-861.34 = 2.7672\,t + 681.644 - \frac{(1444.2\,t + 355750)}{t + 267.13}$$

$$-1542.98 = 2.7672\,t - \frac{(1444.2\,t + 355750)}{t + 267.13}$$

$$-1542.98\,t - 412176 = 2.7672\,t^2 + 739.20\,t - 1444.2\,t - 355750$$

$0 = 2.7672\, t^2 + 838.0\, t + 56442.6$
$0 = t^2 + 303\, t + 20391$

$$t = \frac{-303 \pm \sqrt{(303)^2 - 4(1)\,(2039)}}{2}$$

$$t = \frac{-303 \pm 101}{2}$$

$t = -101$ °C or -202 °C
(-202 °C produces a pressure of 3.22×10^{-13} mmHg, an unrealistic value.)
$T_c = -101$ °C

$$\log p = 6.9379 - \frac{861.34}{t + 246.33}$$

$$\log p = 6.9379 - \frac{861.34}{-101 + 246.33} =$$

$10^{\log p} = p = 10^{1.01} = 10.3 \text{ mmHg} = P_c$

103. Hg(l) 20.0 °C $\rightarrow$ −39 °C
$\Delta H_1 = mC\Delta T$

$$\Delta H_1 = 525 \text{ cm}^3 \text{ Hg} \times \frac{13.6\text{ g}}{\text{cm}^3} \times \frac{0.033 \text{ cal}}{\text{g °C}} \times (-39 - 20)\text{ °C} \times \frac{\text{kcal}}{10^3 \text{ cal}} \times \frac{4.184 \text{ kJ}}{\text{kcal}}$$
$$= -58.2 \text{ kJ}$$

Hg(l) $\rightarrow$ Hg(s) at −39 °C

$$\Delta H_2 = n\Delta H_f = 525 \text{ cm}^3 \times \frac{13.69}{\text{cm}^3} \times \frac{\text{mol}}{200.0\text{ g}} \times \frac{-2.30 \text{ kJ}}{\text{mol}} = -81.9 \text{ kJ}$$

Hg(s) −39 °C $\rightarrow$ −196 °C
$\Delta H_3 = mC\Delta T$

$$\Delta H_3 = 525 \text{ cm}^3 \text{ Hg} \times \frac{13.69}{\text{cm}^3} \times \frac{0.030 \text{ cal}}{\text{g °C}} \times (-196 - (-39))\text{ °C} \times \frac{\text{kcal}}{10^3 \text{ cal}} \times \frac{4.184 \text{ kJ}}{\text{kcal}}$$
$$= -140.7 \text{ kJ}$$

$\Delta H_{total} = \Delta H_1 + \Delta H_2 + \Delta H_3$
$\Delta H_{total} = -58.2 \text{ kJ} - 81.9 \text{ kJ} - 140.7 \text{ kJ} = -280.8 \text{ kJ}$

$$\text{mass N}_2 = 280.8 \text{ kJ} \times \frac{\text{mol}}{5.58 \text{ kJ}} \times \frac{28.01\text{ g}}{\text{mol}} = 1.41 \times 10^3 \text{ g}$$

Chapter 12

Physical Properties of Solutions

Exercises

12.1A $\dfrac{163 \text{ g glucose}}{(163 \text{ g} + 755 \text{ g}) \text{ solution}} \times 100\% = 17.8\%$ glucose by mass

12.1B ? g sucrose = 225 g solution $\times \dfrac{6.25 \text{ g sucrose}}{100 \text{ g solution}} = 14.1$ g sucrose

? g sucrose = 135 g solution $\times \dfrac{8.20 \text{ g sucrose}}{100 \text{ g solution}} = 11.1$ g sucrose

? g sucrose = 14.1 g sucrose + 11.1 g sucrose = 25.2 g sucrose

? g solution = 225 g + 135 g = 360 g solution

mass % $= \dfrac{25.2 \text{ g sucrose}}{360 \text{ g solution}} \times 100\% = 7.00\%$

12.2A volume % toluene $= \dfrac{40.0 \text{ mL toluene x } 100\%}{(40.0 \text{ mL} + 75.0 \text{ mL}) \text{ solution}} = 34.8$ % toluene

12.2B (a) ? g toluene $= \dfrac{0.866 \text{ g}}{\text{mL}} \times 40.0$ mL toluene = 34.6 g toluene

? g benzene $= \dfrac{0.879 \text{ g}}{\text{mL}} \times 75.0$ mL benzene = 65.9 g benzene

mass % toluene $= \dfrac{34.6 \text{ toluene x } 100\%}{(34.6 \text{ g} + 65.9 \text{ g}) \text{ solution}} = 34.4$ % toluene

(b) $d = \dfrac{34.6 \text{ g} + 65.9 \text{ g}}{40.0 \text{ mL} + 75.0 \text{ mL}} = 0.874$ g/mL

12.3A $\dfrac{0.1\ \mu\text{g}}{\text{L}} = \dfrac{0.1\ \mu\text{g}}{1000 \text{ g}} = \dfrac{0.1\ \mu\text{g}}{1000 \text{ g x } \dfrac{\mu\text{g}}{10^{-6} \text{ g}}} = 0.1$ ppb

$\dfrac{0.1\ \mu\text{g}}{\text{L}} = \dfrac{0.1\ \mu\text{g}}{1000 \text{ g}} = \dfrac{0.1\ \mu\text{g x } 10^3}{1000 \text{ g x } \dfrac{\mu\text{g}}{10^{-6} \text{ g}} \text{ x } 10^3} = \dfrac{100\ \mu\text{g}}{10^{12}\ \mu\text{g}} = 100$ ppt

12.3B ppm $Na^+ = \dfrac{1.52 \text{ x } 10^{-3} \text{ mol } Na_2SO_4}{\text{L}} \times \dfrac{10^{-3} \text{ L}}{\text{mL}} \times \dfrac{\text{mL}}{1.0 \text{ g}} \times \dfrac{2 \text{ mol } Na^+}{\text{mol } Na_2SO_4}$

$\times \dfrac{22.99 \text{ g } Na^+}{\text{mol } Na^+} \times \dfrac{10^6}{10^6} = \dfrac{69.9 \text{ g } Na^+}{10^6 \text{ g}} = 69.9$ ppm Na^+

12.4 $? \text{ m} = \frac{225 \text{ mg } C_6H_{12}O_6}{5.00 \text{ mL } C_2H_5OH} \times \frac{\text{mL}}{0.789 \text{ g}} \times \frac{10^3 \text{ g}}{\text{kg}} \times \frac{10^{-3} \text{ g}}{\text{mg}} \times \frac{\text{mol } C_6H_{12}O_6}{180.2 \text{ g } C_6H_{12}O_6}$

$= 0.317 \text{ m } C_6H_{12}O_6$

12.5A $? \text{ mL } C_2H_5OH = 125 \text{ ml } C_6H_6 \times \frac{0.879 \text{ g}}{\text{mL}} \times \frac{\text{kg}}{10^3 \text{ g}} \times \frac{0.0652 \text{ mol } C_2H_5OH}{\text{kg } C_6H_6}$

$\times \frac{46.07 \text{ g } C_2H_5OH}{\text{mol } C_2H_5OH} \times \frac{\text{mL}}{0.789 \text{ g}} = 0.418 \text{ mL}$

12.5B $? \text{ mL } H_2O = 25.0 \text{ g } CO(NH_2)_2 \times \frac{1 \text{ mol } CO(NH_2)_2}{60.06 \text{ g } CO(NH_2)_2} \times \frac{\text{kg } H_2O}{1.65 \text{ mol } CO(NH_2)_2}$

$\times \frac{1000 \text{ g } H_2O}{\text{kg } H_2O} \times \frac{\text{mL } H_2O}{0.998 \text{ g } H_2O} = 253 \text{ mL } H_2O$

12.6A (a) $m \text{ } CH_3OH = \frac{7.50 \text{ g } CH_3OH}{92.5 \text{ g } C_2H_5OH} \times \frac{10^3 \text{ g } C_2H_5OH}{\text{kg } C_2H_5OH} \times \frac{\text{mol } CH_3OH}{32.04 \text{ g } CH_3OH} = 2.53 \, m$

(b) $\text{mole \% } CO(NH_2)_2 = \frac{1.05 \text{ mol } CO(NH_2)_2 \times 100\%}{1.05 \text{ mol } CO(NH_2)_2 + \left(1.00 \text{ kg} \times \frac{10^{-3} \text{ g}}{\text{kg}} \times \frac{\text{mol } H_2O}{18.02 \text{ g}}\right)}$

$= 1.89 \text{ mol \% } CO(NH_2)_2$

12.6B (a) $2.90 \text{ mol} \times 32.04 \text{ g/mol} = 92.9 \text{ g } CH_3OH$

$\frac{92.9 \text{ g}}{(\text{kg x } \frac{1000 \text{ g}}{\text{kg}}) + 93 \text{ g}} \times 100\% = 8.50\% \text{ } CH_3OH \text{ by mass}$

(b) $\frac{2.90 \text{ mol}}{1093 \text{ g soln}} \times \frac{0.984 \text{ g}}{\text{mL}} \times \frac{\text{mL}}{10^{-3} \text{ L}} = 2.61 \text{ M}$

(c) $1000 \text{ g } H_2O \times \frac{\text{mol } H_2O}{18.02 \text{ g } H_2O} = 55.5 \text{ mol}$

$\text{mole \%} = \frac{2.90 \text{ mol}}{55.5 \text{ mol} + 2.9 \text{ mol}} \times 100\% = 4.97\%$

12.7 (b) 5.0% C_2H_5OH by mass is the largest. The solutions are all rather dilute and have densities of approximately 1.0 g/mL. The mole percents of C_2H_5OH are not large, and are expected to be the largest in the solution having the greatest quantity of C_2H_5OH per liter or kilogram of solution. Solution (a) has 0.5 mol C_2H_5OH per liter; (b) has slightly more than 1 mol C_2H_5OH per kilogram of solution; (c) has slightly less than 0.5 mol C_2H_5OH per kilogram of solution; (d) has somewhat less than 1 mol C_2H_5OH per liter.
Another estimation method follows:
(a) is greater than (c) because (a) is 0.50 mole C_2H_5OH in 1 L solution (about 994 g solution - 23 g C_2H_5OH = 971 g H_2O), but (c) is in 1 kg of H_2O. (d) is less than (b) because the 5% mass contains a bigger percent C_2H_5OH than 5% volume, since

the density of C_2H_5OH is lower than water. (b) is greater than (a) because 5 g of C_2H_5OH is about 0.1 mol. 0.1 mol is a larger percent of 5 mol than 0.5 mol is of 50 mol.

Actual calculations -

(a) $? \text{ g solution} = 1 \text{ L solution} \times \frac{\text{mL}}{10^{-3}\text{ L}} \times \frac{0.994 \text{ g solution}}{\text{mL solution}} = 994 \text{ g solution}$

$? \text{ mole } H_2O = (994 \text{ g solution} - 23 \text{ g } C_2H_5OH) \times \frac{\text{mol}}{18.01 \text{ g}} = 53.9 \text{ mol } H_2O$

$? \text{ mole \% } C_2H_5OH = \frac{0.5 \text{ mol}}{0.5 \text{ mol} + 53.9 \text{ mol}} \times 100\% = 0.9\%$

(b) $? \text{ mol } H_2O = 95 \text{ g } H_2O \times \frac{\text{mol}}{18.01 \text{ g}} = 5.27 \text{ mol } H_2O$

$? \text{ mol } C_2H_5OH = 5 \text{ g } C_2H_5OH \times \frac{\text{mol}}{46.0} = 0.11 \text{ mol } C_2H_5OH$

$? \text{ mol \% } C_2H_5OH = \frac{0.11 \text{ mol } C_2H_5OH}{5.38 \text{ mol solution}} \times 100\% = 2.0\%$

(c) $? \text{ mol } H_2O = 1 \text{ kg } H_2O \times \frac{\text{mol}}{18.01 \text{ g}} \times \frac{10^3 \text{ g}}{\text{kg}} = 55.5 \text{ mol } H_2O$

$? \text{ mol \% } C_2H_5OH = \frac{0.5 \text{ mol } C_2H_5OH}{56.0 \text{ mol solution}} \times 100\% = 0.9\%$

(d) $? \text{ g solution} = 1 \text{ L solution} \times \frac{0.991 \text{ g}}{\text{mL}} \times \frac{\text{mL}}{10^{-3}\text{ L}} = 991 \text{ g solution}$

$? \text{ g } C_2H_5OH = \frac{5 \text{ mL}}{100 \text{ mL}} \times 1 \text{ L} \times \frac{\text{mL}}{10^{-3}\text{ L}} \times \frac{0.789 \text{ g}}{\text{mL}} = 39 \text{ g } C_2H_5OH$

$? \text{ g } H_2O = 991 \text{ g} - 39 \text{ g} = 952 \text{ g } H_2O$

$? \text{ mol } H_2O = 952 \text{ g } H_2O \times \frac{\text{mol}}{18.01 \text{ g}} = 52.9 \text{ mol } H_2O$

$? \text{ mol } C_2H_5OH = 39 \text{ g } C_2H_5OH \times \frac{\text{mol}}{46.01 \text{ g}} = 0.85 \text{ mol}$

$? \text{ mol \% } C_2H_5OH = \frac{0.85}{52.9 + 0.9} \times 100\% = 1.7\%$

12.8A Because of the large benzene part of the molecule, nitrobenzene should be more soluble in benzene.

12.8B (a) acetic acid
(b) hexanol
(c) hexane
(d) butanoic acid — least soluble c < b < d < a

12.9A $\frac{\frac{149 \text{ mg } CO_2}{100\text{g } H_2O}}{1 \text{ atm } CO_2} \times 0.00037 \text{ atm } CO_2 = \frac{5.5 \times 10^{-2} \text{ mg}CO_2}{100 \text{ g}H_2O}$

12.9B $\chi_{N_2} = \frac{P_{N_2}}{P_{mixture}} = \frac{P_{N_2}}{10\ atm} = 0.50$

$P_{N_2} = 5.0\ atm$

$\chi_{CH_4} = \frac{P_{CH_4}}{P_{mixture}} = \frac{P_{CH_4}}{10\ atm} = 0.50$

$P_{CH_4} = 5.0\ atm$

at 10 atm N_2 $S = \frac{19\ mg\ N_2}{100\ g\ H_2O}$

CH_4 $S = \frac{23\ mg\ CH_4}{100\ g\ H_2O}$

$k_{N_2} = \frac{S}{P_{N_2}} = \frac{\frac{19\ mg\ N_2}{100\ g\ H_2O}}{10\ atm} = \frac{0.019\ mg\ N_2}{g\ H_2O\ atm}$

$k_{CH_4} = \frac{S}{P_{CH_4}} = \frac{\frac{23\ mg\ CH_4}{100g\ H_2O}}{10\ atm} = \frac{0.023\ mg\ CH_4}{g\ H_2O\ atm}$

? mass $N_2 = \frac{0.019\ mg\ N_2}{g\ atm} \times 5\ atm \times 1.00\ L \times \frac{mL}{10^{-3}\ L} \times \frac{1.00\ g}{mL} = 95\ mg\ N_2$

? mass $CH_4 = \frac{0.023\ mg\ CH_4}{g\ atm} \times 5\ atm \times 1.00\ L \times \frac{mL}{10^{-3}\ L} \times \frac{1.00\ g}{mL} = 115\ mg\ CH_4$

total mass = 95 mg N_2 + 115 mg CH_4 = 2.1×10^2 mg mixture

12.10? mol $C_6H_5COOH = 5.05\ g\ C_6H_5COOH \times \frac{mol\ C_6H_5COOH}{122.1\ g\ C_6H_5COOH}$

$= 0.0414\ mol\ C_6H_5COOH$

? mol $C_6H_6 = 245\ g\ C_6H_6 \times \frac{mol\ C_6H_6}{78.11\ g\ C_6H_6} = 3.14\ mol\ C_6H_6$

$\chi_{C_6H_6} = \frac{3.14\ mol\ C_6H_6}{3.18\ mol\ total} = 0.987$

$P_{C_6H_6} = \chi,C_6H_6)\ P^{\circ}_{C_6H_6} = 0.987 \times 95.1\ mmHg = 93.9\ mmHg$

12.11Assume 100 g

? mol benzene = $50\ g\ C_6H_6 \times \frac{mol\ C_6H_6}{78.11\ g\ C_6H_6} = 0.640\ mol\ C_6H_6$

? mol toluene = $50\ g\ C_7H_8 \times \frac{mol\ C_7H_8}{92.14\ g\ C_7H_8} = 0.543\ mol\ C_7H_8$

$$P_{benzene} = \chi_{benzene} \times P^{\circ}_{benzene} = \frac{0.640 \text{ mol}}{(0.640 + 0.543) \text{ mol}} \times 95.1 \text{ mmHg}$$
$$= 51.4 \text{ mmHg}$$

$$P_{toluene} = \chi_{toluene} \times P^{\circ}_{toluene} = \frac{0.543 \text{ mol}}{(0.640 + 0.543)\text{mol}} \times 28.4 \text{ mmHg}$$
$$= 13.0 \text{ mmHg}$$

$$P_{total} = P_{benzene} + P_{toluene} = 51.4 \text{ mmHg} + 13.0 \text{ mmHg} = 64.4 \text{ mmHg}$$

12.12A The greater mole fraction of benzene in the vapor occurs above the solution with the larger mole fraction of benzene. Since toluene has a larger molecular mass, equal masses of toluene and benzene means a greater mole fraction of benzene and a greater mole fraction of benzene in the vapor.

12.12B $P_{benzene} = \chi_{benzene} \times P^{\circ}_{benzene} = 0.770 \times 95.1 \text{ mmHg} = 73.2 \text{ mmHg}$
$P_{toluene} = \chi_{toluene} \times P^{\circ}_{toluene} = 0.230 \times 28.4 \text{ mmHg} = 6.53 \text{ mmHg}$

12.13 No. The process will continue until the mole fraction of water in B increases to be equal to the mole fraction in A. That is, water vapor passes from the more dilute B to the more concentrated solution A until the concentrations are equal. The evaporation and condensation will continue, but at equal rates in both dishes. There is no net transfer, and the levels will remain constant.

12.14A $$m = \frac{10.0 \text{ g } C_{10}H_8 \text{ x } \frac{\text{mol } C_{10}H_8}{128.2 \text{ g } C_{10}H_8}}{50.0 \text{ g } C_6H_6 \text{ x } \frac{\text{kg}}{10^3 \text{ g}}} = 1.56\ m$$

$$\Delta T = -K_f \times m = \frac{-5.12\,°\text{C}}{\text{m}} \times 1.56\ m = -7.99\ °\text{C}$$

$\Delta T = T_f - T_i$
$-7.99\ °\text{C} = T_f - 5.53\ °\text{C}$
$T_f = -2.46\ °\text{C}$

12.14B $\Delta T = 100.35\ °\text{C} - 100.00\ °\text{C} = 0.35\ °\text{C}$
$\Delta T = K_b m$

$$m = \frac{\Delta T}{K_b} = \frac{0.35\,°\text{C}}{0.512\,°\text{C/m}} = 0.684\ m$$

$$? \text{ g sucrose} = \frac{0.684 \text{ mol sucrose}}{\text{kg } H_2O} \times \frac{\text{kg}}{1000\text{g}} \times 75.0 \text{ g } H_2O \times \frac{342.3 \text{ g}}{\text{mol}} = 18 \text{ g sucrose}$$

12.15 $\Delta T = 4.25\ °\text{C} - 5.53\ °\text{C} = -1.28\ °\text{C}$

$$m = -\frac{\Delta T}{K_f} = \frac{-1.28\,°\text{C}}{-5.12\,°\text{C/m}} = 0.25\ m$$

$$? \text{ mol} = \frac{0.25 \text{ mol cpd}}{\text{kg benz}} \times \frac{\text{kg}}{10^3 \text{ g}} \times 30.00 \text{ g benz} = 0.0075 \text{ mol}$$

$$? \text{ g/mol} = \frac{1.065 \text{ g}}{0.0075 \text{ mol}} = 142 \text{ g/mol}$$

$$? \text{ mol C} = 50.69 \text{ g C} \times \frac{\text{mol}}{12.01 \text{ g}} = 4.221 \text{ mol C} \times \frac{1}{2.818 \text{ mol}} = 1.50 \frac{\text{mol C}}{\text{mol O}}$$

$$? \text{ mol H} = 4.23 \text{ g H} + \frac{\text{mol}}{1.008 \text{ g}} = 4.196 \text{ mol H} \times \frac{1}{2.818 \text{ mol}} = 1.49 \frac{\text{mol H}}{\text{mol O}}$$

$$? \text{ mol O} = 45.08 \text{ g O} \times \frac{\text{mol}}{16.00 \text{ g}} = 2.818 \text{ mol O} \times \frac{1}{2.818 \text{ mol}} = 1.00 \frac{\text{mol O}}{\text{mol O}}$$

empirical formula: $C_3H_3O_2$ 71 u/formula

$$\frac{142 \text{ u}}{\text{molecule}} \times \frac{\text{formula units}}{71 \text{ u}} = \frac{2 \text{ formula units}}{\text{molecule}}$$

molecular formula: $C_6H_6O_4$

12.16A $\pi = MRT$ Replace M with $\frac{n}{V}$. Then replace n with $\frac{m}{\mathcal{M}}$ to get $\pi = \frac{mRT}{\mathcal{M}V}$.

Rearrange.

$$\mathcal{M} = \frac{mRT}{\pi V} = \frac{1.08 \text{ g x } \frac{0.08206 \text{ atm}}{\text{K mol}} \text{ x } 298 \text{ K}}{0.00770 \text{ atm x } 50.0 \text{ mL x } \frac{10^{-3} \text{ L}}{\text{mL}}}$$

$\mathcal{M} = 6.86 \times 10^4$ g/mol

12.16B $\pi = MRT$ Replace M with $\frac{n}{V}$. Then replace n with $\frac{m}{\mathcal{M}}$ to get $\pi = \frac{mRT}{\mathcal{M}V}$.

$$\pi = \frac{mRT}{\mathcal{M}V} = \frac{125 \text{ µg B-12 x } \frac{10^{-6} \text{ g}}{\text{µg}} \text{ x } \frac{62.36 \text{ L mm Hg}}{\text{K mol}}}{\frac{1355 \text{ g B-12}}{\text{mol B12}} \text{ x } 2.50 \text{ mL x } \frac{10^{-3} \text{ L}}{\text{mL}}} \times (273 + 25) \text{ K}$$

$(6.3 \times 12.01) + (88 \times 1.008) + 58.93 + (14 \times 14.01) + (14 \times 16.08) + 30.97 = 1355$

$$\pi = 0.686 \text{ mmHg} \times \frac{13.6 \text{ g/mL}}{1.00 \text{ g/mL}} = 9.33 \text{ mm } H_2O$$

12.17 The lowest freezing point corresponds to the largest molality of ions. The solutions are dilute enough that molarity is essentially equal to molality.

0.0080 M HCl <	0.0050 m $MgCl_2$ ≈	0.0030 M $Al_2(SO_4)_3$<	0.010 m $C_6H_{12}O_6$
0.016 m ions	0.015 m ions	0.015 m ions	0.010 m particles
lowest f.p.			highest f.p.

Review Questions

2. Volume is temperature-dependent, so molarity, mol solute/liter solution, is temperature-dependent. In molality the amount of solute is on a mole basis, and the quantity of solvent is in kilograms. Neither quantity changes with temperature. Mole fraction depends only on the amounts of solution components, in moles. These amounts do not depend on temperature. Volume percent is temperature-

dependent because the volume does change with temperature. A mass percent is not temperature-dependent because mass does not change with temperature.

8. Generally, it would be true only for a solution of liquids having equal vapor pressures. Most solutions are composed of compounds with different vapor pressures and, thus, different vapor composition than the solution composition. Also, solutions of nonvolatile solutes have a vapor phase that is pure solvent–that is, with none of the solute component(s) at all.

14. In the diffusion of gases, the net movement of molecules is from a higher to a lower pressure in a tendency to equilibrate the pressure throughout the mixture. In osmosis, the net movement of solvent molecules is from a region in which they are more abundant and have a higher vapor pressure to one in which they are less abundant and have a lower vapor pressure. This means from a lower concentration of solute (greater mole fraction of solvent) to a higher concentration of solute (smaller mole fraction of solvent). Osmotic flow is similar to gases diffusing from higher to lower pressure.

20. (a) 0.10 M $NaHCO_3$ is a higher molarity than 0.05 $NaHCO_3$.
(b) 1 M NaCl produces more particles per formula unit than 1 M glucose.
(c) 1 M $CaCl_2$ produces more ions per formula unit than 1 M NaCl.
(d) 3 M glucose has more particles per liter than 1 M NaCl.

Problems

21. $$? \text{ g NaCl} = \frac{10.0 \text{ g NaCl}}{100 \text{ g solution}} \times 5.00 \text{ kg} \times \frac{10^3 \text{ g}}{\text{kg}} = 500 \text{ g NaCl}$$
Weigh 500 g NaCl into a container, add 4.50 kg of water.

23. (a) $$? \% = \frac{4.12 \text{ g NaOH}}{(100.00 \text{ g} + 4.12 \text{ g}) \text{ solution}} \times 100\% = 3.96\% \text{ by mass}$$
(b) $? \text{ g ethanol} = 5.00 \text{ mL} \times 0.789 \text{ g/mL} = 3.945 \text{ g ethanol}$
$$? \% = \frac{3.945 \text{ g ethanol}}{(50.00 \text{ g} + 3.945 \text{ g}) \text{ solution}} \times 100\% = 7.31\% \text{ by mass}$$
(c) $? \text{ g glycerol} = 1.50 \text{ mL} \times 1.324 \text{ g/mL} = 1.986 \text{ g glycerol}$
$? \text{ g water} = 22.25 \text{ mL} \times 0.998 \text{ g/mL} = 22.21 \text{ g water}$
$$? \% = \frac{1.986 \text{ g glycerol}}{(1.986 \text{ g} + 22.21 \text{ g}) \text{ solution}} \times 100\% = 8.21\% \text{ by mass}$$

25. (a) $$? \% = \frac{35.0 \text{ mL } H_2O}{725 \text{ mL solution}} \times 100\% = 4.83\% \text{ by volume}$$
(b) $$? \text{ L} = 10.00 \text{ g} \times \frac{\text{mL}}{0.789 \text{ g}} \times \frac{10^{-3} \text{ L}}{\text{mL}} = 0.01267 \text{ L}$$
$$? \% = \frac{0.01267 \text{ L acetone}}{1.55 \text{ L solution}} \times 100\% = 0.817\% \text{ by volume}$$

(c) $? \text{ mol butanol} = 1.05 \text{ g} \times \frac{\text{mL}}{0.810 \text{ g}} = 1.296 \text{ mL 1-butanol}$

$? \text{ mol ethanol} = 98.95 \text{ g} \times \frac{\text{mL}}{0.789 \text{ g}} = 125.4 \text{ mL ethanol}$

$? \% = \frac{1.296 \text{ mL 1-butanol}}{(125.4 \text{ mL} + 1.3 \text{ mL}) \text{ solution}} \times 100\% = 1.02\% \text{ by volume}$

27. $? \text{ mg/dL} = \frac{0.10 \text{ g glucose}}{100 \text{ g blood}} \times \frac{\text{mg}}{10^{-3} \text{ g}} \times \frac{1.0 \text{ g}}{\text{mL}} \times \frac{\text{mL}}{10^{-3} \text{ L}} \times \frac{10^{-1} \text{ L}}{\text{dL}} = 1.0 \times 10^2 \text{ mg/dL}$

29. $1\% = \frac{1 \text{ g}}{100\text{g}} \quad \frac{1 \text{ mg}}{\text{dL}} \times \frac{10^{-3} \text{ g}}{\text{mg}} \times \frac{\text{dL}}{10^{-1} \text{ L}} \times \frac{10^{-3} \text{ L}}{\text{mL}} \times \frac{\text{mL}}{1 \text{ g}} = \frac{1 \text{ g}}{10^5 \text{ g}}$

$\text{ppt} = \frac{1 \text{ g}}{10^{12} \text{ g}} \qquad \text{ppm} = \frac{1 \text{ g}}{10^6 \text{ g}} \qquad \text{ppb} = \frac{1 \text{ g}}{10^9 \text{ g}}$

ppt < ppb < ppm < 1 mg/dL < 1%

31. (a) $? \text{ ppb benzene} = \frac{1 \text{ μg benzene}}{\text{L water}} \times \frac{\text{L}}{10^3 \text{ g}} \times \frac{10^{-6} \text{ g}}{\text{μg}} = \frac{1 \text{ μg}}{10^9 \text{ μg}} = 1 \text{ ppb benzene}$

(b) $? \text{ ppm NaCl} = \frac{0.0035 \text{ g NaCl}}{100 \text{ g solution}} \times \frac{10^4}{10^4} = \frac{35 \text{ g}}{10^6 \text{ g}} = 35 \text{ ppm NaCl}$

(c) $\text{M F}^- = \frac{2.4 \text{ g F}^-}{10^6 \text{ g solution}} \times \frac{\text{mol}}{19.0 \text{ g}} \times \frac{10^3 \text{ g}}{\text{L}} = 1.3 \times 10^{-4} \text{ M F}^-$

33. $m = \frac{18.0 \text{ g glucose}}{80.0 \text{ g solvent}} \times \frac{10^3 \text{ g}}{\text{kg}} \times \frac{\text{mol}}{180.2 \text{ g}} = 1.25\ m$

35. $\text{M} = \frac{75 \text{ g } H_3PO_4}{100 \text{ g solution}} \times \frac{1.57 \text{ g}}{\text{mL}} \times \frac{\text{mL}}{10^{-3} \text{ L}} \times \frac{\text{mol}}{97.99 \text{ g}} = 12 \text{ M}$

$m = \frac{75 \text{ g } H_3PO_4}{25 \text{ g solvent}} \times \frac{10^3 \text{ g}}{\text{kg}} \times \frac{\text{mol}}{97.99 \text{ g}} = 31\ m$

37. $? \text{ g solution} = 375 \text{ mL} \times \frac{1.18 \text{ g}}{\text{mL}} = 443 \text{ g solution}$

$? \text{ g } H_2SO_4 = \frac{3.39 \text{ mol}}{\text{kg } H_2O} \times \frac{98.08 \text{ g}}{\text{mol}} = \frac{333 \text{ g}}{\text{kg } H_2O}$

$? \% \ H_2SO_4 = \frac{333 \text{ g}}{1000 \text{ g } H_2O + 333 \text{ g } H_2SO_4} \times 100\% = 25.0\%$

$? \text{ g } H_2SO_4 = \frac{25.0 \text{ g } H_2SO_4}{100 \text{ g solution}} \times 443 \text{ g solution} = 110 \text{ g } H_2SO_4$

$? \text{ mol } H_2SO_4 = 110 \text{ g } H_2SO_4 \times \frac{\text{mol}}{98.08 \text{ g}} = 1.13 \text{ mol } H_2SO_4$

OR

$$?\text{mol } H_2SO_4 = 443 \text{ g solution} \times \frac{333 \text{ g } H_2SO_4}{1333 \text{ g solution}} \times \frac{1 \text{ mol } H_2SO_4}{98.08 \text{ g } H_2SO_4}$$
$$= 1.13 \text{ mol } H_2SO_4$$

39. (a) $? \text{ mol } C_{10}H_8 = 23.5 \text{ g} \times \frac{\text{mol}}{128.16 \text{ g}} = 0.1834 \text{ mol } C_{10}H_8$

$? \text{ mol } C_6H_6 = 315 \text{ g} \times \frac{\text{mol}}{78.11 \text{ g}} = 4.033 \text{ mol } C_6H_6$

$\chi = \frac{0.183 \text{ mol}}{(4.033 + 0.183) \text{ mol}} = 0.0434$

(b) $? \text{ mol } C_6H_6 = 1 \text{ kg } C_6H_6 \times \frac{10^3 \text{ g}}{\text{kg}} \times \frac{\text{mol}}{78.11 \text{ g}} = 12.80 \text{ mol } C_6H_6$

$\chi = \frac{0.250 \text{ mol}}{(12.80 + 0.250) \text{ mol}} = 0.0192$

41. The solution with the greatest amount of solute per unit mass of solvent has the greatest mole fraction of solute–(b). Solution (a) has 1 mol solute per 1000 g H_2O; (b) has slightly more than 1 mol solute in 950 g H_2O; (c) has about 0.3 mol solute per 900 g H_2O.

Actual -

(a) $\frac{1.00 \text{ mol}}{\text{kg } H_2O}$ $\quad ? \text{ mol} = \text{kg } H_2O \times \frac{10^3 \text{ g}}{\text{kg}} \times \frac{\text{mol}}{18.02 \text{ g}} = 55.50 \text{ mol}$

$\chi = \frac{1.00 \text{ mol}}{(1.00 + 55.50) \text{ mol}} = 0.0177$

(b) $? \text{ mol} = 5.0 \text{ g } C_2H_6O \times \frac{\text{mol}}{46.07 \text{ g}} = 0.1085 \text{ mol}$

$? \text{ mol} = 95.0 \text{ g } H_2O \times \frac{\text{mol}}{18.02 \text{ g}} = 5.272 \text{ mol}$

$\chi = \frac{0.1085 \text{ mol}}{(5.272 + 0.1085) \text{ mol}} = 0.0202$

(c) $? \text{ mol} = 10.0 \text{ g } C_{12}H_{22}O_{11} \times \frac{\text{mol}}{342.3 \text{ g}} = 0.0292 \text{ mol}$

$? \text{ mol} = 90.0 \text{ g } H_2O \times \text{mol}/18.02\text{g} = 4.994 \text{ mol}$

$\chi = \frac{0.0292 \text{ mol}}{(4.994 + 0.029) \text{ mol}} = 0.00582$

43. (a) $CHCl_3$ is insoluble in water. Although $CHCl_3$ is somewhat polar, the primary intermolecular forces in water are hydrogen bonds, which are unimportant in chloroform ($CHCl_3$).
(b) Benzoic acid is slightly soluble in water. There is some hydrogen bonding to the carboxylic acid group, but the benzene ring is very unlike water.
(c) Propylene glycol is highly soluble in water; both solute and solvent have extensive hydrogen bonding.

45. Figure 12.10 shows that a saturated solution at 60 °C has about 55 g NH_4Cl per

100 g H_2O. $\frac{55 \text{ g } NH_4Cl}{100 \text{ g } H_2O} \times 55 \text{ g } H_2O = 30 \text{ g } NH_4Cl$

The given solution is unsaturated.

47. (a) $? \text{ g } H_2O = 35 \text{ g } K_2CrO_4 \times \frac{100\text{g water}}{62.\text{g } K_2CrO_4} = 56 \text{ g water at } 25\ °C$

$? \text{ g } H_2O = 56 \text{ g } H_2O - 35 \text{ g } H_2O = 21 \text{ g } H_2O$ to be added.

(b) $? \text{ g } KNO_3 = \frac{50.0\text{g } KNO_3}{75.0 \text{ g } H_2O} \times 100.0 \text{ g } H_2O = 66.7 \text{ g } KNO_3$

At temperatures above about 44 °C, the solubility of KNO_3 exceeds 66.7 g KNO_3/100.0 g H_2O.

49. The solution could be left open to air so that some of the solvent evaporates. When enough solvent has evaporated, the solute will begin to crystallize.

51. (a) $? \text{ M} = \frac{4.43 \text{ mg } O_2}{100 \text{ g } H_2O} \times \frac{1.0 \text{ g}}{\text{mL}} \times \frac{10^{-3} \text{ g}}{\text{mg}} \times \frac{\text{mL}}{10^{-3} \text{ L}} \times \frac{\text{mol}}{32.00 \text{ g}} = 1.38 \times 10^{-3} \text{ M}$

NOTE: Density of water solution is approximated at 1.0 g/mL.

$? \frac{\text{mg } O_2}{H_2O} = 0.010 \text{ M} \times \frac{10^{-3} \text{ L}}{\text{mL}} \times \frac{32.00 \text{ g}}{\text{mol}} \times \frac{\text{mg}}{10^{-3} \text{ g}} \times \frac{\text{mL}}{1.0 \text{ g}} = \frac{0.32 \text{ mg } O_2}{\text{g } H_2O}$

$? \text{ mg } O_2 = 100 \text{ g } H_2O \times \frac{0.32 \text{ mg } O_2}{\text{g } H_2O} = 32 \text{ mg } O_2$

$k = \frac{S}{P} = \frac{4.43 \text{ mg } O_2}{1.00 \text{ atm}}$

$P = \frac{S}{k} = 32 \text{ mg } O_2 \times \frac{1.00 \text{ atm}}{4.43 \text{ mg } O_2} = 7.2 \text{ atm}$

53. (a) $P_P = \frac{1 \text{ mol}}{5 \text{ mol}} \times 441 \text{ mmHg} = 88.20 \text{ mmHg}$

$P_H = \frac{4 \text{ mol}}{5 \text{ mol}} \times 121 \text{ mmHg} = 96.80 \text{ mmHg}$

(b) $\chi_H = \frac{96.80 \text{ mmHg}}{(96.80 + 88.20) \text{ mmHg}} = 0.523$

$\chi_P = 1.000 - 0.523 = 0.477$

55. $? \text{ mol} = \text{kg} \times \frac{10^3 \text{ g}}{\text{kg}} \times \frac{\text{mol}}{18.02 \text{ g}} = 55.49 \text{ mol}$

$\chi_{H_2O} = \frac{55.49 \text{ mol}}{55.49 + 0.20 \text{ mol}} = 0.9964$

$VP = 0.9964 \times 17.5 \text{ mm Hg} = 17.4 \text{ mmHg}$

57. (a) $\Delta T = -K_f \times m = \frac{-1.86\ °C}{m} \times 0.25 \text{ I} = -0.465\ °C$

$\Delta T = t_f - t_i = t_f - 0.00\ °C = -0.465\ °C$

$t_f = -0.47\ °C$

(b) $m = \dfrac{5.0\ g\ C_6H_4Cl_2}{95.0\ g\ H_2O} \times \dfrac{10^3\ g}{kg} \times \dfrac{mol\ C_6H_4Cl_2}{147.0\ g\ C_6H_4Cl_2} = 0.358\ m$

$\Delta T = \dfrac{-5.12°C}{m} \times 0.358\ m = -1.83\ °C$

$\Delta T = -1.83\ °C = T_f - 5.53\ °C$

$\Delta T = 3.70\ °C$

59. $\Delta T = -K_f \times m = \dfrac{-1.86\ °C}{m} \times 0.55\ m$

$\Delta T = -\ 1.02\ °C = t_f - 0.00\ °C$

$t_f = -1.02\ °C$

$\Delta T = -\ 1.02\ °C - 5.53\ °C = -\ 6.55\ °C$

$m = \dfrac{\Delta T}{-K_f} = \dfrac{-6.55°C}{\dfrac{-5.12°C}{m}}$

$m = 1.28$

61. $m = \dfrac{2.11\ g\ C_{10}H_8}{35.00\ g\ xylene} \times \dfrac{10^3\ g}{kg} \times \dfrac{mole}{128.16\ g} = 0.4704\ m$

$\Delta T = 11.25\ °C - 13.26\ °C = -2.01\ °C = -K_f \times m$

$K_f = \dfrac{2.01°C}{0.4704\ m} = 4.27\ °C/m$

63. $\Delta T = t_f - t_i = 4.70\ °C - 5.53\ °C = -0.83\ °C$

$m = \dfrac{\Delta T}{-K_f} = \dfrac{0.83\ °C}{-5.12\ °C/m} = 0.162\ m$

$?\ mol = 50.00\ mL \times \dfrac{0.874\ g}{mL} \times \dfrac{kg}{10^3\ g} \times 0.162\ m = 0.00708\ mol$

$\mathcal{M} = \dfrac{1.505\ g}{0.00708\ mol} = 213\ g/mol$

$?\ mol\ C = 33.81\ g\ C \times \dfrac{mol}{12.01\ g} = 2.8149\ mol\ C \times \dfrac{1}{1.4080\ mol\ N} = 1.999\ \dfrac{mol\ C}{mol\ N}$

$?\ mol\ H = 1.42\ g\ H \times \dfrac{mol}{1.008\ g} = 1.4087\ mol\ H \times \dfrac{1}{1.4080\ mol\ N} = 1.000\ \dfrac{mol\ H}{mol\ N}$

$?\ mol\ O = 45.05\ g\ O \times \dfrac{mol}{15.999\ g} = 2.8158\ mol\ O \times \dfrac{1}{1.4080\ mol\ N} = 2.000\ \dfrac{mol\ O}{mol\ N}$

$?\ mol\ N = 19.72\ g\ N \times \dfrac{mol}{14.006\ g} = 1.4080\ mol\ N \times \dfrac{1}{1.4080\ mol\ N} = 1.00\ \dfrac{mol\ N}{mol\ N}$

empirical formula C_2HO_2N 71 g/formula

$\dfrac{213\ g}{mole} \times \dfrac{formula}{71\ g} = 3\ formula/mole$

molecular formula $C_6H_3O_6N_3$

65. Because the cucumbers shrivel up, water is leaving the cucumbers, so the salt solution must have a higher osmotic pressure.

67. $$\pi = \frac{mRT}{\mathcal{M}V} = \frac{1.80 \text{ g x } \frac{0.08206 \text{ L atm}}{\text{K mol}} \text{ x } (37 + 273)\text{K}}{46.07 \text{ g/mol x } 100 \text{ mL x } 10^{-3} \text{ L/mL}}$$

$\pi = 9.94$ atm

$$\pi \text{ glucose} = \frac{5.5 \text{ g x } \frac{0.08206 \text{ L atm}}{\text{K mole}} \text{ x } (37 + 273)\text{K}}{180.16 \text{ g/mol x } 100 \text{ mL x } 10^{-3} \text{ L/mL}}$$

$\pi = 7.77$ atm

The CH_3CH_2OH solution is hypertonic.

69. $$M = \frac{5.15 \text{ g urea}}{75.0 \text{ mL } H_2O} \times \frac{\text{mL}}{10^{-3} \text{ L}} \times \frac{\text{mol urea}}{60.06 \text{ g } CO(NH_2)_2} = 1.14 \text{ M}$$

The flow will be to the right to the more concentrated side from A to B. Solution B has the higher osmotic pressure. More molecules are able to leave the dilute side because less-solute molecules block the path.

71. The HCl does not dissociate in benzene. The van't Hoff factor is about 1 for HCl in benzene. $\Delta T_f = -1 \times 5.12 \times 0.01 \approx -0.05$ °C. HCl does ionize in water, producing a van't Hoff factor of about 2: $\Delta T_f = -2 \times 1.86 \times 0.01 \approx -0.04$ °C.

73. NaCl produces two ions per formula unit while glucose is a molecular substance. The total particle molarity is the same in the two solutions.

75. The lowest freezing point will have the largest ΔT. Instead of calculating ΔT, it is easier to compare $i \times m$, as all of the solutions are in the same solvent, water. The largest product ($i \times m$) produces the lowest freezing point.

(b)	$1 \times 0.15 =$	0.15
(a) slightly more than	$1 \times 0.15 =$	slightly more than 0.15
(e)	$2 \times 0.10 =$	0.20
(c) slightly more than	$2 \times 0.10 =$	slightly more than 0.20
(d)	$3 \times 0.10 =$	0.30

(d) is the lowest freezing point.

77. $\Delta T = 100.0$ °C - 99.4 °C = 0.6 °C

$$0.6 \text{ °C} = K_b \times m \times i = \frac{0.512°C}{m} \times m \times 2$$

$m = 0.59\ m$

$$? \text{ g} = 0.59\ m \times 3.50 \text{ kg} \times \frac{58.44 \text{ g}}{\text{mol}} = 1.2 \times 10^2 \text{ g}$$

79. The higher charge on the Al^{3+} ion would more effectively negate the negative charge on the silica and allow coagulation.

Additional Problems

81. (a) $? \% = \dfrac{11.3 \text{ mL } CH_3OH}{75.0 \text{ mL solution}} \times 100\% = 15.1\%$ by volume

(b) $? \% = \dfrac{11.3 \text{ mL } CH_3OH \text{ x } \dfrac{0.793 \text{ g}}{\text{mL}}}{75.0 \text{ mL solution x } \dfrac{0.980 \text{ g}}{\text{mL}}} \times 100\% = 12.2\%$ by mass

(c) $? \% = \dfrac{11.3 \text{ mL } CH_3OH \text{ x } \dfrac{0.793 \text{ g}}{\text{mL}}}{75.0 \text{ mL solution}} \times 100\% = 11.9\ \%$ mass/volume

(d) $? \text{ g } CH_3OH = 11.3 \text{ mL} \times \dfrac{0.793 \text{ g}}{\text{mL}} = 8.961 \text{ g } CH_3OH$

$? \text{ g solution} = 75.0 \text{ mL} \times \dfrac{0.980 \text{ g}}{\text{mL}} = 73.5 \text{ g solution}$

$? \text{ g } H_2O = 73.5 \text{ g} - 9.0 \text{ g} - 64.5 \text{ g water}$

$? \text{ mol} = 8.961 \text{ g } CH_3OH \times \dfrac{\text{mol}}{32.04 \text{ g}} = 0.2797 \text{ mol}$

$? \text{ mol} = 64.5 \text{ g } H_2O \times \dfrac{\text{mol}}{18.02 \text{ g}} = 3.579 \text{ mol}$

$\text{mole } \% = \dfrac{0.2797 \text{ mol}}{(3.579 + 0.280) \text{ mol}} \times 100\% = 7.25\%$

83. $k = \dfrac{S}{P_{gas}} = \dfrac{5.5 \times 10^{-4} \text{ M}}{0.78 \text{ atm}} = 7.1 \times 10^{-4} \dfrac{\text{M}}{\text{atm}}$

$S = k\,P_{gas} = 7.1 \times 10^{-4} \dfrac{\text{M}}{\text{atm}} \times 4.0 \text{ atm} = 2.8 \times 10^{-3} \text{ M}$

$? \text{ mol at } 0.78 \text{ atm} = 5.5 \times 10^{-4} \text{ M} \times 5.0 \text{ L} = 0.0028 \text{ mol}$

$? \text{ mol at } 4.0 \text{ atm} = 2.8 \times 10^{-3} \text{ M} \times 5.0 \text{ L} = 0.014 \text{ mol}$

$? \text{ mol } N_2 \text{ released} = (0.014 \text{ moles} - 0.0028 \text{ mol}) = 0.011 \text{ mol}$

$? \text{ L} = 0.011 \text{ mol} \times \dfrac{0.08206 \text{ L atm}}{\text{K mol}} \times 310 \text{ K} \times \dfrac{1}{0.78 \text{ atm}} = 0.36 \text{ L}$

85. The 25.00 mL H_2O in 25.00 mL CH_3OH has the greater molality. The moles of H_2O (24.95 × mol/18.02 g = 1.385 mol) is greater than the moles of CH_3OH (19.78 g × mol/32.04 g = 0.6172 mol), and the number of kg of solvent is smaller in 25.00 mL of CH_3OH (0.01978 kg) than in 25.00 ml of H_2O (0.02495 kg).

87. $\Delta T = -K \times m$

$$m = \frac{\Delta T}{-K} = \frac{-0.28^\circ C - 0^\circ C}{\dfrac{1.86^\circ C}{m}} = 0.15\ m \text{ of ions}$$

In dilute solutions, molarity and molality are about equal.
(a) has 0.30 M of ions.
(b) has 0.15 M of molecules and could be the solution.
(c) has 0.25 M of solute particles.
(d) has 0.15 M of solute particles and could be the solution.
A few drops of a solution containing silver ion would cause a white precipitate in (d) but not in (b).

89. Because ethanol is less dense than water, the mass percent of ethanol in water is always less than the volume percent. Solutes whose densities are greater than water would have the volume percent less than the mass percent.

91. (a) The acetone solution has the greatest vapor pressure because both the mole fraction in water is as great as in any other solution, and acetone contributes to the vapor pressure.
(b) The saturated NaCl solution has the most ions in solution, so it has the lowest freezing point.
(c) The vapor pressure of the acetone solution and the 0.10 m NaCl will change with time, as some of the vapor completely escapes. The vapor pressure of saturated NaCl will not change, because as water evaporates, some of the NaCl (aq) will become solid, leaving the same concentration of solute to solvent particles.

93. (a) $P\text{solution} = \chi_{\text{benzene}}P_{\text{benzene}} + \chi_{\text{toluene}}P_{\text{toluene}}$

$760.0 \text{ mmHg} = \chi_{\text{benzene}} \times 1351 \text{ mmHg} + (1 - \chi\text{benzene}) \times 556.3 \text{ mmHg}$

$760.0 \text{ mmHg} = 1351\ \chi_{\text{ben}} + 556.3 \text{ mmHg} - 556.3\ \chi_{\text{ben}}$

$203.7 \text{ mmHg} = 795\ \chi_{\text{ben}}$

$\chi_{\text{ben}} = 0.256$

$\chi_{\text{tol}} = 1.000 - \chi_{\text{ben}} = 0.744$

(b) $P_{\text{tol}} = 0.744 \times 556.3 \text{ mmHg} = 414 \text{ mmHg}$

$P_{\text{ben}} = 0.256 \times 1351 \text{ mmHg} = 346 \text{ mmHg}$

$$\chi_{\text{ben}} = \frac{346 \text{ mmHg}}{(346 + 414) \text{ mmHg}} = 0.455$$

$$\chi_{\text{tol}} = \frac{414 \text{ mmHg}}{760 \text{ mmHg}} = 0.545$$

95. $$\frac{0.92 \text{ g NaCl}}{100 \text{ mL solvent}} \times \frac{\text{mL}}{1 \text{ g}} \times \frac{1000 \text{ g}}{\text{kg}} \times \frac{\text{mol}}{58.44 \text{ g}} = 0.157\ m$$

$\Delta T = i \times -K_f \times m = 2 \times (-1.86\ °C/m) \times 0.157\ m = -0.58\ °C$ if $i = 2$.
Actually $i < 2$ and the agreement with -0.52 °C is good.

97. A: $X \text{ mol } CO(NH_2)_2 \times \frac{60.06 \text{ g}}{\text{mol}} + 9X \text{ mol } H_2O \times \frac{18.02 \text{ g}}{\text{mol}} = 200.0 \text{ g}$

$? \text{ g} = 60.06\,X + 162.18\,X = 200.0 \text{ g}$

$X = \frac{200.0 \text{ g}}{222.24 \text{ g/mol}} = 0.900 \text{ mol } CO(NH_2)_2$

$8.100 \text{ mol } H_2O$

B: $X \text{ mol } CO(NH_2)_2 \times \frac{60.06 \text{ g}}{\text{mol}} + 19\,X \text{ mol} \times \frac{18.02 \text{ g}}{\text{mol}} = 100.0 \text{ g}$

$? \text{ g} = 60.06\,X + 342.38\,X = 100.0 \text{ g}$

$402.44\,X = 100$

$X = 0.248 \text{ mol } CO(NH_2)_2$

$4.712 \text{ mol } H_2O$

Final solutions both have the same mole fraction.

$? \text{ mol } H_2O = 8.100 + 4.712 = 12.812 \text{ mol } H_2O$

$? \text{ mol } CO(NH_2)_2 = 0.900 + 0.248 = 1.148 \text{ mol } CO(NH_2)_2$

$\chi_{,CO(NH_2)_2)} = \frac{1.148 \text{ mol}}{(1.148 + 12.812) \text{ mol}} = 0.0822 \qquad \chi_{,H_2O)} = 0.9178$

$\chi = 0.0822 = \frac{0.900 \text{ mol } CO(NH_2)_2}{(0.900 + X)\text{mol}}$

$0.0740 + 0.0822\,X = 0.900$

$X = 10.05 \text{ mol } H_2O$

$? \text{ g } CO(NH_2)_2 = 0.900 \text{ mol } CO(NH_2)_2 \times 60.06 \text{ g/mol} = 54.1 \text{ g } CO(NH_2)_2$ in A

$? \text{ g } H_2O = 10.05 \text{ mol } H_2O \times \frac{18.02 \text{ g}}{\text{mol}} = 181.1 \text{ g } H_2O$ in A

$? \text{ g} = 54.1 \text{ g } CO(NH_2)_2 + 181.1 \text{ g } H_2O = 235.2 \text{ g}$ total in A

$\text{mass fraction} = \frac{54.1 \text{ g } CO(NH_2)2}{235.2 \text{ g total}} = 0.230\ CO(NH_2)_2$

$? \text{ g } CO(NH_2)_2 = 0.248 \text{ mol } CO(NH_2)_2 \times 60.06 \text{ g/mol} = 14.9 \text{ g } CO(NH_2)_2$ in B

$? \text{ g } H_2O = 300.0 \text{ g} - 235.2 \text{ g} - 14.9 \text{ g} = 49.9 \text{ g } H_2O$ in B

or

$0.0822 = \frac{0.248 \text{ mol } CO(NH_2)_2}{(0.248 + X) \text{ mol}}$

$0.0204 + 0.0822X = 0.248$

$X = 2.77 \text{ mol } H_2O \times \frac{18.02 \text{ g}}{\text{mol}} = 49.9 \text{ g } H_2O$

$? \text{ g} = 14.9 \text{ g } CO(NH_2)_2 + 49.9 \text{ g } H_2O = 64.8 \text{ g}$ total in B

$\text{mass fraction} = \frac{54.1 \text{ g } CO(NH_2)2}{235.2 \text{ g total}} = 0.230\ CO(NH_2)_2$

Check: soln A + soln B = total

Initial 200.0 g + 100.0 g = 300.0 g

Final 235.2 g + 64.8 g = 300.0 g

99. $[NaCl] = 0.438 \text{ M} \qquad [MgCl_2] = 0.0512 \text{ M}$

$iM = [\text{ions}] = 2 \times 0.438 \text{ M} + 3 \times 0.0512 \text{ M} = 1.030 \text{ M}$

$$\pi = iMRT = 1.030\text{ M} \times 0.08206\ \frac{\text{L atm}}{\text{K mol}} \times 298\text{ K}$$

$\pi = 25$ atm

The result is of limited accuracy because the i values are based on infinite dilution.

101. $\Delta T = -K_f \times m$

$$m = \frac{\Delta T}{-K_f} = \frac{-2.374^\circ\text{C}}{-14.1\frac{^\circ\text{C}}{m}} = 0.168\ m$$

$$0.168\ m = \frac{\frac{2.58\text{ g}}{\text{molar mass}}}{0.1000\text{ kg}}$$

$$\text{molar mass} = \frac{2.58\text{ g}}{0.1000\text{ kg} \times 0.168\,m} = 154\text{ g/mol}$$

$154\text{ g/mol} = y \times 94.11\text{ g/mol} + (1.00 - y) \times 188.23\text{ g/mol}$

$94.11\ y\text{ g/mol} = 188.23\text{ g/mol} - 154\text{ g/mol} = 34.23\text{ g/mol}$

$$y = \frac{32.23\text{ g/mol}}{94.11\text{ g/mol}} = 0.364$$

fraction as dimers = 1.0000 – 0.364 = 0.636

Apply Your Knowledge

103. The liquid is $CaCl_2(aq)$. $H_2O(g)$ from the air condenses on the solid to form saturated $CaCl_2(aq)$. A solid will not exhibit this phenemenon if the vapor pressure of the saturated solution exceeds the partial pressure of $H_2O(g)$ in the atmosphere. The phenomenon is deliquesence.

105. (a) $? \text{ cm} = 1000\text{ nm} \times \frac{\text{nm}}{10^{-9}\text{ m}} \times \frac{10^{-2}\text{ m}}{\text{cm}} = 10^{-4}\text{ cm}$

$$\left(\frac{1}{2}\right)^n = 1 \times 10^{-4}\text{ cm}$$

$n \log 0.5 = \log 1 \times 10^{-4} = -4$

$$n = \frac{-4}{\log 0.5} = \frac{-4}{0.3} = 13.3 \quad \text{rounds to 14 divisions}$$

(b) $? \text{ size of cube after 14 divisions} = \left(\frac{1}{2}\right)^{14} = 6.10 \times 10^{-5}\text{ cm}$

volume of each cube = $(6.1 \times 10^{-5}\text{ cm})^3 = 2.27 \times 10^{-13}\text{ cm}^3$

number of cubes = $(8)^{14} = 4.40 \times 10^{12}$ cubes

total volume of cubes = 1 cm

$$\text{Total surface area} = 4.40 \times 10^{12}\text{ cubes} \times \frac{6\text{ sides}}{\text{cube}} \times (6.10 \times 10^{-5}\text{ cm})^2$$

$$= 9.82 \times 10^{4}\text{ cm}^2$$

surface area/volume ratio

original cube $\frac{6.0\ cm^2}{cm^3}$

after 14 subdivisions $\frac{9.82 \times 10^4\ cm^2}{cm^3}$

Colloidal particles have a much larger surface arca to volume ratio than does bulk material because as the bulk material is divided, the volume is the same but the surface area increases.

Chapter 13

Chemical Kinetics: Rates and Mechanisms of Chemical Reactions

Exercises

13.1A (a) $\text{rate} = \frac{[D]_2 - [D]_1}{\Delta t} = \frac{0.3546\text{ M} - 0.2885\text{ M}}{2.55\text{ min}} = 0.0259\text{ M/min}$

(b) $\text{rate of formation of C} = 2 \times \text{rate} = 2 \times 0.0259\text{ M/min} \times \frac{\text{min}}{60\text{ sec}}$

$= 8.63 \times 10^{-4}\text{ M/s}$

13.1B (a) $\frac{-\Delta[B]}{\Delta t} = \frac{1}{2} \times \frac{-\Delta[A]}{\Delta t} = \frac{1}{2} \times 2.10 \times 10^{-5}\text{ M/s} = 1.05 \times 10^{-5}\text{ M/s} = \text{rate}$

(b) $\frac{1}{3}\frac{\Delta[C]}{\Delta t} = \frac{-1}{2}\frac{\Delta[A]}{\Delta t}$

$\frac{\Delta[C]}{\Delta t} = \frac{3}{2} \times 2.10 \times 10^{-5}\text{ M/s} = 3.15 \times 10^{-5}\text{ M/s}$

13.2A (a) From tangent line

$\text{rate} = \text{slope} = \frac{-(0 - 0.630)\text{ M}}{(570 - 0)\text{ s}} = 1.11 \times 10^{-3}\text{ M/s}$

(b) $\Delta[H_2O_2] = -10\text{ s} \times 1.11 \times 10^{-3}\text{ M/s} = -1.1 \times 10^{-2}\text{ M}$

$[H_2O_2]_{310} = [H_2O_2]_{300} + \Delta[H_2O_2]$

$= 0.298\text{ M} + (-1.1 \times 10^{-2}\text{ M}) = 0.287\text{ M}$

13.2B A tangent line has to be drawn that is parallel to the purple line. By eyeballing, it appears to be at about 270 s. Values from that line are used to calculate the slope and thus the rate. $\text{rate} = -\text{ slope} = \frac{-(0 - 0.700)\text{M}}{(520 - 0)\text{s}}$

rate = 1.3×10^{-3} M/s

Alternate: Since the purple line is parallel to the tangent line, its slope will be the same as the tangent line. Thus the initial and final values in Table 13.1 can be used.

$\frac{-(0.094 - 0.882)\text{M}}{(600 - 0)\text{s}} = 1.3 \times 10^{-3}\text{ M s}^{-1}$ at about 270 s

Only one tangent line is possible at each point on the curved line.

13.3A rate = $k\,[NO]^2[Cl_2]$

$\text{rate} = \frac{5.70}{\text{M}^2\text{ s}} \times (0.200\text{ M})^2 \times 0.400\text{ M} = 0.0912\text{ M/s}$

13.3B rate = $k[CH_3CHO]^m$

$$\frac{(\text{initial rate})_2}{(\text{initial rate})_1} = \frac{k \times [CH_3CHO]_2^m}{k \times [CH_3CHO]_1^m}$$

$$\frac{2.8 \times (\text{initial rate})_1}{(\text{initial rate})_1} = \frac{k \times 2^m \times [CH_3CHO]_1}{k \times [CH_3CHO]_1}$$

$2.8 = 2^m$

$\log 2.8 = m \log 2$

$$m = \frac{\log 2.8}{\log 2} = \frac{0.45}{0.30} = 1.5$$

Note that the order of this reaction is nonintegral. For m to be integral, the ratio of reaction rates corresponding to a doubling of the initial concentration must be an integral power of 2.

13.4A(a) $\ln \frac{[NH_2NO_2]_t}{[NH_2NO_2]_0} = -kt$

$$\ln \frac{0.0250}{0.105} = -\frac{5.62 \times 10^{-3}}{\text{min}}\, t$$

$t = 255$ min

(b) $\ln \frac{[NH_2NO_2]_t}{0.105\text{ M}} = \frac{-5.62 \times 10^{-3}}{\text{min}} \times 6.00\text{ h} \times \frac{60\text{ min}}{\text{hr}}$

$$\ln \frac{[NH_2NO_2]}{0.105\text{ M}} = -2.02$$

Raise both sides of the equation to a power of e.

$$e^{\ln \frac{[NH_2NO_2]}{0.105\text{ M}}} = \frac{[NH_2NO_2]}{0.105\text{ M}} = e^{-2.02} = 0.133$$

$[NH_2NO_2] = 0.0139$ M

13.4B $\ln \frac{[NH_2NO_2]_t}{[NH_2NO_2]_0} = -kt$

$$\ln \frac{[NH_2NO_2]}{0.0750\text{ M}} = \frac{-5.62 \times 10^{-3}}{\text{min}} \times 35.0\text{ min}$$

$$e^{\ln \frac{[NH_2NO_2]}{0.0750\text{ M}}} = \frac{[NH_2NO_2]}{0.0750\text{ M}} = e^{-0.197} = 0.821$$

$[NH_2NO_2] = 0.0616$ M

$$\text{rate} = k[NH_2NO_2] = \frac{5.62 \times 10^{-3}}{\text{min}} \times 0.0616\text{ M} = 3.46 \times 10^{-4}\text{ M/min}$$

13.5 (a) Because $\frac{1}{16}$ is $\left(\frac{1}{2}\right)^4$, the time required is four half-lives—that is, $4 \times t_{1/2}$.

$4 \times 120\text{ s} = 480\ s$

or

$$\ln \frac{\frac{1}{16}}{1} = \frac{-5.78 \text{ x } 10^{-3}}{s} t$$

$$-2.77 = \frac{-5.78 \text{ x } 10^{-3}}{s} t$$

$t = 480\ s$

(b) $$\ln \frac{m}{m_o} = \frac{-5.78 \text{ x } 10^{-3}}{s} t$$

$$\ln \frac{m}{4.80 \text{ g}} = \frac{-5.78 \text{ x } 10^{-3}}{s} \times 10.0 \text{ min} \times \frac{60\ s}{\text{min}}$$

$$\ln \frac{m}{4.80 \text{ g}} = -3.47$$

Raise both sides of equation to a power of e.

$$e^{\ln \frac{m}{4.80 \text{ g}}} = \frac{m}{4.80 \text{ g}} = e^{-3.50} = 3.02 \times 10^{-2}$$

$m = 0.150$ g

13.6 The $t_{1/2}$ is about 120 s. 370 s is 10 seconds longer than three half–lives.

$$800 \text{ mmHg} \xrightarrow{1} 400 \text{ mmHg} \xrightarrow{2} 200 \text{ mmHg} \xrightarrow{3} 100 \text{ mmHg}.$$

P (N_2O_5) should be slightly less than 100 mmHg, perhaps about 95 mm Hg.

13.7A $$t_{1/2} = \frac{1}{k[A]_o}$$

$$55\ s = \frac{1}{k \text{ x } 0.80 \text{ M}}$$

$$k = \frac{0.023}{M\ s}$$

13.7B $$\frac{1}{[A]_t} = kt + \frac{1}{[A]_0}$$

$$\frac{1}{0.20 \text{ M}} = \frac{0.023}{M\ s} t + \frac{1}{0.80 \text{ M}}$$

$$\frac{5.0}{M} - \frac{1.25}{M} = \frac{0.023}{M \text{ s}} t$$

$t = 1.6 \times 10^2\ s$

$$\frac{1}{0.10 \text{ M}} = \frac{0.023}{M \text{ s}} t + \frac{1}{0.80 \text{ M}}$$

$$\frac{10.0}{M} - \frac{1.25}{M} = \frac{0.023}{M \text{ s}} t$$

$t = 380$ sec

first $t_{1/2} = 55$ s

second $t_{1/2} = 1.6 \times 10^2 - 55 = 1.1 \times 10^2\ s$

third $t_{1/2} = 3.8 \times 10^2 - 1.6 \times 10^2 = 2.2 \times 10^2\ s$

The half-life doubles each half-life because the concentration is halved.

13.8 Method 1

$$\text{rate}_1 = -\left(\frac{0.92\text{ M} - 1.00\text{ M}}{100\text{ sec}}\right) = 8 \times 10^{-4}\text{ M}/s$$

$$\text{rate}_2 = -\left(\frac{12.68\text{ M} - 2.00\text{ M}}{100\text{ sec}}\right) = 3.2 \times 10^{-3}\text{ M}/s$$

$$\frac{\text{rate}_1}{\text{rate}_2} = \frac{k\,[A]_2^n}{k\,[A]_1^n}$$

$$\frac{\dfrac{-3.2 \text{ x } 10^{-3}\text{ M}}{s}}{\dfrac{-8 \text{ x } 10^{-4}\text{ M}}{s}} = \frac{k}{k}\text{ B}\left(\frac{2.00\text{ M}}{1.00\text{ M}}\right)^n$$

$4 = 2^n$

$n = 2$ second order

Method 2

time	[A]	$\frac{1}{[A]}$	change
0	2.00	0.500	.10
100	1.68	0.595	.10
200	1.43	0.699	.10
300	1.26	0.794	.10
400	1.12	0.893	.10
500	1.00	1.00	.09
600	0.92	1.09	.10
700	0.84	1.19	.10
800	0.78	1.29	.10
900	0.72	1.39	.10
1000	0.67	1.49	

The change in $\frac{1}{[A]}$ is constant, so a plot of $\frac{1}{[A]}$ versus time would be a straight line, and the reaction is second order.

13.9A $$\ln\frac{1.0 \text{ x } 10^{-5}}{2.5 \text{ x } 10^{-3}} = \frac{1.0 \text{ x } 10^{2}\text{ kJ /mol x } 1000\text{ J /kJ}}{\dfrac{8.3145\text{ J}}{\text{mol K}}} \times \left(\frac{1}{332\text{ K}} - \frac{1}{T_2}\right)$$

$$-5.52 = 1.20 \times 10^{4}\text{ K}^{-1} \times \left(\frac{1}{332\text{ K}} - \frac{1}{T_2}\right)$$

$$-4.60 \times 10^{-4}\text{ K}^{-1} = \frac{1}{332\text{ K}} - \frac{1}{T_2}$$

$$\frac{1}{T_2} = \frac{1}{332\text{ K}} + 4.60 \times 10^{-4}\text{ K}^{-1} = 3.01 \times 10^{-3}\text{ K}^{-1} + 4.60 \times 10^{-4}\text{ K}^{-1}$$

$$= 3.47 \times 10^{-3}\text{ K}^{-1}$$

$T_2 = 1/3.47 \times 10^{-3}\text{ K}^{-1} = 288\text{ K}$ (15 °C)

13.9B For first order $k = \dfrac{0.693}{t_{\frac{1}{2}}}$

$k_1 = \dfrac{0.693}{17.5\text{ h}} = 0.0396$ $\qquad$ $T_1 = 125\ °C + 273 = 398$ K

$k_2 = \dfrac{0.693}{1.67\text{ h}} = 0.415$ $\qquad$ $T_2 = 145\ °C + 273 = 418$ K

$$\ln\frac{k_2}{k_1} = \frac{Ea}{R}\left(\frac{1}{T_1} - \frac{1}{T_2}\right)$$

$$\ln\frac{0.415}{0.0396} = \frac{Ea \text{ x } 1000\text{ J/kJ}}{8.3145\text{ J mol}^{-1}\text{K}^{-1}}\left(\frac{1}{398} - \frac{1}{418}\right)$$

$\ln 10.5 = 2.35 = E_a \times 0.0145$ kJ

$E_a = 163$ kJ

OR

$$\frac{\ln\dfrac{0.693}{t_{\frac{1}{2}}(2)}}{\ln\dfrac{0.693}{t_{\frac{1}{2}}(1)}} = \ln\frac{t_{\frac{1}{2}}(1)}{t_{\frac{1}{2}}(2)} = \ln\frac{12.5}{1.67} = \ln 10.5$$

$$\ln 10.5 = \frac{E_a}{R}\left(\frac{1}{T_1} - \frac{1}{T_2}\right)$$

$\ln 10.5 = 2.35 = Ea \times 0.0145$ kJ

$Ea = 163$ kJ

13.10A $NOCl \rightarrow NO + Cl$ $\qquad$ slow

$\underline{NOCl + Cl \rightarrow NO + Cl_2}$ $\qquad$ fast

$2\ NOCl \rightarrow 2\ NO + Cl_2$

13.10B (a)Rate of reaction $= k\,[NOCl]$

$NO + O_2 \underset{k_{-1}}{\overset{k_1}{\rightleftharpoons}} NO_3$ $\qquad$ fast

$\underline{NO_3 + NO \xrightarrow{k_2} 2\ NO_2}$ slow

$2\ NO + O_2 \rightarrow 2\ NO_2$

(b) rate $= k\,[NO]^2[O_2]$

from slow reaction rate $= k_2\,[NO_3]\,[NO]$

from fast reaction $\dfrac{k_1}{k_{-1}} = \dfrac{[NO_3]}{[O_2][NO]}$

$[NO_3] = \dfrac{k_1}{k_{-1}}\,[O_2]\,[NO]$

$$\text{rate} = k_2\frac{k_1}{k_{-1}}\ [NO]^2\ [O_2] = k\ [NO]^2\ [O_2]$$

Review Questions

4. The reaction usually goes more rapidly at the very beginning when the concentrations of reactants are at a maximum. An important exception is a zero-order reaction; it has a rate independent of concentration.

5. The average rate is the rate of reaction evaluated over a period of time, not the rate at one particular instant.
The initial rate is the rate of the first few percent of the reaction. The instantaneous rate is the rate at a given time, not necessarily the beginning of the reaction.
The average and instantaneous rates of reaction are the same for a zero-order reaction, and for reactions of other orders, they are nearly the same at the very start of the reaction.
The initial rate and the instantaneous rate are the same if the beginning of the reaction is chosen as the time period.
The three rates are all equal at all times in zero-order reactions.

8. The half-life is the time for 1/2 of the reactants initially present to react.
The half-life for a first-order reaction has a constant value, regardless of the initial concentration.
The half-life for a zero-order reaction depends on the initial concentration. The greater the initial concentration, the longer the half-life.
The half-life of a second-order reaction depends on the inverse of the initial concentration. Each successive half-life period is twice as long as the preceding one.

10. A rise in temperature causes a rise in the average kinetic energy, but the fraction of the molecules with kinetic energies sufficient to form an activated complex increases even more rapidly.

Problems

23. $$\frac{-\Delta A}{\Delta t} = \frac{-(0.1060\text{ M} - 0.1108\text{ M})}{12\text{ s}} = 4.0 \times 10^{-4}\text{ M/s}$$

$$4.0 \times 10^{-4}\text{ M/s} \times \frac{60\text{ sec}}{\text{min}} = 2.4 \times 10^{-2}\text{ M/min}$$

25. (a) 2.2×10^{-4} M/s rate of disappearance of B = rate of disappearance of A
$= -2.2 \times 10^{-4}$ M/s

(b) 1.1×10^{-4} M/s rate of formation of C = $-\frac{1}{2}$ rate of disappearance of A
$= 1.1 \times 10^{-4}$ M/s

(c) general rate of reaction = rate of formation of C = 1.1×10^{-4} M/s

27. zero order $[A]_t = -kt + [A]_0$

29. (a) Statement is true. The rate law is determined by the values of k and the exponents, m, n..., not by concentrations.
(b) The unit for the rate is M/s or M/min. That means the unit for k must be $M^{-1}s^{-1}$ or $M^{-1}min^{-1}$.

31. (a) Assume rate = $k[HgCl_2]^n[C_2O_4^{2-}]^m$
Determine n by using experiments 2 and 3

$$\frac{\text{rate}_2}{\text{rate}_3} = \frac{7.1 \text{ x } 10^{-5}}{3.5 \text{ x } 10^{-5}} = \frac{k\times(0.105\,\text{M})^n\times(0.30\,\text{M})^m}{k\times(0.052\,\text{M})^n\times(0.30\,\text{M})^m}$$

$2.03 = 2.02^n$
$n = 1$ first order in $HgCl_2$
Determine m by using experiments 1 and 2

$$\frac{\text{rate}_2}{\text{rate}_1} = \frac{7.1 \text{ x } 10^{-5}}{1.8 \text{ x } 10^{-5}} = \frac{k\times(0.105\,\text{M})\times(0.30\,\text{M})^m}{k\times(0.105\,\text{M})\times(0.15\,\text{M})^m}$$

$3.9 = 2^m$
$m = 2$ second order in $C_2O_4^{2-}$
Third order overall.
(b) Experiment 1

$$\frac{1.8 \text{ x } 10^{-5}\text{ M}}{\text{min}} = k \text{ x } (0.105\text{ M}) \times (0.15\text{ M})^2$$

$k = 7.6 \times 10^{-3}\ M^{-2}\ min^{-1}$
(c) rate = $7.6 \times 10^{-3}\ M^{-2}\ min^{-1} \times (0.020\text{ M}) \times (0.22\text{ M})^2$
rate = 7.4×10^{-6} M/min

33. $$\frac{\text{rate}_1}{\text{rate}_2} = \frac{k\,[A]_2^n}{k\,[A]_1^n}$$

$$\frac{\text{rate}_1}{0.25\text{ rate}_1} = \frac{k\,[A]_1^n}{k\,(0.50\,[A_1])^n}$$

$4 = 2^n$
n = 2, second order

35. rate = $k\,[H_2O_2] = 3.66 \times \frac{10^{-3}}{s} \times 2.05\text{ M} = 7.50 \times 10^{-3}$ M/s

37. (a) Rate = k [A]

$$k = \frac{0.00250\text{ M/s}}{0.484\text{ M}} = 5.17 \times 10^{-3}/\text{s}$$

(b) For a first-order reaction, the fraction remaining is independent of the initial concentration.

$$\ln\frac{\frac{1}{4}[A]_0}{[A_0]} = -kt$$

$$\ln\frac{1}{4} = -kt_{\frac{3}{4}}$$

$$t_{\frac{3}{4}} = \frac{\ln\frac{1}{4}}{-k} = \text{constant}$$

39. (a) $\ln\frac{1.50\text{ g}}{2.50\text{ g}} = -k\ 109\text{ s}$

$-0.511 = -k\ 109\text{ s}$

$k = 4.69 \times 10^{-3}/\text{s}$

(b) $t_{\frac{1}{2}} = \frac{0.693}{k} = \frac{0.693}{4.69 \text{ x } 10^{-3}/\text{s}} = 148\text{ s}$

(c) $\ln\frac{\text{m}}{2.50\text{ g}} = -4.69 \times 10^{-3}/\text{s} \times 5.0\text{ min} \times \frac{60\text{ s}}{\text{min}} = -1.41$

Raise both sides of the equation to a power of e.

$$e^{\ln\frac{\text{m}}{2.50\text{ g}}} = \frac{\text{m}}{2.50\text{ g}} = e^{-1.41} = 0.244$$

m = 0.61 g

41. $k = \frac{0.693}{32\text{ min}} = 2.17 \times 10^{-2}/\text{min}$

$$\ln\frac{N}{5.0 \text{ x } 10^{14}} = -2.17 \times 10^{-2}/\text{min} \times 1.50\text{ h} \times \frac{60\text{ min}}{\text{h}} = -1.95$$

$$e^{\ln\frac{N}{5.0 \text{ x } 10^{14}}} = \frac{N}{5.0 \text{ x } 10^{14}} = e^{-1.95} = 0.142$$

$N = 7.1 \times 10^{13}$ molecules/L

43. (a) $[NH_3]_{335} = 0.0452\text{ M} - (3.40 \times 10^{-6}\text{ M/s} \times 335) = 0.0441\text{ M}$

(b) $0.0226\text{ M} = 0.0452\text{ M} - 3.40 \times 10^{-6}\text{ M/s} \times t$

$t = 6.65 \times 10^3$ sec = 111 min = 1.85 h

45. For zero order, the half-life gets longer as the initial concentration increases because the rate is constant, and the more molecules present, the longer it takes to consume half of them.
For second order, the half-life is related to the inverse of the concentration—the greater the initial concentration, the faster the initial rate and the sooner the molecules are consumed.

47.

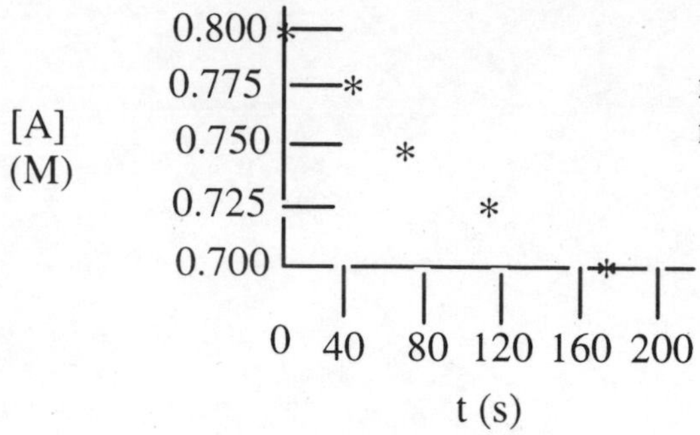

not straight
not zero order

t	[A]	ln [A]	$\frac{1}{[A]}$
0	.800	-0.223	1.25
40	.775	-0.255	1.29
83	.750	-0.288	1.33
129	.725	-0.322	1.38
179	.700	-0.357	1.43

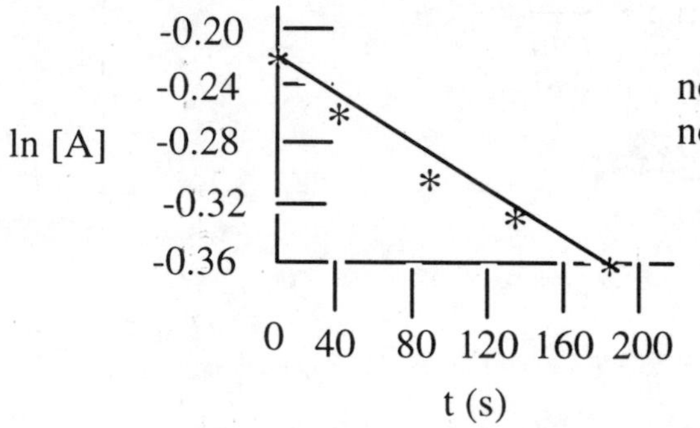

not straight
not first order

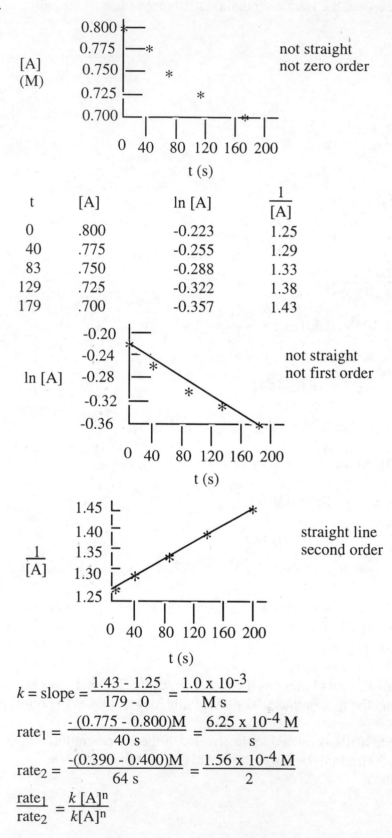

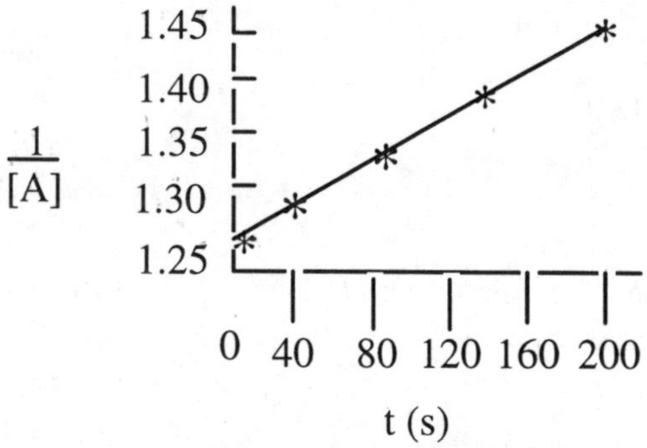

straight line
second order

$$k = \text{slope} = \frac{1.43 - 1.25}{179 - 0} = \frac{1.0 \times 10^{-3}}{\text{M s}}$$

$$\text{rate}_1 = \frac{-(0.775 - 0.800)\text{M}}{40\text{ s}} = \frac{6.25 \times 10^{-4}\text{ M}}{\text{s}}$$

$$\text{rate}_2 = \frac{-(0.390 - 0.400)\text{M}}{64\text{ s}} = \frac{1.56 \times 10^{-4}\text{ M}}{2}$$

$$\frac{\text{rate}_1}{\text{rate}_2} = \frac{k\,[A]^n}{k[A]^n}$$

$$\frac{(6.25) \times 10^{-4} M/s}{1.56 \times 10^{-4} M/s} = \left(\frac{0.800}{0.400}\right)^n$$

$4.00 = (2.00)^n$

$n = 2$ second order

rate $= k\,[A]^2$

$$k = \frac{6.25 \times 10^{-4} M/s}{(.800\ M)^2} = \frac{+9.77 \times 10^{-4}}{M\ s}$$

49. $$t_{1/2} = \frac{1}{k[A_0]_0}$$

$$k = \frac{1}{t_{1/2}[A_0]}$$

$$k = \frac{1}{[A]_0}\,ms$$

$$\frac{1}{\frac{1}{8}[A]_0} = \frac{8}{[A]_0}$$

$$\frac{1}{[A]_t} = kt + \frac{1}{[A]_0}$$

$$\frac{8}{[A]_0} - \frac{1}{[A]_0} = \frac{1}{[A]_0 12\ ms}t$$

$$\frac{7}{[A]_0} \times [A]_0\ 12\ ms = t$$

$t = 84$ ms

Since the half-life is dependent on the concentration and the first half-life is 12 ms, the second half-life concentration is half and the second half-life is twice as long (24 ms). To get to 1/8, the original value requires 12 ms + 24 ms + 48 ms = 84 ms.

51. The calculation involves not only the frequency of molecular collisions but also the fraction of the molecules with sufficient energies to react and a factor to account for the collisions that have a favorable orientation. These latter two quantities are much more difficult to assess than just a collision frequency.

53. E_a is very high. Very few molecules can react at room temperature. In the high-temperature region of the spark, the number of energetic molecules is greatly increased. Reaction starts here; liberated heat raises the temperature (creating more energetic molecules), and the entire mixture reacts with explosive speed.

55. $$\ln\frac{1.63 \times 10^{-3}}{4.75 \times 10^{-4}} = \frac{E_a \times 1000\ J/kJ}{\frac{8.3145\ J}{mol\ K}} \times \left(\frac{1}{293K} - \frac{1}{303K}\right)$$

$1.233 = E_a \times 120.3 \text{ mol/kJ} \times 1.13 \times 10^{-4}$
$E_a = 91 \text{ kJ /mol}$

57. (a) $\ln\dfrac{0.0120}{k_1} = \dfrac{218 \text{ kJ mol x } 1000 \text{ J /kJ}}{8.3145 \text{ J /mol K}} \times \left(\dfrac{1}{525 \text{ K}} - \dfrac{1}{652 \text{ K}}\right)$

$\ln\dfrac{0.0120}{k_1} = \dfrac{218 \text{ x } 1000}{8.3145} \times 3.71 \times 10^{-4} = 9.73$

Raise both sides of the equation to a power of e.

$e^{\ln\frac{0.0120}{k_1}} = \dfrac{0.0120}{k_1} = e^{9.73} = 1.68 \times 10^4$

$k_1 = \dfrac{0.0120}{1.68 \text{ x } 10^4} = 7.14 \times 10^{-7}\text{/min}$

(b) $\ln\dfrac{0.0120}{0.0100} = \dfrac{218 \text{ kJ /mol x } 1000 \text{ J /kJ}}{8.3145 \text{ J /mol K}} \times \left(\dfrac{1}{T_1} - \dfrac{1}{652 \text{ K}}\right)$

$6.954 \times 10^{-6} = \left(\dfrac{1}{T_1} - \dfrac{1}{652 \text{ K}}\right)$

$T_1 = 649 \text{ K}$

59. A molecule acquires enough excess energy through collisions with other molecules to enable it to dissociate. Its dissociation can occur without requiring a further collision.

61. (a) $A + B \rightarrow I$

$\underline{I + B \rightarrow C + D}$

$A + 2B \rightarrow C + D$ net reaction

(b) rate = k [A][B]

63. $2\,NO_2 \rightarrow NO_3 + NO$ slow

$\underline{NO_3 + CO \rightarrow NO_2 + CO_2}$ fast

$NO_2 + CO \rightarrow NO + CO_2$

65. from slow step rate = k [NO][Cl]

from fast step $\dfrac{k_1}{k_{-1}} = \dfrac{[NOCl][Cl]}{[NO][Cl_2]}$ $[Cl] = \dfrac{k_1}{k_{-1}}\dfrac{[NO][Cl_2]}{[NOCl]}$

$\text{rate} = k_2\dfrac{k_1}{k_{-1}}\dfrac{[NO]^2[Cl_2]}{[NOCl]}$

The rate law for the mechanism does not match the exhibited rate law.

67. rate = $k_2[Hg][Tl^{3+}]$

$\dfrac{k_1}{k_{-1}} = \dfrac{[Hg][Hg^{2+}]}{[Hg_2^{2+}]}$

$[Hg] = \dfrac{k_1[Hg_2^{2+}]}{k_{-1}[Hg^{2+}]}$

$$\text{rate} = k_2 \frac{k_1}{k_{-1}} \frac{[Hg_2^{2+}][Tl^{3+}]}{[Hg^{2+}]}$$

69. Because $[I^-]$ remains constant (it is a catalyst), the rate law, rate = $k[I^-][H_2O_2]$, simplifies to rate = $k'[H_2O_2]$. The value of k' depends on $[I^-]$ chosen, but once $[I^-]$ is fixed, the value of k' is fixed.

71. An inhibitor may block the active site of the enzyme, or it may react with the enzyme to change the shape of the active site.

73. Both an enzyme and the surface on which a surface-catalyzed reaction occurs require that reaction occurs at active sites. The kinetics of each type of reaction, then, is governed by the availability of these active sites.

Additional Problems

75. From the data pairs:

t = 0 s	[A] = 0.88 M	t = 100 s	[A] = 0.44 M
t = 25 s	[A] 0.74 M	t = 125 s	[A] = 0.37 M
t = 50 s	[A] = 0.62 M	t = 150 s	[A] = 0.31 M

The half-life is 100 s.

$$k = \frac{0.693}{100\text{ s}} = 6.93 \times 10^{-3}\text{ s}^{-1}$$

$$\text{rate} = k\,[A]_{125} = 6.93 \times 10^{-3}\text{ s}^{-1} \times 0.37\text{ M} = 2.6 \times 10^{-3}\text{ M/s}$$

77.

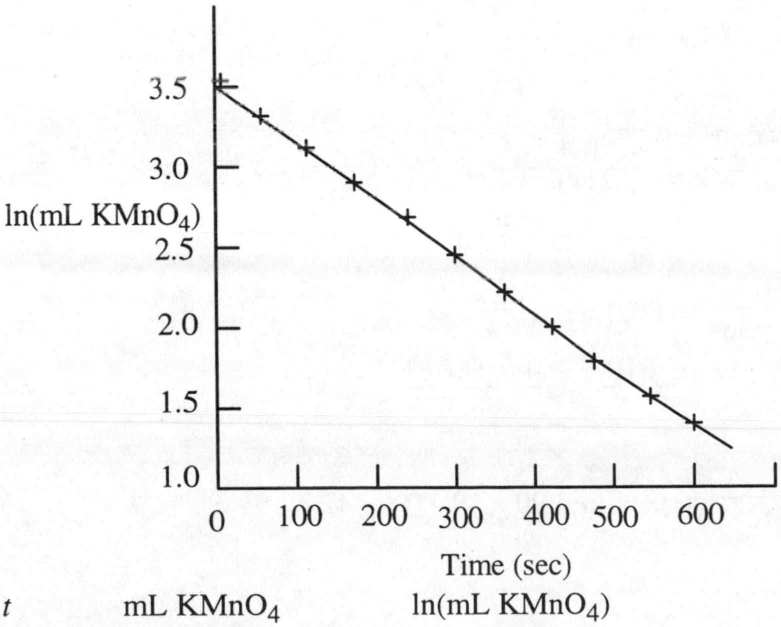

t	mL $KMnO_4$	ln(mL $KMnO_4$)
0	35.3	3.56
60	27.9	3.33
120	22.6	3.12

180	18.3	2.91
240	14.9	2.70
300	11.9	2.48
360	9.44	2.25
420	7.52	2.02
480	6.08	1.81
540	4.80	1.57
600	3.76	1.32

$k = -\text{ slope} = \dfrac{-(1.57-3.33)}{(540-60)\text{s}} 3.67 \times 10^{-3}\ \text{s}^{-1}$ A graphing calculator or computer program can produce a slope value using all of the data points that would be more accurate.

79. (a) $P = \dfrac{mRT}{VM} = \dfrac{4.50\text{ g x } \dfrac{0.08206\text{ L atm}}{\text{K mol}} \text{ x } (147+273)\text{K}}{1.00\text{ L x } 146.2\text{ g/mol}} = 1.06\text{ atm}$

(b) after 1 half-life $P_{DTBP} = 0.53$ atm

$P_{acetone} = 1.06$ atm $P_{ethane} = 0.53$ atm

$P_{total} = 2.12$ atm

(c) $\ln \dfrac{P_t}{P_o} = -kt = -\dfrac{0.693}{t_{1/2}} t$

$\ln \dfrac{P_t}{1.06} = -\dfrac{0.693}{80.0\text{ min}} \times 125\text{ min} = -1.08$

Raise both sides of equation to power of e.

$e^{\ln \frac{P_t}{1.06}} = \dfrac{P_t}{1.06\text{ atm}} = e^{-1.08} = 0.339$

$P_t = 0.36\text{ atm} = P_{DTBP}$

$P_{ethane} = 1.06 - 0.36 = 0.70$ atm

$P_{acetone} = P_{ethane} \times 2 = 1.40$ atm

$P_{total} = 0.36 + 0.70 + 1.40 = 2.46$ atm

81. $t_{1/2} = 1.00\text{ min} \times \dfrac{60\text{ sec}}{\text{min}} = 60.0\text{ sec} = \dfrac{0.693}{k}$

$k = \dfrac{0.693}{t_{1/2}} = \dfrac{0.693}{60.0\text{ s}} = \dfrac{1.16 \text{ x } 10^{-2}}{\text{s}} = \dfrac{4.0 \text{ x } 10^{16}}{\text{s}} e^{-Ea/RT}$

$\ln e^{\frac{-Ea}{RT}} = \dfrac{-Ea}{RT} = \ln\left(\dfrac{1.16\ 10^{-2}}{4.0\ 10^{16}}\right) = \ln(2.90 \times 10^{-19}) = -42.7$

$\dfrac{\dfrac{-262\text{ kJ}}{\text{mol}} \text{ x } \dfrac{10^3\text{J}}{\text{kJ}}}{\dfrac{8.3145\text{ J}}{\text{mol K}}\text{T}} = -42.7$

$T = 738\text{ K} - 273 = 465\ °\text{C}$

83. $\ln \frac{2k}{k} = \frac{E_a \text{ x } 1000 \text{ J /kJ}}{8.3145 \text{ J /mol K}} \times \left(\frac{1}{298} - \frac{1}{308}\right)$

$0.693 = E_a \times 120.3 \text{ mol /kJ} \times 1.09 \times 10^{-4}$

$E_a = 53 \text{ kJ /mol}$

85. (a) For $[OCl^-]$ use rate1/rate 3

$$\frac{\text{rate1}}{\text{rate3}} = \frac{k}{k}\left(\frac{0.0040}{0.0020}\right)^n \times \left(\frac{0.0020}{0.0020}\right)^m \times \frac{[1.00]^p}{[1.00]^p}$$

$$\frac{4.8 \text{ x } 10^{-4}}{2.4 \text{ x } 10^{-4}} = \left(\frac{0.0040}{0.0020}\right)^n$$

$2 = 2^n$

$n = 1$ first order in OCl^-

For $[I^-]$ use rate2/rate 3

$$\frac{\text{rate2}}{\text{rate3}} = \frac{k\,[0.0020]}{k\,[0.0020]} \times \left(\frac{0.0040}{0.0020}\right)^m \times \frac{[1.00]^p}{[1.00]^p}$$

$$\frac{5.0 \text{ x } 10^{-4}}{2.4 \text{ x } 10^{-4}} = \left(\frac{0.0040}{0.0020}\right)^m$$

$2.08 = 2^m$

$m = 1$ first order in I^-

For $[OH^-]$ use rate4/rate 3

$$\frac{\text{rate4}}{\text{rate3}} = \frac{k\,[0.0020]}{k\,[0.0020]} \times \frac{[0.0020]}{[0.0020]} \times \left(\frac{0.50}{1.00}\right)^p$$

$$\frac{4.6 \text{ x } 10^{-4}}{2.4 \text{ x } 10^{-4}} = \left(\frac{0.50}{1.00}\right)^p$$

$$1.92 = \left(\frac{1}{2}\right)^p$$

$\log 1.92 = p \log 0.5$

$$p = \frac{\log 1.92}{\log 0.5} = -0.94$$

$p = -1$ negative first order

Overall order of 1

(b) $\text{rate} = k \frac{[OCl^-][I^-]}{[OH^-]}$

$$\frac{4.8 \text{ x } 10^{-4} \text{ M}}{\text{s}} = k \frac{(0.0040 \text{ M})(0.0020 \text{ M})}{1.00 \text{ M}}$$

$$k = \frac{60}{\text{s}}$$

(c) $OCl^- + H_2O \rightleftharpoons HOCl + OH^-$

$I^- + HOCl \rightarrow HOI + Cl^-$

$HOI + OH^- \rightarrow H_2O + OI^-$

$OCl^- + I^- \rightarrow Cl^- + OI^-$

The first step is a fast reversible reaction. The rapid equilibrium assumption leads to:

$$\frac{k_1}{k_{-1}} = \frac{[HOCl][OH^-]}{[OCl^-][H_2O]}$$

$$[HOCl] = \frac{k_1}{k_{-1}} \frac{[OCl^-][H_2O]}{[OH^-]}$$

The second step is the rate-determining step:

$$\text{rate} = k_2[I^-][HOCl]$$

$$\text{rate} = k_2 \frac{k_1}{k_{-1}} \frac{[I^-][OCl^-][H_2O]}{[OH^-]}$$

Note that $[H_2O]$ is a constant.

$$\text{rate} = k_2 \frac{k_1[H_2O]}{k_{-1}} \frac{[I^-][OCl^-]}{[OH^-]} = k_{total} \frac{[I^-][OCl^-]}{[OH^-]}$$

The third step is a rapid neutralization of the acid HOI by the base OH^-.

(d) A catalyst makes the reaction faster. Increasing the amount of catalyst would either increase the rate of the reaction or have no effect. In this reaction, the OH^- acts as an inhibitor.

87. The NH_2 groups become NH_3^+ groups, and COOH groups retain the H in acidic solutions. Because the enzyme is active in basic solution, it must be that the substituent groups must be present as NH_2 and COO^-, which they are in basic solutions.

89.

Time, hr	mass, g	log mass	1/mass
0	1.71	0.233	0.585
3.58	1.15	0.0607	0.870
7.67	0.87	-0.060	1.15
24.33	0.46	-0.337	2.17
44.58	0.28	-0.553	3.57

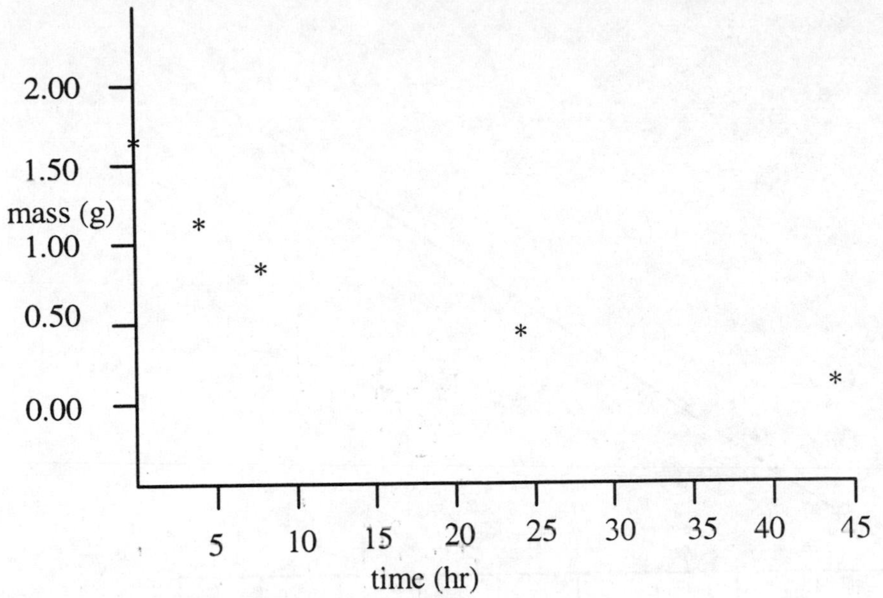

This is not a straight line. It is not zero order.

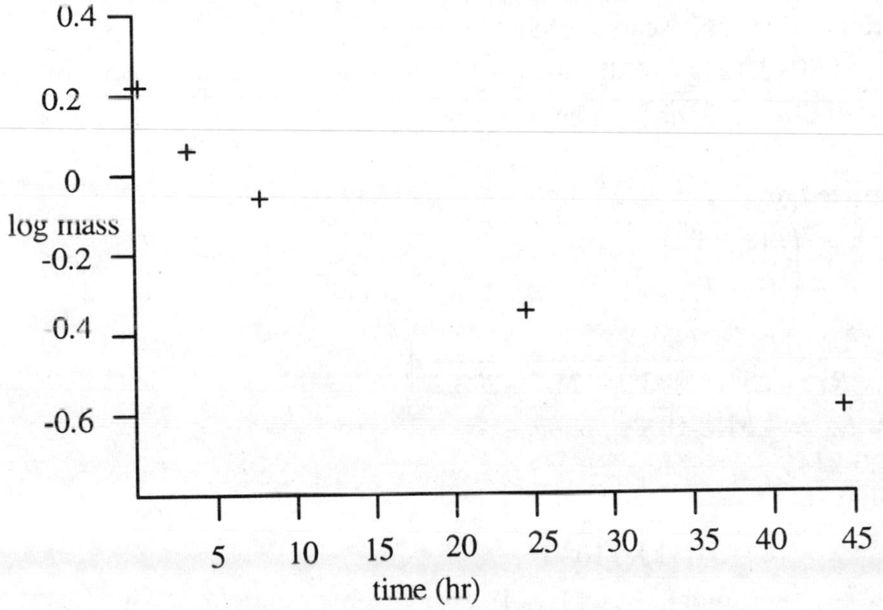

This is not a straight line It is not first order.

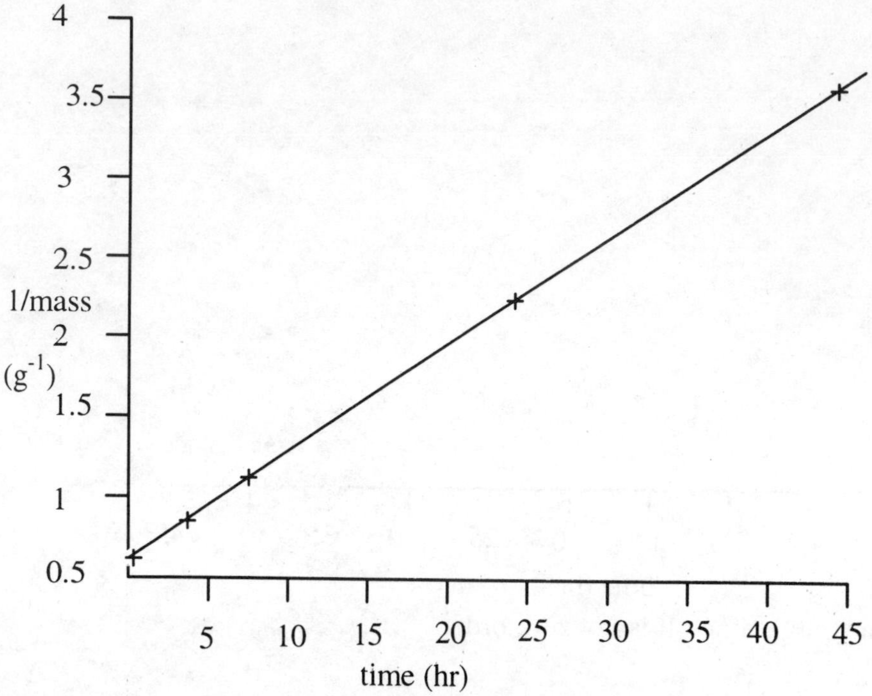

This is a straight line. It is second order.

$$k = \text{slope} = \frac{(4.00 - 1.15)\,\text{g}^{-1}}{(19.08 - 3.58)\,\text{hr}} = \frac{0.184}{\text{g hr}}$$

Apply Your Knowledge

91. (a) $\ln\frac{r2}{r1} = \frac{E_a}{R}\times\left(\frac{1}{T_1}-\frac{1}{T_2}\right)$

$$\ln\frac{142}{179} = \frac{E_a}{R}\times\left(\frac{1}{25.0+273.2}-\frac{1}{21.7+273.2}\right)$$

$-0.232 = Ea \times -4.51 \times 10^{-3}$

$$Ea = \frac{51.3\text{ kJ}}{\text{mol}}$$

(b) $$\ln\frac{r2}{r1} = \ln\frac{r2}{179} = \frac{\frac{-51.3\text{ kJ}}{\text{mol K}}\times\frac{10^3\text{ J}}{\text{kJ}}}{\frac{8.3145\text{ J}}{\text{mol K}}}\times\left(\frac{1}{25.0+273.2}-\frac{1}{20.0+273.2}\right) = -0.353$$

Raise both sides of the equation to a power of e.

$$e^{\ln\frac{r_2}{179}} = \frac{r2}{179} = e^{-0.353} = 0.702$$

$$r_2 = \frac{126\text{ chirps}}{\text{min}}$$

(c) $$\frac{126\text{ chirps}}{\text{min}} = \frac{31.5\text{ chirps}}{15\text{ s}}$$

40 + 31.4 = 71.4 °F

$°F = (20.0\ °C \times \frac{9}{5}) + 32° = 68\ °F$

The rule of thumb is close but not exact.

93.

	cpm	ln cpm
0	648	6.474
3.0	633	6.450
12.0	589	6.378
18.0	562	6.332

$\ln A = -kt + \ln A_0$

$$\text{slope} = -k = \frac{6.474 - 6.378}{12.0 - 0} = -8.00 \times 10^{-3}$$

$k = 8.00 \times 10^{-3}$ (k could also be found using a graphing calculator.)

$$t_{1/2} = \frac{.693}{8.00 \times 10^{-3}} = 86.6 \text{ hr}$$

25 % is after 2 half lives, 173.2 hr.

Chapter 14

Chemical Equilibrium

Exercises

14.1A $K_c = \frac{[COCl_2]}{[CO][Cl_2]} = \frac{[COCl_2]}{[CO]^2}$

There would not be just one value for $[COCl_2]$. There would be a different value of $[COCl_2]$ for each value of CO.

14.1B $[SO_2]$ and $[SO_3]$ do not have unique values, but the ratios $[SO_2]/[SO_3]$, $[SO_2]^2/[SO_3]^2$, and their inverses do have unique values.

$$1.00 \times 10^2 = \frac{[SO_3]^2}{[SO_2]^2[O_2]}$$

When $[O_2]$ is set, then the ratio $\frac{[SO_3]^2}{[SO_2]^2}$ must have a set value. For $[O_2] = 1.00$ M,

$\frac{[SO_3]^2}{[SO_2]^2} = 1.00 \times 10^2$ and $\frac{[SO_3]}{[SO_2]} = 10.$

14.2A $K_c = \left(\frac{1}{K_c}\right)^2 = \left(\frac{1}{20}\right)^2 = 2.5 \times 10^{-3}$

14.2B The reaction is the reverse and four times the first reaction.

$$K_c = \frac{1}{(1.97 \times 10^{-20})^4} = 6.64 \times 10^{78}$$

14.3A $K_c = \left(\frac{1}{1.8x10^{-6}}\right)^{\frac{1}{2}}$ for $NO_2 \rightleftharpoons NO + \frac{1}{2}\ O_2$

$K_c = 7.5 \times 10^2$

$K_p = K_c\,(RT)^{\Delta n} = 7.5 \times 10^2 \times (0.08206 \text{ x } 457)^{1/2}$

$K_p = 4.6 \times 10^3$

14.3B $K_c = \frac{K_p}{(RT)^{\frac{9}{4} - \frac{5}{2}}}$ for $NO + \frac{3}{2}\ H_2O \rightleftharpoons NH_3 + \frac{5}{4}\ O_2$

$$K_c = \frac{2.6 \text{ x } 10^{-16}}{(0.0821 \text{ x } 900)^{-\frac{1}{4}}} = (2.6 \times 10^{-16}) \times (0.0821 \times 900)^{\frac{1}{4}} = 7.6 \times 10^{-16}$$

For $4\ NH_3 + 5\ O_2 \rightleftharpoons 4\ NO + 6\ H_2O$, the chamical equation is four times and the reverse of the starting equation.

$$K_c = \frac{1}{(7.6 \text{ x } 10^{-16})^4} = 3.0 \times 10^{60}$$

14.4A $K_p = \dfrac{P_{CO}\,P_{H_2}}{P_{H_2O}}$

14.4B $K_c = \dfrac{[H_2\ (g)]^4}{[H_2O\ (g)]^4}$ $\qquad K_p = \dfrac{P_{H_2}^{\ 4}}{P_{H_2O}^{\ 4}}$

14.5 A value of K greater than 1 (1.2×10^3) means that the forward rcaction is favored. Because K is not extremely large, the reaction will not go to the point where the reactant concentrations arc essentially zero.

14.6A $Q_c = \dfrac{[H_2][I_2]}{[HI]^2} = \dfrac{0.100 \text{ x } 0.100}{1.00^2} = 1.00 \times 10^{-2}$

$Q_c < K_c$ $\quad 1.00 \times 10^{-2} < 1.84 \times 10^{-2}$ The reaction will proceed to the right.

14.6B $Q_p = \dfrac{P_{HI}^2}{P_{H_2S}} = \dfrac{(0.0010)^2}{(0.010)} = 1.0 \times 10^{-4}$

$Q_p > K_p$ reaction to left.

At equilibrium P_{H_2S} will be greater, P_{HI} will be less; I_2 (s) amount will increase and S (s) amount will dccrease.

14.7 (a) Reaction would go to thc right, using up N_2 and producing more NH_3. The amount of H_2, however, would be greater than in the original equilibrium. The amount of N_2 would be less than in the original equilibrium.

(b) Reaction would go to the left, using up NH_3 and producing more H_2.

(c) Reaction would go to the right by consuming N_2 and H_2. The amount of NH_3 would be less than in the original equilibrium.

14.8A There is no change in the equilibrium amount of HI by changing the pressure or volume, because there are the same number of moles of gas on both sides of the equation.

14.8B Additional NO_2 will cause the reaction to go to the left to use up the excess NO_2. The volume increase causes a pressure decrease; the system reacts to the left to generate more moles of gas and, thus, more pressure. Both the volume change and NO_2 addition push the reaction to the left.

14.9 At low temperature. The reaction is forced to the right to make more heat by its exothermic reaction, and therefore it is more complete at low temperatures.

14.10 (a) The reverse reaction occurs, using up CO_2 and making CO and H_2O. There will be more CO and H_2O and less CO_2. There will be more H_2 because the extra is not all used up.
(b) Because both a reactant (H_2O) and a product (H_2) are added to the equilibrium mixture, we cannot make a qualitative prediction of whether the equilibrium will shift to the left or to the right.
(c) Adding more H_2O favors the forward reaction, as does lowering the temperature (the forward reaction is exothermic). The new equilibrium will have more CO_2, and H_2, and less CO than the original equilibrium. The amount of H_2O is in doubt, because it is not known if the combined effects of the two changes will consume more or less than the 1.00 mol H_2O added.

14.11A $K_c = \dfrac{[PCl_5]}{[PCl_3][Cl_2]}$

	PCl_3 +	Cl_2 $\rightleftharpoons$	PCl_5
Initial, M:	1.00	1.00	0
Changes, M:	-0.82	-0.82	+0.82
Equilibrium, M:	0.18	0.18	0.82

$$K_c = \frac{0.82}{(0.18)^2} = 25$$

14.11B $[H_2] = \dfrac{10.0\text{ g}}{25.0\text{ L}} \times \dfrac{\text{mol}}{2.016\text{ g}} = 0.198\text{ M}$

$$[H_2S] = \frac{72.6\text{ g}}{25.0\text{ L}} \times \frac{\text{mol}}{34.09\text{ g}} = 0.085\text{ M}$$

	Sb_2S_3 (s) + 3 H_2 (g) $\rightleftharpoons$	2 Sb (s) + 3 H_2S (g)
Initial, M:	0.198	--
Changes, M:	–0.085	0.085
Equilibrium, M:	0.113	0.085

$[H_2]_{eq} = 0.1984\text{ M} - [H_2S]_{eq} = 0.198\text{ M} - 0.085\text{ M} = 0.113\text{ M}$

$$K_c = \frac{[H_2S]^3}{[H_2]^3} = \frac{(0.0852)^3}{(0.113)^3}$$

$K_c = 0.429$

$K_p = K_c(RT)^{\Delta n} = K_c(RT)^0 = 0.429$

14.12A (a)

	CO +	H_2O $\rightleftharpoons$	CO_2 +	H_2
Initial, M:	$\dfrac{0.100\text{ mol}}{5.00\text{ L}}$	$\dfrac{0.100\text{ mol}}{5.00\text{ L}}$		
Change, M:	$-X$	$-X$	X	X
Equilibrium, M:	$0.02 - X$	$0.02 - X$	X	X

$$\frac{[CO_2][H_2]}{[CO][H_2O]} = 23.2 = \frac{X^2}{(0.02 - X)^2}$$

$$\sqrt{23.2} = 4.82 = \frac{X}{0.02 - X}$$

$0.0963 - 4.82\,X = X$
$0.0963 = 5.82\,X$
$X = 0.0165$ M
? mol H_2 = 0.0165 M × 5.00 L = 0.0827 mol H_2

(b) $$P = \frac{nRT}{V} = \frac{0.0827\,\text{mol} \times \frac{0.08206\,\text{L atm}}{\text{K mol}} \times 600\,\text{K}}{5.00\,\text{L}}$$

$P = 0.814$ atm

14.12B $$K = \frac{[HI]^2}{[H_2][I_2]} = \frac{\left(\frac{\text{mol HI}}{V}\right)^2}{\left(\frac{\text{mol } H_2}{V}\right) \text{x} \left(\frac{\text{mol } I_2}{V}\right)} = \frac{(\text{mol HI})^2}{\text{mol } H_2 \text{ x mol } I_2}$$

This would be true only for those reactions that have the same number of moles on both sides of the equation, so the volumes cancel.

14.13A	CO	+	Cl_2	$\rightleftharpoons$	$COCl_2$
Initial, M:	$\frac{0.100 \text{ mol}}{25.0 \text{ L}}$		$\frac{0.200 \text{ mol}}{25.0 \text{ L}}$		--
Change, M:	$-X$		$-X$		X
Equilibrium, M:	$0.00400 - X$		$0.00800 - X$		X

$$K_c = 1.2 \times 10^3 = \frac{X}{(0.00400 - X)\times(0.00800 - X)} = \frac{X}{3.20\times10^{-5} - 0.01200X + X^2}$$

$1.2 \times 10^3\,X^2 - 14.4\,X + 3.84 \times 10^{-2} = X$
$1.2 \times 10^3\,X^2 - 15.4\,X + 3.84 \times 10^{-2} = 0$
$X^2 - 1.28 \times 10^{-2}\,X + 3.2 \times 10^{-5} = 0$

$$X = \frac{-b \pm \sqrt{b^2 - 4ac}}{2a}$$

$$= \frac{1.28 \text{ x } 10^{-2} \pm \sqrt{(1.28 \text{ x } 10^{-2})^2 - 4 \text{ x } (1) \text{ x } 3.2 \text{ x } 10^{-5}}}{2}$$

$$= \frac{1.28 \text{ x } 10^{-2} \pm 5.99 \text{ x } 10^{-3}}{2}$$

$X = 3.4 \times 10^{-3}$ or 9.4×10^{-3}
too large, not valid
? mol $COCl_2$ = 3.4×10^{-3} M × 25.0 L = 8.5×10^{-2} mol

14.13B	N_2	+	O_2	$\rightleftharpoons$	2 NO
Initial, mol:	0.78		0.21		--
Change, mol:	$-X$		$-X$		$2X$

Equilibrium, mol: 0.78-X 0.21-X 2X

$$\frac{(2X)^2}{(0.78 - X)(0.21 - X)} = 2.1 \times 10^{-3}$$

$$\frac{4X^2}{0.164 - 0.99X + X^2} = 2.1 \times 10^{-3}$$

$4X^2 = 2.1 \times 10^{-3}\,(0.164 - 0.99X + X^2)$

$4X^2 = 3.44 \times 10^{-4} - 2.08 \times 10^{-3}X + 2.1 \times 10^{-3}X^2$

$4X^2 - 2.1 \times 10^{-3}X^2 \approx 4X^2$

$4X^2 + 2.08 \times 10^{-3}X - 3.44 \times 10^{-4} = 0$

$$X = \frac{-2.08 \text{ x } 10^{-3} \pm \sqrt{(2.08 \text{ x } 10^{-3})^2 + 4 \text{ x } 4 \text{ x } 3.44 \text{ x } 10^{-4}}}{2 \text{ x } 4}$$

$$X = \frac{-2.08 \text{ x } 10^{-3} \pm 7.42 \text{ x } 10^{-2}}{8} = 9.0 \times 10^{-3}$$

at equilibrium moles NO = $2 \times 9.0 \times 10^{-3}$ mol = 0.018 mol

moles N_2 = 0.78 mol – 0.009 mol = 0.77 mol

moles O_2 = 0.21 mol – 0.009 mol = 0.20 mol

$$\text{mol fraction of NO} = \frac{0.018 \text{ mol NO}}{(0.77 + 0.20 + 0.018) \text{ mol}} = 0.018$$

The volume does not matter, because it will cancel out in the equilibrium constant expression. To make concentrations and numbers of moles of reactant numerically equal, a volume of 1.00 L can be assumed.

14.14A H_2 + I_2 $\rightleftharpoons$ 2 HI

0.0100+X X 0.100-2X

The reaction must go to the left because there is no I_2 present initially.

$$\frac{(0.100\text{-}2X)^2}{(0.0100+X)X} = 54.3$$

$0.0100 - 0.400\,X + 4X^2 = 54.3\,X^2 + 0.543\,X$

$50.3\,X^2 + 0.943\,X - 0.0100 = 0$

$$X = \frac{-0.943 \pm \sqrt{(0.943)^2 - 4 \text{ x } 50.3 \text{ x } (-0.0100)}}{2 \text{ x } 50.3}$$

$X = 7.6 \times 10^{-3}$ or -0.26 not valid

H_2 1.76×10^{-2} moles

I_2 7.6×10^{-3} moles

HI $0.1 - 2 \times 7.6 \times 10^{-3} = 0.085$ moles

In this reaction the same answer will be obtained regardless of the volume of the system. To make concentrations and numbers of moles of reactant numerically equal, a volume of 1.00 L can be assumed.

14.14B $$Q = \frac{[HI]^2_{init}}{[H_2]_{init} \text{ x } [I_2]_{init}}$$

$$Q = \frac{(0.100/5.25)^2}{(0.0100/5.25)(0.0100/5.25)} = 100$$

Because $Q > K_c$ (that is, 100 > 54.3), a net reaction must occur in the reverse direction. At equilibrium, the amounts of H_2 and I_2 will be greater than initially, and the amount of HI will be less. This guides us in labeling the changes in amounts as positive or negative.

The reaction:	$H_2(g)$ +	$I_2(g)$	$\rightleftharpoons$	$2\ HI(g)$
Initial amounts:	0.0100 mol	0.0100 mol		0.100 mol
Changes:	$+X$ mol	$+X$ mol		$-2X$ mol
Equil. amounts:	$(0.0100 + X)$ mol	$(0.0100 + X)$ mol		$(0.100 - 2X)$ mol
Equil. concns., M:	$\dfrac{(0.0100 + X)}{5.25}$	$\dfrac{(0.0100 + X)}{5.25}$		$\dfrac{(0.0100 + X)}{5.25}$

$$K_c = \frac{[HI]^2}{[H_2][I_2]} = \frac{\left(\dfrac{0.100 - 2X}{5.25}\right)^2}{\left(\dfrac{0.0100 + X}{5.25}\right)\left(\dfrac{0.0100 + X}{5.25}\right)} = 54.3$$

$$K_c = \frac{(0.100 - 2X)^2}{(0.0100 + X)^2} = 54.3$$

$$\left(\frac{(0.100 - 2X)^2}{(0.0100 + X)^2}\right)^{1/2} = \frac{0.100 - 2X}{0.0100 + X} = (54.3)^{1/2}$$

$0.100 - 2X = (54.3)^{1/2} \times (0.0100 + X)$

$0.100 - 2X = 0.0737 + 7.37X$

$9.37X = 0.0263$

$X = 0.00281$

The equilibrium amounts are:

H_2: $0.0100 + X = 0.0100 + 0.00281 = 0.0128$ mol H_2

I_2: $0.0100 + X = 0.0100 + 0.00281 = 0.0128$ mol I_2

HI: $0.100 - 2X = 0.100 - (2 \times 0.00281) = 0.094$ mol HI

14.15A

	CO +	Cl_2	$\rightleftharpoons$ $COCl_2$ $K_p = 22.5$
Initial, atm:	1.00	1.00	--
Change, atm:	$-X$	$-X$	X
Equilibrium, atm:	$1.00 - X$	$1.00 - X$	X

$$K_p = \frac{P_{COCl_2}}{P_{CO}P_{Cl_2}} = \frac{X}{(1.00 - X)^2} = 22.5$$

$$22.5 = \frac{X}{1.00 - 2.00X + X^2}$$

$22.5 - 45.0 + 22.5X^2 = X$

$22.5X^2 - 46.0X + 22.5 = 0$

$X^2 - 2.04X + 1 = 0$

$$X = \frac{-b \pm \sqrt{b^2 - 4ac}}{2a} = \frac{2.04 \pm \sqrt{(2.04)^2 - 4 X 1 \text{ x } 1}}{2}$$

$$X = \frac{2.04 \pm 0.42}{2}$$

$X = 0.81$ or 1.23 too large, not valid

$P_{COCl_2} = 0.81$ atm

$P_{CO} = 1.00 - 0.81 = 0.19$ atm

$P_{Cl_2} = 1.00 - 0.81 = 0.19$ atm

$P_{total} = P_{COCl_2} + P_{CO} + P_{Cl_2} = 1.19$ atm

14.15B $K_p = 0.108 = P_{NH_3} \times P_{H_2S}$

$P_{NH_3} = P_{H_2S}$

$P_{NH_3} = \sqrt{0.108} = 0.329$ atm

$P_T = P_{NH_3} + P_{H_2S} = 0.329 \text{ atm} + 0.329 \text{ atm} = 0.658$ atm

Review Questions

10. (a) $K_c = \dfrac{[CO]^2}{[CO_2]}$ (b) $K_c = \dfrac{[HI]^2}{[H_2S]}$ (c) $K_c = [O_2]$

11. (a) $2\ H_2S(g) \rightleftharpoons 2\ H_2(g) + S_2(g)$ $\quad K_c = \dfrac{[S_2][H_2]^2}{[H_2S]^2}$

(b) $CS_2(g) + 4\ H_2(g) \rightleftharpoons CH_4(g) + 2\ H_2S(g)$ $K_c = \dfrac{[CH_4][H_2S]^2}{[CS_2][H_2]^4}$

(c) $CO(g) + 3\ H_2(g) \rightleftharpoons CH_4(g) + H_2O(g)$ $\quad K_c = \dfrac{[CH_4][H_2O]}{[CO][H_2]^3}$

12. (a) $K_p = \dfrac{P_{CO_2} P_{H_2}}{P_{CO} P_{H_2O}}$ (b) $K_p = \dfrac{P^2_{NH_3}}{P_{N_2} P^3_{H_2}}$ (c) $K_p = P_{NH_3} P_{H_2S}$

13. (a) $1/2\ N_2(g) + 1/2\ O_2(g) \rightleftharpoons NO(g)$, $K_p = \dfrac{P_{NO}}{(P_{N_2})^{1/2}(P_{O_2})^{1/2}}$

(b) $1/2\ N_2(g) + 3/2\ H_2(g) \rightleftharpoons NH_3(g)$, $K_p = \dfrac{P_{NH_3}}{(P_{N_2})^{1/2}(P_{H_2})^{3/2}}$

(c) $1/2\ N_2(g) + 1/2\ O_2(g) + 1/2\ Cl_2(g) \rightleftharpoons NOCl\ (g)$,

$$K_p = \frac{P_{NOCl}}{(P_{N_2})^{1/2}\ (P_{O_2})^{1/2}(P_{Cl_2})^{1/2}}$$

14. (a) $2\,CO(g) + 2\,NO(g) \rightleftharpoons N_2(g) + 2\,CO_2(g)$ $\quad K_c = \dfrac{[N_2][CO_2]^2}{[CO]^2[NO]^2}$

(b) $5\,O_2(g) + 4\,NH_3(g) \rightleftharpoons 4\,NO(g) + 6\,H_2O(g)$ $\quad K_c = \dfrac{[NO]^4[H_2O]^6}{[NH_3]^4[O_2]^5}$

(c) $2\,NaHCO_3(s) \rightleftharpoons Na_2CO_3(s) + H_2O(g) + CO_2(g)$ $\quad K_c = [H_2O][CO_2]$

Problems

19. (a) $K_p = K_c\,(RT)^{\Delta n}$

Δn = moles gascous products - moles gaseous reactants

$K_p = 1.2 \times 10^3 \times (0.08206 \times 668)^{-1} = 22$

(b) $K_p = 1.32 \times 10^{-2} \times (0.08206 \times 1000)^{-1} = 1.61 \times 10^{-4}$

(c) $K_p = 2.00 \times (0.08206 \times 1273)^0 = 2.00$

21. $K_c = \left(\dfrac{1}{4.08 \text{ x } 10^{-4}}\right)^{\frac{1}{2}} = 49.5$

23. $K_p = K_c(RT)^{4-2}$ for $2\,NH_3 \rightleftharpoons N_2 + 3\,H_2$

$K_p = 6.46 \times 10^{-3} \times \left(\dfrac{0.08206\,\text{L atm}}{\text{K mol}} \times (273 + 300)\,\text{K}\right)^2 = 14.3$

New equation is reverse and one-third of original.

$K_p = \dfrac{1}{(14.3)^{1/3}} = 0.412$

25. $CH_4 + H_2O \rightleftharpoons CO + 3\,H_2$ $\quad K_{p1} = 1.25 \times 10^{-25}$

$\underline{CO + H_2O \rightleftharpoons CO_2 + H_2}$ $\quad \dfrac{1}{K_{p2}} = \dfrac{1}{9.80 \times 10^{-6}}$

$CH_4 + 2\,H_2O \rightleftharpoons CO_2 + 4\,H_2$ $\quad K_{p3} = \dfrac{K_{p1}}{K_{p2}} = \dfrac{1.25 \times 10^{-25}}{9.80 \times 10^{-6}}$

$K_{p3} = 1.28 \times 10^{-20}$

Coefficients of the desired equation are ½ of those in the above equation.

$K_p = \sqrt{K_{p3}} = (1.28 \times 10^{-20})^{1/2} = 1.13 \times 10^{-10}$

27. $\frac{1}{2}\,N_2 + O_2 \rightleftharpoons NO_2$ $\quad K_{p1} = 1.0 \times 10^{-9}$

$NO_2 + \frac{1}{2}\,Cl_2 \rightleftharpoons NO_2Cl$ $\quad K_{p2} = 0.3$

$\underline{NO_2Cl \rightleftharpoons \frac{1}{2}\,O_2 + NOCl}$ $\quad \dfrac{1}{K_{p3}} = \dfrac{1}{1.1 \text{ x } 10^2}$

$\frac{1}{2}\ N_2 + \frac{1}{2}\ O_2 + \frac{1}{2}\ Cl_2 \rightleftharpoons NOCl$ $\qquad \dfrac{K_{p1} \times K_{p2}}{K_{p3}}$

$$Kp = \frac{1.0 \times 10^{-9} \times 0.3}{1.1 \times 10^{2}} = 3 \times 10^{-12}$$

$$K_c = \frac{K_p}{(RT)^{\Delta n}} = \frac{3 \times 10^{-12}}{(0.0821 \times 298\ K)^{1-3/2}}$$

$$K_c = 3 \times 10^{-12} \times (0.0821 \times 298)^{1/2}$$

$$K_c = 1 \times 10^{-11}$$

29. K_p for the reaction 1/2 $N_2O_4 \rightleftharpoons NO_2$ is greater than that for the reaction $N_2O_4 \rightleftharpoons 2\ NO_2$. This is because the square root of a number smaller than 1 is larger than the number. That is, $\sqrt{0.145} = 0.381$. $K_p > 0.145$.

31. $K_c = \dfrac{[C]^2}{[A][B]^2} = \dfrac{(0.55)^2}{(0.025)(0.15)^2} = 5.4 \times 10^2$

33. $K_c = \dfrac{[COCl_2]}{[CO][Cl_2]} = 1.2 \times 10^3$

$$1.2 \times 10^3 = \frac{[COCl_2]}{\frac{1}{2}[COCl_2]\,\frac{1}{4}[COCl_2]} = \frac{8}{[COCl_2]}$$

$$[COCl_2] = 6.7 \times 10^{-3} M$$

35. $K_p = \dfrac{P^2_{CO}}{P_{CO_2}} = 63$

$$63 = \frac{(10P_{CO_2})^2}{P_{CO_2}} = 100\ P_{CO_2}$$

$P_{CO_2} = 0.63$ atm

$P_{CO} = 6.3$ atm

$P_{total} = 6.9$ atm

37. $K_p = \dfrac{P^2_{NO_2}}{P_{N_2O_4}} = \dfrac{(3(P_{N_2O_4})^{1/2})^2}{P_{N_2O_4}} = 9$

39. (a) Not correct. The pressure must decrease as 3 moles of reactant become 2 moles of product.
(b) Not correct. There are the same number of moles of SO_2 as O_2 initially.

Since the mole SO_2 to mole SO_3 ratio is 2 to 2, twice the amount of O_2 initially cannot be produced

(c) Correct. $K_c = \dfrac{[SO_3]^2}{[SO_2]^2[O_2]} = \dfrac{\left(\dfrac{\text{mol } SO_3}{V}\right)^2}{\left(\dfrac{\text{mol } SO_3}{V}\right)^2 \text{x} \left(\dfrac{\text{mol } O_2}{V}\right)} = \dfrac{(\text{mol } SO_2)2\ V}{(\text{mol } SO_3)^2 \text{ x mol } O_2}$

(d) Correct. Two moles of SO_2 react for each mole of O_2.

(e) Not correct. K_c is greater than K_p.

$$K_c = \frac{K_p}{(RT)^{\Delta n}} = \frac{K_p}{(RT)^{2-3}} = K_p \times (RT) = K_p \times (0.0821 \times (1030 + 273)) = K_p \times 107$$

41. SO_2 + Cl_2 → SO_2Cl_2

4	2	12
-x	-x	+x

$Q = \dfrac{12}{(4)(2)} = 1.5$

Reaction goes to right. The number of Cl_2 and SO_2 molecules must drop below 2 and 4, respectively, but not to zero. Drawing b

43. (a) An increase in reaction volume favors the direction in which the number of moles of gas increases–the forward reaction. The percent dissociation is greater than 12.5%.

(b) An increase in temperature favors the endothermic reaction–the forward reaction. The percent dissociation is greater than 12.5%.

(c) No change in percent dissociation. A catalyst does not alter the equilibrium condition.

(d) Neon gas will not change the equilibrium conditions, because the partial pressures of N_2O_4 and NO_2 do not change.

45. These dissociation reactions are endothermic, because they require the breaking of bonds with no new bonds formed. The forward reaction is favored with increasing temperature–dissociation is more extensive when equilibrium is reached.

47. (a) $N_2 + O_2 \rightleftharpoons 2\ NO$

There is no effect, because the same number of moles of gas appear on each side of the equation.

(b) $N_2 + 3\ H_2 \rightleftharpoons 2\ NH_3$

Because the number of molecules of gas produced is less than the number of molecules of reactant gases, this reaction occurs to a greater extent with increased pressure.

(c) $H_2 + I_2 \rightleftharpoons 2\ HI$

There is no effect because the same number of moles of gas appear on each side of the equation.

(d) $2\ H_2 + S_2 \rightleftharpoons 2\ H_2S$

Because fewer molecules of gas are produced, this reaction occurs to a greater extent at high pressure.

49. $Q_c = \frac{[COF_2]^2}{[CO_2][CF_4]} = \frac{(0.75\text{ mol})^2}{(0.525\text{ mol})\times(1.25\text{ mol})} = 0.86$

The volume does not matter.

$Q_c > K_c$ The reaction will be to the left.

51. $$Q_c = \frac{\left(\frac{2.10\text{ mol }CH_4}{7.25\text{ L}}\right) \text{x} \left(\frac{3.15\text{ mol }H_2O}{7.25\text{ L}}\right)}{\left(\frac{0.103\text{ ml CO}}{7.25\text{ L}}\right) \text{x} \left(\frac{0.205\text{ mol}}{7.25\text{ L}}\right)^3} = 3.92 \times 10^5$$

$Q_p = Q_c(RT)^{\Delta n} = 3.92 \times 10^5 \times (0.0821 \times 773\text{ K})^{2-4} = 97.4$

The reaction will go slightly to the right to increase the amount of CH_4 and H_2O because Q_p (97.4) is slightly less than K_p (102).

53. $[PCl_5] = \frac{0.562\text{ g}}{1.75\text{ L}} \times \frac{\text{mol}}{208.22\text{ g}} = 1.54 \times 10^{-3}\text{ M}$

$[PCl_3] = \frac{1.950\text{ g}}{1.75\text{ L}} \times \frac{\text{mol}}{137.32\text{ g}} = 8.11 \times 10^{-3}\text{ M}$

$[Cl_2] = \frac{1.007\text{ g}}{1.75\text{ L}} \times \frac{\text{mol}}{70.91\text{ g}} = 8.11 \times 10^{-3}\text{ M}$

$K_c = \frac{[PCl_3][Cl_2]}{[PCl_5]} = \frac{8.11\times10^{-3}\times8.11\times10^{-3}}{1.54\times10^{-3}} = 4.27 \times 10^{-2}$

$K_p = K_c\,(RT)^1 = 4.27 \times 10^{-2} \times 0.08206 \times (250 + 273)$

$K_p = 1.83$

55. $P_{total} = P_{NH_3} + P_{H_2S} = 0.658\text{ atm}$

$P_{NH_3} = P_{H_2S} = 0.329\text{ atm}$

$K_p = P_{NH_3}\,P_{H_2S} = (0.329)^2 = 0.108$

57.

	2 ICl	$\rightleftharpoons$ I_2	+ Cl_2
Initial, M:	6.72×10^{-3}	--	--
Changes, M:	$-2 \times 2.41\times10^{-4}$	$+2.41 \times 10^{-4}$	$+2.41 \times 10^{-4}$
Equilibrium, M:	6.24×10^{-3}	2.41×10^{-4}	2.41×10^{-4}

$[ICl]_{init} = \frac{0.682\text{ g ICl}}{625\text{ mL}} \times \frac{\text{mol}}{162.4\text{ g}} \times \frac{\text{mL}}{10^{-3}\text{ L}} = 6.72 \times 10^{-3}\text{ M}$

$[I_2]_{eq} = \frac{0.0383\text{ g }I_2}{625\text{ mL}} \times \frac{\text{mol}}{253.8\text{ g}} \times \frac{\text{mL}}{10^{-3}\text{ L}} = 2.41 \times 10^{-4}\text{ M}$

$K_c = \frac{[I_2][Cl_2]}{[ICl]^2} = \frac{(2.41\text{ x }10\text{-}4)^2}{(6.24\text{ x }10^{-3})^2} = 1.49 \times 10^{-3}$

59.

	butane $\rightleftharpoons$	isobutane
Initial, M:	6.882×10^{-3}	--
Change, M:	$-X$	X
Equilibrium, M:	$6.882 \times 10^{-3} - X$	X

$$[\text{butane}] = \frac{5.00\text{ g}}{12.5\text{ L}} \times \frac{\text{mol}}{58.12\text{ g}} = 6.882 \times 10^{-3}\text{ M}$$

$$\frac{X}{6.882 \times 10^{-3} - X} = 7.94$$

$$8.94\,X = 5.465 \times 10^{-2}$$

$$X = 6.11 \times 10^{-3}\text{ M} \times 12.5\text{ L} \times \frac{58.12\text{ g}}{\text{mol}} = 4.44\text{ g}$$

61.

CO	+	H_2O $\rightleftharpoons$	CO_2 +	H_2
0.250		0.250	--	--
$0.250 - X$		$0.250 - X$	X	X

$$\frac{X^2}{(0.250 - X)^2} = 23.2$$

$$\frac{X}{0.250 - X} = 4.82$$

$$5.82\,X = 1.205$$

$X = 0.207$ moles each of CO_2 and H_2

(As long as the same number of terms appear in the numerator and denominator, the volume does not matter.)

63. $$[\text{CO}]_{\text{init}} = \frac{20.0\text{ g CO}}{8.05\text{ L}} \times \frac{\text{mol}}{28.01\text{ g}} = 0.0887\text{ M}$$

$$[\text{Cl}_2]_{\text{init}} = \frac{35.5\text{ g Cl}_2}{8.05\text{ L}} \times \frac{\text{mol}}{70.90\text{ g}} = 6.22 \times 10^{-2}\text{ M}$$

	CO +	Cl_2 $\rightleftharpoons$	$COCl_2$
Initial, M:	8.87×10^{-2}	6.22×10^{-2}	--
Changes, M:	$-X$	$-X$	X
Equilibrium, M:	$8.87 \times 10^{-2} - X$	$6.22 \times 10^{-2} - X$	X

$$\frac{X}{(8.87 \times 10^{-2} - X) \times (6.22 \times 10^{-2} - X)} = 1.2 \times 10^3$$

$$X = 1.2 \times 10^3 X^2 - 1.81 \times 10^2 X + 6.62$$

$$0 = 1.2 \times 10^3 X^2 - 1.82 \times 10^2 X + 6.62$$

$$X = \frac{-b \pm \sqrt{b^2 - 4ac}}{2a}$$

$$X = \frac{1.82 \times 10^2 \pm \sqrt{(1.82 \times 10^2)^2 - 4 \times 1.2 \times 10^3 \times 6.62}}{2 \times 1.2 \times 10^3}$$

$X = 0.0911$ M (too large, not valid) or 0.0605 M

? mol $COCl_2$ = 0.0605 M × 8.05 L = 0.487 mol $COCl_2$

$$?\text{ g COCl}_2 = 0.487\text{ mol COCl}_2 \times \frac{98.91\text{ g}}{\text{mol}} = 48.2\text{ g COCl}_2$$

65. $[PCl_3] = [Cl_2] = \dfrac{0.100 \text{ mol}}{6.40 \text{ L}} = 0.01563 \text{ M}$

$[PCl_5] = \dfrac{0.0100 \text{ mol}}{6.40 \text{ L}} = 0.001563 \text{ M}$

$Q_c = \dfrac{(0.001563)}{(0.01563)(0.01563)} = 6.398$ Since $Q_c < K_c$, the net reaction must go to the right to establish equilibrium.

	PCl_3 +	Cl_2 $\rightleftharpoons$	PCl_5
Initial, M:	0.01563	0.01563	0.001563
Changes, M:	$-X$	$-X$	X
Equilibrium, M:	$0.01563 - X$	$0.01563 - X$	$0.001563 + X$

$$\frac{0.001563 + X}{(0.01563 - X)^2} = 26$$

$0.001563 + X = 26 X^2 - 0.8128 X + 0.006352$

$0 = 26 X^2 - 1.8128 X + 0.004789$

$$X = \frac{-b \pm \sqrt{b^2 - 4ac}}{2a}$$

$$X = \frac{1.8128 \pm \sqrt{(1.8128)^2 - 4 \times 26 \times 0.004789}}{52}$$

$X = 0.0670$ M or 0.00275 M
not valid

PCl_3 and Cl_2 $(0.01563 - 0.00275) \times 6.40 \text{ L} = 0.082$ mol
PCl_5 $(0.001563 + 0.00275) \times 6.40 \text{ L} = 0.028$ mol

67.

	$2\,SO_3$ $\rightleftharpoons$	$2\,SO_2$ +	O_2
Initial, M:	$\dfrac{1.00 \text{ mol}}{5.00 \text{ L}}$	--	--
Changes, M:	$-X$	$+X$	$+\frac{1}{2}X$
Equilibrium, M:	$0.200 - X$	X	$\frac{1}{2}X$

$$K_c = \frac{[SO_2]^2[O_2]}{[SO_3]^2} = \frac{[X]^2[\frac{1}{2}X]}{(0.200 - X)^2} = 0.504$$

$\frac{1}{2}X^3 = 0.0202 - 0.202 X + 0.504 X^2$

$X^3 - 1.01 X^2 + 0.404 X = 0.0404$

Assume $X \approx 0.100$

$(0.100)^3 - 1.01 \times (0.100)^2 + 0.404 \times (0.100) = 0.0313 < 0.0404$

Make X larger $X = 0.120$

$(0.120)^3 - 1.01 \times (0.120)^2 + 0.404 \times (0.120) = 0.0357 < 0.0404$

Make X larger $X = 0.150$

$(0.150)^3 - 1.01 \times (0.150)^2 + 0.404 \times (0.150) = 0.0413 > 0.0404$

Make X smaller $X = 0.145$

$(0.145)^3 - 1.01 \times (0.145)^2 + 0.404 \times 0.145 = 0.0404 = 0.0404$

$X = 0.145$

$[SO_3]_{eq} = 0.200\text{ M} - 0.145\text{ M} = 0.055\text{ M}$

$\%\text{ undissociated} = \dfrac{0.055\text{ M}}{0.200\text{ M}} \times 100\% = 28\%$

To check:

$$\frac{(.145)^2\ 0.073}{(.055)^2} = 0.51$$

69. (a) $K_P = 5.60 = \dfrac{P_{CS_2}}{P_{S_2}} = \dfrac{0.152}{P_{S_2}}$

$P_{S_2} = 0.0271\text{ atm}$

(b) $P_{total} = P_{S_2} + P_{CS_2} = 0.179\text{ atm}$

71. $P_{CH_4} = \dfrac{nRT}{V} = \dfrac{0.100\text{ mol x } \dfrac{0.08206\text{ L atm}}{\text{K mol}} \text{ x } 1273\text{ K}}{4.16\text{ L}}$

$P_{CH_4} = 2.511\text{ atm}$

C (s)	+	2 H_2	⇌	CH_4
		--		2.511 atm
		2X		2.511 - X

$K_P = 0.263 = \dfrac{P_{CH_4}}{P^2_{H_2}} = \dfrac{2.511 - X}{(2X)^2}$

$1.052\ X^2 = 2.511 - X$

$1.052\ X^2 + X - 2.511 = 0$

$X = \dfrac{-b \pm \sqrt{b^2 - 4ac}}{2a}$

$X = \dfrac{-1 \pm \sqrt{1^2 - 4 \text{ x } 1.052 \text{ x } (-2.511)}}{2.104}$

$X = 1.141\text{ atm}$ or -2.09 not valid

$P_{total} = P_{H_2} + P_{CH_4} = 2 \times 1.141 + (2.511 - 1.141) = 3.65\text{ atm}$

73.

	CO +	2 H_2 ⇌	CH_3OH
Initial, atm:	4.0	8.0	--
Changes, atm:	$-P$	$-2P$	P
Equilibrium, atm:	$4.0 - P$	$8.0 - 2P$	P

$12.0\text{ atm} = 2P + P = 3P \qquad P = 4\text{ atm}$

$K_p = \dfrac{[CH_3OH]}{[CO][H_2]^2} = \dfrac{P}{(4.0-P)(8.0-2P)^2} = 9.23 \times 10^{-3}$

$\dfrac{P}{(4.0-P)(64 - 32P + 4P^2)} = 9.23 \times 10^{-3}$

$\dfrac{P}{(256 - 192P + 48P^2 - 4P^3)} = 9.23 \times 10^{-3}$

$P = 2.36 - 1.77P + 0.44P^2 - 0.037P^3$

$0.037P^3 - 0.44P^2 + 2.8P = 2.36$

$X^3 - 12X^2 + 76P = 64$

P must be less than 4, so the important term is 76 P, thus 76.0 P = 64.

$P \approx 0.86$

$(0.86)^3 - 12 \times (0.86)^2 + 76 \times 0.86 = 56 < 64$

Make P larger $P = 0.90$

$(0.90)^3 - 12 \times (0.90)^2 + 76 \times 0.90 = 59 < 64$

Make P larger $P = 1.00$

$(1.00)^3 - 12 \times (1.00)^2 + 76 \times 1.00 = 65 < 64$

Make P smaller $P = 0.99$

$(0.99)^3 - 12 \times (0.99)^2 + 76 \times 0.99 = 64$

$P = P_{CH_3OH} = 0.99$ atm

To check:

$$\frac{(0.99)}{(4.0 - 0.99)(8.0 - 2 \text{ x } 0.99)^2} = \frac{(0.99)}{(3.0)(6.)^2} = 9.2 \times 10^{-3}$$

An alternate method is to plug an estimate of P into the K_p equation.

$P = 3.00$ atm

$$\frac{3.00}{(1.00)(2.00)^2} = 0.75 >> 9.23 \times 10^{-3}$$

$P = 2.00$ atm

$$\frac{2.00}{(2.00)(4.00)^2} = 0.0625 > 9.23 \times 10^{-3}$$

$P = 1.00$ atm

$$\frac{1.00}{(3.00)(6.00)^2} = 9.26 \times 10^{-3} > 9.23 \times 10^{-3}$$

$P = 0.99$ atm

$$\frac{0.99}{(3.01)(6.02)^2} = 9.08 \times 10^{-3} < 9.23 \times 10^{-3}$$

$P = P_{CH_3OH} = 1.00$ atm is closer than 0.99 atm, so the answer is 1.00 atm.

Additional Problems

75. (a) to the left
 (b) 7 moles gas on the left, and 8 moles gas on the right, so to the left.
 (c) Reacts in direction to heat system, so to the right.
 (d) to the right
 (e) Cooling will cause the reaction to generate heat, so it will go to the right. At the temperature at which water is a liquid, an increase in pressure will cause the reaction to go to the right.
 (f) Cooling causes to the right, and at temperatures where water and ammonia are solids, there are 2 moles gas on the right and 3 moles on the left, so the reaction goes to the right.

77. Need to establish the value of K_c from equilibrium concentrations first.

$$K_c = \frac{0.179 \text{ x } 0.079}{0.021 \text{ x } 0.121} = 5.6$$

(a)

	CO	+	H_2O	$\Leftrightarrow$	CO_2	+	H_2
init	0.021		0.121		0.179		0.179

$$Q_c = \frac{0.179 \times 0.179}{0.021 \times 0.121} = 12.6$$

$Q_c > K_c$, so a net reaction occurs to the left.
The same conclusion is reached by considering Le Chatelier's principle. Adding H_2 to a mixture at equilibrium will cause the reaction to go in the direction that uses up H_2–that is, to the left.

(b) $0.021 + X$ $\quad 0.121 + X$ $\quad 0.179 - X$ $\quad 0.179 - X$

$$K_c = 5.6 = \frac{(0.179 - X)^2}{(0.021 + X)(0.121 + X)}$$

$$5.6\,X^2 + 0.795\,X + 0.01423 = X^2 - 0.358\,X + 0.03204$$

$$4.6\,X^2 + 1.153\,X - 0.01781 = 0$$

$$X = \frac{-1.153 \pm \sqrt{(1.153)^2 - 4 \times 4.6 \times (-0.01781)}}{9.2}$$

$X = 0.01460$ mol or -0.26525 not valid

$n_{CO_2} = n_{H_2} = 0.164$ mol

$n_{CO} = 0.036$ mol

$n_{H_2O} = 0.136$ mol

79. Base the calculation of 100.0 g of the gaseous phase at equilibrium

$$13.71 \text{ g C} \times \frac{\text{mol}}{12.011 \text{ g}} = 1.1415 \text{ mol C}$$

All the C in the gas phase is present as CS_2.

The number of moles S in CS_2 = $1.142 \text{ mol } CS_2 \times \frac{2 \text{ mol S}}{1 \text{ mol } CS_2} = 2.284$ mol S.

$$86.29 \text{ g S} \times \frac{\text{mol}}{32.066 \text{ g}} = 2.691 \text{ mol S}$$

The number of moles S_2 = [2.691 mol S (total) - 2.284 mol S in CS_2] $\times \frac{1 \text{ mol } S_2}{2 \text{ mol S}} = 0.204 \text{ mol } S_2$

$$K_c = \frac{n_{CS_2}}{n_{S_2}} = \frac{1.142 \text{ mol } CS_2}{0.204 \text{ mol } S_2} = 5.60$$

81.

	N_2O_4	$\rightleftharpoons$	$2\,NO_2$
Initial, mol:	0.100		--
Changes, mol:	$-\frac{1}{2}X$		$+X$
Equilibrium, mol:	$0.100 - \frac{1}{2}X$		X

$$0.746 = \text{mol fraction } NO_2 = \frac{X}{0.100 - \frac{1}{2}X + X} = \frac{X}{0.100 + \frac{1}{2}X}$$

$X = 0.0746 + 0.373\,X$
$0.0746 = 0.627\,X$
$X = 0.119 \text{ mol} = \text{mol } NO_2$
$\text{mol } N_2O_4 = 0.100 - \frac{1}{2} \times (0.119) = 0.041 \text{ mol}$

$$P_{NO_2} = \frac{nRT}{V} = \frac{0.119 \text{ mol x } \frac{0.08206 \text{ L atm}}{\text{K mol}} \text{ x } 348 \text{ K}}{2.50 \text{ L}} = 1.36 \text{ atm}$$

$$P_{N_2O_4} = \frac{0.041 \text{ mol x } \frac{0.08206 \text{ L atm}}{\text{K mol}} \text{ x } 348 \text{ K}}{2.50 \text{ L}} = 0.47 \text{ atm}$$

$$K_p = \frac{P^2_{NO_2}}{P_{N_2O_4}} = \frac{(1.36)^2}{0.47} = 3.9$$

High pressure and low temperature favor purer N_2O_4.

83. $P_{O_2} = \chi P° = 0.2095 \times 1.0000 = 0.2095 \text{ atm}$

$2\,CaSO_4\,(s) \rightleftharpoons 2\,CaO\,(s) + 2\,SO_2\,(g) + O_2\,(g)$

$X \qquad 0.2095 + \frac{1}{2}X$

$1.45 \times 10^{-5} = P^2_{SO_2} P_{O_2} = X^2 \times (0.2095 + \frac{1}{2}X)$

assume $\frac{1}{2}X << 0.2095$

$1.45 \times 10^{-5} = X^2 \times 0.2095$

$P_{SO_2} = X = 8.32 \times 10^{-3} \text{ atm}$

$(8.32 \times 10^{-3} \text{ atm})^2 \times (0.2095 + 8.32 \times 10^{-3}) = 1.51 \times 10^{-5}$ The approximation produced an error of about 4% so the approximation is justified.

85.

	C_6H_{14} $\rightleftharpoons$	C_6H_6 +	$4\,H_2$
Initial, atm:	1.00	--	--
Changes, atm:	–0.10	+0.10	4×0.10
Equilibrium, atm:	0.90	0.100	0.400

$$K_p = \frac{P_{C_6H_6} P^4_{H_2}}{P_{C_6H_{14}}} = \frac{(0.100) \text{ x } (0.400)^4}{0.90} = 2.84 \times 10^{-3}$$

$$\log_{10} K_p = 23.45 - \frac{13941}{T}$$

$\log_{10}(2.84 \times 10^{-3}) = 23.45 - \dfrac{13941}{T}$

$-25.996 = -\dfrac{13941}{T}$

$T = \dfrac{13941}{25.996} = 536\ K$

$t_C = 536 - 273 = 263\ °C$

87. (a)

	PCl_5	$\rightleftharpoons$	PCl_3	+	Cl_2
Initial, M:	1.00 mol		--		--
Changes,M:	$-\ \alpha$ mol		$+\ \alpha$ mol		$+\ \alpha$ mol
Equilibrium, M:	$1.00 - \alpha$ mol		α mol		α mol

$n_{total} = 1.00 - \alpha + \alpha + \alpha = (1.00 + \alpha)$ mol

equilibrium mol fractions: $\chi_{PCl_3} = \chi_{Cl_2} = \dfrac{\alpha}{1.00+\alpha}$

$\chi_{PCl_5} = \dfrac{1.00-\alpha}{1.00+\alpha}$

equilibrium partial pressures: $P_{PCl_3} = P_{Cl_2} = \dfrac{\alpha \times P_{total}}{1.00+\alpha}$

$P_{PCl_5} = \dfrac{(1.00-\alpha)\times P_{total}}{1.00+\alpha}$

$$K_p = \frac{P_{PCl_3} \times P_{Cl_2}}{P_{PCl_5}} = \frac{(\alpha\, P_{total})^2 \times (1.00+\alpha)}{(1.00+\alpha)^2 \times (1.00-\alpha)\times P_{total}} = \frac{\alpha^2 P_{total}}{(1+\alpha)(1-\alpha)}$$

$$= \frac{\alpha^2 P_{total}}{1-\alpha^2}$$

(b) $K_p = \dfrac{\alpha^2 P}{1-\alpha^2}$

$1.78 = \dfrac{\alpha^2 \text{ x } (2.50 \text{ atm})}{1-\alpha^2}$

$1.78 - 1.78\ \alpha^2 = 2.50\ \alpha^2$

$1.78 = 4.28\ \alpha^2$

$\alpha^2 = 0.416$

$\alpha = 0.645$

(c) $1.78 = \dfrac{(0.100)^2\ P}{1 - (0.100)^2}$

$P = 176$ atm

Apply Your Knowledge

89. (a) $[I_2]_{aq} = X$

$\dfrac{1.33 \text{ x } 10^{-3} - X}{X} = 85.5$

$1.33 \times 10^{-3} - X = 85.5\ X$

$1.33 \times 10^{-3} = 86.5 \, X$

$X = 1.54 \times 10^{-5} = [I_2]_{aq}$

? mg $I_2 = 1.54 \times 10^{-5} \dfrac{mol \, I_2}{L} \times 25.0 \, mL \times \dfrac{253.8 \, g \, I_2}{mol \, I_2} = 9.77 \times 10^{-2}$ mg I_2

(b) The concentration ratio will be the same, but because the volume of the CCl_4 is greater than in (a), there will be fewer milligrams of I_2 left in the aqueous layer than in (a).

(c) Putting the aqueous layer in equilibrium with a second 25.0 mL of CCl_4 means satisfying the 85.5 ratio twice. Using 50.0 mL is like making the ratio double. The two equilibriums will leave much less in the aqueous solution.

91.

CH_4	+	H_2O ⇌	$3 H_2$ +	CO	$\Delta H° = 230$ kJ	$k = 1/190$
CO	+	H_2O ⇌	CO_2 +	H_2	$\Delta H° = -40$ kJ	$k = 1.4$
CH_4	+	$2 H_2O$ ⇌	CO_2 +	$4 H_2$	$k = 1.4/190 = 7.4 \times 10^{-3}$	
0.100 M	0.100 M		--	--	$\Delta H° = 190$ kJ	
$-x$	$-2x$		x	4x		
$-0.100 - x$	$0.100 - 2x$		x	4x		

$7.4 \times 10^{-3} = \dfrac{[CO_2][H_2]^4}{[CH_4][H_2O]^2} = \dfrac{(x/10.0)(4x/10.0)^4}{[(1.00-x)/10.0][(1.00-2x)/10.0]^2}$

$7.4 \times 10^{-3} = \dfrac{x(4x)^4}{100(1.00-x)(1.00-2x)^2}$

$256x^5 = 0.74 \, [(1.00-x)(1.00-2x)^2]$

$256x^5 - 0.74 \, [(1.00-x)(1.00-4x+4x^2)]$

$256x^5 - 0.74 \, (1.00 - 5x + 8x^2 - 4x^3) = 0$

$256x^5 + 2.96x^3 - 5.92x^2 + 3.70x = 0.74$

$x \approx \sqrt[5]{\dfrac{0.74}{256}} = 0.3$

$256 \times (0.3)^5 + 2.96 \times (0.3)^3 - 5.92 \times (0.3)^2 + 3.70 \times (0.3) = 1.28 > 0.74$

x is too large; assume x = 0.2

$256 \times (0.2)^5 + 2.96 \times (0.2)^3 - 5.92 \times (0.2)^2 + 3.70 \times (0.2) = 0.61 < 0.74$

x is too small; assume x = 0.22

$256 \times (0.22)^5 + 2.96 \times (0.22)^3 - 5.92 \times (0.22)^2 + 3.70 \times (0.22) = 0.68 < 0.74$

x is too large; assume x = 0.23

$256 \times (0.23)^5 + 2.96 \times (0.23)^3 - 5.92 \times (0.23)^2 + 3.70 \times (0.23) = 0.74 = 0.74$

x = 0.23 moles

moles $H_2 = 4 \times 0.23 = 0.92$ moles

Because raising the temperature of a reaction shifts the equilibrium toward the endothermic direction, the amount of H_2 will increase.

Chapter 15

Acids, Bases, and Acid-Base Equilibria

Exercises

15.1A (a) $NH_3 + HCO_3^- \rightleftharpoons NH_4^+ + CO_3^{2-}$

base (1) acid (2) acid (1) base (2)

$H_3PO_4 + H_2O \rightleftharpoons H_3O^+ + H_2PO_4^-$

acid (1) base (2) acid (2) base (1)

15.1B HCO_3^- was used as an acid in Exercise 15.1A.

$HCO_3^- + HCl \rightarrow H_2CO_3 + Cl^-$

base acid

H_2O was used as a base in Exercise 15.1A.

$H_2O + NH_3 \rightleftharpoons NH_4^+ + OH^-$

acid base

15.2A (a) H_2Te is the stronger acid. Because the Te atom is larger than the S atom, it is expected that the H-Te bond energy will be less than the H-S bond energy, and the H-Te bond will be more easily broken than the H-S bond.

(b) $CH_3CH_2CH_2CHBrCOOH$ is the stronger acid, because the Br on the second carbon is more electron-withdrawing than the Cl on the fifth carbon.

15.2B (a) < (d) < (b) < (c)

(a) has no extra electron-withdrawing group. (d) has the Br on the far side of the benzene, so it is stronger than (a). (b) has the Cl next to the COOH group and is a stronger withdrawing group than Br. (c) is the strongest because it has two close withdrawing groups.

15.3 (d) < (a) < (c) < (b)

(d) is the weakest because it is aromatic and has two withdrawing groups. (a) is aromatic and has one withdrawing group. (c) has a withdrawing group, so it is weaker than (b), the strongest of these bases.

15.4A $[H^+] = \dfrac{0.0105 \text{ mol } HNO_3}{225 \text{ L}} = 4.67 \times 10^{-5}$ M

$pH = -\log[H^+] = 4.331$

15.4B $[OH^-] = \dfrac{2.65 \text{ g}}{735 \text{ mL}} \times \dfrac{1 \text{ mol } Ba(OH)_2}{171.3 \text{ g } Ba(OH)_2} \times \dfrac{1000 \text{ mL}}{\text{L}} \times \dfrac{2 \text{ mol } OH^-}{\text{mol } Ba(OH)_2} = 0.0421$ M

$pOH = -\log[OH^-] = -\log(0.0421) = 1.376$

$pH = 14.000 - pOH = 12.624$

15.5 The solution is basic. The concentration of OH^- comes from the dissociation of NaOH *and* the self-ionization of water.

15.6A $CH_3CH_2COOH + H_2O \rightleftharpoons CH_3CH_2COO^- + H_3O^+$

$$K_a = 1.3 \times 10^{-5} = \frac{[H_3O^+][CH_3CH_2COO^-]}{[CH_3CH_2COOH]}$$

$[CH_3CH_2COOH] = 0.250 - [H_3O^+]$

Assumption: Self-ionization of water is negligible, so that $[H_3O^+]$ $= [CH_3CH_2COO^-]$.

$$K_a = 1.3 \times 10\text{-}5 = \frac{[H_3O^+]^2}{0.250 - [H_3O^+]}$$

Assume $0.250 >> [H_3O^+]$.

$[H_3O^+] = 1.8 \times 10^{-3}$ M

Assumption is good.

pH = 2.74

15.6B $[C_6H_4NO_2OH] = \frac{2.1\text{ g}}{\text{L}} \times \frac{\text{mole}}{139\text{ g}} = 0.015$ M

$$C_6H_4NO_2OH + H_2O \rightleftharpoons H_3O^+ + C_6H_4NO_2O^-$$

	$C_6H_4NO_2OH$	H_3O^+	$C_6H_4NO_2O^-$
Initial, M:	0.015	≈ 0	–
Charge, M:	-y	+y	+y
Equilibrium, M:	0.015 -y	y	y

$$Ka = \frac{[H_3O^+][C_6H_4NO_2O^-]}{[C_6H_4NO_2OH]} = \frac{y^2}{0.015 - y} = 6.0 \times 10^{-8}$$

Assume $0.015 >> y$.

$y^2 = 9.0 \times 10^{-10}$

Assumption is good.

$y = 3.0 \times 10^{-5} = [H_3O^+]$

$pH = -\log[H_3O^+] = -\log(3.0 \times 10^{-5}) = 4.52$

15.7A $1.4 \times 10^{-3} = \frac{y^2}{0.00200 - y}$

Assume $y << 2 \times 10^{-3}$.

$$1.4 \times 10\text{-}3 = \frac{y^2}{2 \text{ x } 10^{-3}}$$

$y^2 = 2.8 \times 10^{-6}$

$y = 1.7 \times 10^{-3}$

y is not much smaller than 2×10^{-3}.

$$1.4 \times 10^{-3} = \frac{y^3}{0.00200 - y}$$

$y^2 + 1.4 \times 10^{-3}\ y - 2.8 \times 10^{-6} = 0$

$$y = \frac{-1.4 \times 10^{-3} \pm \sqrt{(1.4 \times 10^{-3})^2 - 4 \times 1 \times (-2.8 \times 10^{-6})}}{2}$$

$y = 1.1 \times 10^{-3}$ or -2.9×10^{-3} negative value not valid

$$\% = \frac{1.7 \times 10^{-3} - 1.1 \times 10^{-3}}{1.1 \times 10^{-3}} \times 100\% = 55\%$$

$[H_3O^+] = 1.7 \times 10^{-3}$ M compared to
1.1×10^{-3} M without the assumption, an error of 55%.

15.7B $HClO_2 + H_2O \rightleftharpoons H_3O^+ + ClO_2^-$

	$HClO_2$	H_3O^+	ClO_2^-
Initial, M:	0.0100	≈0	0
Changes, M:	-y	+y	+y
Equilibrium, M:	0.0100 - y	y	y

$$\frac{[H_3O^+][ClO_2^-]}{[HClO_2]} = \frac{y\ y}{[0.0100 - y]} = 1.1 \times 10^{-2}$$

$$\frac{M_{acid}}{K_a} = \frac{0.0100}{1.1 \times 10^{-2}} = 0.91\text{, much less than 100.}$$

The assumption is expected to fail, and the quadratic equation must be used.

$y2 = 1.1 \times 10^{-4} - 1.1 \times 10^{-2}\ y$

$y2 + 1.1 \times 10^{-2}\ y - 1.1 \times 10^{-4} = 0$

$$y = \frac{-1.1 \times 10^{-2} \pm \sqrt{(1.1 \times 10^{-2})^2 - 4 \times 1 \times (-1.1 \times 10^{-4})}}{2}$$

$$y = \frac{-1.1 \times 10^{-2} \pm 2.4 \times 10^{-2}}{2}$$

$y = 6.3 \times 10^{-3}$ 1.7×10^{-2}
Negative values are not valid.

$[H_3O^+] = 6.3 \times 10^{-3}$

$pH = -\log[H_3O^+] = -\log[6.3 \times 10^{-3}] = 2.20$

15.8A $CH_3NH_2 + H_2O \rightleftharpoons CH_3NH_3^+ + OH^-$

	CH_3NH_2	$CH_3NH_3^+$	OH^-
Initial, M:	0.200	0	≈0
Changes, M:	y	+y	+y
Equilibrium, M:	0.200 - y	y	y

$$\frac{[CH_3NH_3^+][OH^-]}{[CH_3NH_2]} = \frac{y^2}{0.200 - y} = 4.2 \times 10^{-4}$$

assume 0.200 >> y

$y^2 = 8.4 \times 10^{-5}$

$y = [OH^-] = 9.2 \times 10^{-3}$

y is 4.6% of 0.200, and the assumption is just within the 5% rule.

$pOH = -\log[OH^-] = -\log[9.2 \times 10^{-3}] = 2.04$

$pH = 14.00 - pOH = 14.00 - 2.04 = 11.96$

15.8B $$[NH_3] = \frac{5.00\text{ mL} \times 0.0100\text{M}}{1.000\text{L}} \times \frac{10^{-3}\text{L}}{\text{mL}} = 5 \times 10^{-5}\text{ M}$$

	$NH_3 + H_2O \rightleftharpoons$	$NH_4^+ +$	OH^-
Initial, M:	5×10^{-5}	0	≈ 0
Changes, M:	-y	y	y
Equilibrium, M:	5×10^{-5} - y	y	y

$$\frac{M_{base}}{K_b} = \frac{5 \text{ x } 10^{-5}}{1.8 \text{ x } 10^{-5}} = 2.7\text{, much less than 100}$$

The assumption will fail, so the quadratic must be used.

$$\frac{[NH_4^+][OH^-]}{[NH_3]} = \frac{y^2}{5 \text{ x } 10^{-5} - y} = 1.8 \times 10^{-5}$$

$y^2 = 9 \times 10^{-10} - 1.8 \times 10^{-5}\ y$

$y^2 + 1.8 \times 10^{-5}\ y - 9 \times 10^{-10} = 0$

$$y = \frac{-1.8 \text{ x } 10^{-5} \pm \sqrt{(1.8 \text{ x } 10^{-5})^2 - 4 \text{ x } 1 \text{ x } (-9 \text{ x } 10^{-10})}}{2}$$

$$y = \frac{-1.8 \text{ x } 10^{-5} \pm 6.3 \text{ x } 10^{-5}}{2}$$

$y = 2.2 \times 10^{-5}$ $\qquad$ -4.0×10^{-5}

Negative values are not valid.

pH = $-\log[OH^-] = -\log(2.2 \times 10^{-5}) = 4.66$

pH = 14.00 - pOH = 14.00 - 4.66 = 9.34

15.9A $K_a = \dfrac{X^2}{0.0100 - X}$

$X = 10^{-3.12} = 7.59 \text{ x } 10^{-4}$ M

$$K_a = \frac{(7.59 \text{ x } 10^{-4})^2}{0.0100 - 7.59 \text{ x } 10^{-4}} = 6.2 \times 10^{-5}$$

$pK_a = 4.21$

15.9B The same pH means the same $[OH^-]$.

$[OH^-] = \sqrt{x} = [(CH_3)_2NH] \times K_b[(CH_3)_2NH]$

$[OH^-] = \sqrt{x} = [NH_3] \times K_b(NH_3)$

$[(CH_3)_2NH] \times K_b[(CH_3)_2NH] = [NH_3] \times K_b\ (NH_3)$

$$[NH_3] = \frac{[(CH_3)_2NH] \text{ x } K_b((CH_3)_2NH)}{K_b(NH_3)}$$

$$[NH_3] = \frac{0.200 \text{ M x } 5.9 \text{ x } 10^{-4}}{1.8 \text{ x } 10^{-5}} = 6.56 \text{ M}$$

15.10 Since both K_b and molarity are larger, it should be obvious that methylamine is more basic.

$$K_b = \frac{[OH^-][NH_4^+]}{[NH_3]} = \frac{[OH^-]^2}{[NH_3]}$$

$[OH^-] = \sqrt{K_b\ [NH_3]}$ $\qquad$ $[OH^-] = \sqrt{K_b\ [CH_3NH_2]}$

$[OH^-] = \sqrt{1.8 \times 10^{-5} \times 0.025}$ $[OH^-] = \sqrt{4.2 \times 10^{-4} \times 0.030}$

$[OH^-] = \sqrt{4.5 \times 10^{-7}}$ $[OH^-] = \sqrt{1.26 \times 10^{-5}}$

$pOH = -\log[OH^-] \approx 3$ actual(3.17) ≈ 2 actual (2.45)

$pH = 10.83$ $= 11.55$

15.11A $C_4H_4O_4 + H_2O \rightleftharpoons H_3O^+ + C_4H_3O_4^-$

	$C_4H_4O_4$	H_3O^+	$C_4H_3O_4^-$
Initial, M:	0.125	≈ 0	0
Changes, M:	-y	+y	+y
Equilibrium, M:	0.125 - y	y	y

$$\frac{[H_3O^+][C_4H_3O_4^-]}{[C_4H_4O_4]} = \frac{y^2}{0.125 - y} = 1.2 \times 10^{-2}$$

$$\frac{M_{acid}}{K_a} = \frac{0.125}{1.2 \times 10^{-2}} = 10.4$$ Assumption fails, use quadratic.

$y^2 = 1.5 \times 10^{-3} - 1.2 \times 10^{-2}\, y$

$y^2 + 1.2 \times 10^{-2} y - 1.5 \times 10^{-3} = 0$

$$y = \frac{-1.2\times10^{-2} \pm \sqrt{(1.2\times10^{-2})^2 - 4\times1\times(-1.5\times10^{-3})}}{2}$$

$$y = \frac{-1.2 \times 10^{-2} \pm 7.8 \times 10^{-2}}{2}$$

$y = 3.3 \times 10^{-2}$ $\quad -4.5 \times 10^{-3}$ Negative values are not valid.

$pH = -\log [H_3O^+] = -\log (3.3 \times 10^{-2}) = 1.48$

Because $Ka_1 >> Ka_2$, essentially all of the H_3O^+ is produced in the Ka_1 reaction.

15.11B $$[H_3PO_4] = \frac{0.057 \text{ g acid solution}}{100 \text{ g cola drink}} \times \frac{75 \text{ g } H_3PO_4}{100 \text{ g acid solution}} \times \frac{\text{mol } H_3PO_4}{98.0 \text{g } H_3PO_4} \times \frac{1 \text{ g cola drink}}{\text{mL cola drink}} \times \frac{\text{mL cola drink}}{10^{-3} \text{ L cola drink}} = 4.4 \times 10^{-3} \text{ M}$$

$$[H_3PO_4] = \frac{0.084 \text{ g acid solution}}{100 \text{ g cola drink}} \times \frac{75 \text{ g } H_3PO_4}{100 \text{ g acid solution}} \times \frac{\text{mol } H_3PO_4}{98.0 \text{g } H_3PO_4} \times \frac{1 \text{ g cola drink}}{\text{mL cola drink}} \times \frac{\text{mL cola drink}}{10^{-3} \text{ L cola drink}} = 6.3 \times 10^{-3} \text{ M}$$

$\frac{4.4 \times 10^{-3}}{7.1 \times 10^{-3}} << 100$ Approximations will not work, so the quadratic equation is needed.

Assume all H_3O^+ is produced in the first acid-dissociation reaction.

$H_3PO_4 + H_2O \rightleftharpoons H_3O^+ + H_2PO_4^-$

H_3PO_4	H_3O^+	$H_2PO_4^-$
4.4×10^{-3} M	≈ 0	0
-y	+y	+y
4.4×10^{-3} - y	y	y

$$K_a = \frac{[H_3O^+][H_2PO_4^-]}{[H_3PO_4]} = \frac{y^2}{4.4 \times 10^{-3} - y} = 7.1 \times 10^{-3}$$

$y^2 = 3.1 \times 10^{-5} \quad -7.1 \times 10^{-3}\, y$

$y^2 + 7.1 \times 10^{-3}\, y - 3.1 \times 10^{-5} = 0$

$$y = \frac{-7.1 \times 10^{-3} \pm \sqrt{(7.1 \times 10^{-3})^2 - 4 \times 1 \times (-3.1 \times 10^{-5})}}{2}$$

$$y = \frac{-7.1 \times 10^{-3} \pm 1.3 \times 10^{-2}}{2}$$

$y = 3.1 \times 10^{-3} = [H_3O^+]$ $\quad -1.0 \times 10^{-2}$

Negative values are not valid.

$pH = -\log[H_3O^+] = -\log(3.0 \times 10^{-3}) = 2.52$

for $[H_3PO_4] = 6.4 \times 10^{-3}$, pH = 2.39.

15.12A For very dilute solutions, reaction goes essentially to completion.

$[H_3O^+] = 2 \times 8.5 \times 10^{-4} = 1.7 \times 10^{-3}$ M

pH = 2.77

15.12B Ionization of 0.020 M H_2SO_4 is complete in the first step and partial in the second: 0.020 M < $[H_3O^+]$ < 0.040 M. Only response (b) fits this requirement: $[H_3O^+]$ = 0.25 M. Response (c) would require almost complete ionization in both steps. A more exact calculation is below.

$$1.1 \times 10^{-2} = \frac{[H_3O^+][SO_4^{2-}]}{[HSO_4^-]} = K_{a2}$$

$$1.1 \times 10^{-2} = \frac{(0.020 + X)\,X}{0.020 - X}$$

$X^2 + 0.031\,X - 2.2 \times 10^{-4} = 0$

$$X = \frac{-0.031 \pm \sqrt{(0.031)^2 - 4 \times 1 \times (-2.2 \times 10^{-4})}}{2}$$

$X = 0.0060$ M or -0.037 M (not valid)

$[H_3O^+] = 0.020 + 0.006 = 0.026$ M

15.13A (a) $NaNO_3$ is neutral. Na^+ is from the strong base, NaOH; NO_3^- is from the strong acid, HNO_3.

(b) $CH_3CH_2CH_2COOK$ is basic. K^+ is from the strong base, KOH. $CH_3CH_2CH_2COO^-$ is from the weak acid, $CH_3CH_2CH_2COOH$.

$CH_3CH_2CH_2COO^- + H_2O \rightleftharpoons CH_3CH_2CH_2COOH + OH^-$

15.13B HCl (aq), NH_4I (aq), NaCl (aq), KNO_2(aq), NaOH(aq)

HCl is a strong acid. NH_4I is the salt of a weak base, so it is slightly acidic. NaCl is a salt of a strong acid–strong base, so it is neutral. KNO_2 is the salt of a weak acid, so it is slightly basic. NaOH is a strong base.

15.14A $NH_4^+ + H_2O \rightleftharpoons H_3O^+ + NH_3$

$$K_a = \frac{[H_3O^+][NH_3]}{[NH_4^+]} = \frac{X \cdot X}{0.052 - X} = \frac{K_w}{K_b} = \frac{10^{-14}}{1.8 \times 10^{-5}} = 5.6 \times 10^{-10}$$

Assume $X << 0.052$.

$X^2 = 0.052 \times 5.6 \times 10^{-10} = 2.9 \times 10^{-11}$

$X = 5.4 \times 10^{-6}$ M

pH = 5.27

15.14B

CH_3COOH +	OH^- $\rightleftharpoons$	$CH_3COO^- + H_2O$
50.00 mL × 0.120 M	18.75 ml × 0.320 M	–
=6.00 mmol	=6.00 mmol	
-6.00mmol	-6.00 mmol	+6.00
—	—	6.00 mmol

$$[CH_3COO^-] = \frac{6.00 \text{ mmol}}{(50.00 + 18.75)\text{mL}} = 0.0873 \text{ M}$$

	$CH_3COO^- + H_2O \rightleftharpoons$	CH_3COOH +	OH^-
Initial, M:	0.0873	0	≈0
Changes, M:	-y	+y	+y
Equilibrium, M:	0.0873 - y	y	y

$$K_b = \frac{K_w}{K_a} = \frac{[CH_3COOH][OH^-]}{[CH_3COO^-]} = \frac{y^2}{(0.0873 - y)} = \frac{10^{-14}}{1.8 \times 10^{-5}}$$

assume 0.0873 >> y

$y^2 = 4.85 \times 10^{-11}$

$y = 6.96 \times 10^{-6}$

pOH = $-\log[OH^-] = -\log(6.96 \times 10^{-6}) = 5.16$

pH = 14.00 - pOH = 8.84

15.15A $CH_3COO^- + H_2O \rightleftharpoons CH_3COOH + OH^-$

$$K_b = \frac{K_w}{K_a} = \frac{10^{-14}}{1.8 \times 10^{-5}} = \frac{[OH^-]^2}{[CH_3COO^-]}$$

pH = 9.10 so pOH = 4.90, and $[OH^-] = 10^{-4.90} = 1.26 \times 10^{-5}$

$$\frac{10^{-14}}{1.8 \times 10^{-5}} = \frac{(1.26 \times 10^{-5})^2}{[CH_3COO^-]}$$

$[CH_3COO^-] = 0.29$ M

15.15B The K_a values are $K_a(HNO_2) = 7.2 \times 10^{-4}$ and $K_a(HCN) = 6.2 \times 10^{-10}$. In comparing K_b values for the conjugate ions, that of CN^- is greater than that of NO_2^-, making NH_4CN (aq) more basic than NH_4NO_2 (aq).

15.16A

NH_3 +	$H_2O \rightleftharpoons$	NH_4^+ +	OH^-
$0.15 - X$		$0.35 + X$	X

$K_b = 1.8 \times 10^{-5} = \dfrac{(0.35 + X)X}{0.15 - X}$

Assume $X << 0.15$.

$1.8 \times 10^{-5} = \dfrac{0.35\ X}{0.15}$

$X = 7.71 \times 10^{-6}$ M = [OH^-]

pOH = 5.11

pH = 8.89

15.16B The common ion is the H^+.

$$CH_3COOH + H_2O \rightleftharpoons CH_3COO^- + H_3O^+$$

	CH_3COOH	CH_3COO^-	H_3O^+
Initial, M:	0.10	-	0.10
Changes, M:	-y	+y	+y
Equilibrium, M:	0.10 - y	y	0.10 + y

$K_a = \dfrac{[CH_3COO^-][H_3O^+]}{[CH_3COOH]} = \dfrac{y\ (0.10 + y)}{0.10 - y} = 1.8 \times 10^{-5}$

Assume 0.10 >> y.

$\dfrac{y\ (0.10)}{0.10} = 1.8 \times 10\text{-}5$ M = [CH_3COO^-]

[H_3O^+] = 0.10 M + 1.8×10^{-5} M = 0.10 M

15.17A

NH_3	+	H^+	$\rightleftharpoons$	NH_4^+
0.24 M		$\dfrac{0.03}{0.50}$ = 0.06 M		0.20M
0.24 - 0.06		--		0.20 + 0.06

$K_b = 1.8 \times 10^{-5} = \dfrac{[NH_4^+][OH^-]}{[NH_3]}$

$K_b = 1.8 \times 10^{-5} = \dfrac{0.26}{0.18} \times [OH^-]$

[OH^-] = 1.25×10^{-5} M

pOH = 4.90

pH = 9.10

or

pK_a for NH_4^+ = 14.00 - pK_b = 14.00 - (- log 1.8×10^{-5}) = 9.26

$pH = pK_a + \log\dfrac{[\text{base}]}{[\text{conj. acid}]}$

$pH = 9.26 + \log\dfrac{0.18}{0.26} = 9.10$

15.17B $9.50 = 9.26 + \log\dfrac{[NH_3]}{[NH_4^+]}$

$0.24 = \log\dfrac{[NH_3]}{[NH_4^+]}$

$$10^{0.24} = 10^{\log\frac{[NH3]}{[NH_4^+]}} = \frac{[NH3]}{[NH_4^+]}$$

$$1.74 = \frac{[NH3]}{[NH_4^+]} = \frac{0.24 + y}{0.20 - y}$$

0.348 - 1.74 y = 0.24 + y
0.108 = 2.74 y
0.039M = y = $[OH^-]$
?mol OH^- = 0.039 M × 0.500L = 0.020 mol NaOH

15.18A $pH = pKa + \log\frac{[CH3COO^-]}{CH3COOH]}$

$$4.50 = 4.74 + \log\frac{(0.250)}{[CH_3COOH]}$$

$$-0.24 = \log\frac{0.250}{[CH_3COOH]}$$

Raise both sides of equation to a power of 10.

$$10^{-0.24} = 10^{\log\frac{0.250}{[CH_3COOH]}} = \frac{0.250}{[CH3COOH]}$$

$$[CH_3COOH] = \frac{0.250\ M}{10^{-0.24}} = \frac{0.250\ M}{0.58}$$

$[CH_3COOH] = 0.43$ M

15.18B $[H_3O^+] = \frac{K_w}{K_b} \times \frac{[NH_4^+]}{[NH_3]}$

$[H_3O^+] = 10^{-9.05} = 8.91 \times 10^{-10}$ M

$$[NH_4^+] = \frac{8.91 \text{ x } 10^{-10} \text{ x } 1.8 \text{ x } 10^{-5} \text{ x } 0.150}{1 \text{ x } 10^{-14}}$$

$[NH_4^+] = 0.241$ M
mass = 0.241 M × 0.250 L × 53.49 g/mol = 3.22 g NH_4Cl
An alternate way to work this problem is shown below:

$$pH = pK_a + \log\frac{[\text{base}]}{[\text{conj. acid}]}$$

$$9.05 = 9.26 + \log\frac{0.150}{M}$$

$$\log\frac{0.150}{M} = -0.21$$

$$\frac{0.150}{M} = 10^{-0.21} = 0.62$$

$M = 0.24$
#g NH_4Cl = 0.24 M × 0.250 L × 53.49 g/mol = 3.2 g NH_4Cl

15.19(a) NH_4Cl pH = 4.63
(b) $NH_4Cl - NH_3$ pH = 9.26

(c) $HCl - HNO_3$ pH = - 0.30
(d) $CH_3COOH - CH_3COO^-$ pH = 4.74

The indicator color shows that the pH is in the range of about 4 to 5.5. Solutions (b) and (c) have pH values outside this range. The 1.00 M NH_4Cl would have a pH in this range due to hydrolysis of NH_4^+. The 1.00 M CH_3COOH-CH_3COONa is a buffer with pH = 4.74 (pK_a of acetic acid). To distinguish between (a) and (d), add a small amount of either an acid or a base. The pH of the buffer solution (d) would not change, and that of the 1.00 M NH_4Cl (solution b) would.

15.20A $H_3O^+ + OH^- \rightleftharpoons H_2O$

(a) $[H_3O^+] = \dfrac{20.00 \text{ mL x } 0.500 \text{ M} - 19.90 \text{ mL x } 0.500 \text{ M}}{(20.00 + 19.90) \text{ mL}}$

$[H_3O^+] = 1.253 \times 10^{-3}$ pH = 2.90

(b) $[H_3O^+] = \dfrac{20.00 \text{ mL x } 0.500 \text{ M} - 19.99 \text{ mL x } 0.500 \text{ M}}{(20.00 + 19.99) \text{ mL}}$

$[H_3O^+] = 1.250 \times 10^{-4}$ pH = 3.90

(c) $[OH^-] = \dfrac{20.01 \text{ mL x } 0.500 \text{ M} - 20.00 \text{ mL x } 0.500 \text{ M}}{(20.00 + 20.01) \text{ mL}}$

$[OH^-] = 1.250 \times 10^{-4}$ pOH = 3.90 pH = 10.10

(d) $[OH^-] = \dfrac{20.10 \text{ mL x } 0.500 \text{ M} - 20.00 \text{ mL x } 0.500 \text{ M}}{(20.00 + 20.10) \text{ mL}}$

$[OH^-] = 1.247 \times 10^{-3}$ pOH = 2.90 pH = 11.10

15.20B $NaOH + HCl \rightleftharpoons H_2O + Na^+ + Cl^-$

NaOH	HCl
25.00 mL × 0.220	8.10 mL × 0.252
5.50mmol	2.04 mmol
-2.04	-2.04
3.46 mmol	—

$[OH^-] = \dfrac{3.46 \text{ mmol}}{33.10 \text{ mL}} = 0.105\text{M}$

pOH = -log[OH] = 0.981

pH = 14.00 - pOH = 13.02

15.21(a) $CH_3COOH + OH^- \rightleftharpoons H_2O + CH_3COO^-$

CH_3COOH	OH^-	$H_2O + CH_3COO^-$
10.00 mmol	12.50 ml × 0.500 M	
-6.25 mmol	-6.25 mmol	+ 6.25 mmol
3.75 mmol	--	6.25 mmol

This is a buffer solution.

$[H_3O^+] = Ka\dfrac{[\text{acid}]}{[\text{base}]} = 1.8 \times 10^{-5} \times \dfrac{3.75 \text{ mmol}/32.50 \text{ mL}}{6.25 \text{ mmol}/32.50 \text{ mL}}$

Notice that since both acid and base are in the same solution, the volume is the same, does cancel out, and can be left out of the calculation.

$[H_3O^+] = 1.08 \times 10^{-5}$

pH = 4.97

(b) CH_3COOH + OH^- $\rightleftharpoons$ $H_2O + CH_3COO^-$

10.00 mmol	20.10 mL × 0.500 M	
-10.00 mmol	-10.00 mmol	+10.00 mmol
0	0.05 mmol	+10.00 mmol

$$[OH^-] = \frac{0.05 \text{ mmol}}{(20.00 + 20.10) \text{ mL}} = 1.25 \times 10^{-3} \text{ M}$$

pOH = 2.90

pH = 11.10

15.22(a) At half neutralization, $[H_3O^+] = \frac{K_w}{K_b}$.

$$10^{-9} = \frac{10^{-14}}{K_b}$$

$K_b = 1.0 \times 10^{-5}$

or

$pH = pK_a \approx 9$

$pK_b = 14.00 - pK_a \approx 5$

$pK_b \approx 1 \times 10^{-5}$

(b) From the graph, pH ≈ 5 at equivalence.
At equivalence, hydrolysis of the cation of the weak base occurs.

$$K_a = \frac{K_w}{K_b} = \frac{10^{-14}}{10^{-5}} = \frac{[H_3O^+]^2}{0.50 \text{ M}}$$

$[H_3O^+] = 2.24 \times 10^{-5}$

pH = 4.7

Review Questions

1. Arrhenius HI (aq) $\rightarrow H^+ + I^-$
Brønsted-Lowry HI (aq) + $H_2O \rightarrow H_3O^+ + I^-$

5. (a) $HClO_2 + H_2O \rightleftharpoons H_3O^+ + ClO_2^-$ $\quad K_a = \frac{[H_3O^+][ClO_2^-]}{[HClO_2]}$

(b) $CH_3CH_2COOH + H_2O \rightleftharpoons CH_3CH_2COO^- + H_3O^+$

$$K_a = \frac{[H_3O^+][CH_3CH_2COO^-]}{[CH_3CH_2COOH]}$$

(c) $HCN + H_2O \rightleftharpoons H_3O^+ + CN^-$ $\quad K_a = \frac{[H_3O^+][CN^-]}{[HCN]}$

(d) $C_6H_5OH + H_2O \rightleftharpoons H_3O^+ + C_6H_5O^-$ $\quad K_a = \frac{[H_3O^+][C_6H_5O^-]}{[C_6H_5OH]}$

13. K_a (acid) × K_b (conjugate base) = K_w

16. The equivalence point in an acid-base titration is the point where the acid and base are in the exact stoichiometric proportions. The end point in an acid-base titration is the point where the indicator color changes.

18. The pH is the highest before an acid is added. The last acid added produces the lowest pH. The equivalence point is that expected for the salt of a strong acid and weak base—below 7.

19. The vertical break is so short and indistinct in a weak base–weak acid titration that it is difficult to determine the equivalence point.

Problems

21. (a) $HOClO_2 + H_2O \rightleftharpoons H_3O^+ + OClO_2^-$
acid 1 base 2 acid 2 base 1
(b) $HSeO_4^- + NH_3 \rightleftharpoons NH_4^+ + SeO_4^{2-}$
acid 1 base 2 acid 2 base 1
(c) $HCO_3^- + OH^- \rightleftharpoons CO_3^{2-} + H_2O$
acid 1 base 2 base 1 acid 2
(d) $C_5H_5NH^+ + H_2O \rightleftharpoons C_5H_5N + H_3O^+$
acid 1 base 2 base 1 acid 2

23. $HSO_3^- + OH^- \rightarrow H_2O + SO_3^{2-}$
$HSO_3^- + HBr \rightarrow H_2SO_3 + Br^-$

25. $Cl^- < NO_3^- < F^- < CO_3^{2-}$ The reaction goes furthest toward completion with CO_3^{2-}, the most basic of these anions.

27. $NH_3 + NH_3 \rightleftharpoons NH_2^- + NH_4^+$
(acid 1) (base 2) (base 1) (acid 2)

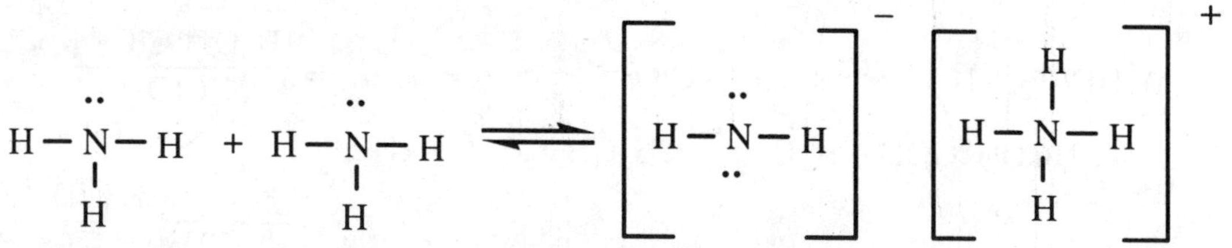

29. (a) $HClO_4$ has three O withdrawing groups pulling electron density from the O-H bond. The H is easily removed. In $HClO_2$ only one O withdrawing group exists, so the H^+ is held to the molecule.
(b) Perchloric acid has three terminal oxygen atoms, and sulfuric acid has only two. Also, the chlorine atom is more electronegative than sulfur. These factors combine to make $HOClO_3$ a better proton donor than $(HO)_2SO_2$.

31. The K_a increased a factor of 6 from 2 chloropropanoic acid to 2,2-dichloropropanoic acid (p. 646). Thus, K_a of dichloroacitic acid should be about $K_a = 1.4 \times 10^{-3} \times 6 = 8.4 \times 10^{-3}$ ($pK_a = 2.08$).

Trichloroacetic acid should have a larger K_a than dichloracetic acid and thus a still smaller pK_a.

33. (a) < (d) < (b) < (c)
Phenol should be weaker than (a). This ranking is determined by the number of Cl atoms in the molecules and their proximity to the -OH group.

35. (a) $[H_3O^+] = 10^{-pH} = 10^{-4.32} = 4.8 \times 10^{-5}$
(b) $[H_3O^+] = 10^{-6.6} = 3 \times 10^{-7}$
(c) $[H_3O^+] = 10^{-1.5} = 3 \times 10^{-2}$
(d) $[H_3O^+] = 10^{-4.3} = 5 \times 10^{-5}$
(e) $[H_3O^+] = 10^{-2.8} = 2 \times 10^{-3}$

37. (a) $[H_3O^+] = 0.0025$ M pH = 2.60
(b) $[OH^-] = 0.055$ M pOH = 1.26 pH = 12.74
(c) $[OH^-] = 2 \times 0.015$ M = 0.030 M pOH = 1.52 pH = 12.48
(d) $[H_3O^+] = 1.6 \times 10^{-3}$ M pH = 2.80

39. $[H_3O^+] = 0.00048 \times 2 = 0.00096$ M pH = 3.02
The vinegar (pH = 2.42) is more acidic. Even if both H^+ ionize, the pH of the $H_2SO_4(aq)$ can be no lower than 3.02.

41. $[H_3O^+] = 10^{-pH} = 10^{-3.60} = 2.5 \times 10^{-4}$ M

$$V_{HCL} = \frac{2.00\text{ L x } 2.5 \text{ x } 10^{-4}\text{M}}{0.100\text{M}} = 5.0 \times 10^{-3}\text{L} \times \frac{\text{mL}}{10^{-3}\text{L}} = 5.0 \text{ mL}$$

Dilute 5.0 mL of 0.100M HCl to 2.00 L.

43. (a) $K_a = 3.2 \times 10^{-7} = \frac{[H^+][SC_6H_5^-]}{[HSC_6H_5]}$

$$3.2 \times 10^{-7} = \frac{[H^+]^2}{1 \times 10^{-3}}$$

$[H^+]^2 = 3.2 \times 10^{-10}$
$[H^+] = 1.8 \times 10^{-5}$
pH = 4.75

(b) $M = \frac{75.0\text{g HCOOH}}{\text{L}} \times \frac{\text{mol HCOOH}}{46.03 \text{ g HCOOH}} = 1.63 \text{ M}$

$$1.8 \times 10^{-4} = \frac{[H_3O^+][HCOO^-]}{[HCOOH]} = \frac{[H_3O^+]^2}{1.63 \text{ M}}$$

$[H_3O^+] = 1.7 \times 10^{-2}$
pH = 1.77

45. $[H_3O^+] = 10^{-3.10} = 7.94 \times 10^{-4}$

$$K_a = \frac{[H_3O^+]^2}{[HN_3]} = \frac{[H_3O^+]^2}{M - 7.94 \text{ x } 10^{-4}} = 1.9 \times 10^{-5}$$

Assume $M >> 7.94 \times 10^{-4}$.

$$\frac{[H_3O^+]^2}{M} = \frac{(7.94 \text{ x } 10^{-4})^2}{M} = 1.9 \times 10^{-5}$$

$M = 3.3 \times 10^{-2}$

47. $[\text{aspirin}] = \frac{1.00 \text{ g}}{0.300 \text{ L}} \times \frac{\text{mol}}{180.15 \text{ g}} = 1.850 \times 10^{-2}$

$[H_3O^+] = 10^{-2.62} = 2.40 \times 10^{-3}$

$$K_a = \frac{[H_3O^+]^2}{M_{aspirin} - [H_3O^+]} = \frac{(2.40 \text{ x } 10^{-3})^2}{(1.85 \text{ x} 10^{-2} - 2.40 \text{ x } 10^{-3})} = 3.6 \times 10^{-4}$$

49. $CCl_3COOH + H_2O \rightleftharpoons H_3O^+ + CCl_3COO^-$

$K_a = 10^{-pKa} = 10^{-0.52} = 3.02 \times 10^{-1}$

K_a is too large for the assumptions to work.

$$0.302 = \frac{[H_3O^+]^2}{0.105-[H_3O^+]}$$

$[H_3O^+]^2 + 0.302 \{H_3O^+] - 0.0317 = 0$

$$[H_3O^+] = \frac{-0.302 \pm \sqrt{(0.302)^2 - 4 \text{ x } 1\text{x } (-0.0317)}}{2}$$

$[H_3O^+] = 0.0825$ M or -0384 M Negative values are not valid.

$[H_3O^+] = 0.0825$ M

pH= 1.08

51. For 0.150 M CH_3 COOH

$$K_a = 1.8 \times 10^{-5} = \frac{[H_3O^+]^2}{0.150 \text{ M}}$$

$[H_3O^+] = 1.6 \times 10^{-3}$

pH = 2.78

For HCOOH

$$K_a = 1.8 \times 10^{-4} = \frac{[H_3O^+]^3}{[HCOOH] - [H_3O^+]}$$

$$1.8 \times 10^{-4} = \frac{(1.6 \text{ x } 10^{-3})^2}{[HCOOH] - 1.6 \text{ x } 10^{-3}}$$

$[HCOOH] - 1.6 \times 10^{-3} = 0.014$

$[HCOOH] = 0.016$ M

53. 0.0045 M H_2SO_4 has the lower pH. H_2SO_4 is a strong acid in its first ionization step; moreover, even K_{a_2} of H_2SO_4 is larger than K_{a_1} of H_3PO_4.

55. (a) $? \text{ M} = \dfrac{83 \text{ g}}{\text{L}} \times \dfrac{\text{mol}}{90.04 \text{ g}} = 0.922 \text{ M}$

$5.4 \times 10^{-2} = \dfrac{[H_3O^+]^2}{0.922 \text{ M} - [H_3O^+]}$

$[H_3O^+]^2 + 5.4 \times 10^{-2}\,[H_3O^+] - 4.98 \times 10^{-2} = 0$

$[H_3O^+] = \dfrac{-5.4 \text{ x } 10^{-2} \pm \sqrt{(5.4 \text{ x } 10^{-2})^2 - 4 \text{ x } 1 \text{ x } (-4.98 \text{ x } 10^{-2})}}{2}$

$[H_3O^+] = 0.198$ M or -0.251 M

Because $K_{a1} >> K_{a2}$, all H_3O^+ is formed in the first ionization step.

$[H_3O^+] = 0.198$ M pH = 0.70

(b) $HOOCCOOH + H_2O \rightleftharpoons H_3O^+ + HOOCCOO^-$

$0.922 - X \qquad X \qquad X$

$HOOCCOO^- + H_2O \rightleftharpoons H_3O^+ + {}^-OOCCOO^-$

$X - Y \qquad X + Y \qquad Y$

$K_{a2} = 5.3 \times 10^{-5} = \dfrac{(X+Y)\left[{}^-OOCCOO^-\right]}{X - Y}$

Assume $X >> Y$.

$K_{a2} = 5.3 \times 10^{-5} = \dfrac{X\left[{}^-OOCCOO^-\right]}{X}$

$[{}^-OOCCOO^-] = 5.3 \times 10^{-5}$ M

57. (a) $H_3PO_4 + H_2O \rightarrow H_3O^+ + H_2PO_4^-$

$Ka = 7.1 \times 10^{-3} = \dfrac{[H_3O^+][H_2PO_4^-]}{[H_3PO_4]} = \dfrac{[H_3O^+]^2}{0.15 - [H_3O^+]}$

$[H_3O^+]^2 + 7.1 \times 10^{-3}\,[H_3O^+] - 1.07 \times 10^{-3} = 0$

$[H_3O^+] = \dfrac{-7.1 \text{ x } 10^{-3} \pm \sqrt{(7.1 \text{ x } 10^{-3})^2 - 4\,(1)\,(-1.07 \text{ x } 10^{-3})}}{2}$

$[H_3O^+] = 0.029$ or -0.036 Negative values are not valid.

$[H_3O^+] = 0.029 \text{ M} = [H_2PO_4^-]$

pH = 1.53

(b) $[H_3PO_4] = 0.15 - 0.029 = 0.12$ M

(c) $H_3PO_4 + H_2O \rightarrow H_3O^+ + H_2PO_4^-$

See (a) above. $[H_2PO_4^-] = 0.029$

(d) $H_2PO_4^- + H_2O \rightarrow H_3O^+ + HPO_4^{2}$

$K_{a2} = 6.3 \times 10^{-8} = \dfrac{[H_3O^+][HPO_4^{2-}]}{[H_2PO_4^-]} = 6.3 \times 10^{-8}$

Assume that $[H_3O^+] \approx [H_2PO_4^-]$ (p. 660), each cancels in the equation and $[HPO_4^{2-}] = K_{a2} = 6.3 \times 10^{-8}$.

(e) $HPO_4^{2-} + H_2O \rightleftharpoons H_3O^+ + PO_4^{3-}$ (see p. 661)

$$4.3 \times 10^{-13} = K_{a3} = \frac{[H_3O^+][PO_4^{3-}]}{[HPO_4^{2-}]}$$

$[HPO_4^{2-}] = K_{a2} = 6.3 \times 10^{-8}$ and $[H_3O^+] = 0.029$ M

$$4.3 \times 10^{-13} = \frac{0.029 \times [PO_4^{3-}]}{6.3 \times 10^{-8}}$$

$9.3 \times 10^{-19} = [PO_4^{3-}]$

OR

$$K_{a3} = \frac{[H_3O^+][PO_4^{3-}]}{[HPO_4^{2-}]}$$

$$K_{a3} = \frac{0.029[PO_4^{3-}]}{K_{a_2}}$$

$$[PO_4^{3-}] = \frac{K_{a_2}K_{a_3}}{0.029} = \frac{6.3 \times 10^{-8} \times 4.3 \times 10^{-13}}{0.029} = 9.3 \times 10^{-19}$$

59. (a) $CH_3CH_2COO^- + H_2O \rightleftharpoons CH_3CH_2COOH + OH^-$
    basic

    (b) $Mg(NO_3)_2$ solution should be neutral because Mg^{2+} would be from a strong base and NO_3^- would be from a strong acid.

    (c) $NH_4^+ + H_2O \rightleftharpoons NH_3 + H_3O^+$ $\quad K_a = \frac{K_w}{K_b} = \frac{10^{-14}}{1.8 \times 10^{-5}} = 5.6 \times 10^{-10}$

    $CN^- + H_2O \rightleftharpoons HCN + OH^-$ $\quad K_b = \frac{K_w}{K_a} = \frac{10^{-14}}{6.2 \times 10^{-10}} = 1.6 \times 10^{-5}$

    Because $K_a(NH_4^+)$ is smaller than $K_b(CN^-)$, hydrolysis of CN^- occurs more extensively than that of NH_4^+, and the solution will be basic.

61. (c) NH_4I. NH_4^+ is the conjugate acid of the base NH_3 and produces an acidic solution by hydrolysis. Of the other solutions, (a) is neutral and (b) and (d) are basic due to hydrolysis of the anions.

63. (a) $OCl^- + H_2O \rightleftharpoons HOCl + OH^-$

    (b) $K_b = \frac{[HOCl][OH^-]}{[OCl^-]} = \frac{[OH^-]^2}{0.080 - [OH^-]} \approx \frac{[OH^-]^2}{0.080} = \frac{K_w}{K_a} = \frac{10^{-14}}{2.9 \times 10^{-8}}$

    $= 3.4 \times 10^{-7}$

    $\frac{[OH^-]^2}{0.080} = 3.4 \times 10^{-7}$

    $[OH^-] = 1.66 \times 10^{-4}$ M

    pOH = 3.78

    (c) pH = 10.22

65. $[OH^-] = 10^{-(14.00 - 9.05)} = 1.12 \times 10^{-5}$

$CH_3CO_2^- + H_2O \rightleftharpoons CH_3CO_2H + OH^-$

$$K_b = \frac{K_w}{K_a} = \frac{10^{-14}}{1.8 \text{ x } 10^{-5}} = \frac{[OH^-]^2}{[CH_3CO_2^-]} = \frac{(1.12 \text{ x } 10^{-5})^2}{[CH_3CO_2^-]}$$

$[CH_3CO_2^-] = 0.23$ M

67. $HCOOH + H_2O \rightleftharpoons HCOO^- + H_3O^+$
(c) or (d). The H_3O^+ ion from HNO_3 or the $HCOO^-$ ion from $(HCOO)_2Ca$ will suppress the reaction.

69. $NH_3 + H_2O \rightleftharpoons NH_4^+ + OH^-$
0.15 M 0.015 M

$$K_b = \frac{[NH_4^+][OH^-]}{[NH_3]} = 1.8 \times 10^{-5}$$

$$1.8 \times 10^{-5} = \frac{X\,(0.015 + X)}{0.15 - X}$$

Assuming that $0.015 >> X$, then

$$1.8 \times 10^{-5} = \frac{X0.015}{0.15}.$$

$X = 1.8 \times 10^{-4}$ M $= [NH_4^+]$

71. $$K_a = \frac{[H_3O^+][CH_3CH_2COO^-]}{[CH_3CH_2COOH]}$$

Assume $[H_3O^+] << CH_3CH_2COO^-$.

$$1.3 \times 10^{-5} = \frac{0.0786X}{0.350}$$

$X = 5.79 \times 10^{-5} = [H_3O^+]$

pH = 4.24

73. (a) C_6H_5COOH is an acid. pH < 7
(b) NH_3 is a weak base. pH > 7
(c) CH_3NH_2 is a weak base; $CH_3NH_3^+$ is a salt of a weak base. pH < 7
(d) $K_{b1} = \frac{Kw}{K_{a1}} = \frac{10^{-14}}{7.1 \text{ x } 10^{-3}} = 1.4 \times 10^{-12}$. $K_{a2} = 6.3 \times 10^{-8}$ Because K_{a2} is larger than K_{b1}, KH_2PO_4 is more likely to lose a H^+ than to gain one. The solution is acidic. pH < 7
(e) $Ba(OH)_2$ (aq) will be basic.
(f) NO_2^- is the salt of a weak acid. The solution is basic.

75. $HPO_4^{2-} + H_2O \rightleftharpoons PO_4^{3-} + H_3O^+$ K_{a3}
$HPO_4^{2-} + H_2O \rightleftharpoons H_2PO_4^- + OH^-$ K_b

$K_{a3} = 4.3 \times 10^{-13}$ $\qquad K_b = \dfrac{K_w}{K_{a2}} = \dfrac{10^{-14}}{6.3 \text{ x } 10^{-8}} = 1.6 \times 10^{-7}$

Basic, because the hydrolysis of HPO_4^{2-} occurs more extensively than its further ionization as an acid.

77. $[H_3O^+] = K_a \times \dfrac{[\text{acid}]}{[\text{base}]} = 1.8 \times 10^{-4} \times \dfrac{0.405}{0.326}$
$[H_3O^+] = 2.24 \times 10^{-4}$ M
pH = 3.65

79. $[H^+] = 10^{-10.05} = 8.91 \times 10^{-11}$ M
$8.91 \times 10^{-11} = \dfrac{10^{-14}}{1.8 \text{ x } 10^{-5}} \dfrac{[\text{acid}]}{[0.350]}$
$[\text{Acid}] = 5.6 \times 10^{-2}$ M

$? \text{ g}(NH_4)_2SO_4 = 5.6 \times 10^{-2}\text{M} \times 0.100\text{L} \times \dfrac{\text{mol}(NH_4)_2SO_4}{2\text{mol}NH_4^+} \times \dfrac{132 \text{ g}(NH_4)_2SO_4}{\text{mol}(NH_4)_2SO_4}$
$= 0.37 \text{ g}(NH_4)_2SO_4$

The solution is so basic that a negligible amount of SO_4^{2-} will become HSO_4^-, and that reaction can be ignored.

81.

	$HCOO^-$	+	H_3O^+	$\rightleftharpoons$	$HCOOH$	+	H_2O
	0.326 M × 50.0 mL		1.00 mL × 0.250 M		0.405 M × 50.0 mL		
initial:	16.3 mmol		0.25 mmol		20.3 mmol		
equil:	16.0		--		20.6		

$[H_3O^+] = 1.8 \times 10^{-4} \times \dfrac{20.6}{16.0} = 2.32 \times 10^{-4}$

pH = 3.63

83. No. The components need to be a base and its *conjugate* acid or an acid and its *conjugate* base. The HCl and NaOH would neutralize one another, leaving NaCl (aq) with either excess HCl or excess NaOH, not a buffer solution.

85. The pH range from just slightly before to just slightly beyond the equivalence point is longer for a strong acid–strong base titration than for a weak acid–strong base titration. More indicators are available that change color in this longer pH range.

87. (a) NH_4Cl: $\qquad NH_4^+ + H_2O \rightleftharpoons NH_3 + H_3O^+$

$K_a = \dfrac{K_w}{K_b} = \dfrac{1 \text{ x } 10^{-14}}{1.8 \text{ x } 10^{-5}} = \dfrac{[H_3O^+]^2}{[NH_4^+]} = \dfrac{[H_3O^+]^2}{(0.10 \text{ M})}$

$[H_3O^+] = 7.45 \times 10^{-6}$ M
pH = 5.13 $\qquad$ yellow

(b) CH_3COOK: $\qquad CH_3COO^- + H_2O \rightleftharpoons CH_3COOH + OH^-$

$K_b = \dfrac{K_w}{K_a} = \dfrac{10^{-14}}{1.8 \text{ x } 10^{-5}} = \dfrac{[OH^-]^2}{[CH_3COO^-]} = \dfrac{[OH^-]^2}{(0.10 \text{ M})}$

$[OH^-] = 7.45 \times 10^{-6}$ M
pOH = 5.13
pH = 8.87 blue

(c) Na_2CO_3: $CO_3^{2-} + H_2O \rightleftharpoons HCO_3^- + OH^-$

$$K_b = \frac{K_w}{K_a} = \frac{10^{-14}}{4.7 \times 10^{-11}} = 2.1 \times 10^{-4} = \frac{[OH^-]^2}{[CO_3^{2-}]} = \frac{[OH^-]^2}{(0.10\ M)}$$

$[OH^-] = 4.58 \times 10^{-3}$ M
pOH = 2.34
pH = 11.66 red

(d) CH_3COOH: $CH_3COOH + H_2O \rightleftharpoons CH_3COO^- + H_3O^+$

$$K_a = 1.8 \times 10^{-5} = \frac{[H_3O^+]^2}{[CH_3COOH]} = \frac{[H_3O^+]^2}{(0.10\ M)}$$

$[H_3O^+] = 1.34 \times 10^{-3}$ M
pH = 2.87 violet

89. The new solution will be pH = 7. The new color will be yellow because the phenolphthalein will be colorless and the thymol blue will be yellow.

91. In a strong base–strong acid titration; (1) the initial pH is high because the base is completely ionized; (2) at the half-neutralization point, the pH depends on the concentration of the base remaining (half has been neutralized); (3) at the equivalence point, the pH is 7.00 because neither cation nor anion ionize; (4) the steep portion of the curve is over a wide range; (5) the choice of indicators is extensive. Any indicator with a color change in the pH range of 4 to 10 will work. The same indicator can be used for either titration because the region of rapid pH change (steep portion of titration curve) is the same in either case; it is just appoached from a different direction. (Something not discussed in the text is that since it is easier to see the color change from a light to a dark color, usually the same indicator is not used for both types of titrations.)

93.

CH_3COOH	+ OH^-	$\rightleftharpoons$ $CH_3COO^- + H_2O$
20.00 mL × 0.500 M	7.45 mL × 0.500 M	--
10.00 mmol	3.73 mmol	
-3.73 mmol	-3.73 mmol	+3.73 mmol
6.27 mmol	--	3.73 mmol

$$[H_3O^+] = K_a \times \frac{[acid]}{[base]} = 1.8 \times 10^{-5} \times \frac{6.27}{3.73} = 3.03 \times 10^{-5}$$

pH = 4.52

95. $NaOH + HCl \rightarrow Na^+ + Cl^- + H_2O$

NaOH	HCl
25.00 mol × 0.324M	V × 0.250M
8.10mmol	0.250V

$[OH^-] = 10^{-(14.00 - 11.50)} = 3.16 \times 10^{-3}$ M

$$3.16 \times 10^{-3} = \frac{(8.10\ \text{mmol} - 0.250\ V)\ \text{mmol}}{(V + 25.00)\ \text{mL}}$$

$3.16 \times 10^{-3}\ V + 7.9 \times 10^{-2} = 8.10 - 0.250\ V$
$0.253\ V = 8.02$
$V = 31.7\ mL$

97. (a) $NH_3 + HCl \rightarrow NH_4^+ + Cl^-$

$$? \text{ mL HCl} = 40.00 \text{ mL} \times 0.200M\ NH_3 \times \frac{1 \text{ mol HCl}}{1 \text{ mol } NH_3} \times \frac{L}{0.500 \text{ mol HCl}} = 16 \text{ mL HCl}$$

(b) $NH_3 + H_2O \rightleftharpoons NH_4^+ + OH^-$

$$K_b = 1.8 \times 10^{-5} = \frac{[OH^-]^2}{0.200}$$

$[OH^-] = 1.90 \times 10^{-3}$ M pOH = 2.72
pH = 11.28

(c)

	NH_3	+	H^+	$\rightarrow$	NH_4^+
	40.00 mL 0.200 M		5.00 mL × 0.500 M		
I	8.00 mmol		2.50 mmol		
C	- 2.50		- 2.50		+ 2.50 mmol
E	5.50 mmol		0		2.50 mmol

$$[H_3O^+] = \frac{10^{-14}}{1.8 \times 10^{-5}} \times \frac{(2.50)}{(5.50)} = 2.53 \times 10^{-10} \text{ M}$$

pH = 9.60

(d) $$[H_3O^+] = \frac{10^{-14}}{1.8 \times 10^{-5}} = 5.56 \times 10^{-10} \text{ M}$$

pH = 9.26
or
$pH = pK_a(NH_4^+) = 14.00 - pK_b(NH_3)$
pH = 14.00- 4.74 = 9.26

(e)

	NH_3	+	H^+	$\rightarrow$	NH_4^+
I	8.00 mmol		10.00 mL × 0.500 M		
C	- 5.00		- 5.00		+ 5.00
E	3.00				5.00

$$[H_3O^+] = \frac{1 \times 10^{-14}}{1.8 \times 10^{-5}} \times \frac{(5.00)}{(3.00)} = 9.26 \times 10^{-10} \text{ M}$$

pH = 9.03

(f) $$K_a = \frac{K_w}{K_b} = \frac{10^{-14}}{1.8 \times 10^{-5}} = \frac{[H_3O^+]^2}{[NH_4^+]}$$

$$[NH_4^+] = \frac{8.00 \text{ mmol}}{(40.00 + 16.00) \text{ mL}} = 0.143 \text{ M}$$

$$\frac{10^{-14}}{1.8 \times 10^{-5}} = \frac{[H_3O^+]^2}{0.143}$$

$[H_3O^+] = 8.91 \times 10^{-6}$
pH = 5.05

(g)

NH_3	+	H^+	$\rightarrow$	NH_4^+
8.00 mmol		20.00 mL × 0.500 M		
		10.00 mmol		
- 8.00		- 8.00		+ 8.00
--		2.00 mmol		8.00

$[H_3O^+] = \frac{2.00\ \text{mmol}}{60.00\ \text{mL}} = 3.33 \times 10^{-2}\ \text{M}$

pH = 1.48

99.

	Acid	Base
	e^- pair acceptor	e^- pair donor
(a)	$Al(OH)_3$	OH^-
(b)	Cu^{2+}	NH_3
(c)	CO_2	OH^-

Additional Problems

101. (a) $[H^+] = K_a \frac{[\text{acid}]}{[\text{base}]}$

$[H^+] = 10^{-4.0} = 1.4 \times 10^{-4} \frac{[\text{acid}]}{[\text{base}]}$

$\frac{1.0\times10^{-4}}{1.4\times10^{-4}} = 0.71 = \frac{[\text{acid}]}{[\text{base}]}$

? moles NaOH = 2.0 g NaOH $\frac{\text{mol}}{40.0\ \text{g}}$ = 0.050 mol NaOH

$CH_3CHOHCOOH + OH^- \rightarrow CH_3CHOHCOO^- + H_2O$

0.050 mol NaOH will make 0.050 mol of conjugate base.

? mol acid = 0.71 × 0.050 mol NaOH = 0.036 mol acid

? mol acid total = mol reacted with NaOH + mol as acid

? mol acid = 0.050 mol + 0.036 mol = 0.086 mol

? g lactic acid = 0.086 mol × $\frac{90.0\ \text{g}}{\text{mol}}$ = 7.7 g lactic acid

(b) $\text{pH} = \text{p}K\text{a} + \log \frac{[\text{base}]}{[\text{acid}]}$

$3.80 = 4.74 + \log \frac{[\text{OAc}^-]}{[\text{HOAC}]}$

$-0.94 = \log \frac{[\text{OAc}^-]}{[\text{HOAC}]}$

$0.115 = \frac{[\text{OAc}^-]}{[\text{HOAC}]}$

Also $[\text{HOAc}] + [\text{OAc}^-] = 0.0500$ M

$$0.115 = \frac{[OAc^-]}{[0.0500 - [OAc]}$$

$0.00575 - 0.115\,[OAc^-] = [OAc^-]$

$1.115 = 0.00575$

$[OAc^-] = 0.0052\text{ M}$

$[HOAc] = 0.0448\text{ M}$

$$0.0052\text{ mol }OAc^- \times \frac{\text{mol NaOH}}{\text{mol }OAc^-} \times \frac{40.00\text{ g}}{\text{mol}} = 0.21\text{ g NaOH}$$

$$0.0500\text{ M} \times 1.00\text{ L} \times \frac{60.05\text{ g}}{\text{mol}} = 3.00\text{ g acetic acid}$$

103. (a) $\text{pH} = \text{p}K_a + \log \frac{[\text{base}]}{[\text{acid}]}$

$$7.4 = 12.4 + \log \frac{[\text{base}]}{[\text{acid}]}$$

$$-5.0 = \log \frac{[\text{base}]}{[\text{acid}]} = \log \frac{[PO_4^{3-}]}{[HPO_4^{2-}]}$$

$$10^{-5.0} = 1.0 \times 10^{-5} = \frac{[PO_4^{3-}]}{[HPO_4^{2-}]}$$

(b) $7.4 = 7.2 + \log \frac{[HPO_4^{2-}]}{[H_2PO_4^-]}$

$$10^{0.2} = \frac{[HPO_4^{2-}]}{[H_2PO_4^-]} = 1.6$$

(c) $7.4 = 2.1 + \log \frac{[H_2PO_4^-]}{[H_3PO_4]}$

$$10^{5.3} = \frac{[H_2PO_4^-]}{[H_3PO_4]} = 2 \times 10^5$$

(d) $7.4 = 10.3 + \log \frac{[CO_3^{2-}]}{[HCO_3^-]}$

$$10^{-2.9} = \frac{[CO_3^{2-}]}{[HCO_3^-]} = 1.3 \times 10^{-3}$$

(e) $7.4 = 6.4 + \log \frac{[HCO_3^-]}{[H_2CO_3]}$

$$10^{1.0} = \frac{[HCO_3^-]}{[H_2CO_3]} = 10$$

The HCO_3^-/H_2CO_3 ratio is the most responsible for maintaining the normal pH buffer, although usually the buffer system with K_a closest to the pH is the best. The blood pH depends heavily on the bicarbonate ion/carbonic acid system.

105. (a) $pH = pK_a + \log \frac{[HPO_4^{2-}]}{[H_2PO_4^-]}$

$$7.20 = 7.2 + \log \frac{[HPO_4^{2-}]}{[H_2PO_4^-]}$$

$$10^0 = 1.0 = \frac{[HPO_4^{2-}]}{[H_2PO_4^-]}$$

$[HPO_4^{2-}] = [H_2PO_4^-] = 0.10$ M

Add 100 mL each of 1.00 M Na_2HPO_4 and NaH_2PO_4 to a volumetric flask and dilute to 1.000 L.

(b) $pH = pKa + \log \frac{[HPO_4^{2-}]}{[H_2PO_4^-]}$

$$6.8 = 7.2 + \log \frac{[HPO_4^{2-}]}{[H_2PO_4^-]}$$

$$-0.4 = \log \frac{[HPO_4^{2-}]}{[H_2PO_4^-]}$$

$$10^{-0.4} = 0.398 = \frac{[HPO_4^{2-}]}{[H_2PO_4^-]}$$

Also $[H_2PO_4^-] = 0.25$ M

$[HPO_4^{2-}] = 0.398 \times [H_2PO_4^-] = 0.398 \times 0.25$ M

$[HPO_4^{2-}] = 0.0995$ M

Mix 250 mL of 1.00 M NaH_2PO_4 with 99.5 mL of 1.00 M Na_2HPO_4 and dilute to 1.00 L.

OR

If $[HPO_4^{2-}] = 0.25$ M

$$[H_2PO_4^-] = \frac{[HPO_4^{2-}]}{0.398} = \frac{0.25}{0.398} = 0.628 \text{ M}$$

Mix 628 mL of 1.00 M NaH_2PO_4 with 250 mL of 1.00 M Na_2HPO_4 and dilute to 1.00 L.

107. $CaO\ (s) + H_2O\ (l) \rightarrow Ca(OH)_2\ (aq) \rightarrow Ca^{2+}\ (aq) + 2\ OH^-\ (aq)$

$$1.0 \text{ ton} \times \frac{2000 \text{ lb}}{\text{ton}} \times \frac{454 \text{ g}}{\text{lb}} \times \frac{72 \text{ g}}{100 \text{ g}} \times \frac{\text{mL}}{\text{g}} \times \frac{\text{L}}{1000 \text{ mL}} = 654 \text{ L solution}$$

$[H_3O^+] = 1 \times 10^{-7}$ in pure water. The amount of H_3O^+ that must be neutralized to raise the pH from 6.0 to 7.0 is:

$$654\ L \times \left(\frac{1 \times 10^{-6}\ mol\ H_3O^+}{L} - \frac{1 \times 10^{-7}\ mol\ H_3O^+}{L} \right) = 5.9 \times 10^{-4}\ mol\ H_3O^+$$

(a negligible amount)

To raise the pH of water from 7.0 to 12.0:

$[OH^-] = 10^{-(14.00 - 12.00)} = 10^{-2}$ M

$$10^{-2}\ M \times 654\ L = 6.54\ moles\ OH^- \times \frac{mole\ CaO}{2\ mole\ OH^-} \times \frac{56.08\ g}{mole} = 1.8 \times 10^2\ g\ CaO$$

109. $HCl + H_2O \rightarrow H_3O^+ + Cl^-$ $\qquad [H_3O^+]_{HCl} = 10^{-8}$

$2\ H_2O \rightarrow H_3O^+ + OH^-$

$\qquad 10^{-8} + X \qquad X$

$K_w = 10^{-14} = (10^{-8} + X)\,X$

$10^{-14} = 10^{-8}\,X + X^2$

$X^2 + 10^{-8}\,X - 10^{-14} = 0$

$$X = \frac{-10^{-8} + \sqrt{(10^{-8})^2 - 4 \times 1 \times (-10^{-14})}}{2}$$

$[H_3O^+]_{H_2O} = X = 9.51 \times 10^{-8}$ or -8.5×10^{-8} not valid

$[H_3O^+] = [H_3O^+]_{HCl} + [H_3O^+]_{H_2O}$

$[H_3O^+] = 1.05 \times 10^{-7}$ $\quad$ pH = 6.98

111. Thymol blue changes to blue at about pH 9.5.

$[H^+] = 10^{-9.5} = 3.16 \times 10^{-10}$

$$?\ mol\ NaOH\ excess = 3.16 \times 10^{-10} \times 100\ mL \times \frac{10^{-3} L}{mL} = 3.16 \times 10^{-11}\ mol\ NaOH$$

$$100.0\ mL \times 0.200 M \times \frac{10^{-3} L}{mL} = 2.00 \times 10^{-2}\ mol\ NaOH\ to\ neutralize\ buffer.$$

$?mol\ NaOH_{total} = 3.16 \times 10^{-11}\ mol + 2.00 \times 10^{-2}\ mol = 2.00 \times 10^{-2}\ mol\ NaOH$

113. $\Delta H^\circ = \Delta H_f^\circ(H^+) + \Delta H_f^\circ(OH^-) - \Delta H_f^\circ(H_2O)$

$\Delta H^\circ = 0 - 230.0 kJ - (-285.8\ kJ)$

$H_2O \rightleftharpoons H^+ + OH^-$ $\qquad \Delta H^\circ = 55.8$ kJ

$K_w = [H^+][OH^-]$

Equilibrium shifts in the direction of the endothermic reaction at a high temperature. Thus, the numerical value of K_w increases with temperature.

115 $$?\ g\ NaHSO_4 = 36.56\ mL \times 0.225\ M \times \frac{10^{-3}\ mol}{mmol} \times \frac{mol\ NaHSO_4}{mol\ NaOH} \times \frac{120.1\ g}{mol} = 0.988\ g$$

$$? \text{ \% } NaHSO_4 = \frac{0.988 \text{ g}}{1.016 \text{ g}} \times 100 \text{ \%} = 97.2 \text{ \% } NaHSO_4$$

$? \text{ \% } NaCl = 100.0 \text{ \%} - 97.2 \text{ \%} = 2.8 \text{ \% } NaCl$

Bromthymol blue, phenol red, and other indicators that change close to pH 7 are suitable. With a K_a of 1.1×10^{-2} a titration of HSO_4^- will reach the equivalence point close to 7.

117.

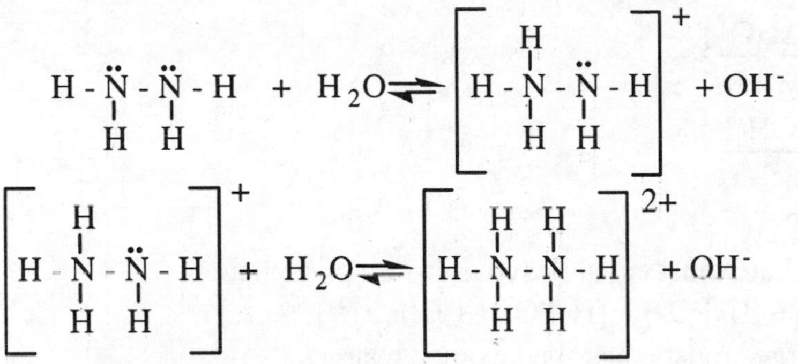

The second H^+ is repelled by the charge on the $NH_3NH_2^+$ ion. This makes $K_{b2} < K_{b1}$ and $pK_{b2} > pK_{b1}$.

$$K_{b1} = \frac{[N_2H_5^+][OH^-]}{[N_2H_4]} = 10^{-6.07} = 8.5 \times 10^{-7}$$

$[OH^-]^2 = 0.150 \times 8.5 \times 10^{-7}$

$[OH^-] = 3.6 \times 10^{-4}$

pOH = 3.45

pH = 14.00 – 3.45 = 10.55

119. (a) The pH starts high and decreases during the titration. The solution being titrated is a base, and the titrant is an acid. The base must be a strong base; the initial pH is 14. The equivalence point is pH ≈ 8.8, so the acid is a weak acid. Also, the final pH remains fairly high (about 4), indicating that the titrant is a weak acid.

(b) At twice the equivalence point the concentration of weak acid and conjugate base will be equal and pH = $pK_a \approx 3.8$. $Ka \approx 1.6 \times 10^{-4}$.

(c) From the graph the 2< pH <9 at the equivalence point. By calculation the pH = 8.8 (assuming $pK_a = 3.8$ for the titrant).

$A^- + H_2O \rightleftharpoons HA + OH^-$

$$\frac{K_w}{K_a} = \frac{[OH^-]^2}{[A^-]}$$

$$\frac{10^{-14}}{1.6 \times 10^{-4}} = \frac{[OH^-]^2}{0.50}$$

$[A^-] = 0.50$ M because equal volumes of 1.00 M acid and 1.00 M base are required to reach the equivalence point.

$[OH^-] = 5.6 \times 10^{-6}$

pOH = 5.25

$pH = 14 - 5.25 = 8.75$

121. (a) $H_2PO_4^- + H_2O \rightleftharpoons HPO_4^{2-} + H_3O^+$ [1]

$H_2PO_4^- + H_2O \rightleftharpoons H_3PO_4 + OH^-$ [2]

(b) The traditional method of deriving the equation as shown in many analytical books is shown below. There are five unknowns: $[HPO_4^{2-}]$, $[H_3PO_4]$, $[H_2PO_4^-]$, $[H_3O^+]$, and $[OH^-]$, so five equations are needed.

$$K_{a2} = \frac{[HPO_4^{2-}][H_3O^+]}{[H_2PO_4^-]}$$

$$\frac{K_w}{K_{a1}} = \frac{[H_3PO_4][OH^-]}{[H_2PO_4^-]}$$

$$K_w = [OH^-][H_3O^+]$$

The initial phosphate must equal the equilibrium phosphate.

$$[NaH_2PO_4]_{initial} = [H_3PO_4] + [H_2PO_4^-] + [HPO_4^{2-}]$$

The positive charges must equal the negative charges.

$$[Na^+] + [H_3O^+] = [H_2PO_4^-] + 2\,[HPO_4^{2-}] + [OH^-]$$

All Na^+ comes from NaH_2PO_4. $[NaH_2PO_4]_{initial} = [Na^+]$

subtract $[Na^+] = [H_3PO_4] + [H_2PO_4^-] + [HPO_4^{2-}]$

from $\underline{[Na^+] + [H_3O^+] = [H_2PO_4^-] + 2[HPO_4^{2-}] + OH^-}$

to get $[H_3O^+] = [HPO_4^{2-}] + [OH^-] - [H_3PO_4]$ {1}

Rearrange from above. $[H_3PO_4] = \dfrac{[H_2O^+][H_2PO_4^-]}{K_{a1}}$

$$[HPO_4^{2-}] = \frac{K_{a2}\,[H_2PO_4^-]}{[H_3O^+]} \qquad [OH^-] = \frac{K_w}{[H_3O^+]}$$

Plug these into {1} to get

$$[H_3O^+] = \frac{K_{a2}[H_2PO_4^-]}{[H_3O^+]} + \frac{K_w}{[H_3O^+]} - \frac{[H_3O^+][H_2PO_4^-]}{K_{a1}}$$

$$[H_3O^+]^2 = K_{a2}[H_2PO_4^-] + K_w - \frac{[H_3O^+]^2[H_2PO_4^-]}{K_{a1}}$$

$$[H_3O^+]^2 \left(\frac{[H_2PO_4^-]}{K_{a1}} + 1\right) = K_{a2}[H_2PO_4^-] + K_w$$

$$[H_3O^+]^2 = \frac{K_{a2}[H_2PO_4^-] + K_w}{\dfrac{[H_2PO_4^-]}{K_{a1}} + 1}$$

$K_{a2}[H_2PO_4^-]$ will be much larger than K_w.

$\dfrac{[H_2PO_4^-]}{K_{a1}}$ will be much larger than 1.

$$[H_3O^+]^2 = \frac{K_{a2}[H_2PO_4^-]}{[H_2PO_4^-]} K_{a1}$$

$[H_3O^+] = \sqrt{K_{a2}K_{a1}}$

$pH = \frac{1}{2}(pK_{a1} + pK_{a2})$

An alternate simpler, more straightforward method as described in the answer appendix is shown below.
The equation for the two reactions are:

$[1]\ K_{a2} = \frac{[HPO_4^{2-}][H_3O^+]}{[H_2PO_4^-]}$

$[2]\ K_b = \frac{K_w}{K_{a1}} = \frac{[H_3PO_4][OH^-]}{[H_2PO_4^-]}$

Assume $[HPO_4^{2-}] = [H_3O^+]$ in [1] and $[H_3PO_4] = [OH^-]$ in [2]

$[1]\ K_{a2} = \frac{[H_3O^+]^2}{[H_2PO_4^-]}$

$[2]\ K_b = \frac{K_w}{K_{a1}} = \frac{[OH^-]^2}{[H_2PO_4^-]}$

and $[OH-] = \frac{K_w}{[H_3O^+]}$

Assuming H_3O^+ from [1] neutralizes $[OH^-]$ from [2], this means that

$[H_3O+] = [HPO_4^{2-}] - [H_3PO_4]$

Substitute for $[HPO_4^{2-}]$ and $[H_3PO_4]$ with [1] and [2] to get

$[H_3O^+] = \frac{K_{a2}[H_2PO_4^-]}{[H_3O^+]} - \frac{[K_bH_2PO_4^-]}{[OH-]}$

$[H_3O^+] = \frac{K_{a2}H_2PO_4^-]}{[H_3O^+]} - \frac{[H_2PO_4^-][H_3O^+]}{K_w}\frac{K_w}{K_{a1}}$

$[H_3O^+]^2K_{a1} = K_{a1}K_{a2}[H_2PO_4^-] - [H_2PO_4^-][H_3O^+]^2$

$[H_3O^+]^2 (K_{a1} + [H_2PO_4^-]) = K_{a1}K_{a2}[H_2PO_4^-]$

$[H_3O^+]^2 = \frac{K_{a1}K_{a2}[H_2PO_4^-]}{K_{a1} + [H_2PO_4^-]}$

Since $[H_2PO_4^-] \approx 0.01$, $[H_2PO_4^-] >> K_{a1}$ (7.1×10^{-3})

$[H_3O^+]^2 = \frac{K_{a1}K_{a2}[H_2PO_4^-]}{[H_2PO_4^-]}$

$[H_3O^+]^2 = K_{a1}K_{a2}$

$[H_3O^+] = \sqrt{K_{a1}K_{a2}}$

$pH = \frac{1}{2}(pK_{a1} + pK_{a2})$

(c) $pH = ½(pK_{a1} + pK_{a2})$

$K_{a1} = 4.4 \times 10^{-7}$ $\qquad pK_{a1} = 6.36$

$K_{a2} = 4.7 \times 10^{-11}$ $\qquad pK_{a2} = 10.33$

$pH = ½(6.36 + 10.33)$
$pH = 8.35$

123. $\left(\dfrac{0.759\,mL\,CO_2}{mL\,H_2O} = \dfrac{0.759\,L\,CO_2}{L\,H_2O}\right)$

$$?\,M = \frac{0.759\,L\,CO_2}{L\,H_2O} \times \frac{mol}{22.4L} = 0.034\ M$$

$$K = \frac{c}{p} = \frac{0.034\ M}{1\ atm} = \frac{[H_2CO_3]}{0.00037\,atm}$$

$[H_2CO_3] = 1.26 \times 10^{-5}$ M

$$K_{a1} = 4.4 \times 10^{-7} = \frac{[H_3O^+]^2}{1.26{\times}10^{-5} - [H_3O^+]}$$

$$[H_3O^+]^2 + 4.4 \times 10^{-7}[H_3O^+] - 5.5 \times 10^{-12} = 0$$

$$[H_3O^+] = \frac{-4.4{\times}10^{-7} + \sqrt{(4.4{\times}10^{-7})^2 - 4{\times}1{\times}(-5.5{\times}10^{-12})}}{2}$$

$[H_3O^+] = 2.1 \times 10^{-6}$ M or -2.51×10^{-6} not valid
pH = 5.68

125. For dilute solutions, molarity is approximately equal to molality.
$0.0500M \approx 0.0500m$
$\Delta T = iKm$

$$(-0.096-0)\ °C = -i \times \frac{1.86\ °C}{m} \times 0.0500\ M$$

$i = 1.03$
3% of the molecules have dissociated.
$[H_2C{=}CHCH_2COO^-] = [H_3O^+] = 0.03 \times 0.0500 = 1.5 \times 10^{-3}$ M
$[H_2C{=}CHCH_2COOH] = 0.97 \times 0.0500\ M = 4.85 \times 10^{-2}$ M

$$K_a = \frac{(1.5 \text{ x } 10^{-3})^2}{4.85 \text{ x } 10^{-2}} = 4.6 \times 10^{-5}$$

127. (a) 2 moles acetic acid and 1 mole of sodium acetate in 2.00 L of solution.
(b) 3 moles acetic acid and 1 mole of NaOH in 2.00 L of solution.
(c) 3 moles sodium acetate and 2 moles hydrochloric acid in 2.00 L of solution.
In (a) and (b) there are acetic acid molecules, acetate ions, sodium ions, and water. In (c) there is also 1 M NaCl, and the high ionic strength may have a small effect on the pH.

129. $[HCl] = \dfrac{0.10\,mL \times 0.05\,M}{100.0\,mL} = 5 \times 10^{-5}$
pH = 4.3 instead of 7.

Any residual drops of solution on the electrode would change the pH of the water. A buffer at & would react with the residual drops and keep the pH at 7. One appropriate buffer would be $H_2PO_4^-$ and HPO_4^{2-} in the ratio of $\frac{[HPO_4^{2-}]}{[H_2PO_4^-]} = 0.63$.

$$pH = pK_{a2} + \log \frac{[HPO_4^{2-}]}{[H_2PO_4^-]} = 7.00$$

$$pH = 7.2 + \log \frac{[HPO_4^{2-}]}{[H_2PO_4^-]} = 7.00$$

$$\frac{[HPO_4^{2-}]}{[H_2PO_4^-]} = 10^{-0.20} = 0.63$$

Apply Your Knowledge

131. $CH_3COOH + H_2O \rightleftharpoons CH_3COO^- + H_3O^+$

$$K_a = 1.8 \times 10^{-5} = \frac{[CH_3COO^-][H_3O^+]}{0.250M}$$

$HCOOH + H_2O \rightleftharpoons HCOO^- + H_3O^+$

$$K_a = 1.8 \times 10^{-4} = \frac{[HCOO^-][H_3O+]}{0.150M}$$

$$\frac{4.5 \text{ x } 10^{6}}{[H_3O^+]} = [CH_3COO^-]$$

$$\frac{2.7 \text{ x } 10^{-5}}{[H_3O^+]} = [HCOO^-]$$

$$[CH_3COO^-] + [HCOO^-] = [H_3O+] = \frac{4.5 \text{ x } 10^{-6}}{[H_3O^+]} + \frac{2.7 \text{ x } 10^{-5}}{[H_3O^+]}$$

$[H_3O^+]^2 = 3.15 \times 10^{-5}$

$[H_3O^+] = 5.61 \times 10^{-3}$

$pH = 2.25$

133. At the equivalence point. The solution consists of the salt (NaP) of the weak monoprotic acid (HP). The value of $[P^-]$ at the equivalence point could be calculated from the titration and dilution data, but this is not necessary. Addition of a second 25.00-mL portion of the weak acid at the equivalence point produces a solution in which $[HP] = [P^-]$, no matter what their particular value. In a solution with $[HP] = [P^-]$, $pH = pK_a + \log \frac{[P^-]}{[HP]} = pK_a + \log 1 = pK_a = 3.84$. Thus,

$[H^+] = K_a = 10^{-3.84} = 1.4 \times 10^{-4}$.

135. (a) TRIS is best. Aniline's K_b is small, and its reaction with acid is less complete. Sodium phosphate is tribasic and would give a less distinct end point (see graph

on Problem 132). Ammonia is volatile, causing its concentration to change and thus not be known accurately.

(b) The solution at the equivalence point is $Na^+(aq) + Cl^- + H_2O(l) + CO_2(aq)$, but the solution is likely to be supersaturated in CO_2. This enhances the ionization of CO_2 as a weak acid, leads to a slightly lower pH, and slightly affects the titration volume. Boiling the solution, followed by cooling to room temperature, reduces $[CO_2]$ to its normal equilibrium value. The solution does not need to be boiled with sodium phosphate, as no CO_2 is produced in the itration reaction. Some of the NH_3 is lost to vaporization.

(c) $$30\text{ mL} \times 1\text{M} \times \frac{10^{-3}\text{ mole}}{\text{mmole}} \times \frac{\text{mole TRIS}}{\text{mole HCl}} \times \frac{121\text{ g}}{\text{mole TRIS}} = 3.6\text{ g}$$

$$45\text{ mL} \times 1\text{M} \times \frac{10^{-3}\text{ mole}}{\text{mmole}} \times \frac{\text{mole TRIS}}{\text{mole HCl}} \times \frac{121\text{ g}}{\text{mole TRIS}} = 5.4\text{ g}$$

$$30\text{ mL} \times 1\text{ M} \times \frac{10^{-3}\text{ mol}}{\text{mmol}} \times \frac{\text{mol Na}_2\text{CO}_3}{2\text{ mol HCl}} \times \frac{106\text{ g}}{\text{mol Na}_2\text{CO}_3} = 1.6\text{ g}$$

$$45\text{ mL} \times 1\text{ M} \times \frac{10^{-3}\text{ mol}}{\text{mmol}} \times \frac{\text{mol Na}_2\text{CO}_3}{2\text{ mol HCl}} \times \frac{106\text{ g}}{\text{mol Na}_2\text{CO}_3} = 2.4$$

(d) ? moles HCl = 1 M × 45 mL = 45 mmol

$$[\text{TRIS H}^+] = \frac{45\text{ moles}}{90\text{ mL}} = 0.5\text{ M}$$

$\text{TRIS} \rightarrow H^+ + \text{TRIS}$

$$K = \frac{K_w}{K_b} = \frac{10^{-14}}{1.20 \times 10^{-6}} = \frac{[H^+]^2}{0.5}$$

$[H^+]^2 = 4.2 \times 10^{-9}$

$[H^+] = 6.5 \times 10^{-5}$

pH = 4.2

Methyl orange changes light to dark as pH decreases. Bromphenol blue would also work at that pH range.

Chapter 16

More Equilibria in Aqueous Solutions: Slightly Soluble Salts and Complex Ions

Exercises

16.1A (a) $MgF_2(s) \rightleftharpoons Mg^{2+}(aq) + 2\,F^-(aq)$ $\quad K_{sp} = [Mg^{2+}][F^-]^2$

(b) $Li_2CO_3(s) \rightleftharpoons 2\,Li^+(aq) + CO_3^{2-}(aq)$ $\quad K_{sp} = [Li^+]^2[CO_3^{2-}]$

(c) $Cu_3(AsO_4)_2(s) \rightleftharpoons 3\,Cu^{2+}(aq) + 2\,AsO_4^{3-}(aq)$ $\quad K_{sp} = [Cu^{2+}]^3[AsO_4^{3-}]^2$

16.1B (a) $Mg(OH)_2(s) \rightleftharpoons Mg^{+2}(aq) + 2\,OH^-(aq)$ $\quad K_{sp} = [Mg^{2+}][OH^-]^2$

(b) $ScF_3(s) \rightleftharpoons Sc^{3+}(aq) + 3\,F^-(aq)$ $\quad K_{sp} = [Sc^{3+}][F^-]^3$

(c) $Zn_3(PO_4)_2(s) \rightleftharpoons 3\,Zn^{2+}(aq) + 2\,PO_4^{3-}(aq)$ $\quad K_{sp} = [Zn^{2+}]^3[PO_4^{3-}]^2$

16.2A $s = 1.7 \times 10^{-4}$ M $Mg(OH)_2$

$$Mg(OH)_2(s) \rightleftharpoons \underset{s}{Mg^{+2}(aq)} + \underset{2s}{2\,OH^-(aq)}$$

$$K_{sp} = [Mg^{+2}][OH^-]^2 = s\,(2s)^2 = 4s^3$$

$$K_{sp} = 4 \times (1.7 \times 10^{-4})^3 = 2.0 \times 10^{-11}$$

16.2B $[Ag^+] = \dfrac{14\text{ g }Ag^+}{10^6\text{g solution}} \times \dfrac{1\text{ g}}{\text{mL}} \times \dfrac{\text{mL}}{10^{-3}\text{L}} \times \dfrac{\text{mol}}{108\text{ g}} = 1.3 \times 10^{-4}$ M

$$Ag_2CrO_4(s) \rightleftharpoons 2\,Ag^+(aq) + CrO_4^{2-}(aq)$$

$$[CrO_4^{2-}] = \tfrac{1}{2}[Ag^+] = 6.5 \times 10^{-5}$$

$$K_{sp} = [Ag^+]^2[CrO_4^{2-}] = (1.3 \times 10^{-4})^2(6.5 \times 10^{-5})$$

$$K_{sp} = 1.1 \times 10^{-12}$$

16.3A $Ag_3AsO_4(s) \rightleftharpoons \underset{3s}{3Ag^+(aq)} + \underset{s}{AsO_4^{3-}(aq)}$

$$K_{sp} = [Ag^+]^3[AsO_4^{3-}] = 1.0 \times 10^{-22}$$

$$27\,s^4 = 1.0 \times 10^{-22}$$

$$s = 1.4 \times 10^{-6}\text{ M}$$

16.3B $PbI_2(s) \rightleftharpoons \underset{\frac{1}{2}s}{Pb^{2+}(aq)} + \underset{s}{2\,I^-(aq)}$

$$K_{sp} = 7.1 \times 10^{-9} = \tfrac{1}{2}\,s\,(s)^2 = \tfrac{1}{2}\,s^3$$

$s = 2.4 \times 10^{-3}$ M I^-

$$? \text{ ppm } I^- = \frac{2.4 \text{ x } 10^{-3} \text{ mol}}{\text{L}} \times \frac{10^{-3} \text{ L}}{\text{mL}} \times \frac{\text{mL}}{1 \text{ g}} \times \frac{127 \text{ g } I^-}{\text{mol}} \times \frac{10^6 \text{ g ppm}}{\text{g}}$$

$= 3.0 \times 10^2$ ppm

OR

$$? \text{ ppm } I^- = \frac{2.4 \text{ x } 10^{-3} \text{ mol}}{\text{L}} \times \frac{1000}{1000} \times \frac{127 \text{ g } I^-}{\text{mol}} \times \frac{1 \text{ L}}{1000 \text{ g}} = \frac{305 \text{ g } I^-}{1000000 \text{ g}}$$

$= 3.0 \times 10^2$ ppm

16.4 The solutes are all of the same type, MX_2. Their molar solubilities parallel their K_{sp} values:

CaF_2	<	PbI_2	<	MgF_2	<	$PbCl_2$
5.3×10^{-9}		7.1×10^{-9}		3.7×10^{-8}		1.6×10^{-5}

16.5A $Ag_2SO_4(s) \rightleftharpoons 2\ Ag^+(aq) + SO_4^{2-}(aq)$

$1.00 + 2s \qquad s$

$K_{sp} = 1.4 \times 10^{-5} = (1.00 + 2s)^2\ s$

assume $1.00 >> 2s$

$1.4 \times 10^{-5} = 1.00\ s$

$s = 1.4 \times 10^{-5}$ M

16.5B $Ag_2SO_4(s) \rightleftharpoons 2\ Ag^+(aq) + SO_4^{2-}(aq)$

$2\ s + y \qquad +s$

$s = 1.0 \times 10^{-3}$ M

$K_{sp} = 1.4 \times 10^{-5} = (2.0 \times 10^{-3} + y)^2\ (1.0 \times 10^{-3})$

$1.4 \times 10^{-5} = (4.0 \times 10^{-6} + 4.0 \times 10^{-3}\ y + y^2)\ (1.0 \times 10^{-3})$

$1.4 \times 10^{-5} = 4.0 \times 10^{-9} + 4.0 \times 10^{-6}\ y + 1.0 \times 10^{-3}\ y^2$

$10 \times 10^{-3}\ y^2 - 4.0 \times 10^{-6}\ y - 1.4 \times 10^{-5} = 0$

$y^2 + 4.0 \times 10^{-3}\ y - 1.4 \times 10^{-2} = 0$

$$y = \frac{-4.0 \times 10^{-3} \pm \sqrt{(4.0 \times 10^{-3})^2 - 4 \times 1 \times (-1.4 \times 10^{-2})}}{2}$$

$$y = \frac{-4x10^{-3} \pm 0.237}{2}$$

$y = 0.117$ M or -0.12 M (Negative values are not valid.)

$$? \text{ g } AgNO_3 = 0.117 \times 250.0 \text{ mL} \times \frac{10^{-3} \text{ L}}{\text{mL}} \times \frac{170 \text{ g}}{\text{mol}} = 5.0 \text{ g } AgNO_3$$

16.6A $$[Pb^{2+}] = \frac{1.00 \text{ g}}{1.50 \text{ L}} \times \frac{\text{mol}}{331.2 \text{ g}} \times \frac{1 \text{ mol } Pb^{2+}}{\text{mol } Pb(NO_3)_2} = 2.01 \times 10^{-3} \text{ M}$$

$$[I^-] = \frac{1.00 \text{ g}}{1.50 \text{ L}} \times \frac{\text{mol}}{278.1 \text{ g}} \times \frac{2 \text{ mol } I^-}{\text{mol } MgI_2} = 4.79 \times 10^{-3} \text{ M}$$

$Q_{ip} = (2.01 \times 10^{-3})(4.79 \times 10^{-3})^2 = 4.61 \times 10^{-8}$

$Q_{ip} > K_{sp}$ A precipitate should form.

16.6B $[OH^-] = 10^{-(14.00 - 10.35)} = 2.2 \times 10^{-4}$

$$[Mg^{2+}] = \frac{2.5 \text{ g } MgCl_2}{10^6 \text{ g solution}} \times \frac{1 \text{ g}}{\text{mL}} \times \frac{\text{mL}}{10^{-3} \text{ L}} \times \frac{\text{mol } MgCl_2}{95.2 \text{ g } MgCl_2} \times \frac{\text{mol } Mg^{2+}}{\text{mol } MgCl_2}$$

$$= 2.6 \times 10^{-5} \text{ M}$$

$Q_{ip} = [Mg^{2+}][OH^-]^2 = (2.6 \times 10^{-5})(2.2 \times 10^{-4})^2$

$Q_{ip} = 1.3 \times 10^{-12}$

$Q_{ip} = < K_{sp}$ No precipitate.

16.7 Add KI(aq) dropwise from a buret to a known volume of solution of known $[Pb^{2+}]$. Stir after each drop is added, observing first the appearance and then disapperance of PbI_2(s). Continue until a single drop produces a lasting precipitate. Now, $Q_{IP} = K_{sp}$. Calculate K_{sp} from the $[Pb^{2+}]$ and $[I^-]$.

16.8A Adding equal volumes will mean that the ion concentrations are halved.

$$[Mg^{2+}] = \tfrac{1}{2} \times \frac{0.0010 \text{ mol } MgCl_2}{\text{L}} \times \frac{1 \text{ mol } Mg^{2+}}{1 \text{ mol } MgCl_2} = 5.0 \times 10^{-4} \text{ M}$$

$$[F^-] = \tfrac{1}{2} \times \frac{0.020 \text{ mol NaF}}{\text{L}} \times \frac{1 \text{ mol } F^-}{1 \text{ mol NaF}} = 1.0 \times 10^{-2} \text{ M}$$

$Q_{ip} = (5 \times 10^{-4})(1 \times 10^{-2})^2 = 5 \times 10^{-8} > 3.7 \times 10^{-8}\ K_{sp}$

Yes, precipitation should just occur.

16.8B $K_{sp} = [Pb^{2+}][I^-]^2 = 7.1 \times 10^{-9} = 2 \times 10^{-3} \times [I^-]^2$

$[I^-] = 1.9 \times 10^{-3}$ M

$$? \text{ g KI} = 10.5 \text{ L} \times 1.9 \times 10^{-3} \text{ M} \times \frac{166 \text{ g}}{\text{mol}} \times \frac{\text{mol KI}}{\text{mol } I^-} = 3.3 \text{ g KI}$$

16.9

	$Ca^{2+}(aq)$ +	$C_2O_4^{2-}(aq)$ $\rightleftharpoons$	$CaC_2O_4(s)$
Init.	0.0050	0.0100	
Change	– 0.0050	– 0.0050	
Eq.	--	0.0050	

	$CaC_2O_4(s)$ $\rightleftharpoons$	$Ca^{2+}(aq)$	+ $C_2O_4^{2-}(aq)$
Init.		0	0.0050
Change		s	s
Eq.		s	$0.0050 + s$

$K_{sp} = 4 \times 10^{-9} = s \times (0.0050 + s)$

Assume $s << 0.0050$.

$4 \times 10^{-9} = s \times 0.0050$

$s = 8 \times 10^{-7}$ M

$\frac{8 \text{ x } 10^{-7}}{0.0050} \times 100\% = 0.02\%$

Yes, precipitation is complete.

16.10A $AgCl(s) \rightleftharpoons Ag^+(aq) + Cl^-(aq)$ $K_{sp} = 1.6 \times 10^{-10}$

$AgBr(s) \rightleftharpoons Ag^+(aq) + Br^-(aq)$ $K_{sp} = 5.0 \times 10^{-13}$

(a) Br- precipitates first.

$$[Ag^+] = \frac{K_{sp}}{[Cl^-]} = \frac{1.8\times10^{-10}}{0.0100} = 1.8 \times 10^{-8} \text{ M}$$

$$[Ag^+] = \frac{K_{sp}}{[Br^-]} = \frac{5.0\times10^{-13}}{0.0100} = 5 \times 10^{-11} \text{ M}$$

(b) Cl^- begins to precipitate when

$[Ag^+] = 1.8 \times 10^{-8}$

$$[Br^-] = \frac{K_{sp}}{[Ag^+]} = \frac{5.0\times10^{-13}}{1.8\times10^{-8}} = 2.8 \times 10^{-5} \text{ M}$$

(c) Percent of Br left in solution

$$? \,\%\, Br^- = \frac{2.8\times10^{-5}}{0.0100} \times 100\% = 0.28\%$$

Because 0.28% is greater than 0.1%, the precipitation is not complete, but it is close.

16.10B $AgBr(s) \rightleftharpoons Ag^+(aq) + Br^-(aq)$ $K_{sp} = 5.0 \times 10^{-13}$

$AgI(s) \rightleftharpoons Ag^+(aq) + I^-(aq)$ $K_{sp} = 8.5 \times 10^{-17}$

The selective precipitation of Br^- and Cl^- barely failed. There is a greater difference between the K_{sp} values of AgBr and AgI than between AgCl and AgBr; thus, the selective precipitation will probably work.

16.11A $Fe(OH)_2(s) \rightleftharpoons Fe^{2+}(aq) + 2\,OH^-(aq)$

s $\quad 10^{-(14.00 - 6.50)}$

$K_{sp} = 8.0 \times 10^{-16} = (s)(3.16 \times 10^{-8})^2$

$s = [Fe^{2+}] = 0.80$ M

16.11B $[H^+] = K_a \frac{\text{acid}}{\text{base}} = 1.8 \times 10^{-5} \times \frac{0.520}{0.180} = 5.2 \times 10^{-5}$

$$[OH^-] = \frac{K_w}{[H^+]} = \frac{10^{-14}}{5.2\times10^{-5}} = 1.9 \times 10^{-10}$$

$Q_{ip} = [Fe^{3+}][OH^-]^3 = (1.0 \times 10^{-3}) \times (1.9 \times 10^{-10})^3$

$Q_{ip} = 7.1 \times 10^{-33} > 4 \times 10^{-38}\ K_{sp}$

No, Fe^{3+} cannot exist in the buffer solution without precipitation occurring.

16.12 $Mg(OH)_2(s) \rightarrow Mg^{2+}(aq) + 2\,OH^-(aq)$

$2\,NH_4^+(aq) + 2\,H_2O(l) \rightarrow 2\,NH_3(aq) + 2\,H_3O^+(aq)$

$\underline{2\,H_3O^+(aq) + 2\,OH^-(aq) \rightarrow 4\,H_2O(l)}$

$Mg(OH)_2(s) + 2NH_4Cl(aq) \rightarrow MgCl_2(aq) + 2H_2O(l) + 2NH_3(aq)$

16.13A $Ag^+(aq) + 2\,S_2O_3^{2-}(aq) \rightleftharpoons [Ag(S_2O_3)_2]^{3-}(aq)$

$X \qquad 1.0- 2(0.10 - X) \qquad 0.10 - X$

Setting $[Ag^+] = X$ is the equivalent of the formation of $[Ag(S_2O_3)_2]^{3-}$ going to completion and the amount X dissociating.

$$K_f = 1.7 \times 10^{13} = \frac{0.10\text{-}X}{X(0.80 + 2X)^2}$$

Assume $X << 0.10$

$$1.7 \times 10^{13} = \frac{0.10}{X(0.80)^2}$$

$X = [Ag^+] = 9.2 \times 10^{-15}$ M

16.13B $$K_f = \frac{[Ag(NH_3)_2]^+}{[Ag^+][NH_3]^2} = 1.6 \times 10^7$$

$$\frac{(0.050 - 1.0\times10^{-8})}{(1.0\times10^{-8})[NH_3]^2} = 1.6 \times 10^7$$

$[NH_3] = 0.56$ M

$[NH_3]_{total} = [NH_3]_{free} + [NH_3]_{complex}$

$$[NH_3]_{total} = 0.56\text{ M} + 0.050\text{ M} \times \frac{2\text{ mol }NH_3}{\text{mol complex}} = 0.66\text{ M}$$

16.14A $$[I^-] = \frac{1.00\text{ g}}{1.00\text{ L}} \times \frac{\text{mol}}{166.0\text{ g}} = 6.024 \times 10^{-3}\text{ M}$$

$Q_{ip} = [Ag^+][I^-] = 9.2 \times 10^{-15} \times 6.024 \times 10^{-3} = 5.5 \times 10^{-17}$

K_{sp} $8.5 \times 10^{-17} > Q_{ip}$ No, precipitation does not occur.

16.14B $$K_f = \frac{[Ag(NH_3)_2]^+}{[Ag^+][NH_3]^2} = 1.6 \times 10^7 = \frac{0.050}{[Ag^+](1.00)^2}$$

$[Ag^+] = 3.1 \times 10^{-9}$ M

$$[Br^-] = \frac{K_{sp}}{[Ag^+]} = \frac{5.0 \text{ x } 10^{-13}}{3.1 \text{ x } 10^{-9}} = 1.6 \times 10^{-4}\text{ M}$$

$$?\text{ g KBr} = 250.0\text{ mL} \times \frac{10^{-3}\text{ L}}{\text{mL}} \times 1.6 \times 10^{-4}\text{ M} \times \frac{\text{mol KBr}}{\text{mol Br}^-} \times \frac{119\text{ g KBr}}{\text{mol KBr}}$$

$= 4.8 \times 10^{-3}$ g KBr

16.15A $AgBr(s) + 2\ S_2O_3^{2-}(aq) \rightleftharpoons [Ag(S_2O_3)_2]^{3-}(aq) + Br^-(aq)$

	$S_2O_3^{2-}$	$[Ag(S_2O_3)_2]^{3-}$	Br^-
	0.500	0	0
	-2s	+s	+s
	0.500-2s	s	s

$$K_c = 8.5 = \frac{s^2}{(0.500 - 2s)^2}$$

$$2.9 = \frac{s}{0.500 - 2s}$$

1.5 - 5.8 s = s

1.5 = 6.8 s

s = 0.22 M   molar solubility

16.15B Because $[Ag(CN)_2]^-$ has a larger K_f than the K_f of $[Ag(S_2O_3)_2]^{3-}$ or $[Ag(NH_3)_2]^+$, AgI is most soluble in 0.100 M NaCN.

16.16 No. NH_4^+ does not react with NO_3^-. The complex ion, $[Ag(NH_3)_2]^+$, is not destroyed, and the concentration of free Ag^+ remains too low for AgCl(s) to precipitate.

Review Questions

1. $K_{sp} = [Fe^{3+}][OH^-]^3$
   $K_{sp} = [Au^{3+}]^2[C_2O_4^{2-}]^3$

2. $Zn_3(PO_4)_2(s) \rightleftharpoons 3\ Zn^{2+}(aq) + 2\ PO_4^{3-}(aq) \quad K_{sp} = 9.0 \times 10^{-33}$

4. (a) $PbSO_4$ is more soluble because its K_{sp} value is larger and the two solutes are of the same type: MX.
   (b) PbI_2 is more soluble because its K_{sp} is larger and the two compounds are of the same type: MX_2.

8. The washing is to remove ions that are adhered to the precipitate but are not ions of the precipitate. The salt solution should contain one ion common to the precipitate because the solubility of the precipitate is greatly reduced in this solution compared to pure water.

10. The solubility of $Fe(OH)_3(s)$ is increased by HCl(aq) and $CH_3COOH(aq)$, both acids, because the H^+ would react with the OH^- from the $Fe(OH)_3(s)$. Its solubility is lowered by NaOH(aq) and $NH_3(aq)$, because of the common ion OH^-.

13. $Pb(NO_3)_2(aq)$, through the common ion Pb^{2+}, reduces the solubility of $PbCl_2(s)$; but because Pb^{2+} forms the complex ion $[PbCl_3]^-$, HCl(aq) increases the solubility of $PbCl_2(s)$.

14. HCl and NH_3 will dissolve more $Cu(OH)_2$. H^+ from HCl will react with OH^- to form H_2O, and the NH_3 will complex the Cu^{2+} to form $[Cu(NH_3)_4]^{2+}$. Adding more H_2O would not increase the solubility of $Cu(OH)_2$, and adding NaOH would reduce the solubility through the common-ion effect.

15. $[Al(H_2O)_6]^{3+} + H_2O \rightarrow H_3O^+ + [Al(OH)(H_2O)_5]^{2+}$

Problems

19. (a) $Hg_2(CN)_2(s) \rightleftharpoons Hg_2^{2+}(aq) + 2\ CN^-(aq)$

$K_{sp} = 5 \times 10^{-40} = [Hg^{2+}][CN^-]^2$

(b) $Ag_3AsO_4(s) \rightleftharpoons 3\ Ag^+(aq) + AsO_4^{3-}(aq)$

$K_{sp} = 1 \times 10^{-22} = [Ag^+]^3[AsO_4^{3-}]$

21. No, the molar solubility and K_{sp} cannot have the same value. The molar solubility must be raised to a power and generally multiplied by a factor to obtain K_{sp}. The molar solubility is larger than K_{sp}, because the solubility is generally much smaller than 1M, and raising such a number to a power (2, 3, 4 ...) produces a result that is smaller still.

23. (a) $CaF_2(s) \rightleftharpoons Ca^{2+}(aq) + 2F^-(aq)$

s 2s

$K_{sp} = (s)(2s)^2 = 4s^3$

$K_{sp} = 4 \times (3.32 \times 10^{-4})^3$

$K_{sp} = 1.46 \times 10^{-10}$

(b) $Hg_2SO_4(s) \rightleftharpoons Hg_2^{2+}(aq) + SO_4^{2-}(aq)$

s s

$K_{sp} = s^2 = (8.9 \times 10^{-4})^2$

$K_{sp} = 7.9 \times 10^{-7}$

(c) $Ce(IO_3)_4(s) \rightleftharpoons Ce^{4+}(aq) + 4\ IO_3^-(aq)$

s 4s

$K_{sp} = (s)(4s)^4 = 256s^5$

$K_{sp} = 256 \times (1.8 \times 10^{-4})^5$

$K_{sp} = 4.8 \times 10^{-17}$

25. (a)

$SrCO_3$	$K_{sp} = 1.1 \times 10^{-10}$
$SrSO_4$	$K_{sp} = 3.2 \times 10^{-7}$
SrF_2	$K_{sp} = 2.5 \times 10^{-9}$

$SrSO_4$ is more soluble than $SrCO_3$ $\sqrt{3.2\times10^{-7}} = (5.7 \times 10^{-4})$ compared to

$\sqrt{1.1\times10^{-10}} = (1.0 \times 10^{-5})$

$SrF_2(s) \rightleftharpoons Sr^{2+}(aq) + 2F^-(aq)$

s   2s

$2.5 \times 10^{-9} = 4s^3$

$s = 8.5 \times 10^{-4}$

SrF_2 is more soluble than $SrSO_4$ and is the most soluble of the three.

8.5×10^{-4} compared to 5.7×10^{-4}.

(b) $Mg_3(AsO_4)_2(s) \rightleftharpoons 3Mg^{2+}(aq) + 2AsO_4^{3-}(aq)$

$$K_{sp} = 2.0 \times 10^{-20} = (s)^3\left(\frac{2}{3}s\right)^2 = \frac{4}{9}s^5$$

$s = 1.4 \times 10^{-4}$ M

$Mg(OH)_2(s) \rightleftharpoons Mg^{2+}(aq) + 2OH^-(aq)$

s   2s

$K_{sp} = 1.8 \times 10^{-11} = (s)(2s)^2 = 4s^3$

$s = 1.7 \times 10^{-4}$ M

$MgC_2O_4(s) \rightleftharpoons Mg^{2+}(aq) + C_2O_4^{2-}(aq)$

s   s

$K_{sp} = s^2 = 9.3 \times 10^{-3}$

$s = 9.6 \times 10^{-2}$ M

MgC_2O_4 is more soluble than $Mg(OH)_2$ or $Mg_3(AsO_4)_2$.

27. $$[OH^-] = 2.3\text{ mL} \times 0.0010\text{ M HCl} \times \frac{\text{mol OH}^-}{\text{mol HCl}} \times \frac{1}{250.0\text{ mL}} = 9.2 \times 10^{-6}\text{ M OH}^-$$

$$[La^{3+}] = \frac{1\text{ mol La}^{3+}}{3\text{ mol OH}^-} \times 9.2 \times 10^{-6}\text{ M OH}^- = 3.1 \times 10^{-6}\text{ M La}^{3+}$$

$$K_{sp} = [La^{3+}][OH^-]^3 = (3.1 \times 10^{-6})(9.2 \times 10^{-6})^3 = 2.4 \times 10^{-21}$$

29.

	$CaCO_3$	$CaSO_4$	CaF_2
K_{sp}	2.8×10^{-9}	9.1×10^{-6}	5.3×10^{-9}

$CaSO_4$ has more $[Ca^{2+}]$ than $CaCO_3$ because of K_{sp} values. To determine whether CaF_2 or $CaSO_4$ has a higher $[Ca^{2+}]$ requires at least an estimation. That is, $s^2 = 9.1 \times 10^{-6}$ and $s \approx 3 \times 10^{-3}$ for $CaSO_4$, and $s^3 \approx 1 \times 10^{-9}$ and $s \approx 1 \times 10^{-3}$ for CaF_2. $CaSO_4$ also has higher $[Ca^{2+}]$ than CaF_2.

actual

$CaSO_4(s) \rightleftharpoons Ca^{2+}(aq) + SO_4^{2-}(aq)$

s s

$s^2 = 9.1 \times 10^{-6}$

$s = 3.0 \times 10^{-3}$ M

$CaF_2(s) \rightleftharpoons Ca^{2+}(aq) + 2\,F^-(aq)$

s $2s$

$4s^3 = 5.3 \times 10^{-9}$

$s = 1.1 \times 10^{-3}$ M

31. $K_{sp} = 7.6 \times 10^{-36}$

$Cu_3(AsO_4)_2(s) \rightleftharpoons 3\ Cu^{2+}(aq) + 2\ AsO_4^{3-}(aq)$

$3s \quad 2s$

$K_{sp} = 7.6 \times 10^{-36} = (3s)^3(2s)^2 = 108s^5$

$s = 3.71 \times 10^{-8}$ M

$[Cu^{2+}] = 3s = 1.11 \times 10^{-7}$ M

$$\frac{1.11 \times 10^{-7}\ \text{mol}}{\text{L}} \times \frac{10^{-3}\ \text{L}}{\text{mL}} \times \frac{63.55\ \text{g Cu}}{\text{mol Cu}} \times \frac{\text{mL}}{\text{g}} \times \frac{10^6}{10^6} = \frac{7.1\ \text{g Cu}}{10^9\ \text{g}} = 7.1\ \text{ppb}$$

33. $PbCl_2(s) \rightleftharpoons Pb^{2+}(aq) + 2\ Cl^-(aq)$

$s \quad 2s$

K_{sp} (80 °C) $= 3.3 \times 10^{-3} = 4s^3$

$s = 0.094$ M

K_{sp} (25 °C) $= 1.6 \times 10^{-5} = 4s^3$

$s = 0.016$ M

$0.094 - 0.016 = 0.078$ M

$$\frac{0.078\ \text{mmol}}{\text{mL}} \times 1.00\ \text{mL} \times \frac{278.1\ \text{mg}}{\text{mmol}} = 22\ \text{mg}$$

Yes, it is visible.

35. AgBr will be most soluble in water, as both of the other solutions contain common ions that reduce solubility.

37. (a) $[OH^-] = 0.25$ M

$Mg(OH)_2(s) \rightleftharpoons Mg^{2+}(aq) + 2\ OH^-(aq)$

$s \quad 0.25 + 2s$

$1.8 \times 10^{-11} = s\,(0.25 + 2s)^2$

Assume $0.25 >> 2s$

$s = 2.9 \times 10^{-10}$ M

(b) $BaSO_4(s) \rightleftharpoons Ba^{2+}(aq) + SO_4^{2-}(aq)$

$s \quad 0.10 + s$

$K_{sp} = 1.1 \times 10^{-10} = (s)(0.010 + s)$

Assume $0.010 >> s$

$1.1 \times 10^{-10} = 0.010\,s$

$s = 1.1 \times 10^{-8}$

39. $$[I^-] = \frac{15.0\ \text{g}}{0.250\ \text{L}} \times \frac{\text{mol}}{149.89} = 0.400\ \text{M}$$

$PbI_2(s) \rightleftharpoons Pb^{2+}(aq) + 2\ I^-(aq)$

$s \quad s \quad 0.400 + 2s$

$K_{sp} = 7.1 \times 10^{-9} = s\,(0.400 + 2s)^2$

Assume $0.400 >> 2s$

$s \times (0.400)^2 = 7.1 \times 10^{-9}$
$s = 4.4 \times 10^{-8}$ M = $[Pb^{2+}]$
$[I^-] = 0.400$ M

41. $Ag_2SO_4(s) \rightleftharpoons 2\ Ag^+(aq) + SO_4^{2-}(aq)$
$X \qquad \frac{1}{2}X$

$K_{sp} = 1.4 \times 10^{-5} = X^2(\frac{1}{2}X) = \frac{1}{2}X^3$

(a) $[Ag^+] = X = 3.04 \times 10^{-2}$ M
$[SO_4^{2-}] = \frac{1}{2}X = 1.52 \times 10^{-2}$ M

(b) $[SO_4^{2-}] = \dfrac{K_{sp}}{[Ag^+]^2} = \dfrac{1.4 \text{ x } 10^{-5}}{(4.0 \text{ x } 10^{-3}\text{ M})^2} = 0.88$ M

0.88M	$- 1.5 \times 10^{-2}$M	= 0.86 M
needed in sol.	already in sol.	to be added

$0.86\text{ M } SO_4^{2-} \times 0.500\text{ L} \times \dfrac{\text{mol } Na_2SO_4}{\text{mol } SO_4^{2-}} \times \dfrac{142.05\text{ g}}{\text{mol } Na_2SO_4} = 61\text{ g } Na_2SO_4$ added

An alternate method to reach this number is shown below.

$Ag_2SO_4(s) \rightleftharpoons$	$2\ Ag^+(aq)$	$+\ SO_4^{2-}(aq)$	
Initial	$2X$	X	
Change		M	
Equil	$2X$	$X + M$	$2X = 4.0 \times 10^{-3}$
	4.0×10^{-3}	$2.0 \times 10^{-3} + M$	

$1.4 \times 10^{-5} = (4.0 \times 10^{-3})^2 \times (2.0 \times 10^{-3} + M)$
$0.88 = 2.0 \times 10^{-3} + M$
$M = 0.88$ M SO_4^{2-} needed in solution.

0.88M	$- 1.5 \times 10^{-2}$M	= 0.86 M
needed in sol.	already in sol.	to be added

$0.86\text{ M } SO_4^{2-} \times 0.500\text{ L} \times \dfrac{\text{mol } Na_2SO_4}{\text{mol } SO_4^{2-}} \times \dfrac{142.05\text{ g}}{\text{mol } Na_2SO_4} = 61\text{ g } Na_2SO_4$ added

43. $SrCrO_4(s) \rightarrow Sr^{2+}(aq) + CrO_4^{2-}(aq)$
$s \qquad 0.0025\text{ M} + s$

$K_{sp} = 2.2 \times 10^{-5} = s \times (0.0025 + s)$

$2.2 \times 10^{-5} = 0.0025\ s + s^2$

$s^2 + 0.0025\ s - 2.2 \times 10^{-5} = 0$

$$s = \frac{-0.0025 \pm \sqrt{(0.0025)^2 - 4 \text{ x } 1 \text{ x } (-2.2 \text{ x } 10^{-5})}}{2}$$

$$s = \frac{-0.0025 \pm 9.7 \text{ x } 10^{-3}}{2}$$

$s = 3.6 \times 10^{-3}$ M or -6.1×10^{-3} Negative values are not valid.

45. $[CrO_4^{2-}] = \dfrac{K_{sp}}{[Ag^+]^2} = \dfrac{1.1\times10^{-12}}{(1.05\times10^{-3})^2} = 1.0 \times 10^{-6}$ M

47. (a) $Q_{ip} = [Hg_2^{2+}][Cl^-]^2 = \left(\dfrac{1.0\times10^{-3}}{20}\right)\left(\dfrac{1.0\times10^{-3}}{20}\right)^2$

$Q_{ip} = 1.25 \times 10^{-13}$

$K_{sp} = 1.3 \times 10^{-18}$ $Q_{ip} > K_{sp}$ Precipitation occurs.

(b) $K_{sp} = 3.5 \times 10^{-8}$

$[Mg^{2+}] = \dfrac{0.48 \text{ mg}}{225 \text{ mL}} \times \dfrac{\text{mmol } Mg^{2+}}{\text{mmol } MgCl_2} \times \dfrac{\text{mmol } MgCl_2}{95.21 \text{ mg}} = 2.24 \times 10^{-5}$ M

$[CO_3^{2-}] = \dfrac{12.2 \text{ mg}}{225 \text{ mL}} \times \dfrac{\text{mmol } CO_3^{2-}}{\text{mmol } Na_2CO_3} \times \dfrac{\text{mmol } Na_2CO_3}{105.99 \text{ mg}} = 5.12 \times 10^{-4}$ M

$Q_{ip} = [Mg^{2+}][CO_3^{2-}] = 1.15 \times 10^{-8}$ No precipitation occurs.

49. $[F^-] = \dfrac{1.0 \text{ g}}{10^3 \text{ L}} \times \dfrac{\text{mol}}{19.00 \text{ g}} = 5.26 \times 10^{-5}$ M

$Q_{ip} = [Ca^{2+}][F^-]^2 = 2.0 \times 10^{-3} \times (5.26 \times 10^{-5})^2$

$Q_{ip} = 5.5 \times 10^{-12}$

$K_{sp} = 5.3 \times 10^{-9}$ No precipitation occurs.

51. $CaCO_3$ $K_{sp} = 2.8 \times 10^{-9}$ $CaSO_4$ $K_{sp} = 9.1 \times 10^{-6}$

$CaCO_3$ will precipitate first.

$[Ca^{2+}] = \dfrac{K_{sp}}{[SO_4^{2-}]} = \dfrac{9.14\times10^{-6}}{0.010} = 9.1 \times 10^{-4}$ M

$[CO_3^{2-}] = \dfrac{K_{sp}}{[Ca^{2+}]} = \dfrac{2.8\times10^{-9}}{9.1\times10^{-4}} = 3.1 \times 10^{-6}$ M

53. (a) $[Mg^{2+}] = \dfrac{K_{sp}}{[OH^-]^2} = \dfrac{1.8 \text{ x } 10^{-11}}{(2.0 \text{ x } 10^{-3})^2} = 4.5 \times 10^{-6}$ M

(b) $\dfrac{4.5 \text{ x } 10^{-6}}{0.059} \times 100\% = 7.6 \times 10^{-3}\%$

Yes, precipitation is complete.

55. $[Mg^{2+}] = \dfrac{0.010 \text{ mol}}{100 \text{ mol}} \times 0.360 \text{ M} = 3.6 \times 10^{-5}$ M

$[OH^-] = \sqrt{\dfrac{K_{sp}}{[Mg^{2+}]}} = \sqrt{\dfrac{1.8 \text{ x } 10^{-11}}{3.6 \text{ x } 10^{-5}}} = 7.1 \times 10^{-4}$ M

pOH = 3.15

pH = 10.85

57. K_{sp} CaF_2 5.3×10^{-9} K_{sp} MgF_2 3.7×10^{-8}

(a) CaF_2 will precipitate first, as it has a smaller K_{sp}.

(b) $[F^-] = \sqrt{\dfrac{K_{sp}}{[Mg^{2+}]}} = \sqrt{\dfrac{3.7 \text{ x } 10^{-8}}{0.010 \text{ M}}} = \sqrt{3.7 \text{ x } 10^{-6}} \text{ M} = 1.9 \times 10^{-3} \text{ M}$

(c) No, the K_{sp} values are too close together.

$$[Ca^{2+}] = \frac{K_{sp}}{[F^-]^2} = \frac{5.3 \text{ x } 10^{-9}}{(1.9 \text{ x } 10^{-3})^2} = 1.47 \times 10^{-3}$$

$$\% \text{ Ca remaining} = \frac{1.47 \times 10^{-3} \text{ M}}{0.010 \text{ M}} \times 100\% = 15\%$$

To be a successful separation, the % remaining should be < 0.1%.

59. (a) $[Ag^+] = \dfrac{10.00 \text{ mL x } 0.50 \text{ M}}{110.0 \text{ mL}} = 4.5 \times 10^{-2} \text{ M}$

$[Cl^-] = \dfrac{100.0 \text{ mL x } 0.010 \text{ M}}{110.0 \text{ mL}} = 9.1 \times 10^{-3} \text{ M}$

$Q_{ip} = [Ag^+][Cl^-] = 4.5 \times 10^{-2} \times 9.1 \times 10^{-3} = 4.1 \times 10^{-4} > K_{sp} = 1.8 \times 10^{-10}$

K_{sp} Precipitation will occur.

$Ag^+(aq)$	+	$Cl^-(aq)$	$\rightleftharpoons$	$AgCl(s)$
4.5×10^{-2} M		9.1×10^{-3} M		
-9.1×10^{-3}		-9.1×10^{-3}		
3.6×10^{-2} M		≈ 0		

$$? \text{ mol AgCl} = 9.1 \times 10^{-3} \text{ M} \times 110 \text{ mL} \times \frac{10^{-3} \text{ L}}{\text{mL}} = 1.0 \times 10^{-3} \text{ moles}$$

All 9.1×10^{-3} M can precipitate and leave a solution with $[Ag^+] = 0.036$ M.

(b) $[SO_4^{2-}] = \dfrac{100.0 \text{ mL x } 0.010 \text{ M}}{110.0 \text{ mL}} = 9.1 \times 10^{-3} \text{ M}$

$Q_{ip} = [Ag^+]^2[SO_4^{2-}] = (3.6 \times 10^{-2})^2(9.1 \times 10^{-3})$

$Q_{ip} = 1.2 \times 10^{-5} < 1.4 \times 10^{-5} = K_{sp}$

Precipitation will not occur after the AgCl precipitates.

61. $[OH^-] = 10^{-(14.0 - 4.5)} = 3.2 \times 10^{-10}$ M

$Al(OH)_3(s) \rightleftharpoons Al^{3+}(aq) + 3\ OH^-(aq)$

s $\quad$ 3.2×10^{-10}

In a buffer solution, the $[OH^-]$ is constant.

$K_{sp} = 1.3 \times 10^{-33} = s\,(3.2 \times 10^{-10})^3$

$s = 4.0 \times 10^{-5}$ M

63. $NaHSO_4$ is acidic. The H^+ will react to form HCO_3^- and increase the molar solubility.

65. $CaCO_3(s) + 2\ H_3O^+(aq) \rightarrow Ca^{2+}(aq) + H_2CO_3\ (aq) + 2\ H_2O(l) \rightarrow$
$Ca^{2+}(aq) + 3\ H_2O(l) + CO_2(g)$

$CaCO_3(s) + 2\ CH_3COOH(vinegar) \rightarrow$
$Ca^{2+}(aq) + 2\ CH_3COO^-(aq) + H_2O(l) + CO_2(g)$

67. $[CrCl_2(NH_3)_4]^+$

69. $[Ag(CN)_2]^-$ has the largest K_f; its formation will go the farthest toward completion, leaving the lowest $[Ag^+]$ in solution.

71. $SO_4^{2-}(aq) + 2\ Ag^+(aq) \rightleftharpoons Ag_2SO_4(s)$
$Ag_2SO_4(s) + 4\ NH_3(aq) \rightleftharpoons 2[Ag(NH_3)_2]^+\ (aq) + SO_4^{2-}(aq)$
$4\ HNO_3(aq) + 2\ [Ag(NH_3)_2]^+(aq) + SO_4^{2-}(aq) \rightarrow Ag_2SO_4(s) + 4\ NH_4^+(aq)$
$+ 4\ NO_3^-(aq)$

73. (a) $Zn^{2+}(aq) + 4\ OH^-(aq) \rightarrow [Zn(OH)_4]^{2-}(aq)$
(b) $Al_2O_3(s) + 3\ H_2O(l) + 6\ H_3O^+(aq) \rightarrow 2\ [Al(H_2O)_6]^{3+}(aq)$
(c) $Fe(OH)_3(s) + OH^-(aq) \rightarrow$ N.R.

75. $K_f = 4.1 \times 10^8 = \dfrac{0.25}{[Zn^{2+}](1.50)^4}$
$[Zn^{2+}] = 1.2 \times 10^{-10}$ M

77. $K_f = 2.4 \times 10^1 = \dfrac{[[PbCl_3]^-]}{[Cl^-]^3[Pb^{2+}]} = \dfrac{0.010}{(1.50)^3[Pb^{2+}]}$

$[Pb^{2+}] = 1.2 \times 10^{-4}$

$[I^-] = \dfrac{2.00\text{ mL x } 2.00\text{ M}}{0.302\text{ L}} \times \dfrac{10^{-3}\text{ L}}{\text{mL}} = 1.3 \times 10^{-2}$ M

$[Pb^{2+}][I^-]^2 = (1.2 \times 10^{-4}) \times (1.3 \times 10^2)^2 = 2.1 \times 10^{-8} > 7.1 \times 10^{-9}$
Yes, a precipitate should form.

79. $K_f = 1.6 \times 10^7 = \dfrac{[[Ag(NH_3)_2]^+]}{[NH_3]^2[Ag^+]} = \dfrac{0.220}{(0.805)^2[Ag^+]}$

$[Ag^+] = 2.1 \times 10^{-8}$ M

$[Br^-] = \dfrac{K_{sp}}{[Ag^+]} = \dfrac{5.0\text{ x }10^{-13}}{2.1\text{ x }10^{-8}} = 2.4 \times 10^{-5}$ M

? mg KBr $= 2.4 \times 10^{-5}\text{ M} \times 1.42\text{ L} \times \dfrac{119\text{ g KBr}}{\text{mol KBr}} \times \dfrac{\text{mg}}{10^{-3}\text{ g}} = 4.0$ mg KBr

81. $AgBr(s) \rightleftharpoons Ag^{+}(aq) + Br^{-}(aq)$ $K_{sp} = 5.0 \times 10^{-13}$

$\underline{Ag^{+}(aq) + 2\ S_2O_3^{2-}(aq) \rightleftharpoons [Ag(S_2O_3)_2]^{3-}(aq)}$ $K_f = 1.7 \times 10^{13}$

$AgBr(s) + 2\ S_2O_3^{2-}(aq) \rightleftharpoons [Ag(S_2O_3)_2]^{3-}(aq) + Br^{-}(aq)$ $K_c = K_{sp}K_f = 8.5$

0.100 M		
- 2s	+s	+s
0.100 - 2s	s	s

$$8.5 = \frac{s^2}{(.100 - 2\ s)^2}$$

$$2.9 = \frac{s}{0.100 - 2s}$$

$0.29 - 5.8s = s$

$0.29 = 6.8s$

$s = 4.3 \times 10^{-2}$ M = molar solubility

83. $PbCl_2$ is soluble enough that $[Pb^{2+}]$ remaining in solution is sufficiently high that K_{sp} of PbS is exceeded in Group 2 after the Group 1 precipitation. AgCl is so insoluble that $[Ag^{+}]$ remaining in solution after the Group 1 precipitation is not enough to yield a detectable precipitate of Ag_2S in Group 2.

85. Only Hg_2^{2+} is proven to be present, based on the gray color produced when the Group 1 precipitate is treated with $NH_3(aq)$. The presence of Pb^{2+} and Ag^{+} remains uncertain. Treatment of the Group 1 precipitate with hot water and a subsequent test for Pb^{2+} was not performed, and the $NH_3(aq)$ was not tested for the presence of Ag^{+}.

Additional Problems

87. (a) $BaSO_4$ is insoluble enough that little dissolves and therefore will not be dangerous.

(b) $BaSO_4(s) \rightleftharpoons Ba^{2+}(aq) + SO_4^{2-}(aq)$

s s s

$K_{sp} = 1.1 \times 10^{-10} = s^2$

$$s = 1.05 \times 10^{-5}\ \text{M} \times \frac{137.33\ \text{g}}{\text{mol}} \times \frac{\text{mg}}{10^{-3}\ \text{g}} = 1.4\ \text{mg/L}\ Ba^{2+}$$

(c) SO_4^{2-} from $MgSO_4$ reduces the solubility of $BaSO_4$ through the common-ion effect.

89. $K_b = 1.8 \times 10^{-5} = \dfrac{[OH^{-}]^2}{0.750\ \text{M}}$

$[OH^{-}] = 3.67 \times 10^{-3}$ M

$Q_{ip} = (0.350)(3.67 \times 10^{-3})^2 = 4.73 \times 10^{-6}$

$Q_{ip} > K_{sp}$ $K_{sp} = 1.8 \times 10^{-11}$ A precipitate should form.

Adding sufficient NH_4Cl will lower the concentration of OH^- to the point where no precipitation can occur.

$[OH^-] = \sqrt{\frac{1.8 \times 10^{-11}}{0.350}} = 7.17 \times 10^{-6}$ M   This is the maximum possible $[OH^-]$ allowable before $Mg(OH)_2(s)$ precipitates.

$[H_3O^+] = \frac{K_w}{7.17 \times 10^{-6}} = \frac{10^{-14}}{7.17 \times 10^{-6}} = 1.39 \times 10^{-9}$ M When $[H_3O^+]$ falls below this value, $Mg(OH)_2(s)$ precipitates.

$NH_3(g) + H_2O(l) \rightleftharpoons NH_4^+(aq) + OH^-(aq)$

$K_b = 1.8 \times 10^{-5} = \frac{[OH^-][NH_4^+]}{[NH_3]} = \frac{[NH_4^+] \times 7.17 \times 10^{-6}}{0.750}$

$[NH_4^+] = 1.88$ M

$1.88 \text{ M} \times 0.500 \text{ L} \times \frac{53.5 \text{ g}}{\text{mol}} = 50 \text{ g } NH_4Cl$

91. $5\, C_2O_4^{2-}(aq) + 2\, MnO_4^-(aq) + 16\, H^+(aq) \rightarrow 10\, CO_2(g) + 2\, Mn^{2+}(aq) + 8\, H_2O(l)$

$\frac{0.00500 \text{ M } MnO_4^- \times 3.22 \text{ mL}}{100.0 \text{ mL}} \times \frac{5 \text{ mmol } C_2O_4^{2-}}{2 \text{ mmol } MnO_4^-} = 4.03 \times 10^{-4} \text{ M } C_2O_4^{2-}$

or

$$\frac{\frac{5.00\times10^{-3} \text{ mmol } MnO_4^-}{\text{mL}} \times 3.22 \text{ mL} \times \frac{5 \text{ mmol } C_2O_4^{2-}}{2 \text{ mmol } MnO_4^-}}{100.0 \text{ mL}} = 4.03 \times 10^{-4} \text{ M } C_2O_4^{2-}$$

$K_{sp} = [Sr^{2+}][C_2O_4^{2-}] = [C_2O_4^{2-}]^2 = (4.03 \times 10^{-4})^2$

$K_{sp} = 1.62 \times 10^{-7}$

93. $Pb(N_3)_2(s) \rightleftharpoons Pb^{2+}(aq) + 2\, N_3^-(aq)$ $K_{sp} = 2.5 \times 10^{-9}$

$2\, N_3^-(aq) + 2\, H_3O^+(aq) \rightleftharpoons 2\, HN_3(aq) + 2\, H_2O(l)$ $\frac{1}{K_a^2} = \frac{1}{(1.9 \times 10^{-5})^2}$

$Pb(N_3)_2(s) + 2\, H_3O^+(aq) \rightleftharpoons Pb^{2+}(aq) + 2\, HN_3(aq) + H_2O(l)$ $K_c = \frac{K_{sp}}{(K_a)^2} = 6.9$

s $2s$

$[H_3O^+] = 10^{-2.85} = 1.4 \times 10^{-3}$

$K_c = \frac{[Pb^{2+}][HN_3]^2}{[H_3O^+]^2} = \frac{s\,(2s)^2}{(1.4 \times 10^{-3})^2} = \frac{4\,s^3}{(1.4 \times 10^{-3})^2}$

$s = 0.015$ M molar solubility

95. $M^{2+}(aq) + HS^-(aq) + OH^-(aq) \rightleftharpoons MS(s) + H_2O(l)$ $\frac{1}{K_{sp}}$

$2\, H_2O(l) \rightleftharpoons OH^-(aq) + H_3O^+(aq)$ K_w

$$M^{2+}(aq) + HS^-(aq) + H_2O(l) \rightleftharpoons MS(s) + H_3O^+(aq) \qquad \frac{K_w}{K_{sp}}$$

$$\underline{H_2S(aq) + H_2O(l) \rightleftharpoons H_3O^+(aq) + HS^-(aq)} \qquad K_{a1}$$

$$M^{2+}(aq) + H_2S(aq) + 2\ H_2O(l) \rightleftharpoons MS(s) + 2\ H_3O^+(aq) \qquad K_c = \frac{K_w K_{a_1}}{K_{sp}}$$

$$K_c = \frac{1\times10^{-14} \times 1\times10^{-7}}{6\times10^{-19}} = 2 \times 10^{-3}$$

97. $K_f = 1.6 \times 10^7 = \dfrac{[[Ag(NH_3)_2]^+]}{[NH_3]^2[Ag^+]} = \dfrac{0.62}{(0.50)^2[Ag^+]}$

$[Ag^+] = 1.6 \times 10^{-7}$ M

$[Br^-] = \dfrac{K_{sp}}{[Ag^+]} = \dfrac{5.0\times10^{-13}}{1.6\times10^{-7}} = 3.1 \times 10^{-6}$ M

$? \text{ mL NaBr} = \dfrac{3.1\times10^{-6} \text{ mol Br}^-}{\text{L}} \times 165 \text{ mL} \times \dfrac{\text{L NaBr(aq)}}{0.0050 \text{ mol Br}^-} = 0.10 \text{ mL NaBr(aq)}$

99. $Li_3PO_4(s) \rightleftharpoons 3Li^+(aq) + PO_4^{3-}(aq) \qquad K_{sp} = 3.2 \times 10^{-9}$

$$\underline{PO_4^{3-}(aq) + H_2O(l) \rightleftharpoons HPO_4^{2-}(aq) + OH^-(aq)} \qquad \frac{K_w}{K_{a_3}} = \frac{1\times10^{-14}}{4.2\times10^{-13}}$$

$$Li_3PO_4(s) + H_2O(l) \rightleftharpoons 3Li^+(aq) + HPO_4^{2-}(aq) + OH^-(aq) \qquad K = \frac{K_{sp}K_w}{K_{a_3}}$$

3*s* *s* *s*

$$K = \frac{K_{sp}K_w}{K_{a_3}} = \frac{3.2\times10^{-9} \times 1\times10^{-14}}{4.2\times10^{-13}} = 7.6 \times 10^{-11}$$

$K = 7.6 \times 10^{-11} = [Li^+]^3\ [HPO_4^{2-}\ ][\ OH^-]$

$K = (3s)^3(s)\ (s) = 27s^5$

$27s^5 = 7.6 \times 10^{-11}$

$s = [OH^-] = 4.90 \times 10^{-3}$ M

pOH = 2.31

pH = 11.69

101. To simplify this solution, the possible formation of the complex ion $[Pb(S_2O_3)_3]^{4-}$ is ignored.

The expression required are the two K_{sp} expressions:

$PbSO_4(s) \rightleftharpoons Pb^{2+}(aq) + SO_4^{2-}(aq) \qquad K_{sp1} = [Pb^{2+}][SO_4^{2-}] = 1.6 \times 10^{-8}$

$PbS_2O_3(s) \rightleftharpoons Pb^{2+}(aq) + SO_3^{2-}(aq)$ $\qquad K_{sp2} = [Pb^{2+}][S_2O_3^{2-}] = 4.0 \times 10^{-7}$

together with an electronegativity condition, which requires the total concentration of positive and negative charge to be equal.

Positive charge = negative charge

$2 \times [Pb^{2+}] = 2 \times [SO_4^{2-}] + 2 \times [S_2O_3^{2-}]$

$[Pb^{2+}] = [SO_4^{2-}] + [S_2O_3^{2-}]$

(Note: The concentrations of H_3O^+ and OH^- are assumed to be neglible.)

$$[Pb^{2+}] = \frac{1.6\times10^{-8}}{[Pb^{2+}]} + \frac{4.0\times10^{-7}}{[Pb^{2+}]}$$

$[Pb^{2+}]^2 = 1.6 \times 10^{-8} + 4.0 \times 10^{-7} = 4.2 \times 10^{-7}$

$[Pb^{2+}] = 6.5 \times 10^{-4}$ M

103. $Cu^{2+} + NH_3 \rightleftharpoons [Cu(NH_3)]^{2+}$ $\qquad K_1 = 1.9 \times 10^4$

$[Cu(NH_3)]^{2+} + NH_3 \rightleftharpoons [Cu(NH_3)_2]^{2+}$ $\qquad K_2 = 3.9 \times 10^3$

$[Cu(NH_3)_2]^{2+} + NH_3 \rightleftharpoons [Cu(NH_3)_3]^{2+}$ $\qquad K_3 = 1.0 \times 10^3$

$\underline{[Cu(NH_3)_3]^{2+} + NH_3 \rightleftharpoons [Cu(NH_3)_4]^{2+}}$ $\qquad K_4 = ?$

$Cu^{2+} + 4NH_3 \rightleftharpoons [Cu(NH_3)_4]^{2+}$ $\qquad K_f = 1.1 \times 10^{13}$

$K_f = 1.1 \times 10^{13} = K_1\, K_2\, K_3\, K_4$

$$K_4 = \frac{1.1\times10^{13}}{1.9\times10^4 \times 3.9\times10^3 \times 1.0\times10^3} = 1.5 \times 10^2$$

Apply Your Knowledge

105. (a) $Ca(palm)_2(s) \rightleftharpoons Ca^{2+}(aq) + 2\ palm^-(aq)$

$\qquad\qquad s \qquad\qquad 2s$

$K_{sp} = [Ca^{2+}][palm^-]^2 = s\,(2s)^2 = 4s^3$

$$\text{molar solubility} = \frac{0.003\ g}{100\ mL} \times \frac{mL}{10^{-3}\ L} \times \frac{mol}{550.9\ g} = 5 \times 10^{-5}\ M$$

$K_{sp} = (5 \times 10^{-5})(2 \times 5 \times 10^{-5})^2 = 5 \times 10^{-13}$

$$[Ca^{2+}] = \frac{25\ g}{10^6\ g} \times \frac{mol}{40.0\ g} \times \frac{g}{mL} \times \frac{mL}{10^{-3}\ L} = 6.3 \times 10^{-4}\ M$$

$Q_{ip} = [Ca^{2+}][palm]^2 = (6.3 \times 10^{-4})(0.10)^2 = 6 \times 10^{-6} > K_{sp} = 5 \times 10^{-13}$

Yes, soap scum will form.

(b) ? g scum = 6.3×10^{-4} M × 6.5 L × 550.9 g/mol = 2.3 g calcium palmitate

107. (a) $BaSO_4(s) \rightleftharpoons Ba^{2+}(aq) + SO_4^{2-}(aq)$ $\qquad K_{sp} = 1.1 \times 10^{-11}$

$\underline{Ba^{2+}(aq) + CO_3^{2-}(aq) \rightleftharpoons BaCO_3(s)}$ $\qquad \dfrac{1}{K_{sp}} = \dfrac{1}{5.1\times10^{-9}}$

$BaSO_4(s) + CO_3^{2-}(aq) \rightleftharpoons BaCO_3(s) + SO_4^{2-}(aq)$ $K = \dfrac{K_{sp}}{K_{sp}} = \ = 2.2\times 10^{-2}$

$$K = \frac{1.1\times10^{-10}}{5.1\times10^{-9}} = 2.2\times10^{-2} = \frac{[SO_4^{\ 2-}]}{[CO_3^{\ 2-}]}$$

$[SO_4^{2-}] = 2.2\times10^{-2}\,[CO_3^{2-}] = 2.2\times10^{-2}\times 3\text{ M} = 6.6\times10^{-2}\text{ M}$

Yes, this procedure works.

(b) $Mg(OH)_2(s) \rightleftharpoons 2OH^-(aq) + Mg^{2+}(aq)$ $K_{sp} = 1.8\times10^{-11}$

$\underline{Mg^{2+}(aq) + CO_3^{2-}(aq) \rightleftharpoons MgCO_3(s)}$ $\dfrac{1}{K_{sp}} = \dfrac{1}{3.5\times10^{-8}}$

$Mg(OH)_2(s) + CO_3^{2-}(aq) \rightleftharpoons MgCO_3(s) + 2OH^-(aq)$ $K = \dfrac{1.8\times10^{-11}}{3.5\times10^{-8}} = 5.1\times10^{-4}$

$$K = 5.1\times10^{-4} = \frac{[OH^-]^2}{[CO_3^{\ 2-}]}$$

$[OH^-] = \sqrt{5.1\times10^{-4}\times 3} = 3.9\times10^{-2}\text{ M}$ No, this procedure does not work

(c) $2AgCl(s) \rightleftharpoons 2Ag^+(aq) + 2Cl^-(aq)$ $K' = (K_{sp})^2 = (1.8\times10^{-10})^2$

$\underline{2Ag^+(aq) + CO_3^{2-}(aq) \rightleftharpoons Ag_2CO_3(s)}$ $\dfrac{1}{K_{sp}} = 8.5\times10^{-12}$

$2AgCl(s) + CO_3^{2-}(aq) \rightleftharpoons Ag_2CO_3(s) + 2Cl^-(aq)$ $K = \dfrac{(1.8\times10^{-10})^2}{8.5\times10^{-12}} = 3.8\times10^{-9}$

$$K = 3.8\times10^{-9} = \frac{[Cl^-]^2}{[CO_3^{\ 2-}]}$$

$[Cl^-] = \sqrt{3\times3.8\times10^{-9}} = 1.1\times10^{-4}\text{ M}$

No, this procedure does not work.

Chapter 17

Thermodynamics: Spontaneity, Entropy, and Free Energy

Exercises

17.1 (a) Spontaneous. The molecules in the wood (principally cellulose, a carbohydrate) would eventually oxidize to CO_2 and H_2O. The decay is greatly enhanced by microorganisms.

(b) Nonspontaneous. Stirring NaCl(aq) cannot supply the energy input required to dissociate NaCl into its elements.

(c) Indeterminate. $CaCO_3$(s) should decompose on heating, but whether the decomposition is sufficient to produce CO_2(g) at 1 atm at 650 °C cannot be determined without more information.

(d) Spontaneous. In water, hydrogen chloride becomes a strong acid and completely dissociates.

17.2A (a) This is a decrease in entropy because two moles of gas go to one mole of a solid.

(b) This is an increase in entropy because two moles of solid go to five moles (two moles of solid and three moles of gas).

(c) There is no prediction here. All compounds are gaseous. The same number of moles are on both sides of the equation, and the molecules have the same total number of atoms.

17.2B The dish of water is enclosed. Once enough molecules of water have vaporized, there will be an equilibrium between the vapor molecules and the liquid molecules. That is, vapor molecules will become liquid at the same rate that liquid molecules become vapor. The molecules in the vapor phase produce the vapor pressure. If the dish were not enclosed, the liquid would disappear because the vapor molecules would escape from the vicinity of the dish.

17.3A $\Delta S° = \Sigma \nu_p \times S°$ products - $\Sigma \nu_r \times S°$ reactants

$\Delta S° =$ 1 mol $CO_2 \times 213.6$ J /mol K + 1 mol $H_2 \times 130.6$ J/mol K - 1 mol CO $\times$ 197.6 J /mol K - 1 mol $H_2O \times 188.7$ J /mol K

$\Delta S° = -\,42.1$ J /K

17.3B $4\,NH_3(g) + 3\,O_2(g) \rightarrow 2\,N_2(g) + 6\,H_2O(g)$

$\Delta S° = \Sigma \nu_p \times S°$ products - $\Sigma \nu_r \times S°$ reactants

$\Delta S° =$ 2 mol $N_2 \times 191.5$ J /mol K + 6 mol $H_2O \times 188.7$ J/mol K

– 4 mole $NH_3 \times 192.3$ J/mol K – 3 mol $O_2 \times 205.0$ J/mol K = 131.0 J/K

17.4A (a) $\Delta S < 0$, $\Delta H < 0$, case 2
(b) $\Delta S > 0$, $\Delta H > 0$, case 3

17.4B $\Delta H^\circ_{rxn} = \Sigma\nu_p \times \Delta H^\circ_f$ product - $\Sigma\nu_r \times \Delta H^\circ_f$ reactant

$$\Delta H^\circ_{rxn} = 3 \text{ mol BrF} \times \frac{-93.85 \text{ kJ}}{\text{mol}} - 1 \text{ mol Br}_2 \times \frac{30.91 \text{ kJ}}{\text{mol}}$$

$$- 1 \text{ mol BrF}_3 \times \frac{-255.6 \text{ kJ}}{\text{mol}} = -56.9 \text{ kJ}$$

Because 2 moles of gas go to 3 moles of gas, $\Delta S > 0$ is expected. For a negative ΔH and positive ΔS, the reaction should be spontaneous at all temperatures.

17.5 The phase diagram (Figure 11.12) shows that at one atm the equilibrium of $CO_2(s) \rightleftharpoons CO_2(g)$ occurs at -78.5 °C. This is the temperature at which the line, $P = 1$ atm, intersects the sublimation curve of CO_2 (s). At higher T, vaporization goes to completion. At lower T, the vaporization does not occur at 1 atm pressure.

17.6A (a) $\Delta G^\circ = \Delta H^\circ - T\Delta S^\circ$

$$\Delta G^\circ = -114.1 \text{ kJ} - \left(298 \text{ K x } (-146.2 \text{ J/K}) \text{ x } \frac{\text{kJ}}{10^3 \text{ J}}\right)$$

$$\Delta G^\circ = -70.5 \text{ kJ}$$

(b) $\Delta G^\circ = \Delta H^\circ - T\Delta S^\circ$

$$\Delta G^\circ = 409.0 \text{ kJ} - \left(298 \text{ K x} \left(\frac{-129.1 \text{ J}}{\text{mol K}}\right) \text{x } \frac{\text{kJ}}{10^3 \text{ J}}\right)$$

$$\Delta G^\circ = 447.5 \text{ kJ}$$

17.6B (a) $\Delta G^\circ = \Sigma\nu_p \times \Delta G^\circ_f$ products - $\Sigma\nu_r \times \Delta G^\circ_f$ reactants

$$\Delta G^\circ = 1 \text{ mol CCl}_4 \text{ (l)} \times (-65.27 \text{ kJ/mol}) + 6 \text{ mol S} \times 0$$

$$- 1 \text{ mol CS}_2 \times (+65.27 \text{ kJ/mol}) - 2 \text{ mol S}_2\text{Cl}_2 \times (-31.8 \text{ kJ/mol})$$

$$\Delta G^\circ = -66.9 \text{ kJ}$$

(b) $4 NH_3(g) + 3 O_2(g) \rightarrow 2 N_2(g) + 6 H_2O(g)$

$\Delta G^\circ = \Sigma\nu_p \times \Delta G^\circ_f$ products - $\Sigma\nu_r \times \Delta G^\circ_f$ reactants

$$\Delta G^\circ = 2 \text{ mol N}_2 \times 0 \text{ kJ/mol} + 6 \text{ mol H}_2\text{O} \times (-228.6 \text{ kJ/mol})$$

$$-3 \text{ mol O}_2 \times 0 \text{ kJ/mol} - 4 \text{ mol NH}_3 \times (-16.48 \text{ kJ/mol})$$

$$\Delta G^\circ = -1305.7 \text{ kJ}$$

17.7A $$\Delta S^\circ_{vapor} = \frac{\Delta H^\circ_{vap}}{(145.1 + 273.2)\text{K}} \sim \frac{87 \text{ J}}{\text{mol K}}$$

$$\Delta H^\circ_{vapor} = \frac{3.6 \text{ x } 10^4 \text{ J}}{\text{mol}} \times \frac{\text{kJ}}{10^3 \text{ J}} = \frac{36 \text{ kJ}}{\text{mol}}$$

17.7B Methanol will hydrogen-bond extensively. Vaporization will involve breaking the hydrogen bonds and creating a disordered vapor from an ordered liquid. The ΔS° vapor should be much larger than 87 J/mol K.

$CH_3OH(l) \rightarrow CH_3OH(g)$

$$\Delta S^\circ_{vapor} = \frac{239.7\ J}{mol\ K} - \frac{126.8\ J}{mol\ K} = \frac{112.9\ J}{mol\ K}$$

or

$$\Delta H^\circ_{vapor} = \frac{-200.7\ kJ}{mol} - \frac{(-238.7\ kJ)}{mol} = \frac{+38\ kJ}{mol}$$

$$\Delta S^\circ_{vapor} = \frac{\Delta H^\circ_{vap}}{T_{bp}} = \frac{\frac{+38\ kJ}{mol} \times \frac{10^3\ J}{kJ}}{(64.7^\circ + 273.2)\ K} = \frac{112.5\ J}{mol}$$

17.8A $2\ Al(s) + 6\ H^+(aq) \rightarrow 2\ Al^{3+}(aq) + 3\ H_2(g)$

$$K_{eq} = \frac{P^3_{H_2}\,[Al^{3+}]^2}{[H^+]^6}$$

17.8B $Mg(OH)_2(s) + 2\ H_3O^+(aq) \rightarrow Mg^{2+}(aq) + 4\ H_2O(l)$

$$K_{eq} = \frac{[Mg^{2+}]}{[H_3O^+]^2}$$

17.9A $\Delta G^\circ = 2\ mol\ Hg \times 0 + 1\ mol\ O_2 \times 0 - 2\ mol\ HgO \times -58.56\ kJ/mol$

$\Delta G^\circ = 117.12\ kJ/mol$

$$\ln K_{eq} = \frac{\Delta G^\circ}{-RT} = \frac{117.12\ kJ/mol \times 10^3\ J/kJ}{-8.3145\ J/mol\ K \times (273.15 + 25.00)\ K} = -47.25$$

$e^{\ln K_{eq}} = K_{eq} = e^{-47.25} = 3.02 \times 10^{-21}$

17.9B $\Delta G^\circ = \Sigma\nu_p \times \Delta G^\circ_f\ \text{products} - \Sigma\nu_r \times \Delta G^\circ_f\ \text{reactants}$

$\Delta G^\circ = 2\ mol\ NOBr \times 82.4\ kJ/mol - 2\ mol\ NO \times 86.57\ kJ/mol - 1\ mol\ Br_2 \times 0\ kJ/mol = -8.3\ kJ/mol$

$\Delta G^\circ = -RT \ln K_{eq}$

$$\ln K_{eq} = \frac{\Delta G^\circ}{-RT} = \frac{\frac{-8.3\ kJ}{mol} \times \frac{10^3\ J}{kJ}}{\frac{-8.3145\ J}{mol\ K} \times 298.2\ K} = 3.3$$

$e^{\ln K_{eq}} = K_{eq} = e^{3.3} = 27$

$2\ NO + Br_2$	$\rightleftharpoons$	$2\ NOBr$
1 atm		--
$1-X$		X

$$27 = \frac{P^2_{NOBr}}{P^2_{NO}} = \frac{X^2}{(1-X)^2}$$

$$5.2 = \frac{X}{1-X}$$

$5.2 - 5.2\,X = X$

$X = 0.84$

$P_{NOBr} = 0.84$ atm

$P_{NO} = 1.00 - 0.84 = 0.16$ atm

17.10A $\Delta H°_{rxn} = \Sigma\Delta H°_f$ product - $\Sigma\Delta H°_f$ reactant

$\Delta H°_{rxn} = 1$ mol $N_2O_4 \times 9.16$ kJ /mol - 2 mol $NO_2 \times 33.18$ kJ /mol

$\Delta H°_{rxn} = -57.20$ kJ/mol

$$\ln\frac{K_2}{K_1} = \frac{\Delta H°}{R}\left(\frac{1}{T_1} - \frac{1}{T_2}\right) = \frac{\frac{-57.20\,\text{kJ}}{\text{mol}} \times \frac{10^3\,\text{J}}{\text{kJ}}}{\frac{8.3145\,\text{J}}{\text{mol K}}}\left(\frac{1}{298\text{K}} - \frac{1}{338\text{K}}\right)$$

$$\ln\frac{K_2}{6.9} = -2.732$$

$$\frac{K_2}{6.9} = e^{-2.732} = 6.51 \times 10^{-2}$$

$K_2 = 0.45$

17.10B $\ln\frac{P_2}{P_1} = \frac{\Delta H°}{R}\left(\frac{1}{T_1} - \frac{1}{T_2}\right)$

$$\ln\frac{P_2}{23.8\text{ mmHg}} = \frac{\frac{44.0\,\text{kJ}}{\text{mol}} \times \frac{10^3\,\text{J}}{\text{kJ}}}{8.3145\,\text{J/mol K}}\left(\frac{1}{298.2\,\text{K}} - \frac{1}{313.2\,\text{K}}\right)$$

$$\ln\frac{P_2}{23.8\text{ mmHg}} = = 0.850$$

Raise both sides of the equation to a power of e.

$$e^{\ln\frac{P_2}{23.8\text{ mmHg}}} = \frac{P_2}{23.8\text{ mmHg}} = e^{0.850} = 2.34$$

$P_2 = 55.7$ mmHg

From the table V.P. (40°) = 55.3 mmHg

17.11 $\ln\frac{K_2}{K_1} = \frac{\Delta H°}{R}\left(\frac{1}{T_1} - \frac{1}{T_2}\right)$

$$\ln\frac{K_2}{1.00} = \frac{283\,\text{kJ} \times \frac{10^3\,\text{J}}{\text{kJ}}}{\frac{8.3145\,\text{J}}{\text{mol K}}}\left(\frac{1}{1395\text{K}} - \frac{1}{298\,\text{K}}\right)$$

$$\ln\frac{K_2}{1.00} = -89.8$$

$$e^{\ln\frac{K_2}{1.00}} = \frac{K_2}{1.00} = e^{-89.8} = 9.8 \times 10^{-40}$$

$$K_2 = 9.8 \times 10^{-40}$$

$$\Delta G° = -RT \ln K_{eq}$$

$$\Delta G° = \frac{-8.3145\,\text{J}}{\text{mol K}} \times 298\text{ K} \times \ln 9.8 \times 10^{-40}$$

$$\Delta G° = 2.23 \times 10^5\text{ J/mol} \times \frac{\text{kJ}}{10^3\text{ J}} = 223\text{ kJ/mol}$$

$$\Delta G° = \Delta H° - T\,\Delta S°$$

$$\Delta S° = \frac{\Delta G° - \Delta H°}{-T} = \frac{223\text{ kJ - }283\text{ kJ}}{-298\text{ K}}$$

$$\Delta S° = 0.201\text{ kJ/K}$$

This is very close to the calculation in Example 17.11. It should be; it is the same value but calculated by another method. This method is more exact, because it has fewer approximations.

Review Questions

2. (a) Spontaneous. Microorganisms that lead to its souring are present in the milk; no further intervention is needed.
   (b) Nonspontaneous. The extraction of copper metal from copper ores requires a great deal of external intervention.
   (c) Spontaneous. The corrosion of iron in moist air cannot be prevented without external intervention.

3. (a) Decrease in entropy. A liquid is converted to a solid.
   (b) Increase in entropy. A solid is converted to a gas.
   (c) Increase in entropy. A liquid combines with oxygen gas to produce an even greater amount of gaseous products.

9. NOF_3 has more atoms than NO_2F, more modes of vibration, and a greater entropy.

10. Low temperature. At low temperatures the ΔH term dominates in the Gibbs equation, and a $\Delta H < 0$ produces a $\Delta G < 0$.

11. No. Vaporization of water will occur spontaneously, but not to produce a vapor at one atm (which occurs only at 100 °C). A vapor pressure of less than one atm will be produced.

13. ΔH zero. There is no chemical reaction; no heat is involved.
    ΔS positive. Mixing gases produces greater disorder.
    ΔG negative. $\Delta G = \Delta H - T\Delta S < 0$. The mixing of gases is spontaneous, and $\Delta G < 0$ for a spontaneous process.

18. For $\Delta G° = 0$, $K_{eq} = 1$ because $\Delta G° = - RT \ln K_{eq}$. There is no free-energy gain in either direction ($\Delta G° = 0$), and at equilibrium reactants and products are in their standard states.

Problems

19. (a) Increase. A gas is less ordered than a liquid.
    (b) Increase. Production of a greater number of molecules leads to more disorder.
    (c) One cannot tell, because there are the same number of molecules of about the same complexity on both sides of the equation.
    (d) Increase. A large amount of gas (a disordered phase) is produced from a solid (a more ordered phase).
    (e) Decrease. A solid is more ordered than the liquid phase from which it is frozen.
    (f) Indeterminate. The gases are all diatomic and the same number of moles of gas appear on each side of the equation.
    (g) Increase. A liquid decomposes to produce a large amount of gas.
    (h) Decrease. The conversion of a gas to an aqueous solution should produce a more ordered state.

21. Water is extensively hydrogen-bonded. To break all of the hydrogen bonds would require a large, positive ΔH that would not be offset by the increase in entropy, ΔS. Two separate liquids remain (octane floating on liquid water); there is very little mixing.

23. The correct statement would be: For a process to occur spontaneously, the total entropy or the entropy of the universe must increase; that is, the entropy of the system and of the surroundings together must increase.
    Errors in other statements are: (a) Entropy of the system may increase in some cases and decrease in others; (b) Entropy of the surroundings may also increase or decrease; (c) Entropy of the system and surroundings need not both increase, as long as the increase in one exceeds the decrease in the other—that is, the total entropy increases.

25. Since the numbers of gas molecules are the same on both sides of the equation, the first guidelines are not very helpful. The second set of guidelines indicate that the more complex molecules have more entropy, so entropy should increase in this reaction.

    $\Delta S° = 1 \text{ mol } S_2Cl_2 \times 331.5 \text{ J/mol K} - 2 \text{ mol S} \times 31.8 \text{ J/mol K} - 1 \text{ mol } Cl_2 \times 223.0 \text{ J/mol K}$

    $\Delta S° = 44.9 \text{ J/K}$

27. In pollution processes, disorder increases as molecules break down and/or are spread out over a large area. $\Delta S°_{universe} > 0$. To clean up pollution, the dispersed material must be collected and processed. $\Delta S°_{system} < 0$. It requires considerable outside action to gather them up again. $\Delta S°_{surroundings} > 0$. Pollution is a

spontaneous process, and pollution cleanup is nonspontaneous. To remove tert-butyl ether would require distillation, aeration, or a specialized filtration.

29. To use entropy change alone as a criterion for spontaneous change requires evaluating entropy changes in the system and in the surroundings. Free-energy change requires only measurements in the system: $\Delta G = \Delta H - T\Delta S$.

31.

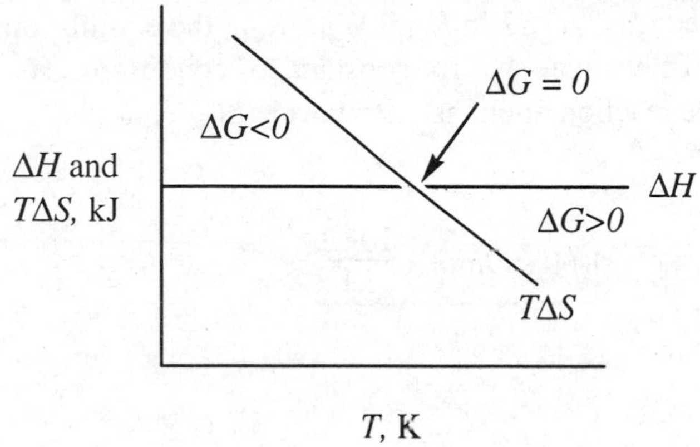

At low temperature $\Delta G < 0$, and the reaction is spontaneous. At higher temperatures the difference $\Delta G > 0$, and the reaction is nonspontaneous.

33. (a) The melting of a solid is nonspontaneous below its melting point and spontaneous above its melting point. For water, this temperature is 0 °C.
(b) The condensation of a vapor at 1 atm pressure to liquid is spontaneous below the normal boiling point and nonspontaneous above the normal boiling point. For water, this temperature is 100 °C.

35. $\Delta S > 0$ because two moles of gas are produced from one mole of gas. Because the reaction is nonspontaneous at room temperature, $\Delta H > 0$. This corresponds to case 3 of Table 19.1. The reaction should become spontaneous at a higher temperature.

37. (a) $\Delta G° = 1 \text{ mol } C_2H_6 \times (-32.89 \text{ kJ/mol}) - 1 \text{ mol } C_2H_4 \times 68.12 \text{ kJ/mol} - 1 \text{ mol } H_2 \times 0$

$\Delta G° = -101.01 \text{ kJ}$

(b) $\Delta G° = 1 \text{ mol } CaSO_4 \times (-1322 \text{ kJ/mol}) - 1 \text{ mol } CaO \times (-604.0 \text{ kJ/mol}) - 1 \text{ mol } SO_3 \times (-371.1 \text{ kJ/mol})$

$\Delta G° = -346.9 \text{ kJ}$

39. $\Delta H° = 1 \text{ mol } CO \times (-110.5 \text{ kJ/mol}) + 1 \text{ mol } H_2 \times 0 - 1 \text{ mol } H_2O\text{ (g)} \times (-241.8 \text{ kJ/mol}) - 1 \text{ mol C} \times 0$

$\Delta H° = 131.3 \text{ kJ}$

$\Delta S° = 1 \text{ mol } CO \times 197.6 \text{ J/mol K} + 1 \text{ mol } H_2 \times 130.6 \text{ J/mol K} - 1 \text{ mol } H_2O \times 188.7 \text{ J/mol K} - 1 \text{ mol C} \times 5.74 \text{ J/mol K}$

$\Delta S° = 133.8 \text{ J/K}$

$$\Delta G° = \Delta H° - T\Delta S° = 131.3\text{ kJ} - 298\text{ K} \times 133.8\text{ J/K} \times \frac{\text{kJ}}{10^3\text{ J}}$$

$\Delta G° = 91.4$ kJ

$\Delta G° = 1$ mol CO $\times$ (-137.2 kJ /mol) + 1 mol $H_2 \times 0$

- 1 mol $H_2O \times$ (-228.6 kJ /mol) - 1 mol C $\times$ 0

$\Delta G° = 91.4$ kJ

The results are the same.

41. With $\Delta G°$, K_{eq} can be evaluated: $\Delta G° = -RT \ln K_{eq}$. With K_{eq}, the equilibrium condition can be determined. To evaluate ΔG for nonstandard conditions, $\Delta G = \Delta G° + RT \ln Q$ (where Q is the reaction quotient). However, ΔG cannot be obtained without knowing $\Delta G°$.

43. $$\Delta S°_{vap} = 87\text{ J/mol K} = \frac{\Delta H°_{vap}}{T_{bp}} = \frac{31.69\text{ kJ/mol x } \frac{10^3\text{ J}}{\text{kJ}}}{T_{bp}}$$

$T_{bp} = 364$ K - 273° = 91 °C

45. $$\Delta S°_{vap} = 87\text{ J/k} = \frac{\Delta H°_{vap}}{T_{bp}} = \frac{\Delta H°_{vap} \times 10^3\text{ J/kJ}}{(59.47 + 273.15)\text{K}}$$

$\Delta H°_{vap} = 29$ kJ /mol

Br_2 (l) $\rightarrow$ Br_2 (g)

$\Delta H°_{vap} = 1 \times 30.91$ kJ /mol - $1 \times 0 = 30.91$ kJ /mol

The agreement between 29 kJ/mol and 30.91 kJ/mol is rather good considering that Trouton's rule is not exact.

47. (a) $\Delta G = 0$

(b) $\Delta G° = - RT \ln K_p$

$\Delta G° = - 8.3145$ J /mol K $\times$ 765 K $\times$ ln 46.0

$$\Delta G° = -2.44 \times 10^4\text{ J/mol} \times \frac{\text{kJ}}{10^3\text{ J}} = -24.4\text{ kJ/mol}$$

49. (a) $$K_{eq} = \frac{P^2_{NO_2}}{P^2_{NO} P_{O_2}} = K_p$$

(b) $K_{eq} = P_{SO_2} = K_p$

(c) $$K_{eq} = \frac{[H_3O^+][CN^-]}{[HCN]} = K_a$$

51. (a) $$\Delta G° = 4\text{ mol NO} \times \frac{86.57\text{ kJ}}{\text{mol}} - 2\text{ mol } N_2O \times \frac{104.2\text{ kJ}}{\text{mol}} - 1\text{ mol } O_2 \times 0 = 137.9\text{ kJ}$$

$\Delta G° = - RT \ln K_p$

Note that in this equation the unit of $\Delta G°$ is changed to the basis of a "mole of reaction" to provide for the proper cancellation of units.

$$\ln K_p = \frac{137.9 \text{ kJ /mol x } \frac{10^3 \text{ J}}{\text{kJ}}}{-8.3145 \text{ J /mol K x 298 K}} = -55.7$$

$K_p = e^{-55.7} = 6.5 \times 10^{-25}$

(b) $\Delta G° = 3 \text{ mol } H_2O \times (-228.6 \text{ kJ /mol}) + 1 \text{ mol } N_2O \times 104.2 \text{ kJ /mol}$
$- 2 \text{ mol } NH_3 \times (-16.48 \text{ kJ /mol}) - 2 \text{ mol } O_2 \times 0 = -548.6 \text{ kJ}$

$\Delta G° = - RT \ln K_p$

$$\ln K_p = \frac{-548.6 \text{ kJ /mol x } 10^3 \text{ J /kJ}}{-8.3145 \text{ J /mol K x 298 K}} = 221 \qquad K_p = e^{221} = 1 \times 10^{96}$$

53. $\Delta G° = \Delta H° - T\Delta S°$

$$\Delta G° = 38.0 \text{ kJ /mol} - 298 \text{ K} \times 99.7 \text{ J /mol K} \times \frac{\text{kJ}}{10^3 \text{ J}}$$

$\Delta G° = 8.3 \text{ kJ /mol} = - RT \ln K_{eq}$

$$\ln K_{eq} = \frac{8.3 \text{ kJ /mol x } 10^3 \text{ J /kJ}}{-8.3145 \text{ J /mol K x 298 K}} = -3.35$$

$K_{eq} = 0.035 = P_{C_6H_5CH_3}$

$$P_{C_6H_5CH_3} = 0.035 \text{ atm} \times \frac{760 \text{ mmHg}}{\text{atm}} = 27 \text{ mmHg}$$

55. (a)
$$\frac{X_{CO_2}X_{H_2}}{X_{CO}X_{H_2O}} = \frac{\frac{P_{CO_2}}{P_T} \times \frac{P_{H2}}{P_T}}{\frac{P_{CO}}{P_T} \times \frac{P_{H_2O}}{P_T}} = \frac{P_{CO_2}P_{H_2}}{P_{CO}P_{H_2O}} = K_p$$

$$K_p = \frac{(0.320)^2}{0.0133 \text{ x } 0.347} = 22.2$$

$$\Delta G° = -RT \ln K_p = -8.3145 \text{ J/mol K} \times (345 + 273) \text{ K} \times \ln 22.2 \times \frac{\text{kJ}}{10^3 \text{ J}}$$

$\Delta G° = -15.9 \text{ kJ/mol}$

(b) $$Q_p = \frac{0.145 \text{ x } 0.226}{0.085 \text{ x } 0.112} = 3.44 \qquad Q_p < K_p \ (22.2)$$

The reaction will go to the right.

(c)

	CO	+	H_2O	⇌	CO_2	+	H_2
I	0.085		0.112		0.145		0.226
C	$-X$		$-X$		$+X$		$+X$
E	$0.085-X$		$0.112-X$		$0.145+X$		$0.226+X$

$$K = 22.2 = \frac{(0.145+X)(0.226+X)}{(0.085-X)(0.112-X)}$$

$$22.2 = \frac{0.0328 + 0.371X + X^2}{0.00952 - 0.197X + X^2}$$

$22.2X^2 - 4.37X + 0.211 = X^2 + 0.371X + 0.0328$

$21.2X^2 - 4.74X + 0.178 = 0$

$$X = \frac{4.74 \pm \sqrt{(4.74)^2 - 4 \text{ x } 21.1 \text{ x } 0.178}}{2 \text{ x } 21.2}$$

$$X = \frac{+4.74 \pm 2.73}{42.4}$$

$X = 0.176$ or 0.048 Too large, not valid.

? mol CO = 0.085 - 0.048 = 0.037 mol CO

? mol H_2O = 0.112 - 0.048 = 0.064 mol H_2O

? mol CO_2 = 0.145 + 0.048 = 0.193 mol CO_2

? mol H_2 = 0.226 + 0.048 = 0.274 mol H_2

57. (a) C(graphite) + $\frac{1}{2}O_2 \rightarrow$ CO

$\Delta G°$ = 1 mol CO × (-137.2 kJ/mol) - 1 mol C × 0 - $\frac{1}{2}$ mol O_2 × 0 = -137.2 kJ No

(b) 1 $H_2 + \frac{1}{2}O_2 \rightarrow H_2O$ (g)

$\Delta G°$ =1 mol H_2O×(-228.6 kJ/mol)-1 mol H_2 × 0 -$\frac{1}{2}$ mol O_2×0 = -228.6 kJ No

(c) CO + $\frac{1}{2}O_2 \rightarrow CO_2$

$\Delta G°$ = 1 mol CO_2×(-394.4 kJ/mol)-1 mol CO×(-137.2 kJ/mol)-$\frac{1}{2}$ mol O_2 × 0

= -257.2 kJ Yes

59. $\Delta H°_{rxn}$ = 1 mol SO_2 × (-296.8 kJ/mol) + 1 mol $OSCl_2$ × (-245.6 kJ/mol) -

1 mol SO_3 × (-395.7 kJ/mol) - 1 mol SCl_2 (-50.0 kJ/mol)

$\Delta H°_{rxn}$ = -96.7 kJ

$$\Delta S_{rxn} = 1 \text{ mol } SO_2 \times \frac{248.1 \text{ J}}{\text{mol K}} + 1 \text{ mol } OSCl_2 \times \left(\frac{121 \text{ J}}{\text{mol K}}\right) - 1 \text{ mol } SO_3 \times \frac{256.6 \text{ J}}{\text{mol K}}$$

$$- 1 \text{ mol } SCl_2 \times \frac{184 \text{ J}}{\text{mol K}} = \frac{-72 \text{ J}}{\text{K}}$$

$\Delta G° = \Delta H° - T\Delta S°$

$\Delta G°$ = -96.7 kJ - T(-72 J/K) = $-RT$ ln K

$$-9.67 \times 10^4 \text{ J} + \frac{72T \text{ J}}{\text{K}} = \frac{-8.3145 \text{ J}}{\text{mol K}} T \ln 1.0 \times 10^{15}$$

$$-9.67 \times 10^4 \text{ J} + \frac{72T \text{ J}}{\text{K}} = \frac{-287 T \text{ J}}{\text{K}}$$

$$\frac{359 T \text{ J}}{\text{K}} = 9.67 \times 10^4 \text{ J}$$

T = 269 K - 273 = -4 °C

61. $\ln\frac{P_2}{P_1} = \frac{\Delta H^\circ}{R}\left(\frac{1}{T_1} - \frac{1}{T_2}\right)$

$$\ln\frac{747\text{ mmHg}}{760\text{ mmHg}} = \frac{\frac{43.82\text{ kJ}}{\text{mol}} \times \frac{10^3\text{ J}}{\text{kJ}}}{\frac{8.3145\text{ J}}{\text{mol K}}}\left(\frac{1}{(117.8+273.2)\text{ K}} - \frac{1}{T_2}\right)$$

$$-0.0173 = 5.270 \times 10^3 \left(\frac{1}{391.0\text{ K}} - \frac{1}{T_2}\right)$$

$$-3.28 \times 10^{-6} = 0.002558\text{ K}^{-1} - \frac{1}{T_2}$$

$$\frac{1}{T_2} = 2.561 \times 10^{-3}\text{ K}^{-1}$$

$$T_2 = 390.4\text{ K} - 273.2 = 117.2\ ^\circ\text{C}$$

63. $\Delta H^\circ_{rxn} =$ 2 mol $NO_2 \times 33.18$ kJ/mol - 2 mol NO $\times$ 90.25 kJ/mol - 1 mol $O_2 \times 0$

$= -114.14$ kJ

$$\Delta S^\circ_{rxn} = 2\text{ mol NO}_2 \times \frac{240.0\text{ J}}{\text{mol K}} - 2\text{ mol NO} \times \frac{210.6\text{ J}}{\text{mol K}} - 1\text{ mol O}_2 \times \frac{205.0\text{ J}}{\text{mol K}}$$

$$= \frac{-146.2\text{ J}}{\text{K}}$$

$$\Delta G^\circ = \Delta H^\circ - T\Delta S^\circ = -114.14\text{ kJ} \times \frac{10^3\text{ J}}{\text{kJ}} - (155\ ^\circ\text{C} + 273)\text{ K} \times \left(\frac{-146.2\text{ J}}{\text{K}}\right)$$

$$\Delta G^\circ = -5.16 \times 10^4\text{ J/mol} = -RT\ln K_{eq}$$

$$-5.16 \times 10^4\text{ J/mol} = \frac{-8.3145\text{J}}{\text{K}} \times 428\text{ K}\ln K$$

$$e^{\ln K} = K = e^{14.5} = 2.0 \times 10^6$$

65. $CCl_4(l) \rightarrow CCl_4(g)$

$$\Delta H^\circ_{rxn} = \Delta H^\circ_f(CCl_4\ (g)) - \Delta H^\circ_f(CCl_4\ (l))$$

$$\Delta H^\circ_{rxn} = \frac{-102.9\text{ kJ}}{\text{mol}} - \left(\frac{-135.4\text{ kJ}}{\text{mol}}\right) = \frac{32.5\text{ kJ}}{\text{mol}}$$

$$\Delta G^\circ_{rxn} = \Delta G^\circ_f(CCl_4\ (g)) - \Delta G^\circ_f(CCl_4\ (l))$$

$$\Delta G^\circ_{rxn} = \frac{-60.63\text{ kJ}}{\text{mol}} - (\frac{-65.27\text{ kJ}}{\text{mol}}) = \frac{4.64\text{ kJ}}{\text{mol}}$$

$$\Delta G^\circ = \frac{4.64\text{ kJ}}{\text{mol}} \times \frac{10^3\text{ J}}{\text{kJ}} = -RT\ln K_p = \frac{-8.3145\text{ J}}{\text{mol K}} \times 298\text{ K} \times \ln K_p$$

$$\frac{4.64 \text{ x } 10^3\text{ J}}{\text{mol}} = -\frac{2.48 \text{ x } 10^3\text{ J}}{\text{mol}}\ \ln K_p$$

$$\ln K_p = -1.87$$

$$e^{\ln K_p} = K_p = e^{-1.87} = 1.54 \times 10^{-1}$$

$$\ln\frac{K_2}{K_1} = \frac{\Delta H^\circ}{R}\left(\frac{1}{T_1} - \frac{1}{T_2}\right)$$

$$\ln\frac{0.154}{1.0} = \frac{\frac{32.5\text{ kJ}}{\text{mol}} \text{ x } \frac{10^3\text{ J}}{\text{kJ}}}{\frac{8.3145\text{ J}}{\text{mol K}}} \times \left(\frac{1}{T_1} - \frac{1}{298}\right)$$

$$-1.87 = 3.909 \times 10^3 \times \left(\frac{1}{T_1} - \frac{1}{298}\right)$$

$$-4.78 \times 10^{-4} = \left(\frac{1}{T_1} - \frac{1}{298}\right)$$

$$\frac{1}{T} = 2.88 \times 10^{-3}$$

$$T = 347\text{ K}$$

67. $K_{eq} = P_{O_2} = 0.21$ atm

$$\Delta G^\circ = -RT\ln K_{eq} = \frac{-8.3145\ JT}{\text{mol K}} \times \ln 0.21 = \frac{13.0\ JT}{\text{K}}$$

$$\Delta H^\circ = 1\text{ mol } O_2 \times 0\text{ kJ/mol} + 4\text{ mol Ag} \times 0\text{ kJ/mol} - 2\text{ mol } Ag_2O \times (-31.0\text{ kJ/mol}) = 62.0\text{ kJ}$$

$$\Delta S^\circ = 1\text{ mol } O_2 \times \frac{205.0\text{ J}}{\text{mol K}} + 4\text{ mol Ag} \times \frac{42.55\text{ J}}{\text{mol K}} - 2\text{ mol } Ag_2O \times \frac{121\text{ J}}{\text{mol K}} = 133.2\text{ J/K}$$

$$\Delta G^\circ = 62.0\text{ kJ} \times \frac{10^3\text{ J}}{\text{kJ}} - T \times \frac{133.2\text{ J}}{\text{K}} = \frac{13.0\text{ J } T}{\text{K}}$$

$$6.20 \times 10^4\text{ J} = \frac{146.2\text{ J } T}{\text{K}}$$

$$T = 424\text{ K} - 273$$

$$T = 151\ ^\circ\text{C}$$

An alternate method is to determine K_p at 298K and then use van't Hoff's equation to calculate the temperature at which $K_p = 0.21$.

69. The expansion in Figure 6.5 is not reversible because the process cannot be reversed by an infinitesimal change. The weights are too large; the change occurs too quickly.

71. $CH_3(CH_2)_6CH_3$ (c) is nonpolar and should follow Trouton's rule. CH_3CH_2OH (b) has the most extensive hydrogen bonding and will deviate the most from Trouton's rule.

73. In Chapter 14, it was seen that an endothermic reaction would be forced to the right by an increase in temperature. The effect of being forced to the right increases the K_{eq}. From the van't Hoff equation, for positive ΔH (endothermic) an increase in

temperature increases the K_{eq}. That is, $\ln \frac{K_2}{K_1} = \frac{\Delta H°}{R}\left(\frac{1}{T_1} - \frac{1}{T_2}\right)$, and with $\Delta H > 0$ and $T_2 > T_1$, $K_2 > K_1$.

75. (a) $\Delta H° = 1 \text{ mole } CO_2 \times (-393.5 \text{ kJ/mol}) + 1 \text{ mol } H_2O \times (-241.8 \text{ kJ/mol})$
$+ 1 \text{ mol } Na_2CO_3 \times (-1131 \text{ kJ/mol}) - 2 \text{ mol } NaHCO_3 \times (-950.8 \text{ kJ/mol})$
$\Delta H° = 135 \text{ kJ/mol}$
$\Delta S° = 1 \text{ mol } CO_2 \times 213.6 \text{ J/mol K} + 1 \text{ mol } H_2O \times 188.7 \text{ J/mol K}$
$+1 \text{ mol } Na_2CO_3 \times 135.0 \text{ J/mol K} - 2 \text{ mol } NaHCO_3 \times (102 \text{ J/mol K})$
$\Delta S° = 333 \text{ J/mol K}$
(b) $\Delta G° = 135.3 \text{ kJ} - 298 \text{ K} \times 333 \text{ J/K} \times \text{kJ}/10^3 \text{ J}$

$$\ln K_p = \frac{\Delta G°}{-RT} = \frac{36.0 \text{ kJ/mol} \times 10^3 \text{ J/kJ}}{-8.3145 \text{ J/mol K} \times 298 \text{ K}}$$

$\ln K_p = -14.5$
$K_p = 5.0 \times 10^{-7}$

$K_p = P_{CO_2} \times P_{H_2O} = 0.50 \text{ atm} \times 0.50 \text{ atm} = 0.25$

$$\ln \frac{0.25}{5.0 \times 10^{-7}} = \frac{135 \text{ kJ/mol}}{8.3145 \text{ J/mol K}} \times \frac{10^3 \text{ J}}{\text{kJ}} \times \left(\frac{1}{298 \text{ K}} - \frac{1}{T}\right)$$

$$13.1 = 1.62 \times 10^4 \times \left(\frac{1}{298 \text{ K}} - \frac{1}{T}\right)$$

$$8.09 \times 10^{-4} = \frac{1}{298 \text{ K}} - \frac{1}{T}$$

$T = 393 \text{ K}$

77. (a) $K_c = \frac{[CO][H_2O]}{[CO_2][H_2]} = \frac{0.224 \times 0.224}{0.276 \times 0.276} = 0.659$

$\Delta G° = -RT \ln K_c$

$$\Delta G° = \frac{-8.3145 \text{ J}}{\text{mol K}} \times 1000 \text{ K} \times \ln 0.659 \times \frac{\text{kJ}}{10^3 \text{ J}}$$

$\Delta G° = 3.47 \text{ kJ}$
(b) $CO_2 + H_2 \rightleftharpoons CO + H_2O$

CO_2	H_2	CO	H_2O
0.0500	0.0700	0.0400	0.0850

$$Q = \frac{0.0400 \times 0.0850}{0.0500 \times 0.0700} = 0.971$$

The reaction will go to the left.
(c)

CO_2	H_2	CO	H_2O
$+x$	$+x$	$-x$	$-x$
$0.0500 + x$	$0.0700 + x$	$0.0400 - x$	$0.0850 - x$

$$0.659 = \frac{(0.0400 - x)(0.0850 - x)}{(0.0500 + x)(0.0700 + x)}$$

$0.659x^2 + 0.0791x + 2.31 \times 10^{-3} = x^2 - 0.1250x + 3.40 \times 10^{-3}$
$0.341x^2 - 0.2041x + 1.09 \times 10^{-3} = 0$

$$x = \frac{0.2041 \pm \sqrt{(0.2041)^2 - 4 \times 0.341 \times 1.09 \times 10^{-3})}}{0.682}$$

$$x = \frac{0.2041 \pm 0.2004}{0.682}$$

$x = 0.593$ $\quad$ 5.4×10^{-3}

Too Large

Not Valid

? moles CO_2 = 0.0500 + 0.0054 = 0.0554 moles CO_2

? moles H_2 = 0.0700 + 0.0054 = 0.0754 moles H_2

? moles CO = 0.0400 – 0.0054 = 0.0346 moles CO

? moles H_2O = 0.0850 – 0.0054 = 0.0796 moles H_2O

79.

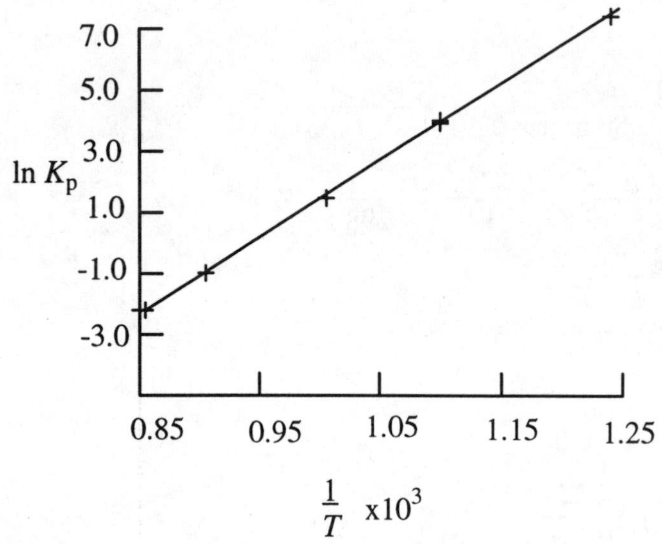

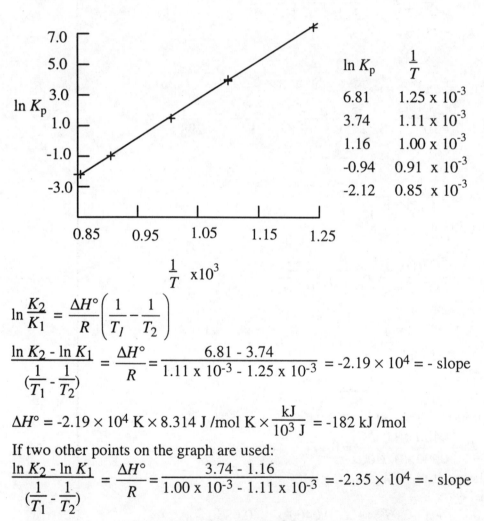

$\ln K_p$	$\frac{1}{T}$
6.81	1.25×10^{-3}
3.74	1.11×10^{-3}
1.16	1.00×10^{-3}
-0.94	0.91×10^{-3}
-2.12	0.85×10^{-3}

$$\ln\frac{K_2}{K_1} = \frac{\Delta H°}{R}\left(\frac{1}{T_1} - \frac{1}{T_2}\right)$$

$$\frac{\ln K_2 - \ln K_1}{(\frac{1}{T_1} - \frac{1}{T_2})} = \frac{\Delta H°}{R} = \frac{6.81 - 3.74}{1.11 \text{ x } 10^{-3} - 1.25 \text{ x } 10^{-3}} = -2.19 \times 10^4 = -\text{ slope}$$

$$\Delta H° = -2.19 \times 10^4 \text{ K} \times 8.314 \text{ J /mol K} \times \frac{\text{kJ}}{10^3 \text{ J}} = -182 \text{ kJ /mol}$$

If two other points on the graph are used:

$$\frac{\ln K_2 - \ln K_1}{(\frac{1}{T_1} - \frac{1}{T_2})} = \frac{\Delta H°}{R} = \frac{3.74 - 1.16}{1.00 \text{ x } 10^{-3} - 1.11 \text{ x } 10^{-3}} = -2.35 \times 10^4 = -\text{ slope}$$

$$\Delta H° = -2.35 \times 10^4 \text{ K} \times 8.314 \text{ J /mol K} \times \frac{\text{kJ}}{10^3 \text{ J}} = -195 \text{ kJ /mol.}$$

It appears that the slope is about 2.3×10^4 and $\Delta H° \approx -190$ kJ/mol.

(There are computer and calculator programs to find the best straight line that would include all of the data points and provide a more accurate answer.)

81. $Hg(s) \rightarrow Hg(l)$ $\qquad \Delta H°_{fus} = 2.30$ kJ

$\underline{Hg(l) \rightarrow Hg(g)}$ $\qquad \Delta H°_{vapor} = 61.32$ kJ

$Hg(s) \rightarrow Hg(g)$ $\qquad \Delta H°_{sub} = 63.62$ kJ

$$\Delta S°_{fus} = \frac{\Delta H°_{fus}}{T} = \frac{\Delta H°_{fus}}{(-38.86+273.15)\,K} = \frac{2.30 \text{ kJ x } \frac{10^3 \text{ J}}{\text{kJ}}}{(-38.86° + 273.15)} = \frac{9.82 \text{ J}}{\text{K}}$$

$$\Delta S°_{vapor} = 1 \text{ mol Hg(g)} \times \frac{174.9 \text{ J}}{\text{mol K}} - 1 \text{ mol Hg(l)} \times \frac{76.02 \text{ J}}{\text{mol K}} = \frac{98.9 \text{ J}}{\text{K}}$$

$$\Delta S°_{sub} = \Delta S°_{fus} + \Delta S°_{vap} = \frac{9.8 \text{ J}}{\text{K}} + \frac{98.9 \text{ J}}{\text{K}} = \frac{108.7 \text{ J}}{\text{K}}$$

$$\Delta G°_{sub} = \Delta H°_{sub} - T\,\Delta S°_{sub} = 63.62 \text{ kJ} \times \frac{10^3 \text{ J}}{\text{kJ}} - (-78.5 \text{ °C} + 273.2) \times \frac{108.7 \text{ J}}{\text{K}}$$

$$= 4.246 \times 10^4 \text{ J/mol} = -RT \ln K = \frac{-8.3145 \text{ J}}{\text{mol K}} \times 194.7 \times \ln K$$

$-26.23 = \ln K$

$e^{\ln K} = K = e^{-26.23} = 4.06 \times 10^{-12}$

$$K = P_{Hg} = 4.06 \times 10^{-12} \text{ atm} \times \frac{760 \text{ mmHg}}{\text{atm}} = 3 \times 10^{-9} \text{ mmHg}$$

83. $$\Delta H°_{rxn} = 1 \text{ mole CO} \times \frac{-110.5 \text{ kJ}}{\text{mol}} + 1 \text{ mole } Cl_2 \times \frac{0 \text{ kJ}}{\text{mol}} - 1 \text{ mol } COCl_2 \times \frac{-220.9 \text{ kJ}}{\text{mol}}$$

$$= 110.4 \text{ kJ}$$

$$\Delta S°_{rxn} = 1 \text{ mole CO} \times \frac{197.6 \text{ J}}{\text{mole K}} + 1 \text{ mole } Cl_2 \frac{223.0 \text{ kJ}}{\text{mole}} - 1 \text{ mol } COCl_2 \times \frac{283.8 \text{ kJ}}{\text{mol}}$$

$$= \frac{136.8 \text{ J}}{\text{K}}$$

$$\Delta G° = 110.4 \text{ kJ} \times \frac{10^3 \text{ J}}{\text{kJ}} - T \times \frac{136.8 \text{ J}}{\text{K}}$$

$$COCl_2 \rightleftharpoons CO + Cl_2$$

$$K_p = \frac{\alpha^2 P}{1-\alpha^2} = \frac{(0.100)^2 \times 1.00 \text{ atm}}{1-(0.100)^2}$$

$K_p = 1.01 \times 10^{-2}$

$$\Delta G° = -RT \ln K_p = \frac{-8.3145 \text{ J}}{\text{mol K}}\, T \ln 1.01 \times 10^{-2} = \frac{3.82 \text{ x } 10^1 \text{ J}}{\text{K}}\, T$$

$$\Delta G° = 1.104 \times 10^5 \text{ J} - \frac{136.8 \text{ J}}{\text{K}}\, T = \frac{3.82 \text{ x } 10^1 \text{ J}}{\text{K}}\, T$$

$$\Delta G° = 1.104 \times 10^5 \text{ J} = \frac{175.0 \text{ J}}{\text{K}}\, T$$

$T = 631$ K $- 273 = 358$ °C

85. $Ag_2SO_4(s) \rightarrow 2\,Ag^+(aq) + SO_4^{2-}(aq)$

$$\Delta G° = 2\text{ mol }Ag^+ \times 77.11\text{ kJ/mol} + 1\text{ mole }SO_4^{2-} \times \left(\frac{-744.5\text{ kJ}}{\text{mol}}\right) - 1\text{ mol }Ag_2SO_4 \times (-618.5\text{ kJ/mol}) = 28.2\text{ kJ}$$

$\Delta G° = -RT \ln K_{sp}$

$28.2\text{ kJ/mol} \times \frac{10^3\text{ J}}{\text{kJ}} = -8.3145 \times 298\text{ K} \times \ln K_{sp}$

$e^{\ln K_{sp}} = K_{sp} = e^{-11.4} = 1.1 \times 10^{-5}$

Table $K_{sp} = 1.4 \times 10^{-5}$ The values are close.

87. glucose-1-phosphate → glucose-6-phosphate $\Delta G° = -7.28\ \frac{\text{kJ}}{\text{mol}}$

glucose-6-phosphate → fructose-6-phosphate $\Delta G° = 1.67\ \frac{\text{kJ}}{\text{mol}}$

glucose-1-phosphate → fructose-6-phosphate $\Delta G° = -5.61\ \frac{\text{kJ}}{\text{mol}}$

Assume body temperature = 37 °C = 310 K

$$\ln K_{eq} = \frac{\Delta G°}{-RT} = \frac{-5.61\frac{\text{kJ}}{\text{mol}} \times \frac{10^3\text{ J}}{\text{kJ}}}{\frac{-8.3145\text{ J}}{\text{mol K}} \times 310\text{ K}}$$

$e^{\ln K_{eq}} = K_{eq} = e^{2.177} = 8.82$

89. $ATP + H_2O \rightarrow ADP + HPO_4^{2-}$ $\Delta G° = -30.5\ \frac{\text{kJ}}{\text{mol}}$

HPO_4^{2-} + glucose → H_2O + glucose-6-phosphate $\Delta G° = 13.9\ \frac{\text{kJ}}{\text{mol}}$

ATP + glucose → ADP + glucose-6-phosphate $\Delta G° = -16.6\ \frac{\text{kJ}}{\text{mol}}$

$$\ln K = \frac{\Delta G°}{-RT} = \frac{\frac{-16.6\text{ kJ}}{\text{mol}} \times \frac{10^3\text{ J}}{\text{kJ}}}{\frac{-8.314\text{ J}}{\text{mol K}} \times 310\text{ K}} = 6.44$$

$e^{\ln K} = K = e^{6.44} = 627$

A concentration of glucose-6-phosphate that is 627 times the glucose concentration would cause the reverse reaction to be spontaneous.

91. (a) True. With $H_2O(l)$ and $H_2O(g)$ in their standard states, equilibrium is at 100 °C.

(b) False. $H_2O(g)$ is not in its standard state, so the system is not at equilibrium.

(c) False. At 100 °C, $\Delta G°$ must be zero.

(d) True. The process is nonspontaneous.
$\Delta G = 0$ applies only to equilibrium, and $\Delta G°$ applies only to standard-state conditions. If the standard-state conditions are also equilibrium conditions, $\Delta G°$ = 0.

93. $PCl_5 \rightleftharpoons PCl_3 + Cl_2$

$$\Delta G° = 1 \text{ mol } PCl_3 \times \left(\frac{-267.8\text{ kJ}}{\text{mol}}\right) - 2 \text{ mol } Cl_2 \times 0 - 1 \text{ mol } PCl_3 \times \left(\frac{-305\text{ kJ}}{\text{mol}}\right)$$
$$= 37.2 \text{ kJ}$$

$$\ln K = \frac{\Delta G°}{-RT} = \frac{\frac{37.2\text{ kJ}}{mol} \times \frac{10^3\text{ J}}{\text{kJ}}}{\frac{-8.3145\text{ J}}{\text{mol K}} \times 298\text{ K}} = -15.0$$

$e^{\ln K} = K = e^{-15.0} = 3.06 \times 10^{-7}$

$$\Delta H° = 1 \text{ mol } PCl_3 \times \left(\frac{-287\text{ kJ}}{\text{mol}}\right) - 1 \text{ mol } Cl_2 \times 0 - 1 \text{ mol } PCl_3 \times \left(\frac{-374.9\text{ kJ}}{\text{mol}}\right)$$
$$= 87.9 \text{ kJ}$$

$$\ln\frac{K_2}{K_1} = \frac{\Delta H°}{R}\left(\frac{1}{T_1} - \frac{1}{T_2}\right)$$

$$\ln\frac{K_2}{3.06\times10^{-7}} = \frac{\frac{87.9\text{ kJ}}{\text{mol}} \times \frac{10^3\text{ J}}{\text{kJ}}}{\frac{8.3145\text{ J}}{\text{K mol}}}\left(\frac{1}{298.15\text{ K}} - \frac{1}{500.15\text{ K}}\right)$$

$$\ln\frac{K_2}{3.06\times10^{-7}} = 14.32$$

$$\frac{K_2}{3.06\times10^{-7}} = e^{14.32} = 1.66 \times 10^6$$

$K_2 = 0.508$

$$P_{\text{Init}} = \frac{nRT}{V} = \frac{0.100\text{ mol} \times \frac{0.08206\text{ L atm}}{\text{K mol}} \times 500\text{ K}}{1.50\text{ L}}$$

$P_{\text{Init}} = 2.74$ atm

PCl_5	$\rightleftharpoons$	PCl_3	+	Cl_2
2.74				
$-x$		x		x
$2.74-x$		x		x

$$0.508 = \frac{x^2}{2.74 - x}$$

$x^2 + 0.508\,x - 1.39 = 0$

$$x = \frac{-0.508 \pm \sqrt{(0.508)^2 - 4 \times 1 \times (-1.39)}}{2}$$

$$x = \frac{-0.508 \pm 2.41}{2}$$

$x = -1.46$ 0.95 atm

Not Valid

$P_{PCl_5} = 2.74 - x = 1.79$ atm

$P_{Cl_2} = P_{PCl_3} = 0.95$ atm

$P_{total} = 3.69$ atm

95. (a) $P_{O_2} = 0.25$ atm

$(P_{O_2})^{\frac{1}{2}} = 0.50 = K_{eq2}$

At 25 °C: $\Delta G^\circ = 58.56 \times 10^3$ J/mol $= -RT\ln K_{eq}$

58.56×10^3 J/mol $= -8.3145$ J/mol K $\times$ 298 K $\times \ln K_{eq}$

$\ln K_{eq} = -23.62$

$K_{eq1} = 5.44 \times 10^{-11}$

$$\ln \frac{K_2}{K_1} = \frac{-90.83 \times 10^3 \text{ J/mol}}{8.3145 \text{ J/mol K}} \left(\frac{1}{T_2} - \frac{1}{T_1} \right)$$

$$\ln \frac{0.50}{5.44 \times 10^{-11}} = \frac{-90.83 \times 10^3 \text{ J/mol}}{8.3145 \text{ J/mol K}} \left(\frac{1}{T_2} - \frac{1}{298 \text{ K}} \right)$$

$$-2.10 \times 10^{-3} \text{ K}^{-1} = \left(\frac{1}{T_2} - \frac{1}{298 \text{ K}} \right)$$

$$\frac{1}{T_2} = -2.10 \times 10^{-3} \text{ K}^{-1} + 3.36 \times 10^{-3} \text{ K}^{-1}$$

$T_2 = 796$ K $- 273 = 523$°C

(b) In order for the reaction

$HgO(s) \rightarrow Hg(l) + ½ O_2(g)$ $\Delta G^\circ = 58.56$ kJ/mol

to be coupled with another, the ΔG° for the sum of the two reactions must be negative. If the reaction to be coupled with the one above is based on ½ $O_2(g)$, $\Delta G^\circ_{tot} < 0$ or the coupling reaction must have $\Delta G^\circ < -58.56$ kJ/mol. If the coupling reaction is based on something other than the ½ $O_2(g)$, the two reactions must be combined as in the case for aluminum shown below. If the principal reactant in the coupling reaction is already in its highest oxidation state, that reactant cannot be oxidized further.

Yes $H_2(g) + ½ O_2(g) \rightarrow H_2O(l)$ $\Delta G^\circ = -237.2$ kJ/mol

Yes C (graphite) + ½ $O_2(g) \rightarrow CO(g)$ $\Delta G^\circ = -137.2$ kJ/mol

Yes $CO(g) + ½ O_2(g) \rightarrow CO_2(g)$ $\Delta G^\circ = -257.2$ kJ/mol

No $CO_2(g)$ (CO_2 cannot be oxidized.)

Yes $Cu_2O(s) + ½ O_2(g) \rightarrow 2\ CuO(s)$ $\Delta G^\circ = -113.4$ kJ/mol

No $CuO(s)$ (CuO cannot be oxidized.)
No N_2 (g) (All oxides of nitrogen have $\Delta G°_f > 0$.)
No $N_2O(g) + ½\,O_2(g) \rightarrow 2NO(g)$ $\Delta G° = +68.94$ kJ/mol
No $NO(g) + ½\,O_2(g) \rightarrow NO_2(g)$ $\Delta G° = -35.27$ kJ/mol
Yes $SO_2(g) + ½\,O_2(g) \rightarrow SO_3(g)$ $\Delta G° = -70.9$ kJ/mol
No $SO_3(g)$ (SO_3 cannot be oxidized.)
No $Cl_2(g) + ½\,O_2(g) \rightarrow Cl_2O(g)$ $\Delta G° = +97.49$ kJ/mol
Yes $Al(s) + 3/4\,O_2(g) \rightarrow ½\,Al_2O_3(s)$ $\Delta G° = -791$ kJ/mol
$3\,HgO(s) \rightarrow 3\,Hg(l) + 3/2\,O_2(g)$ $\Delta G° = +175.7$ kJ/mol
$\Delta G_{tot} = -615$ kJ/mol
No $2\,Ag(g) + ½\,O_2(g) \rightarrow Ag_2O(s)$ $\Delta G° = -11.2$ kJ/mol
No $MgO(s)$ (MgO cannot be oxidized

97. $CO_2(g) + 4\,H_2(g) \rightarrow CH_4(g) + 2\,H_2O\,(l)$

$$\Delta H°_{rxn} = 1 \text{ mole } CH_4 \times \left(\frac{-74.81 \text{ kJ}}{\text{mol}}\right) + 2 \text{ mol } H_2O \times \left(\frac{-285.8 \text{ kJ}}{\text{mole}}\right) - 1 \text{ mole } CO_2 \times \frac{-393.5 \text{ kJ}}{\text{mol}} - 4 \text{ mol } H_2 \times \frac{0 \text{ kJ}}{\text{mol}} = -252.9 \text{ kJ}$$

$$\Delta S°_{rxn} = 1 \text{ mole } CH_4 \times \frac{186.2 \text{ J}}{\text{K mol}} + 2 \text{ mol } H_2O \times \frac{69.91 \text{ J}}{\text{K mol}} - 1 \text{ mol } CO_2 \times \frac{213.6 \text{ J}}{\text{mole K}} - 4 \text{ mole } H_2 \times \frac{130.6 \text{ J}}{\text{K mol}} = \frac{-410.0 \text{ J}}{\text{K}}$$

$2\,H_2O\,(l) \xrightarrow{\text{electrolysis}} 2\,H_2 + O_2$
$CH_4 + 2\,O_2 \rightarrow CO_2 + 2\,H_2O$
$\Delta G°_{rxn} = \Delta H°_{rxn} - T\,\Delta S°_{rxn}$
The first reaction would be spontaneous (with reactants and products in their standard states) at a T lower than 345 °C (617 K).

$$-252.9 \text{ kJ} \times \frac{10^3 \text{ J}}{\text{kJ}} = T \times \frac{-410.0 \text{ J}}{\text{K}}$$

T = 617 K - 273 = 345 °C
The electrolysis will require a lot of energy to form H_2 and O_2 from water. Solar cells can be used to generate the necessary electricity. The combustion of methane is a viable source of energy. This proposal seems feasible if solar energy is used.

Chapter 18

Electrochemistry

Exercises

18.1A $Zn(s) \rightarrow Zn^{2+}$ $\qquad$ $NO_3^- \rightarrow N_2O$

$2\ NO_3^- \rightarrow N_2O$

$2\ NO_3^- \rightarrow N_2O + 5\ H_2O$

$2\ NO_3^- + 10\ H^+ \rightarrow N_2O + 5\ H_2O$

$Zn(s) \rightarrow Zn^{2+} + 2\ e^-$ $\qquad$ $2\ NO_3^- + 10\ H^+ + 8\ e^- \rightarrow N_2O + 5\ H_2O$

$4\ Zn \rightarrow 4\ Zn^{2+} + 8\ e^-$

$4\ Zn + 2\ NO_3^- + 10\ H^+ \rightarrow 4\ Zn^{2+} + N_2O + 5\ H_2O$

18.1B $P_4 \rightarrow H_2PO_4^-$ $\qquad$ $NO_3^- \rightarrow NO$

$P_4 \rightarrow 4\ H_2PO_4^-$

$P_4 + 16\ H_2O \rightarrow 4\ H_2PO_4^-$ $\qquad$ $NO_3^- \rightarrow NO + 2\ H_2O$

$P_4 + 16\ H_2O \rightarrow 4\ H_2PO_4^- + 24\ H^+$ $\qquad$ $NO_3^- + 4\ H^+ \rightarrow NO + 2\ H_2O$

$P_4 + 16\ H_2O \rightarrow 4\ H_2PO_4^- + 24\ H^+ + 20\ e^-$ $\qquad$ $NO_3^- + 4\ H^+ + 3\ e^- \rightarrow NO + 2\ H_2O$

$3\ P_4 + 48\ H_2O \rightarrow 12\ H_2PO_4^- + 72\ H^+ + 60\ e^-$

$20\ NO_3^- + 80\ H^+ + 60\ e^- \rightarrow 20\ NO + 40\ H_2O$

$3\ P_4 + 20\ NO_3^- + 8\ H_2O + 8\ H^+ \rightarrow 12\ H_2PO_4^- + 20\ NO$

18.2A $OCN^- \rightarrow CO_3^{2-} + N_2$ $\qquad$ $OCl^- \rightarrow Cl^-$

$2\ OCN^- \rightarrow 2\ CO_3^{2-} + N_2$

$2\ OCN^- + 4\ H_2O \rightarrow 2\ CO_3^{2-} + N_2$ $\qquad$ $OCl^- \rightarrow Cl^- + H_2O$

$2\ OCN^- + 4\ H_2O \rightarrow 2\ CO_3^{2-} + N_2 + 8\ H^+$ $\qquad$ $OCl^- + 2H^+ \rightarrow Cl^- + H_2O$

$2\ OCN^- + 4\ H_2O \rightarrow 2\ CO_3^{2-} + N_2 + 8\ H^+ + 6\ e^-$

$OCl^- + 2H^+ + 2\ e^- \rightarrow Cl^- + H_2O$

$3\ OCl^- + 6\ H^+ + 6\ e^- \rightarrow 3\ Cl^- + 3\ H_2O$

$2\ OCN^- + 3\ OCl^- + H_2O \rightarrow 2\ CO_3^{2-} + N_2 + 2\ H^+ + 3\ Cl^-$

$2\ OCN^- + 3\ OCl^- + H_2O + 2\ OH^- \rightarrow 2\ CO_3^{2-} + N_2 + 2\ H_2O + 3\ Cl^-$

$2\ OCN^- + 3\ OCl^- + 2\ OH^- \rightarrow 2\ CO_3^{2-} + N_2 + H_2O + 3\ Cl^-$

18.2B $MnO_4^- \rightarrow MnO_2$

$CH_3CH_2OH \rightarrow CH_3CO_2^-$

$MnO_4^- \rightarrow MnO_2 + 2\ H_2O$

$C_2H_6O + H_2O \rightarrow C_2H_3O_2^-$

$MnO_4^- + 4\ H^+ \rightarrow MnO_2 + 2\ H_2O$

$C_2H_6O + H_2O \rightarrow C_2H_3O_2^- + 5\ H^+$

$MnO_4^- + 4\ H^+ + 3\ e^- \rightarrow MnO_2 + 2\ H_2O$

$C_2H_6O + H_2O \rightarrow C_2H_3O_2^- + 5\ H^+ + 4\ e^-$

$4\ MnO_4^- + 16\ H^+ + 12\ e^- \rightarrow 4\ MnO_2 + 8\ H_2O$

$3\ C_2H_6O + 3\ H_2O \rightarrow 3\ C_2H_3O_2^- + 15\ H^+ + 12\ e^-$

$4\ MnO_4^- + 3\ C_2H_6O + H^+ \rightarrow 4\ MnO_2 + 3\ C_2H_3O_2^- + 5\ H_2O$

$4\ MnO_4^- + 3\ C_2H_6O + H_2O \rightarrow 4\ MnO_2 + 3\ C_2H_3O_2^- + 5\ H_2O + OH^-$

$4\ MnO_4^- + 3\ C_2H_6O \rightarrow 4\ MnO_2 + 3\ C_2H_3O_2^- + 4\ H_2O + OH^-$

18.3A anode $\{Al(s) \rightarrow Al^{3+} + 3\ e^-\} \times 2$

cathode $\underline{\{2\ H^+ + 2\ e^- \rightarrow H_2(g)\} \times 3}$

$2\ Al(s) + 6\ H^+(aq) \rightarrow 3\ H_2(g) + 2\ Al^{3+}(aq)$

18.3B anode $Cu(s) \rightarrow Cu^{2+} + 2\ e^-$

cathode $\{Ag^+ + e^- \rightarrow Ag(s)\} \times 2$

Cell reaction: $Cu(s) + 2\ Ag^+ \rightarrow Cu^{2+} + 2\ Ag(s)$

Cell diagram: $Cu(s) \mid Cu^{2+}(aq) \parallel Ag^+(aq) \mid Ag(s)$

18.4A $I_2 + 2\ e^- \rightarrow 2\ I^-$ $\qquad E^\circ_{I_2/2I^-} = 0.535\ V$

$E^\circ_{cell} = E^\circ(\text{right}) - E^\circ(\text{left}) = E^\circ(\text{cathode}) - E^\circ(\text{anode}) = E^\circ_{I_2/2I^-} - E^\circ_{Sm^{2+}/Sm}$

$3.21\ V = 0.535 - E^\circ_{Sm^{2+}/Sm}$

$E^\circ_{Sm^{2+}/Sm} = 0.535\ V - 3.21\ V = -2.67\ V$

18.4B $Cl_2 + 2\ e^- \rightarrow 2\ Cl^-$ $\qquad E^\circ_{Cl_2/2Cl^-} = 1.358\ V$

$E^\circ_{cell} = E^\circ(\text{right}) - E^\circ(\text{left}) = E^\circ(\text{cathode}) - E^\circ(\text{anode})$

$= E^\circ_{Cl_2/2Cl^-} - E^\circ_{ClO_4^-/Cl_2}$

$-0.034\ V = 1.358\ V - E^\circ_{ClO_4^-/Cl_2}$

$E°_{ClO_4^-/Cl_2} = 1.392\text{ V}$

18.5 (a) $\{Al \rightarrow Al^{3+} + 3\ e^-\} \times 2$ oxidation anode

$[Cu^{2+} + 2\ e^- \rightarrow Cu\} \times 3$ reduction cathode

$E°_{cell} = E°(\text{cathode}) - E°(\text{anode})$

$E°_{cell} = E°_{Cu^{2+}/Cu} - E°_{Al^{3+}/Al}$

$E°_{cell} = 0.340\text{ V} - (-1.676\text{ V})$

$E°_{cell} = 2.016\text{ V}$

(b) $S_2O_8^{2-} + 2\ e^- \rightarrow 2\ SO_4^{2-}$ reduction cathode

$Mn^{2+} + 4\ H_2O \rightarrow MnO_4^- + 8\ H^+ + 5\ e^-$ oxidation anode

$E°_{cell} = E°(\text{cathode}) - E°(\text{anode})$

$E°_{cell} = E°_{S_2O_8^{2-}/SO_4^{2-}} - E°_{MnO_4^-/Mn^{2+}}$

$E°_{cell} = 2.01\text{ V} - 1.51\text{ V}$

$E°_{cell} = 0.50\text{ V}$

(c) $NO_3^- + 4\ H^+ + 3\ e^- \rightarrow NO + 2\ H_2O$ reduction cathode

$Pb^{2+} + 2\ H_2O \rightarrow PbO_2 + 4\ H^+ + 2\ e^-$ oxidation anode

$E°_{cell} = E°(\text{cathode}) - E°(\text{anode})$

$E°_{cell} = E°_{NO_3^-/NO} - E°_{PbO_2/Pb^{2+}}$

$E°_{cell} = 0.956\text{ V} - 1.455\text{ V}$

$E°_{cell} = -0.499\text{ V}$

18.6A $Cu^{2+} + 2\ e^- \rightarrow Cu(s)$ $E°_{Cu^{2+}/Cu} = 0.340\text{ V}$

$\{Fe^{2+} \rightarrow Fe^{3+} + e-\} \times 2$ $E°_{Fe^{3+}/Fe^{2+}} = 0.771\text{ V}$

$E°_{cell} = E°(\text{cathode}) - E°(\text{anode})$

$E°_{cell} = 0.340\text{ V} - 0.771\text{ V} = -0.431\text{ V}$

Since $E°_{cell}$ is negative, the reaction is not spontaneous in the forward direction.

18.6B $Mn^{2+} + 4\ H_2O \rightarrow MnO_4^- + 8\ H^+ + 5\ e^-$ oxidation anode

$2\ IO_3^- + 12\ H^+ + 10\ e^- \rightarrow I_2 + 6\ H_2O$ reduction cathode

$E°_{cell} = E°(\text{cathode}) - E°(\text{anode})$

$E°_{cell} \;= E°_{IO_3^-/I_2} - E°_{MnO_4^-/Mn^{2+}}$

$E°_{cell} \;= 1.20\ V - 1.51\ V = -0.31\ V$

Since $E°_{cell}$ is negative, the reverse direction is spontaneous.

18.7 Where the zinc strip touches the citric acid in the lemon, zinc metal atoms give up electrons (oxidation of Zn to Zn^{2+} occurs on the zinc electrode), because zinc is above SHE on the table of standard electrode potentials. The electrons go through the zinc metal, the voltmeter, and the copper metal to where it touches the juice of the lemon. At that point, H^+ is changed to $H_2(g)$ (the reduction half-reaction is that of $H^+(aq)$ in citric acid to $H_2(g)$). The lemon juice supplies the H^+ and acts as an electrolyte. (Notice that both strips have to be in the same section of the lemon.) In part this reduction occurs directly on the zinc electrode, but also some electrons pass through the electric measuring circuit to the copper electrode, where reduction of $H^+(aq)$ also occurs.

18.8A $\{Mg \rightarrow Mg^{2+} + 2\ e^-\} \times 3$ oxidation anode

$\{Al^3 + 3\ e^- \rightarrow Al\} \times 2$ reduction cathode

$E°_{cell} = E°(\text{cathode}) - E°(\text{anode})$

$E°_{cell} = E°_{Al^{3+}/Al} - E°_{Mg^{2+}/Mg} = -1.676\ V - (-2.356\ V)$

$E°_{cell} = 0.680\ V$

$\Delta G° = -nFE° = -6\ \text{moles}\ e^- \times 96485\ C\ (\text{mol}\ e^-)^{-1} \times 0.680\ V$

$\Delta G° = -3.94 \times 10^5\ J \times \frac{kJ}{10^3\ J} = -394\ kJ$

$$E°_{cell} = \frac{0.0592}{n} \log K_{eq}$$

$$\log K_{eq} = \frac{nE°_{cell}}{0.0592} = \frac{6 \times 0.680}{0.0592} = 69$$

Raise both sides of the equation to a power of e.

$10^{\log K_{eq}} = K_{eq} = 10^{68.9} = 1 \times 10^{69}$

18.8B $Ag \rightarrow Ag^+ + e^-$ oxidation anode

$NO_3^- + 4\ H^+ + 3\ e^- \rightarrow NO + 2\ H_2O$ reduction cathode

$$E°_{cell} = \frac{0.0592}{n} \log K_{eq}$$

$E°_{cell} = E°_{NO_3^-/NO} - E°_{Ag^+/Ag} = 0.956\ V - 0.800\ V$

$$E°_{cell} = 0.156\ V = \frac{0.0592}{n} \log K_{eq}$$

$$\log K_{eq} = \frac{3 \times 0.156}{0.0592} = 7.91$$

Raise both sides of the equation to a power of e.

$$10^{\log K_{eq}} = K_{eq} = 10^{7.91} = 8.0 \times 10^{7}$$

18.9 (a) $Zn (s) \rightarrow Zn^{2+} + 2\ e^{-}$ $\qquad E^{\circ}_{Zn/Zn^{2+}} = -0.763\ V$

$Cu^{2+} + 2\ e^{-} \rightarrow Cu (s)$ $\qquad E^{\circ}_{Cu^{2+}/Cu} = 0.340\ V$

$E^{\circ}_{cell} = E^{\circ}(\text{cathode}) - E^{\circ}(\text{anode}) = 0.340 - (-0.763\ V) = 1.103\ V$

$$E_{cell} = E^{\circ}_{cell} - \frac{0.0592}{n} \log \frac{[Zn^{2+}]}{[Cu^{2+}]}$$

$$E_{cell} = 1.103\ V - \frac{0.0592}{n} \log \frac{(2.0)}{(0.050)} = 1.103\ V - 0.047\ V$$

$E_{cell} = 1.056\ V$

(b) $$E_{cell} = E^{\circ}_{cell} - \frac{0.0592}{2} \log \frac{(0.050)}{(2.0)}$$

$E_{cell} = 1.103\ V + 0.047\ V$

$E_{cell} = 1.150\ V$

(c) $Cu (s) \rightarrow Cu^{2+} + 2\ e^{-}$ $\qquad E^{\circ}_{Cu^{2+}/Cu} = 0.340\ V$

$Cl_2 (g) + 2\ e^{-} \rightarrow 2\ Cl^{-}$ $\qquad E^{\circ}_{Cl_2/Cl^{-}} = 1.358\ V$

$E^{\circ}_{cell} = E^{\circ}(\text{cathode}) - E^{\circ}(\text{anode}) = 1.358\ V - 0.340\ V = 1.018\ V$

$$E_{cell} = E^{\circ}_{cell} - \frac{0.0592}{2} \log \frac{(1.0)(0.25)^2}{(0.50)}$$

$E_{cell} = 1.018\ V + 0.027\ V$

$E_{cell} = 1.045\ V$

18.10 The metal in the bend in the nail is more strained and thus more energetic, so it, along with the head and tail of the nail, is preferentially oxidized. Reduction occurs along the rest of the body of the nail.

18.11A The principal species in the solution are K^{+} and Br^{-} ions and H_2O molecules.

$K^{+} + e^{-} \rightarrow K$ $\qquad E^{\circ}_{K^{+}/K} = -2.924\ V$

It is expected that the reduction half-reaction is:

$2\ H_2O + 2\ e^{-} \rightarrow H_2 (g) + 2\ OH^{-}(aq)$ $\qquad E^{\circ}_{H_2O/H_2} = -0.828\ V$

The two possibilities for oxidation are that of water molecules to $O_2(g)$ and $Br^{-}(aq)$ to $Br_2(l)$. These oxidation half-reactions and their E°_{ox} values are

$2\ Br^-(aq) \rightarrow Br_2(l) + 2\ e^-$ $\quad E^\circ_{Br^-/Br_2} = 1.065\ V$

$2\ H_2O \rightarrow O_2(g) + 4\ H^+ + 4\ e^-$ $\quad E^\circ_{H_2O/O_2} = 1.229\ V$

We should expect the oxidation of $Br^-(aq)$ to predominate and the net electrolysis reaction to be

$2\ Br^-(aq) + 2\ H_2O \xrightarrow{\text{electrolysis}} Br_2(l) + H_2(g) + 2\ OH^-(aq)$ $\quad E^\circ_{cell} = -1.893\ V$

18.11B The electrolysis reaction needs to be:

$Ag^+(aq) + e^- \rightarrow Ag\ (s)$ $E^\circ_{Ag^+/Ag} = 0.800\ V$

This reduction should occur exclusively over that of water molecules.

$2\ H_2O + 2\ e^- \rightarrow H_2\ (g) + 2\ OH^-(aq)$ $\quad E^\circ_{red} = -0.828\ V$

To cause this with a silver anode and a copper cathode means that the reaction that can occur, $Cu\ (s) + 2\ Ag^+ \longrightarrow 2\ Ag\ (s) + Cu^{2+}$, must be overcome.

$Cu\ (s) \rightarrow Cu^{2+} + 2\ e^-$ $\quad E^\circ_{Cu^{2+}/Cu} = 0.340\ V$

$\underline{Ag^+(aq) + e^- \rightarrow Ag\ (s)}$ $\quad E^\circ_{Ag^+/Ag} = 0.800\ V$

$Cu\ (s) + 2\ Ag^+ \longrightarrow 2\ Ag\ (s) + Cu^{2+}$

$E^\circ_{cell} = E^\circ(\text{cathode}) - E^\circ(\text{anode}) = 0.800\ V - 0.340\ V = 0.460\ V$

To force the electrolysis, Ag (s, anode) $\xrightarrow{\text{electrolysis}}$ Ag(s, cathode) requires the external voltage > 0.460 V. Note that Cu(s) will spontaneously deposit Ag(s) from $AgNO_3(aq)$ (recall Figure 18.1). What is required here, however, is that the electrodeposited Ag(s) be derived from oxidation of the silver electrode.

18.12 Cell A $\quad Zn^{2+} + Cu\ (s) \rightarrow Zn\ (s) + Cu^{2+}$

Cell B $\quad Zn\ (s) + Cu^{2+} \rightarrow Zn^{2+} + Cu\ (s)$

The current will continue to flow only as long as the concentrations are unequal—that is, as long as there is a force to push the electrons.

18.13A $Cu^{2+} + 2e^- \rightarrow Cu\ (s)$

$$\frac{1.00\ g\ Cu}{2.25\ A} \times \frac{mol}{63.55\ g} \times \frac{2\ mol\ e^-}{mol\ Cu} \times \frac{96485\ C}{mol\ e^-} \times \frac{1\ A \times 1\ s}{1\ C} \times \frac{min}{60\ s} = 22.5\ min$$

18.13B(a) $t = 21\ min \times \dfrac{60\ s}{min} = 1260\ s + 12\ s = 1272\ s$

$$?\ C = 2.175\ g\ Ag \times \frac{mol}{107.87\ g} \times \frac{mol\ e^-}{mol\ Ag} \times \frac{96485\ C}{mol\ e^-} = 1.945 \times 10^3\ C$$

(b) $?\ amp = \dfrac{1.945 \times 10^3\ C}{1272\ s} = 1.529\ amp$

18.14 $2\ Cl^- \rightarrow 2\ e^- + Cl_2(g)$ $\qquad E^\circ_{Cl_2/Cl^-} = 1.398\ V$

$2\ H_2O \rightarrow O_2(g) + 4\ H^+ + 4\ e^-$ $\qquad E^\circ_{O_2/H_2O} = 1.229\ V$

$2\ I^- \rightarrow 2\ e^- + I_2(s)$ $\qquad E^\circ_{I_2/I^-} = 0.535\ V$

Nitrate ion cannot be oxidized, so the solution of magnesium nitrate will produce O_2 at the anode. For each mole of electrons, there will be one-fourth of a mole of gas. Although it would seem that O_2 would be produced from a solution of sodium chloride, Cl_2 is produced due to the over-voltage. For each mole of electrons, there will be one-half of a mole of gas.
$I_2(s)$ would be produced at the anode from a potassium iodide solution. That is a solid—no gas is produced.
The sodium chloride solution would produce the most gas for the same amount of charge.

Review Questions

6. No. The basis of an electrochemical cell reaction must be an oxidation-reduction reaction, and a Brønsted-Lowry acid–base reaction does not involve changes in oxidation states.

8. Standard electrode potentials are based on an arbitrary value of zero assigned to the half-reaction, $2\ H^+(1\ M) + 2\ e^- \rightarrow H_2(g, 1\ atm)$. If the reduction most easily achieved is assigned $E^\circ = 0$, all others will have $E^\circ < 0$. If the reduction most difficult to achieve is assigned $E^\circ = 0$, all others will have $E^\circ > 0$.

9. E° is a standard electrode potential that describes the tendency for a particular reduction half-reaction to occur when reactants and products are in their standard states. E°_{cell} is a difference in E° values for two standard half-reactions, and its value depends on the specific half-reactions. E_{cell} is the cell potential when the conditions are not standard state.

11. Au and Ag have positive reduction potentials and do not react with HCl (aq).

12. Redox reactions outside of an electrochemical cell are still reactions of an oxidation and a reduction. The sign of E°_{cell} of those reactions is still applicable. The prediction of the direction of spontaneous change based on the sign of E_{cell} is essentially a prediction based on the sign of ΔG ($\Delta G = -nFE_{cell}$). Because ΔG is a function of state, the prediction of spontaneous change is independent of the path of the reaction, whether in a voltaic cell or directly between the reactants.
The electrode potentials can be used to predict spontaneity for all redox reactions, but they are not applicable to other types, for example, acid-base reactions. .

13. $E^\circ_{cell} < 0$ means that a reaction will not occur spontaneously with reactants and products at standard-state conditions. Often it can be made to occur at other conditions.

23. Standard electrode potentials are used to calculate the minimum voltage required to cause electrolysis.

24. The metal needs to be the cathode, so that metal ions in solution are reduced to metal atoms, which become attached to the plated metal.

Problems

25. (a) $ClO_2 + H_2O \rightarrow ClO_3^- + 2\ H^+ + e^-$ oxidation

(b) $MnO_4^- + 4\ H^+ + 3\ e^- \rightarrow MnO_2 + 2\ H_2O$ reduction

(c) $SbH_3 \rightarrow Sb + 3\ H^+ + 3\ e^-$

$SbH_3 + 3\ OH^- \rightarrow Sb + 3\ H^+ + 3\ OH^- + 3\ e^-$

$SbH_3 + 3\ OH^- \rightarrow Sb + 3\ H_2O + 3\ e^-$ oxidation

27. The solution to the balancing of redox equations is listed in this order: First, the two balanced half-reactions are listed side by side. These are followed by the final balanced equation for acidic solution. For basic solutions, two more equations are listed. One is to show the addition of the OH^- ions, and the other, the final balanced equation in basic solution.

(a) $Ag \rightarrow Ag^+ + e^-$ $NO_3^- + 4\ H^+ + 3\ e^- \rightarrow NO\ (g) + 2\ H_2O$

$3\ Ag + NO_3^- + 4\ H^+ \rightarrow 3\ Ag^+ + NO(g) + 2\ H_2O$

(b) $H_2O_2 \rightarrow O_2 + 2\ H^+ + 2\ e^-$ $MnO_4^- + 8\ H^+ + 5\ e^- \rightarrow Mn^{2+} + 4\ H_2O$

$2\ MnO_4^- + 5\ H_2O_2 + 6\ H^+ \rightarrow 5\ O_2 + 2\ Mn^{2+} + 8\ H_2O$

(c) $Cl_2 + 2\ e^- \rightarrow 2\ Cl^-$ $I^- + 3\ H_2O \rightarrow IO_3^- + 6\ H^+ + 6\ e^-$

$I^- + 3\ Cl_2 + 3\ H_2O \rightarrow 6\ Cl^- + IO_3^- + 6\ H^+$

(d) $Fe(OH)2 + H_2O \rightarrow Fe(OH)3 + H^+ + e^-$ $O_2 + 4\ H^+ + 4\ e^- \rightarrow 2\ H_2O$

$4\ Fe(OH)2 + 2\ H_2O + O_2 \rightarrow 4\ Fe(OH)3$

(e) $S_8 + 12\ H_2O \rightarrow 4\ S_2O_3^{2-} + 24\ H^+ + 16\ e^-$ $S_8 + 16\ e^- \rightarrow 8\ S^{2-}$

$2\ S_8 + 12\ H_2O \rightarrow 4\ S_2O_3^{2-} + 8\ S^{2-} + 24\ H^+$

$2\ S_8 + 12\ H_2O + 24\ OH^- \rightarrow 4\ S_2O_3^{2-} + 8\ S^{2-} + 24\ H^+ + 24\ OH^-$

$S_8 + 12\ OH^- \rightarrow 2\ S_2O_3^{2-} + 4\ S^{2-} + 6\ H_2O$

(f) $CrI_3 + 16\ H_2O \rightarrow CrO_4^{2-} + 3\ IO_4^- + 32\ H^+ + 27\ e^-$

$H_2O_2 + 2\ H^+ + 2\ e^- \rightarrow 2\ H_2O$

$2\ CrI_3 + 27\ H_2O_2 \rightarrow 2\ CrO_4^{2-} + 6\ IO_4^- + 10\ H^+ + 22\ H_2O$

$2\ CrI_3 + 27\ H_2O_2 + 10\ OH^- \rightarrow$

$2\ CrO_4^{2-} + 6\ IO_4^- + 10\ H^+ + 10OH^- + 22\ H_2O$

$2\ CrI_3 + 27\ H_2O_2 + 10\ OH^- \rightarrow 2\ CrO_4^{2-} + 6\ IO_4^- + 32\ H_2O$

29. (a) $MnO_4^- + 8\,H^+ + 5\,e^- \rightarrow Mn^{2+} + 4\,H_2O$ $C_2H_2O_4 \rightarrow 2\,CO_2 + 2\,H^+ + 2\,e^-$

$2\,MnO_4^- + 5\,C_2H_2O_4 + 6\,H^+ \rightarrow 2\,Mn^{2+} + 10\,CO_2 + 8\,H_2O$

(b) $Cr_2O_7^{2-} + 14\,H^+ + 6\,e^- \rightarrow 2\,Cr^{3+} + 7\,H_2O$

$UO^{2+} + H_2O \rightarrow UO_2^{2+} + 2\,H^+ + 2\,e^-$

$Cr_2O_7^{2-} + 3\,UO^{2+} + 8\,H^+ \rightarrow 2\,Cr^{3+} + 3\,UO_2^{2+} + 4\,H_2O$

(c) $Zn \rightarrow Zn^{2+} + 2\,e^-$ $NO_3^- + 9\,H^+ + 8\,e^- \rightarrow NH_3(g) + 3\,H_2O$

$4\,Zn + NO_3^- + 9\,H^+ \rightarrow 4\,Zn^{2+} + NH_3(g) + 3\,H_2O$

$4\,Zn + NO_3^- + 9\,H^+ + 9\,OH^- \rightarrow 4\,Zn^{2+} + NH_3(g) + 3\,H_2O + 9\,OH^-$

$4\,Zn + NO_3^- + 6\,H_2O \rightarrow 4\,Zn^{2+} + NH_3(g) + 9\,OH^-$

31. (a) $V^{2+} \rightarrow V^{3+} + e^-$ oxidation

$Sn^{2+} + 2\,e^- \rightarrow Sn(s)$ reduction

$2\,V^{2+} + Sn^{2+} \rightarrow 2\,V^{3+} + Sn(s)$

$E°_{cell} = E°(\text{cathode}) - E°(\text{anode}) = E°(\text{reduction}) - E°(\text{oxidation})$

$E°_{cell} = E°_{Sn^{2+}/Sn} - E°_{V^{3+}/V^{2+}} = -0.137\text{ V} - (-0.255\text{ V})$

$E°_{cell} = 0.118\text{ V}$

(b) $V^{2+} \rightarrow V^{3+} + e^-$ oxidation

$Zn^{2+} + 2\,e^- \rightarrow Zn$ reduction

$2\,V^{2+} + Zn^{2+} \rightarrow 2\,V^{3+} + Zn(s)$

$E°_{cell} = E°(\text{cathode}) - E°(\text{anode}) = E°(\text{reduction}) - E°(\text{oxidation})$

$E°_{cell} = E°_{Zn^{2+}/Zn} - E°_{V^{3+}/V^{2+}} = -0.763\text{ V} - (-0.255\text{ V})$

$E°_{cell} = -0.508\text{ V}$

(c) $V^{2+} \rightarrow V^{3+} + e^-$ oxidation

$Cu^{2+} + 2\,e^- \rightarrow Cu$ reduction

$2\,V^{2+} + Cu^{2+} \rightarrow 2\,V^{3+} + Cu(s)$

$E°_{cell} = E°(\text{cathode}) - E°(\text{anode}) = E°(\text{reduction}) - E°(\text{oxidation})$

$E°_{cell} = E°_{Cu^{2+}/Cu} - E°_{V^{3+}/V^{2+}} = 0.340\text{ V} - (-0.255\text{ V})$

$E°_{cell} = 0.595\text{ V}$

33. $2\,CuI + 2\,e^- \rightarrow 2\,Cu(s) + 2\,I^-$ $E°_{cathode} = ?$

$Cd(s) \rightarrow Cd^{2+} + 2\,e^-$ $E°_{anode} = -0.403\text{ V}$

$E°_{cell} = E°_{cathode} - E°_{anode}$

$0.23\text{ V} = E°_{cathode} + 0.403\text{ V}$

$E° = 0.23\text{ V} - 0.403\text{ V} = -0.17\text{ V}$

35. $V(s) \rightarrow V^{2+} + 2e^-$ $E°_{anode} = ?$

$Cu^{2+} + 2e^- \rightarrow Cu$ $E°_{cathode} = 0.340$ V

$E°_{cell} = E°_{cathode} - E°_{anode} = 1.47\text{ V} = 0.340\text{ V} - E°_{anode}$

$E°_{anode} = -1.13\text{ V} = E°_{V^{2+}/V}$

37. (a) $2\,I^-(aq) \rightarrow I_2(s) + 2e^-$ $E°_{anode} = 0.535$ V

$Cl_2(g) + 2e^- \rightarrow 2\,Cl^-(aq)$ $E°_{cathode} = 1.358$ V

$2\,I^-(aq) + Cl_2(g) \rightarrow I_2(s) + 2\,Cl^-(aq)$

$E°_{cell} = E°_{cathode} - E°_{anode} = 1.358\text{ V} - 0.535\text{ V} = 0.823\text{ V}$

(b) $S_2O_8^{2-} + 2e^- \rightarrow 2\,SO_4^{2-}$ $E°_{cathode} = 2.01$ V

$Pb^{2+} + 2\,H_2O \rightarrow 2e^- + 4\,H^+ + PbO_2$ $E°_{anode} = 1.455$ V

$Pb^{2+} + S_2O_8^{2-} + 2\,H_2O \rightarrow 2\,SO_4^{2-} + PbO_2 + 4\,H^+$

$E°_{cell} = E°_{cathode} - E°_{anode} = 2.01\text{ V} - 1.455\text{ V} = 0.55\text{ V}$

39. (a) $Zn(s) \rightarrow Zn^{2+} + 2e^-$ $E°_{anode} = -0.763$ V

$\{Ag^+ + e^- \rightarrow Ag\} \times 2$ $E°_{cathode} = 0.800$ V

$Zn(s) + 2\,Ag^+ \rightarrow 2\,Ag(s) + Zn^{2+}$

$Zn(s)\,|\,Zn^{2+}(aq)\,||\,Ag^+(aq)\,|\,Ag(s)$

$E°_{cell} = E°_{cathode} - E°_{anode}\ 0.800\text{ V} - (-0.763\text{ V}) = 1.563\text{ V}$

(b) $\{Fe^{2+} \rightarrow Fe^{3+} + e^-\} \times 4$ $E°_{anode} = 0.771$ V

$O_2(g) + 4\,H^+ + 4e^- \rightarrow 2\,H_2O$ $E°_{cathode} = 1.229$ V

$4\,Fe^{2+} + O_2(g) + 4\,H^+(aq) \rightarrow 4\,Fe^{3+} + 2\,H_2O(l)$

$Pt\,|\,Fe^{2+}(aq), Fe^{3+}(aq)\,||\,H_2O(l), H^+(aq)\,|\,O_2(g), Pt$

$E°_{cell} = E°_{cathode} - E°_{anode} = 1.229\text{ V} - 0.771\text{ V} = 0.458\text{ V}$

41. (a) $Sn(s) \rightarrow Sn^{2+} + 2e^-$ $E°_{anode} = -0.137$ V

$Co^{2+} + 2e^- \rightarrow Co(s)$ $E°_{cathode} = -0.277$ V

$E°_{cell} = E°_{cathode} - E°_{anode} = -0.277\text{ V} - (-0.137\text{ V}) = -0.140\text{ V}$

not spontaneous

(b) $2\,Br^- \rightarrow Br_2(l) + 2e^-$ $E°_{anode} = 1.065$ V

$Cr_2O_7^{2-} + 14\,H^+ + 6e^- \rightarrow 2\,Cr^{3+} + 7\,H_2O$ $E°_{cathode} = 1.33$ V

$E°_{cell} = E°_{cathode} - E°_{anode} = 1.33\text{ V} - 1.065\text{ V} = 0.26\text{ V}$

spontaneous

43. (a) $Cd^{2+} + 2e^- \rightarrow Cd(s)$ $E°_{cathode} = -0.403$ V

$Al(s) \rightarrow Al^{3+} + 3e^-$ $E°_{anode} = -1.676$ V

$E°_{cell} = E°_{cathode} - E°_{anode} = -0.403\text{ V} - (-1.676\text{ V}) = 1.273\text{ V}$

spontaneous

(b) $2\ Cl^- \rightarrow Cl_2 + 2\ e^-$ $E°_{anode}$ = 1.358 V

$Br_2\ (l) + 2\ e^- \rightarrow 2\ Br^-$ $E°_{cathode}$ = 1.065 V

$E°_{cell} = E°_{cathode} - E°_{anode} = 1.065\ V - 1.358\ V = -0.293\ V$

not spontaneous

(c) $Cl^- + 6\ OH^- \rightarrow ClO_3^- + 3\ H_2O + 6\ e^-$ $E°_{anode}$ = 0.622 V

$HO_2^- + H_2O + 2\ e^- \rightarrow 3\ OH^-$ $E°_{cathode}$ = 0.88 V

$E°_{cell} = E°_{cathode} - E°_{anode} = 0.88\ V - 0.622\ V = 0.26\ V$

spontaneous

45. (a) Silver does not react with HCl (aq) because H^+ (aq) is not a good enough oxidizing agent to oxidize Ag (s) to Ag^+ (aq). $E°_{cell}$ for the reaction is – 0.800 V.

(b) Nitrate ion in acidic solution is a good enough oxidizing agent to oxidize Ag (s) to Ag+ (aq). $E°_{cell}$ for the reaction is 0.156 V.

$3\ Ag\ (s) + 4\ H^+ + NO_3^- \rightarrow 3\ Ag^+ + NO\ (g) + 2\ H_2O$

47. Because Cu displaces $Rh^{3+}(aq)$, $E°_{cathode}$ for the half-reaction $Rh^{3+} + 3\ e^- \rightarrow Rh(s)$ must be greater than 0.340 V. Because Ag does not displace $Rh^{3+}(aq)$, $E°_{cathode}$ for the half-reaction $Rh^{3+} + 3\ e^- \rightarrow Rh(s)$ must be less than 0.800 V. Because Rh (s) is oxidized to $Rh^{3+}(aq)$ by $HNO_3(aq)$, $E°_{anode}$ for $Rh(s) \rightarrow Rh^{3+} + 2\ e^-$ must be less positive than 0.956 V. Thus, for the half-reaction $Rh^{3+} + 3\ e^- \rightarrow Rh(s)$ $0.340\ V < E°_{Rh^{3+}/Rh} < 0.800\ V$.

49. (a) $Al(s) \rightarrow Al^{3+} + 3\ e^-$ $E°_{anode}$ = –1.676 V

$\{Ag^+ + e^- \rightarrow Ag(s)\} \times 3$ $E°_{cathode}$ = 0.800 V

$E°_{cell} = E°_{cathode} - E°_{anode} = 0.800\ V - (-1.676\ V) = 2.476\ V$

$\Delta G° = -nFE° = -3\ \text{moles}\ e^- \times 96485\ \text{C/mol}\ e^- \times 2.476\ V$

$\Delta G° = -716700\ J = -716.7\ kJ$

(b) $\{2\ IO_3^- + 12\ H^+ + 10\ e^- \rightarrow I_2 + 6\ H_2O\} \times 2$ $E°_{cathode}$ = 1.20 V

$\{2\ H_2O \rightarrow 4\ e^- + 4\ H^+ + O_2\} \times 5$ $E°_{anode}$ = 1.229 V

$E°_{cell} = E°_{cathode} - E°_{anode} = 1.20\ V - 1.229\ V = -0.03\ V$

$\Delta G° = -nFE° = -20\ \text{mol}\ e^- \times 96485\ \text{C/mol}\ e^- \times (-0.03\ V)$

$\Delta G° = 5.8 \times 10^4\ J = 6 \times 10^1\ kJ$

51. (a) $Ag^+ + e^- \rightarrow Ag\ (s)$ $E°_{cathode}$ = 0.800

$Fe^{2+} \rightarrow Fe^{3+} + e^-$ $E°_{anode}$ = 0.771

$E°_{cell} = E°_{cathode} - E°_{anode} = 0.800\ V - 0.771\ V = 0.029\ V$

$$K_{eq} = \frac{[Fe^{3+}]}{[Fe^{2+}][Ag^+]}$$

$$E^\circ_{cell} = 0.029\text{ V} = \frac{0.0592}{n}\log K_{eq} = \frac{0.0592}{n}\log K_{eq}$$

$$\log K_{eq} = \frac{nE^\circ_{cell}}{0.0592} = \frac{1\times 0.029}{0.0592} = 0.49$$

$$10^{\log Keq} = K_{eq} = 10^{0.49} = 3.1$$

(b) $MnO_2 + 4\ H^+ + 2\ e^- \rightarrow Mn^{2+} + 2\ H_2O$ $\qquad E^\circ_{cathode} = 1.23$ V

$2\ Cl^- \rightarrow Cl_2 + 2\ e^-$ $\qquad E^\circ_{anode} = 1.358$ V

$E^\circ_{cell} = E^\circ_{cathode} - E^\circ_{anode} = 1.23\text{ V} - 1.358\text{ V} = -0.13\text{ V}$

$$K_{eq} = \frac{[Mn^{2+}][P_{Cl_2}]}{[H^+]^4[Cl^-]^2}$$

$$E^\circ_{cell} = -0.13\text{ V} = \frac{0.0592}{n}\log K_{eq} = \frac{0.0592}{2}\log K_{eq}$$

$$\log K_{eq} = \frac{2\times(-0.13)}{0.0592} = -4.39$$

$$10^{\log Keq} = K_{eq} = 10^{-4.39} = 4\times 10^{-5}$$

(c) $\{OCl^- + H_2O + 2\ e^- \rightarrow Cl^- + 2\ OH^-\}\times 2$ $\qquad E^\circ_{cathode} = 0.890$ V

$4\ OH^- \rightarrow 4\ e^- + 2\ H_2O + O_2$ $\qquad E^\circ_{anode} = 0.401$ V

$E^\circ_{cell} = E^\circ_{cathode} - E^\circ_{anode} = 0.890\text{ V} - 0.401\text{ V} = 0.489\text{ V}$

$$K_{eq} = \frac{P_{O_2}[Cl^-]^2}{[OCl^-]^2}$$

$$E^\circ_{cell} = 0.489\text{ V} = \frac{0.0592}{2}\log K_{eq}$$

$$\log K_{eq} = \frac{4\times 0.489}{0.0592} = 33.04$$

$$10^{\log Keq} = K_{eq} = 10^{33.04} = 1\times 10^{33}$$

53. $Sn(s) + Pb^{2+} \rightarrow Sn^{2+} + Pb(s)$

$Sn \rightarrow Sn^{2+} + 2\ e^-$ $\qquad E^\circ_{anode} = -0.137$ V

$Pb^{2+} + 2\ e^- \rightarrow Pb$ $\qquad E^\circ_{cathode} = -0.125$ V

$E^\circ_{cell} = E^\circ_{cathode} - E^\circ_{anode} = -0.125\text{ V} - (-0.137\text{ V}) = 0.012\text{ V}$

$$\log K_{eq} = \frac{2\times 0.012}{0.0592} = 0.405$$

$$10^{\log Keq} = K_{eq} = 10^{0.405} = 2.5 = \frac{[Sn^{2+}]}{[Pb^{2+}]}$$

	$Sn(s) + Pb^{2+}$	$\rightleftharpoons$	$Sn^{2+} + Pb(s)$
init	1.00 M		—
change	$-X$		$+X$
eq.	$1.00 - X$		X

$$\frac{X}{1.00 - X} = 2.5$$
$X = 2.5 - 2.5\ X$
$3.5\ X = 2.5$
$X = 0.71\ M = [Sn^{2+}]$
$[Pb^{2+}] = 0.29\ M$

55. (a) $E_{cell} = E°_{cell} - \frac{0.0592}{n} \log Q$

$Fe^{2+} + 2\ e^- \rightarrow Fe(s)$ $\quad E°_{anode} = -0.440\ V$

$Ag^+ + e^- \rightarrow Ag(s)$ $\quad E°_{cathode} = 0.800\ V$

$E°_{cell} = E°_{cathode} - E°_{anode} = 0.800\ V - (-0.440\ V) = 1.240\ V$

$n = 2$

$$Q = \frac{[Fe^{2+}]}{[Ag^+]^2}$$

$$E_{cell} = 1.240\ V - \frac{0.0592}{2} \log \frac{1.33}{(0.0015)^2}$$

$= 1.240\ V - 0.171\ V$
$= 1.069\ V$

(b) $E_{cell} = E°_{cell} - \frac{0.0592}{n} \log Q$

$VO^{2+} + H_2O \rightarrow VO_2^+ + 2\ H^+ + e^-$ $\quad E°_{anode} = 1.000\ V$

$O_2 + 4\ H^+ + 4\ e^- \rightarrow 2\ H_2O(l)$ $\quad E°_{cathode} = 1.229\ V$

$E°_{cell} = E°_{cathode} - E°_{anode} = 1.229\ V - 1.000\ V = 0.229\ V$

$n = 4$

$$Q = \frac{[VO_2^+]^4[H^+]^4}{[VO^{2+}]^4 P_{O_2}}$$

$$E_{cell} = 0.229\ V - \frac{0.0592}{n} \log \frac{(0.75)^4 (0.30)^4}{(0.050)^4 (0.25)}$$

$E_{cell} = 0.229\ V - 0.048\ V = 0.181\ V$

57. $H_2(g, 1\ atm) + 2\ H^+(1\ M) \rightarrow 2\ H^+(0.0025\ M) + H_2\ (g, 1\ atm)$

$$E_{cell} = 0.00 - \frac{0.0592}{2} \log \frac{[H^+]^2 P_{H_2}}{[H^+]^2 P_{H_2}}$$

$$E_{cell} = 0.00 - \frac{0.0592}{2} \log \frac{(0.0025)^2\ (1)}{(1)^2\ (1)} = 0.15\ V$$

OR

concentration cell: $2\ H^+\ (1M) \rightarrow 2\ H^+\ (0.0025\ M)$

$$E_{cell} = -\frac{0.0592}{n}\log\frac{[H^+]^2_{prod}}{[H^+]^2_{reac}}$$

$$E_{cell} = -\frac{0.0592}{2}\log\frac{(0.0025)^2}{(1)^2} = 0.15\ V$$

59. $H_2(g, 1\ atm) + 2\ H^+(1\ M) \rightarrow 2\ H^+(X\ M) + H_2(g, 1\ atm)$

$$Q = \frac{[H^+]^2 P_{H_2}}{[H^+]^2 P_{H_2}}$$

$n = 2$

$$E_{cell} = E°_{cell} - \frac{0.0592}{2}\log\frac{[H^+]^2\ (1)}{(1)^2\ (1)}$$

$$0.108 = 0 - \frac{0.0592}{2}\log\ [H^+]^2$$

$-8.407 = \log\ [H^+]^2$ or $-8.407 = 2\log\ [H^+]$

$[H^+]^2 = 10^{-8.407}$ | $-4.204 = \log\ [H^+]$

$[H^+]^2 = 2.23 \times 10^{-4}$ | $[H^+] = 10^{-4.204}$

$[H^+] = 1.5 \times 10^{-2}$ | $[H^+] = 1.5 \times 10^{-2}$

$pH = 1.82$

61. $Cu \rightarrow Cu^{2+} + 2\ e^-$ $E°_{anode} = 0.340\ V$

$Ag^+ + e^- \rightarrow Ag$ $E°_{cathode} = 0.800\ V$

$Cu(s) + 2\ Ag^+ \rightarrow Cu^{2+} + 2\ Ag(s)$

$E°_{cell} = E°_{cathode} - E°_{anode} = 0.800\ V - 0.340\ V = 0.460\ V$

$n = 2$

$$Q = \frac{[Cu^{2+}]}{[Ag^+]^2}$$

$$E_{cell} = E°_{cell} - \frac{0.0592}{n}\log\frac{[Cu^{2+}]}{[Ag^+]^2}$$

The higher cell voltage will be the cell with the lower ratio $[Cu^{2+}]/[Ag^+]^2$.

(a) $\frac{1.25}{(0.55)^2} = 4.13$ $E_{cell} = 0.445$

(b) $\frac{0.12}{(0.60)^2} = 0.33$ $E_{cell} = 0.476$

In this case the $[Ag^+]$ is about the same. So the lower value of $[Cu^{2+}]$ in cell (b) means a greater tendency for oxidation to occur there and the higher cell voltage.

63. $Zn(s) \rightarrow Zn^{2+} + 2\ e^-$ $E°_{anode} = -0.763$ V
$Cl_2(g) + 2\ e^- \rightarrow 2\ Cl^-$ $E°_{cathode} = 1.358$ V
$Zn(s) + Cl_2(g) \rightarrow Zn^{2+} + 2\ Cl^-$
$E°_{cell} = E°_{cathode} - E°_{anode} = 1.358\ V - (-0.763\ V) = 2.121\ V$

65. anode: $Zn(s) + 2\ OH^- \rightarrow ZnO(s) + H_2O + 2\ e^-$
cathode: $Ag_2O + H_2O + 2\ e^- \rightarrow 2\ Ag + 2\ OH^-$
$Zn(s) + Ag_2O(s) \rightarrow ZnO(s) + 2\ Ag(s)$

67. Oxygen is the oxidizing agent required to oxidize Fe (s) to Fe^{2+} and then to Fe^{3+}. Water is a reactant in the reduction half-reaction, in which O_2 (g) is reduced to OH^- (aq). Water is also a reactant in the conversion of $Fe(OH)_2$ to $Fe_2O_3 \cdot xH_2O$ (rust). The electrolyte completes the electrical circuit between the cathodic and anodic areas.

69. A sacrificial anode is used up—that is, the anode metal is oxidized to metal ions at the same time that a cathodic half-reaction occurs on the protected metal.

71. Zinc protects the iron from corrosion. Zinc is oxidized instead of the iron. The faint white precipitate is zinc ferricyanide.

73. (a) anode $Ni(s) \rightarrow Ni^{2+} + 2\ e^-$
cathode $Ni^{2+} + 2\ e^- \rightarrow Ni(s)$
$Ni(s,anode) \rightarrow Ni(s,cathode)$
This is similar to the electrorefining of copper.
(b) anode $Ni(s) \rightarrow Ni^{2+} + 2\ e^-$
cathode $Ni^{2+} + 2\ e^- \rightarrow Ni(s)$
$Ni(s,anode) \rightarrow Ni(s,cathode)$
(c) anode $2\ H_2O \rightarrow 4\ H^+ + O_2 + 4\ e^-$
cathode $Ni^{2+} + 2\ e^- \rightarrow Ni(s)$
$2\ H_2O + 2\ Ni^{2+} \rightarrow 2\ Ni(s) + 4\ H^+ + O_2(g)$

75. (a) anode $2\ Cl^- \rightarrow Cl_2 + 2\ e^-$ $E°_{anode} = 1.358$ V
cathode $Ba^{2+} + 2\ e^- \rightarrow Ba(l)$ $E°_{cathode} = -2.92$ V
$BaCl_2(l) \xrightarrow{electrolysis} Ba(l) + Cl_2(g)$
probable products Ba(l) and $Cl_2(g)$
$E°_{cell}$ (voltaic) $= E°_{cathode} - E°_{anode} = -2.92\ V - 1.358\ V = -4.28\ V$
$E°_{cell}$ (electrolytic) ≥ 4.28 V
(b) anode $2\ Br^- \rightarrow Br_2(l) + 2\ e^-$ $E°_{anode} = 1.065$ V
cathode $2\ H^+ + 2\ e^- \rightarrow H_2(g)$ $E°_{cathode} = 0.00$ V
$2\ Br^- + 2\ H^+ \rightarrow Br_2(l) + H_2(g)$

probable products $Br_2(l)$ and $H_2(g)$

$E°_{cell}$ (voltaic) = $E°_{cathode} - E°_{anode}$ = 0.00 V – 1.065 V = – 1.065 V

$E°_{cell}$ (electrolytic) ≥ 1.065 V

(c) anode $\quad 2\ H_2O \rightarrow 4\ H^+ + O_2(g) + 4\ e^-$ $\quad E°_{anode}$ = 1.229 V

cathode $\quad \{2\ H_2O + 2\ e^- \rightarrow H_2(g) + 2\ OH^-\} \times 2$ $\quad E°_{cathode}$ = –0.828 V

net $\quad 2\ H_2O \rightarrow 2\ H_2(g) + O_2(g)$

probable products $O_2(g)$ and 2 $H_2(g)$

$E°_{cell}$ (voltaic) = $E°_{cathode} - E°_{anode}$ = – 0.828 V – 1.229 V = – 2.057 V

$E°_{cell}$ (electrolytic) ≥ 2.057 V

77. $Ag^+ + e^- \rightarrow Ag(s)$

$$? \text{ g Ag} = 1.73\ \text{A} \times \frac{\text{C}}{\text{A s}} \times 2.05\ \text{h} \times \frac{3600\ \text{s}}{\text{h}} \times \frac{\text{mol e}^-}{96485\ \text{C}} \times \frac{\text{mol Ag}}{\text{mol e}^-} \times \frac{107.9\ \text{g Ag}}{\text{mol Ag}}$$

$$= 14.3\ \text{g Ag}$$

79. $Cu^{2+} + 2\ e^- \rightarrow Cu$

$$? \text{ C} = 25.0\ \text{g Cu} \times \frac{\text{mol Cu}}{63.55\ \text{g Cu}} \times \frac{2\ \text{mol e}^-}{\text{mol Cu}} \times \frac{96485\ \text{C}}{\text{mol e}^-} = 7.59 \times 10^4\ \text{C}$$

81. Na (s) does not electrodeposit from $NaNO_3$ (aq). Of the remaining solutions, $AgNO_3$ (aq) yields the greatest number of moles of deposit. One mole of silver is formed for every mole of electrons ($Ag^+ + e^- \rightarrow Ag$), whereas only one-half mole of copper and zinc is formed ($M^{2+} + 2\ e^- \rightarrow M$). Also, given that the atomic weight of Ag is greater than those of Cu and Zn, $AgNO_3$(aq) (solution c) yields the greatest mass of metal deposit.

$$1.00\ \text{A} \times 1\ \text{h} \times \frac{3600\text{s}}{\text{h}} \times \frac{\text{C}}{\text{A s}} \times \frac{\text{mol e}^-}{96485\ \text{C}} \times \frac{\text{mol M}}{n\ \text{mol e}^-} \times \frac{\text{molar mass g}}{\text{mol M}} =$$

Additional Problems

83. $P_4 + 12\ H^+ + 12\ e^- \rightarrow 4\ PH_3$ $\qquad P_4 + 16\ H_2O \rightarrow 4\ H_3PO_4 + 20\ H^+ + 20\ e^-$

$5\ P_4 + 3\ P_4 + 48\ H_2O \rightarrow 20\ PH_3 + 12\ H_3PO_4$

$2\ P_4 + 12\ H_2O \rightarrow 5\ PH_3 + 3\ H_3PO_4$

85. (a) $B_2Cl_4 + 4\ H_2O \rightarrow 2\ BO_2^- + 4\ Cl^- + 8\ H^+ + 2\ e^-$ $\qquad 2\ e^- + 2\ H^+ \rightarrow H_2$

$B_2Cl_4 + 4\ H_2O \rightarrow 2\ BO_2^- + 4\ Cl^- + 6\ H^+ + H_2$

$B_2Cl_4 + 6\ OH^- \rightarrow 2\ BO_2^- + 4\ Cl^- + 2\ H_2O + H_2$

(b) $CH_3CH_2ONO_2 + 6\ H^+ + 6\ e^- \rightarrow H_2O + CH_3CH_2OH + NH_2OH$

$Sn \rightarrow Sn^{2+} + 2\ e^-$

$CH_3CH_2ONO_2 + 3\ Sn + 6\ H^+ \rightarrow CH_3CH_2OH + NH_2OH + 3\ Sn^{2+} + H_2O$

(c) $F_5SeOF + 3\ H_2O + 2\ e^- \rightarrow SeO_4^{2-} + 6\ F^- + 6\ H^+$

$2\ H_2O \rightarrow 4\ H^+ + O_2 + 4\ e^-$

$2\ F_5SeOF + 8\ H_2O \rightarrow 2\ SeO_4^{2-} + 12\ F^- + 16\ H^+ + O_2$

$2\ F_5SeOF + 16\ OH^- \rightarrow 2\ SeO_4^{2-} + 12\ F^- + 8\ H_2O + O_2$

(d) $As_2S_3 + 20\ H_2O \rightarrow 2\ AsO_4^{3-} + 3\ SO_4^{2-} + 40\ H^+ + 28\ e^-$

$H_2O_2 + 2\ H^+ + 2\ e^- \rightarrow 2\ H_2O$

$As_2S_3 + 14\ H_2O_2 \rightarrow 2\ AsO_4^{3-} + 3\ SO_4^{2-} + 12\ H^+ + 8\ H_2O$

$As_2S_3 + 14\ H_2O_2 + 12\ OH^- \rightarrow 2\ AsO_4^{3-} + 3\ SO_4^{2-} + 20\ H_2O$

(e) $XeF_6 + 6\ e^- \rightarrow Xe + 6\ F^-$ $\quad$ $XeF_6 + 6\ H_2O \rightarrow XeO_6^{4-} + 6\ F^- + 12\ H^+ + 2\ e^-$

$2\ H_2O \rightarrow O_2 + 4\ H^+ + 4\ e^-$

$2\ XeF_6 + 8\ H_2O \rightarrow Xe + XeO_6^{4-} + 12\ F^- + O_2 + 16\ H^+$

$2\ XeF_6 + 16\ OH^- \rightarrow Xe + XeO_6^{4-} + 12\ F^- + O_2 + 8\ H_2O$

87. Sodium metal is very reactive with water. The sodium will become Na^+ while water is reduced to OH^- and $H_2(g)$. Instead of just establishing a half-reaction electrode equilibrium, a piece of sodium enters into a complete redox reaction with water.

89. $Ag^+ + e^- \rightarrow Ag\ (s)$

$O^{2-} + 4\ e^- \rightarrow O_2$

$P_{total} = P_{O_2} + VP_{H_2O}$

$P_{O_2} = 761.5\ \text{mmHg} - 17.5\ \text{mmHg}$

$P_{O_2} = 744.0\ \text{mmHg}$

$$?\ \text{mL } O_2 = 1.02\ \text{g Ag} \times \frac{\text{mol Ag}}{107.9\ \text{g Ag}} \times \frac{\text{mol e}^-}{\text{mol Ag}} \times \frac{\text{mol } O_2}{4\ \text{mol e}^-} \times \frac{\frac{62.4\ \text{L mmHg}}{\text{K mol}} \times 293.2\ \text{K}}{744.0\ \text{mmHg}} \times \frac{\text{mL}}{10^{-3}\ \text{L}} = 58.1\ \text{mL } O_2$$

91. $MnO_2(s) + 4\ H^+ + 2\ e^- \rightarrow Mn^{2+} + 2\ H_2O$ $\quad$ $E°_{cathode} = 1.23\ \text{V}$

$2\ Cl^- \rightarrow Cl_2 + 2\ e^-$ $\quad$ $E°_{anode} = 1.358\ \text{V}$

$MnO_2(s) + 4\ H^+ + 2\ Cl^- \rightarrow Cl_2(g) + Mn^{2+} + 2\ H_2O$

$E°_{cell} = E°_{cathode} - E°_{anode} = 1.23\ \text{V} - 1.358\ \text{V} = -0.13\ \text{V}$

The Nernst equation will show that if the concentration of Mn^{2+} is small and the concentrations of Cl^- and H^+ are high, it can make the log term large enough to overcome the negative $E°_{cell}$. More importantly, the escape of $Cl_2(g)$ will push the reaction to the right.

93. $H_2(g) \rightarrow 2\ H^+ + 2\ e^-$ $\qquad E°_{anode} = 0.00\ V$

$2\ H^+ + 2\ e^- \rightarrow H_2(g)$ $\qquad E°_{cathode} = 0.00\ V$

$$E_{cell} = E°_{cell} - \frac{0.0592}{2}\log\frac{[H^+]^2}{[H^+]^2}$$

$$E_{cell} = 0.00 - \frac{0.0592}{2}\log\frac{[H^+]^2 \text{ from acid}}{(0.010)^2}$$

$$K_a = 1.8 \times 10^{-5} = \frac{[H^+]^2}{1\times10^{-4} - [H^+]}$$

$$[H^+]^2 + 1.8 \times 10^{-5}\,[H^+] - 1.8 \times 10^{-9} = 0$$

$$[H^+] = \frac{-1.8\times10^{-5} \pm \sqrt{(1.8\times10^{-5})^2 - 4\times(-1.8\times10^{-9})}}{2}$$

$$[H^+] = \frac{-1.8\times10^{-5} \pm 8.7\times10^{-5}}{2}$$

$$[H^+] = 3.4 \times 10^{-5}$$

$$E_{cell} = 0.00 - \frac{0.0592}{2}\log\frac{(3.4\times10^{-5})^2}{(0.010)^2}$$

$$E_{cell} = 0.15\ V$$

95. $Cu\ (s) \rightarrow Cu^{2+} + 2\ e^-$ $\qquad E°_{anode} = 0.340\ V$

$Ag^+ + e^- \rightarrow Ag$ $\qquad E°_{cathode} = 0.800\ V$

$E°cell = E°_{cathode} - E°_{anode} = 0.800\ V - 0.340\ V = 0.460\ V$

$Ag_2CrO_4 \rightarrow 2\ Ag^+ + CrO_4^{2-}$

$\qquad\qquad\qquad 2s \qquad s$

$K_{sp} = 1.1 \times 10^{-12} = 4\ s^3$

$s = 6.50 \times 10^{-5}$

$[Ag^+] = 2\ s = 1.30 \times 10^{-4}\ M$

$$E_{cell} = 0.460\ V - \frac{0.0592}{2}\log\frac{[Cu^{2+}]}{[Ag^+]^2}$$

$$E_{cell} = 0.460\ V - \frac{0.0592}{2}\log\frac{(0.10)}{(1.30 \times 10^{-4})^2} = 0.260\ V$$

97. $Cu^{2+} + 2\ e^- \rightarrow Cu\ (s)$ $\qquad$ cathode

$Cu\ (s) \rightarrow Cu^{2+} + 2\ e^-$ $\qquad$ anode

$$E_{cell} = E°_{cell} - \frac{0.0592}{2}\log\frac{[Cu^{2+}]\text{ anode}}{[Cu^{2+}]\text{ cathode}}$$

$$E_{cell} = 0.00 - \frac{0.0592}{2}\log\frac{0.025}{1.50}$$

$E_{cell} = 0.0526$ V

As the cell operates, the concentration of the 1.50 M solution will decrease, and that of the 0.025 M solution will increase, causing the voltage to decrease. When the concentrations of the solutions become equal (0.76 M in each half-cell), E_{cell} will fall to zero.

99. $Cu^{2+} + 2e^- \rightarrow Cu$ $\quad E°_{cathode} = 0.340$ V

$Fe \rightarrow 2e^- + Fe^{2+}$ $\quad E°_{anode} = -0.440$ V

$E°_{cell} = E°_{cathode} - E°_{anode} = 0.340\text{ V} - (-0.440)\text{ V} = 0.780\text{ V}$

When the asbestos insulation wore away, voltaic cells could form, having iron as a sacrificial anode protecting the underlying copper cathode.

101. $Cu^{2+} + 2\,e- \rightarrow Cu$

$$?\text{ mol } Cu^{2+}_{initial} = 250.0\text{ mL} \times 0.1000\text{ M} \times \frac{10^{-3}\text{ L}}{\text{mL}} = 2.500 \times 10^{-2}\text{ mol } Cu^{2+}$$

$$\text{mol } Cu^{2+}\text{ electrolyzed} = 3.512\text{ A} \times 1368\text{ s} \times \frac{\text{C}}{\text{A s}} \times \frac{\text{mol e}^-}{96485\text{C}} \times \frac{\text{mol Cu}^{2+}}{2\text{ mol e}^-}$$

$$= 2.490 \times 10^{-2}\text{ mol } Cu^{2+}$$

$$\text{mol } Cu^{2+}\text{ in solution} = 2.500 \times 10^{-2}\text{ mol} - 2.490 \times 10^{-2}\text{ mol} = 0.010 \times 10^{-2}\text{ mol}$$

$$= 1.0 \times 10^{-4}\text{ mol}$$

$$[Cu^{2+}]_{init} = \frac{1.0 \text{ x } 10^{-4}\text{ mol}}{250.0\text{ mL}} \times \frac{\text{ml}}{10^{-3}\text{ L}} = 4 \times 10^{-4}\text{ M}$$

To simplify the calculation, it is assumed that the formation reaction goes to completion. Then $[Cu^{2+}]$ formed in the reverse reaction is determined to see if $[Cu^{2+}]$ is indeed negligible.

	Cu^{2+}	+	$4\ NH_3$	$\rightleftharpoons$	$[Cu(NH_3)_4]^{2+}$
Init			0.10 M		4×10^{-4} M
Change	$+X$		no change		$-X$
Eq	X		0.10 M		$4 \times 10^{-4}\text{ M} - X$

$$K_f = 1.1 \times 10^{13} = \frac{4\times10^{-4} - X}{(0.10)^4 \times X}$$

Assume $X << 4.0 \times 10^{-4}$

$$X = \frac{4.0\times10^{14}}{(0.10)^4 \times 1.1\times10^{13}} = 3.6 \times 10^{-13}$$

$[Cu(NH_3)_4]^{2+} = 4.0 \times 10^{-4} - 3.6 \times 10^{-13} = 4.0 \times 10^{-4}$ M

Yes, the blue color will be seen.

103. $Hg^{2+} + 2\,e^- \rightarrow Hg\,(l)$ $\qquad E^\circ_{cathode} = 0.854\text{ V}$

$[Fe^{2+} \rightarrow Fe^{3+} + e^-] \times 2$ $\qquad E^\circ_{anode} = -0.771\text{ V}$

$Hg^{2+} + 2\,Fe^{2+} \rightarrow 2\,Fe^{3+} + Hg\,(l)$

$E^\circ_{cell} = E^\circ_{cathode} - E^\circ_{anode} = 0.854\text{ V} - (-0.771\text{ V}) = 0.083\text{ V}$

At $E_{cell} = 0$

$$0 = 0.083 - \frac{0.0592}{2}\log\frac{[Fe^{3+}]^2}{[Fe^{2+}]^2[Hg^{2+}]}$$

$$6.4 \times 10^2 = K = \frac{[Fe^{3+}]^2}{[Fe^{2+}]^2[Hg^{2+}]}$$

	Hg^{2+}	+	$2\,Fe^{2+}$	$\rightarrow$	$2\,Fe^{3+}$	+	Hg(l)
Init	0.250		0.180		0.210		
Change	$-X$		$-2X$		$+2X$		
Eq	$0.250 - X$		$0.180 - 2X$		$0.210 + 2X$		

$$6.4 \times 10^2 = \frac{(0.210 + 2X)^2}{(0.250 - X)(0.180 - 2X)^2}$$

The maximum value of X must be less than 0.090.

Assume $X = 0.050$ $\quad \dfrac{(0.310)^2}{(0.200)(0.080)^2} = 75 < 640$

Assume $X = 0.060$ $\quad \dfrac{(0.330)^2}{(0.190)(0.060)^2} = 159 < 640$

Assume $X = 0.080$ $\quad \dfrac{(0.370)^2}{(0.170)(0.020)^2} = 2013 > 640$

Assume $X = 0.070$ $\quad \dfrac{(0.350)^2}{(0.180)(0.040)^2} = 425 < 640$

Assume $X = 0.075$ $\quad \dfrac{(0.360)^2}{(0.175)(0.030)^2} = 8227 > 640$

Assume $X = 0.071$ $\quad \dfrac{(0.352)^2}{(0.179)(0.038)^2} = 479 < 640$

Assume $X = 0.072$ $\quad \dfrac{(0.354)^2}{(0.178)(0.036)^2} = 543 < 640$

Assume $X = 0.073$ $\quad \dfrac{(0.356)^2}{(0.177)(0.034)^2} = 619 < 640$

$X = 0.073$

$[Hg^{2+}] = 0.250 - X = 0.177$ M

$[Fe^{2+}] = 0.180 - 2X = 0.034$ M

$[Fe^{3+}] = 0.210 + 2X = 0.356$ M

105. $Cu^{2+} + 2e^- \rightarrow Cu$ $\quad E^\circ_{cathode} = 0.340$

$H_2 \rightarrow 2e^- + 2H^+$ $\quad E^\circ_{anode} = 0.00$

$$E = E^\circ - \frac{0.0592}{2}\log\frac{[H^+]^2}{P_{H_2}[Cu^{2+}]}$$

Plot potential (E_{cell}) versus log $\frac{1}{[Cu^{2+}]}$. The slope will be $-\frac{0.0592\text{ V}}{2}$.

For silver, plot potential (E_{cell}) versus log $\frac{1}{[Ag^+]}$. The slope will be –0.0592. The difference in the two plots results from n = 2 mol e⁻ at the copper electrode and n = 1mol e⁻ at the silver electrode.

107. $? \text{ g Pb} = 30 \text{ s} \times 80 \text{ A} \times \frac{\text{C}}{\text{A s}} \times \frac{\text{mol e}^-}{96485\text{C}} \times \frac{\text{mol Pb}}{2\text{ mol e}^-} \times \frac{207.2\text{ g}}{\text{mol}} = 2.6 \text{ g Pb}$

109. $\text{g cyclohexene} = 9.65 \text{ mA} \times \frac{10^{-3}\text{ A}}{\text{mA}} \times 235 \text{ s} \times \frac{\text{C}}{\text{A s}} \times \frac{\text{mol e}^-}{96485\text{C}} \times \frac{\text{mol Br}_2}{2\text{ mol e}^-}$

$$\times \frac{\text{mole C}_6\text{H}_{10}}{\text{mole Br}_2} \times \frac{82.14\text{ g}}{\text{mole C}_6\text{H}_{10}} = 9.65 \times 10^{-4}\text{ g}$$

Apply Your Knowledge

111. anode $\quad 2 H_2O \rightarrow 4 H^+ + O_2 + 4 e^-$ $\quad E^\circ(\text{anode}) = -1.229$ V

cathode $\quad \underline{\{2 H_2O + 2 e^- \rightarrow H_2 + 2 OH^-\} \times 2}$ $\quad E^\circ(\text{cathode}) = -0.828$ V

net $\quad 2 H_2O \rightarrow 2 H_2 (g) + O_2 (g)$

$E^\circ_{cell} = E^\circ(\text{cathode}) - E^\circ(\text{anode}) = -0.828 \text{ V} - (-1.229 \text{ V}) = -2.057 \text{ V}$

(a) Left $\quad 2 H_2O \rightarrow 4 H^+ + O_2 + 4 e^-$ $\quad$ anode

Right $\quad 2 H_2O + 2 e^- \rightarrow H_2 + 2 OH^-$ $\quad$ cathode

Pink will appear around the cathode and color the solution.

(b) It requires more than 2.06 V to make the reaction occur. A 1.5-V battery will not work.

(c) When the two solutions are mixed, the result is simply Na_2SO_4 (aq) at its characteristic pH (about 7). The H^+ produced at the anode neutralizes the OH^- produced at the cathode. The neutral solution is colorless.

(d) The H^+ (aq) produced in the anode compartment and the OH^- (aq) in the cathode compartment are produced in exactly equal molar amounts, regardless of the concentration of Na_2SO_4, the electrolysis time, or the current used.

(e) $? \text{ moles OH}^- = \left[\left(8\,\text{min} \times \frac{60\,\text{s}}{\text{min}}\right) + 22\,\text{s}\right] \times 23.2\,\text{mA} \times \frac{10^{-3}\,\text{A}}{\text{mA}} \times \frac{\text{C}}{\text{A s}} \times \frac{\text{mol e}^-}{96485\,\text{C}}$

$$\times \frac{2\,\text{mol OH}^-}{2\,\text{mol e}^-} = 1.21 \times 10^{-4}\ \text{moles OH}^-$$

$\text{moles}_{\text{acid}} = \text{moles}_{\text{base}}$ at color change.

$$[\text{HCl}] = [\text{H}^+] = \frac{1.21 \times 10^{-4}\ \text{moles}}{10\ \text{mL}} \times \frac{\text{mL}}{10^{-3}\ \text{L}} = 1.21 \times 10^{-2}\ \text{M}$$

113. Cell half-reactions:

anode: $H_2(g) + 2\,OH^-(aq) \rightarrow 2\,H_2O(l) + 2e^-$

cathode: $Ag_2O(s) + H_2O(l) + 2\,e^- \rightarrow 2\,Ag(s) + 2OH^-(aq)$

overall: $H_2(g) + Ag_2O(s) \rightarrow 2\,Ag(s) + H_2O(l)$

$$\Delta G^\circ = -nFE^\circ_{\text{cell}} = -2\ \text{mol e}^- \times \frac{96485\,\text{C}}{\text{mol e}^-} \times 1.172\ \text{V} \times \frac{\text{kJ}}{10^3\ \text{J}} = -226.2\ \text{kJ}$$

Combine the overall cell reaction with $H_2O(l) \rightarrow H_2(g) + \frac{1}{2}\,O_2\,(g)$ for which $\Delta G^\circ = -\Delta G^\circ_f[H_2O(l)] = -(-237.2) = 237.2$ kJ to obtain:

$2Ag_2O\,(s) \rightarrow 4Ag\,(s) + O_2\,(g)$ $\qquad \Delta G^\circ = -226.2\ \text{kJ} + 237.2\ \text{kJ} = 11.0\ \text{kJ}$

$$\ln K_{eq} = \frac{-\Delta G^\circ}{RT} = \frac{-11{,}000\,\text{J/mol}}{\frac{8.3145\,\text{J}}{\text{mol K}} \times 298\,\text{K}} = -4.44$$

$$e^{\ln K_{eq}} = K_{eq} = e^{-4.44} = 0.012$$

$$K_{eq} = (P_{O_2})^{1/2}$$

$$P_{O_2} = (K_{eq})^2 = (0.012)^2 = 1.4 \times 10^{-4}\ \text{atm} \times \frac{760\ \text{mmHg}}{\text{atm}} = 0.11\ \text{mmHg}$$

Because P_{O_2} in equilibrium with $Ag_2O(s)$ and $Ag(s)$ at 298 K is much less than P_{O_2} in air (0.21 atm), Ag(s) oxides spontaneously in air. That is, the reverse reaction is favored.

Chapter 19

Nuclear Chemistry

Exercises

19.1A (a) $^{212}_{86}\text{Rn} \rightarrow {}^{4}_{2}\text{He} + ?$

$212 = 4 + A \quad A = 208$

$86 = 2 + Z \quad Z = 84$ Polonium

$^{212}_{86}\text{Rn} \rightarrow {}^{4}_{2}\text{He} + {}^{208}_{84}\text{Po}$

(b) $^{37}_{18}\text{Ar} + {}^{0}_{-1}\text{e} \rightarrow ?$

$37 + 0 = A \quad A = 37$

$18 - 1 = Z \quad Z = 17$

$^{37}_{18}\text{Ar} + {}^{0}_{-1}\text{e} \rightarrow {}^{37}_{17}\text{Cl}$

(c) $^{60m}_{27}\text{Co} \rightarrow \gamma + {}^{60}_{27}\text{Co}$

19.1B (a) The sum of the mass numbers of $^{4}_{2}\text{He}$ and $^{214}_{82}\text{Pb}$ equals the mass number of the starting nucleus (218). The sum of the atomic numbers of He (2) and Pb(82) equals the atomic number of the parent nucleus (84): $^{218}_{84}\text{Po} \rightarrow {}^{214}_{82}\text{Pb} + {}^{4}_{2}\text{He}$.

(b) The sum of the mass numbers of $^{36}_{16}\text{S}$ and $^{0}_{1}\text{e}$ equals the mass number of the starting nucleus (36). The sum of the atomic numbers of S (16) and a positron (+1) equals the parent nucleus's atomic number (17): $^{36}_{17}\text{Cl} \rightarrow {}^{36}_{16}\text{S} + {}^{0}_{1}\text{e}$

19.2A $\ln \dfrac{A_t}{A_o} = -\lambda t \qquad t_{1/2} = 14.3\text{d} = \dfrac{0.693}{\lambda}$

$$\lambda = \frac{0.693}{t_{1/2}} = \frac{0.693}{14.3\,\text{d}} = 0.0485\ \text{d}^{-1}$$

$$\ln \frac{A_t}{2.50x10^{10}\ \frac{\text{atom}}{\text{sec}}} = -\,0.0485\ \text{d}^{-1} \times 365\text{d} = -17.7$$

$$e^{\ln \frac{A_t}{2.50x10^{10}\ \frac{\text{atom}}{\text{sec}}}} = \frac{A_t}{2.50x10^{10}\ \frac{\text{atom}}{\text{sec}}} = e^{-17.7} = 2.08 \times 10^{-8}$$

$$A_t = 520\ \frac{\text{atoms}}{\text{sec}}$$

19.2B $\ln \frac{A_t}{A_o} = -\lambda t \qquad \lambda = \frac{0.693}{t_{1/2}} = \frac{0.693}{2.411 \times 10^4 \text{ y}} = \frac{2.874 \times 10^{-5}}{\text{y}}$

$\ln \frac{1}{100} = \frac{-2.874 \times 10^{-5}}{\text{y}} \; t = -4.61$

$t = 1.60 \times 10^5$ y

19.3 (b). There are about 10 times more ^{24}Na atoms than ^{11}C atoms, but ^{11}C atoms disintegrate about 50 times as fast. There are about $10^4 - 10^5$ times as many ^{238}U atoms as ^{11}C atoms, but $t_{1/2}$ of ^{238}U exceeds that of ^{11}C by a factor of more than 10^9.

An alternate approach to the problem using activity follows.

$A = \lambda N \qquad \lambda = \frac{0.693}{t_{\frac{1}{2}}}$

(a) $^{24}\text{Na } A = \frac{0.693}{14.659 \text{ h}} \times \frac{\text{h}}{60 \text{ min}} \times 1\ \mu\text{mol} \times \frac{1 \times 10^{-6} \text{ mol}}{\mu\text{mol}} \times \frac{6.02 \times 10^{23} \text{ atoms}}{\text{mol}}$

(b) $^{11}\text{C } A = \frac{0.693}{20.39 \text{ m}} \times 1 \times 10^{-6} \text{ g} \times \frac{1 \text{ mol}}{11 \text{ g}} \times \frac{6.02 \times 10^{23} \text{ atoms}}{\text{mol}}$

(c) $^{238}\text{U } A = \frac{0.693}{4.51 \times 10^9 \text{ y}} \times \frac{\text{y}}{365 \text{ d}} \times \frac{\text{d}}{24 \text{ h}} \times \frac{\text{h}}{60 \text{ min}} \times 1 \text{ g} \times \frac{1 \text{ mol}}{238 \text{ g}} \times \frac{6.02 \times 10^{23} \text{ atoms}}{\text{mol}}$

By ignoring the factors that are the same, such as 0.693, 10^{-6}, and 6.02×10^{23}, it is evident that C-11 has the greatest rate of decay because (b) $\frac{1}{20 \times 11}$ is more than (a) $\frac{1}{15 \times 60}$ and much more than (c) $\frac{1}{10^9 \times 10^2 \times 10^1 \times 10^1 \times 10^2}$.

Actual ^{24}Na $\quad A = \frac{4.7 \times 10^{14} \text{ atoms}}{\text{min}}$

^{11}C $\quad A = \frac{1.9 \times 10^{15} \text{ atoms}}{\text{min}}$

^{238}U $\quad A = \frac{7.4 \times 10^{5} \text{ atoms}}{\text{min}}$

19.4A $\ln \frac{A_t}{A_o} = -\lambda t = \ln \frac{A_t}{\frac{7.2 \text{ dis}}{\text{min}}} = -(1.21 \times 10^{-4} \text{ y}) \times 1500 \text{ y} = -0.182$

$e^{\ln \frac{A_t}{A_o}} = \frac{A_t}{A_o} = \frac{A_t}{\frac{7.2 \text{ dis}}{\text{min}}} = e^{-0.182} = 0.834$

$A_t = \frac{6.0 \text{ dis}}{\text{min}}$

19.4B The activity falls to one-half its initial value in one half-life period—$t_{1/2}$ = 12.26 y. The brandy is 12.26 y old, not 25 years. The claimed age is not authentic.

19.5A The sum of the mass numbers of $^{35}_{17}Cl$ and $^{1}_{0}n$ equals the sum of the mass numbers of $^{35}_{16}S$ and the product (1). The sum of the atomic numbers of Cl (17) and ^{1}n (0) equals the sum of the atomic number of S (16) and the product (1). The product must be a hydrogen nucleus. $^{35}_{17}Cl + ^{1}_{0}n \rightarrow ^{35}_{16}S + ^{1}_{1}H$

19.5B $^{249}_{98}Cf + ^{15}_{7}N \rightarrow ^{260}_{105}Db + ?^{1}_{0}n$

$249 + 15 = 260 + A \quad A = 4$

$98 + 7 = 105$

$^{249}_{98}Cf + ^{15}_{7}N \rightarrow ^{260}_{105}Db + 4\,^{1}_{0}n$

19.6 $^{40}_{20}Ca$ is not radioactive because it is an even nuclide, and 20 is a magic number. $^{48}_{20}Ca$ is not radioactive because it is an even-even nuclide, and 20 and 28 are magic numbers. $^{39}_{20}Ca$ is radioactive because it is an odd-even nuclide that is below the ratio of stability. $\frac{n}{p+} = 0.95 < 1.0.$

19.7 To bring $^{84}_{39}Y$ closer to the belt of stability, the ratio of neutrons to protons needs to increase ($\frac{45}{39} = 1.15$ stable Y is $\frac{50}{39} = 1.28$). This means either positron emission or electron capture. The resultant isotope would be $^{84}_{38}Sr$ ($\frac{46}{38} = 1.21$). The product is barely inside the belt of stability and actually is stable.

19.8A $^{237}_{93}Np \rightarrow ^{4}_{2}He + ^{233}_{91}Pa$

$\Delta E = 232.9901\ u + 4.0015\ u - 236.9970\ u = -0.0054\ u$

$?MeV = -0.0054\ u \times \frac{931.5 MeV}{u} = -5.0\ MeV$

19.8B Mass e$^-$ $= 9.1093897 \times 10^{-31}\ kg \times \frac{u}{1.6605402 \times 10^{-27}\ kg} = 5.485799 \times 10^{-4}\ u$

nuclear masses

$^{27}_{13}Al$ $\quad 26.9815\ u - 13 \times 5.485799 \times 10^{-4}\ u = 26.9744\ u$

$^{4}_{2}He$ $\quad 4.0026\ u - 2 \times 5.485799 \times 10^{-4}\ u = 4.0015\ u$

$^{30}_{15}P$ $\quad 29.9783\text{ u} - 15 \times 5.485799 \times 10^{-4}\text{ u} = 29.9701\text{ u}$

$\Delta E = 29.9701\text{ u} + 1.0087\text{ u} - 26.9744\text{ u} - 4.0015\text{ u} = 0.0029\text{ u}$

$$\Delta E = 0.0029\text{ u} \times \frac{931.5\text{ MeV}}{\text{u}} = 2.7\text{ MeV}$$

ΔE from atomic masses

$\Delta E = 29.9783\text{ u} + 1.0087\text{ u} - 26.9815\text{ u} - 4.0026\text{ u} = 0.0029\text{ u}$

$$\Delta E = 0.0029\text{ u} \times \frac{931.5\text{ MeV}}{\text{u}} = 2.7\text{ MeV}$$

Both calculations are the same because the difference is the difference in masses of nuclei. Electrons are not involved in the reaction.

Review Question

1. An α is a helium nucleus with a mass of 4 u and a +2 charge. A β^- is an electron with a mass of 0.0055u and a -1 charge. A γ is energy having neither mass nor charge.
   γ has the greatest penetrating power through matter. α has the greatest ionizing power. β^- suffers the greatest deflection in a magnetic field.
   The main difference between a γ ray and an X-ray is that a γ ray is emitted from the nucleus while an X-ray is produced in the electron shell. The γ ray is usually more energetic.

5. In electron capture an electron is absorbed by the nucleus. This leaves a hole in the electron shell of the atom. An electron from a higher shell will fall to the lower shell to fill the hole. The energy lost by going to a lower energy shell is an X-ray.

7. $A = \lambda N$ Activity A is equal to the decay constant times the number of atoms.
   $\ln \frac{A}{A_0} = -\lambda t$ The activity decreases with time as the number of atoms decreases.
   The activity declines by an exponential decay curve.

10. Nuclear binding energy relates to the mass difference between a nuclide and the sum of the masses of protons and neutrons that form the nucleus of that nuclide. This mass difference is expressed as an energy.
    The nuclear binding energy divided by the number of protons and neutrons is the binding energy per nucleon.
    The binding energy per nucleon is at a maximum at about Z =26 and A = 56. $^{56}_{26}Fe$ has the greatest binding energy per nucleon.

12. Nuclear reactions are more energetic than chemical reactions because changing matter into energy produces much more energy than simply rearranging electrons in chemical bonds. Chemical reactions involve bond breakage and bond formation, which involve changes in electrostatic forces. Nuclear reactions involve changes in the much stronger nuclear forces that bind nucleons together.

17. Primary electrons are the electrons produced by the direct impact of ionizing radiation on atoms and molecules. Secondary electrons are those released by the impact of primary electrons on atoms and molecules.

Problems

21. (a) ${}^{32}_{16}S \rightarrow {}^{32}_{17}Cl + {}^{0}_{-1}e$

(b) ${}^{14}_{8}O \rightarrow {}^{14}_{7}N + {}^{0}_{+1}e$

(c) ${}^{238}_{92}U \rightarrow {}^{234}_{90}Th + {}^{4}_{2}He$

23. (a) ${}^{210}_{84}Po \rightarrow {}^{4}_{2}He + {}^{206}_{82}Pb$

(b) ${}^{59}_{29}Cu \rightarrow {}^{0}_{1}e + {}^{59}_{28}Ni$

(c) ${}^{212}_{82}Pb \rightarrow {}^{0}_{-1}e + {}^{212}_{83}Bi$

${}^{212}_{83}Bi \rightarrow {}^{4}_{2}He + {}^{208}_{81}Tl$

${}^{208}_{81}Tl \rightarrow {}^{0}_{-1}e + {}^{208}_{82}Pb$

25.

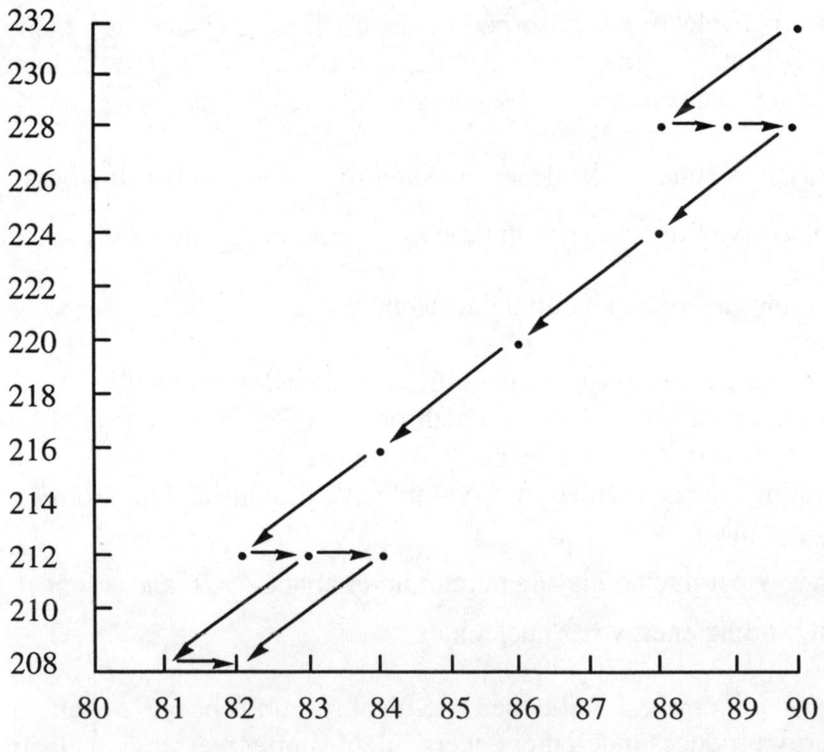

27. Chemical reactions are based on collision and a series of elementary steps that can lead to reactions of different orders. Radioactive decay is a nuclear process unrelated to chemical reactions. Although the decay may involve different types of

particles (α, β, and γ rays), all radioactive decay follows the same rate law, which happens to be first-order.

29. (a) $A = \lambda N = \dfrac{0.693}{t_{1/2}} N$

The shortest half-life would be the greatest activity. ^{15}O.

(b) 25% left means two half-lives. If two days is two half-lives, then one half-life is one day. ^{28}Mg is close to one day.

31. For the same mass, there are about four times as many atoms of ^{32}P as of ^{131}I and the decay constant (λ) of ^{32}P is about one-half that of ^{131}I. The activity is proportional to the number of atoms by the factor of the decay constant: $A = \lambda \cdot N$. The activity of ^{32}P is about twice that of ^{131}I ($4 \times ½ = 2$).

33. $A = \lambda N = \dfrac{0.693}{t_{1/2}} \times 1 \times 10^6 \text{ atoms} = \dfrac{0.693}{8.55 \text{ h}} \times \dfrac{\text{hr}}{3600 \text{ s}} \times 1 \times 10^6 \text{ atoms} = 22.5 \dfrac{\text{dis}}{\text{s}}$

35. $\ln \dfrac{118}{138} = -\lambda \times 20.0 \text{ d}$

$-0.157 = -\lambda \times 20.0 \text{ d}$

$\lambda = 7.83 \times 10^{-3}/\text{d}$

$t_{1/2} = \dfrac{0.693}{7.83 \text{ x } 10^{-3}/\text{d}} = 88.5 \text{ d}$

37. $1 \rightarrow \frac{1}{2} \rightarrow \frac{1}{4} \rightarrow \frac{1}{8} \rightarrow \frac{1}{16}$

50% 25% 12.5% 6.25%

8 + 8 + 8 + 8

More than 24 d and less than 32.

$\ln 0.1 = \dfrac{-0.693}{8.04 \text{ d}} \times t \qquad t = 26.7 \text{ d}$

The sample size does not matter.

39. $\ln \dfrac{A_t}{A_0} = -\lambda t = -\dfrac{0.693}{5730 \text{ y}} \times 2.80 \times 10^3 \text{ y} = -0.339$

$e^{\ln \frac{A_t}{A_0}} = \dfrac{A_t}{A_0} = \dfrac{A_t}{15 \frac{\text{dis}}{\text{min g}}} = e^{-0.339} = 0.71$

$A_t = 11 \dfrac{\text{dis}}{\text{min g}}$

41. (a) ${}^{10}_{5}B + {}^{1}_{0}n \rightarrow {}^{10}_{4}Be + {}^{1}_{1}H$

(b) $^{121}_{51}\text{Sb} + ^{1}_{1}\text{H} \rightarrow ^{121}_{52}\text{Te} + ^{1}_{0}\text{n}$

(c) $^{59}_{27}\text{Co} + ^{1}_{0}\text{n} \rightarrow ^{56}_{25}\text{Mn} + ^{4}_{2}\text{He}$

43. (a) $2\,^{2}_{1}\text{H} \rightarrow ^{3}_{2}\text{He} + ^{1}_{0}\text{n}$

(b) $^{241}_{95}\text{Am} + ^{4}_{2}\text{He} \rightarrow ^{243}_{97}\text{Bk} + 2\,^{1}_{0}\text{n}$

(c) $^{238}_{92}\text{U} + ^{4}_{2}\text{He} \rightarrow ^{239}_{94}\text{Pu} + 3\,^{1}_{0}\text{n}$

45. $^{99}_{43}\text{Tc} + ^{1}_{0}\text{n} \rightarrow ^{100}_{43}\text{Tc} \rightarrow ^{100}_{44}\text{Ru} + ^{0}_{-1}\text{e}$

47. mass of protons $1.0073 \times 26 = 26.1898$ u
mass of neutrons $1.0087 \times (56\text{-}26) = \underline{30.2610\text{ u}}$
56.4508 u
mass defect = 56.4508 - 55.92068 = 0.5301 u

49. The nucleus of $^{40}_{20}\text{Ca}$ has 20 protons and 20 neutrons. This is 10 times the mass of the two protons and two neutrons to form a $^{4}_{2}\text{He}$ nucleus.
Mass defect = $10 \times 4.0320\text{ u} - 39.95162\text{ u} = 0.368\text{ u}$

$$\frac{\text{binding energy}}{\text{nucleon}} = \frac{0.368\text{ u}}{40\text{ nucleons}} \times \frac{931.5\text{ MeV}}{1\text{ u}} = 8.57\text{ MeV/nucleon}$$

51. $^{224}_{88}\text{Ra} \rightarrow ^{220}_{80}\text{Rn} + ^{4}_{2}\text{He}$
energy = (219.9642 u + 4.00150 u - 223.9719 u)

$$= -0.0062\text{ u} \times \frac{931.5\text{ MeV}}{\text{u}} = -5.8\text{ MeV}$$

53. $^{229}\text{Ra} \rightarrow ^{0}_{1}\text{e} + ^{229}\text{Ac}$

$$\text{mass defect} = -1.14\text{ MeV} \times \frac{\text{u}}{931.5\text{ MeV}} = -0.00122\text{ u}$$

mass Ac -229 + 0.00055 u - 228.9866 u = – 0.00122 u
nuclear mass Ac -229 = 228.9848 u
atomic mass = 228.9848 u + 89 × 5.486 × 10^{-4} u = 229.0336 u

55. $$\text{energy} = (235.0439\text{ u} + 1.0087\text{ u} - 4.0026\text{ u} - 232.0382\text{ u}) \times \frac{931.5\text{ MeV}}{\text{u}}$$
$$= 11.0\text{ MeV}$$

57. β^- emission effectively changes a neutron into a proton and thus decreases the neutron/proton ratio. β^+ emission increases the neutron/proton ratio. The isotope

with the larger neutron/proton ratio should decay by β^- emission; this is $^{22}_{9}F$. $^{17}_{9}F$ decays by β^+ emission.

59. Cu–64 and Ga–70, isotopes with mass numbers closest to the weighted-average atomic mass do not occur naturally because they are odd-odd isotopes. Isotopes on each side of the weighted-average atomic mass are stable and average to the tabulated atomic mass.

61. Calcium-40 hass a doubly magic number and is in the region where nuclides with the same number of neutrons and protons are the most stable.

63. $$\Delta E = (138.8891\text{ u} + 94.8985\text{ u} + 2 \times (1.0087\text{ u}) - 234.9934\text{ u} - 1.0087\text{ u})$$
$$= -0.1971\text{u} \times \frac{931.5\text{ MeV}}{\text{u}} = -183.6\text{ MeV}$$

65. A short half-life nuclide will decay to a smaller number of radioactive atoms quickly and will no longer be a hazard. A long half-life has a low activity and is less of a hazard than a nuclide of intermediate half-life. The intermediate half-life nuclide will have a fairly strong activity and will last for a considerable length of time.

67. First determine the decay constant. $\lambda = \frac{0.693}{8.04\text{ d}} = 0.0862\text{ d}^{-1}$

Activity = $A = \lambda N$

Now determine the number of ^{131}I atoms to produce one curie.

$$3.7 \times 10^{10}\text{ s}^{-1} = 0.0862\text{ d}^{-1} \times \frac{1\text{ d}}{24\text{ h}} \times \frac{1\text{ h}}{3600\text{ s}} \times N$$

$$N = \frac{3.7\times10^{10}\text{ s}^{-1}}{1.0\times10^{6}\text{ s}^{-1}} = 3.7 \times 10^{16}\text{ I-131 atoms/curie}$$

$$^{131}\text{I mass} = 10 \times 10^{6}\text{ curie} \times \frac{3.7\times10^{16}\ ^{131}\text{I atoms}}{\text{curie}} \times \frac{\text{mol }^{131}\text{I}}{6.022\times10^{23}\text{ atoms}}$$
$$\times \frac{131\text{ g I-131}}{1\text{ mol }^{131}\text{I}} = 80\text{ g }^{131}\text{I minimum.}$$

$$^{131}\text{I mass} = 50 \times 10^{6}\text{ curie} \times \frac{3.7\times10^{16}\ ^{131}\text{I atoms}}{\text{curie}} \times \frac{\text{mol }^{131}\text{I}}{6.022\times10^{23}\text{ atoms}}$$
$$\times \frac{131\text{ g I-131}}{1\text{ mol }^{131}\text{I}} = 400\text{ g }^{131}\text{I maximum.}$$

The amount of ^{131}I released ranges from 80 to 400 g

Additional Problems

69. The deflection of a positively-charged positron will be to the right and larger than the alpha ray. The alpha particle has more mass so is deflected less than a beta

minus or a positron. A large enough field will cause the particles to circle. The α will make a larger circle than the less massive β^+ or β^-. β^+ and β^- annihilate each other on contact, producing two gamma rays.

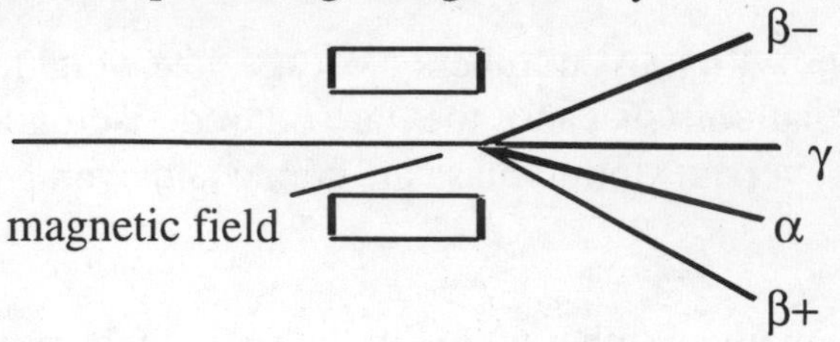

71. (a) Radioactivity will appear in the NH_3. When the HCl reacts with NH_3 to make NH_4Cl, the hydrogen atoms become equivalent. When the second reaction occurs, three out of four times the radioactive hydrogen atoms stays with the nitrogen.
(b) Radioactivity will appear in the NaCl solution because some of the radioactive hydrogens will become part of the water molecules. So the water is radioactive, making the solution radioactive.
(c) No radioactivity will be in the solid NaCl. It is only hydrogen that is radioactive.

73. First determine the energy per photon in joules, then the wavelength.

$$0.050 \text{ MeV} \times \frac{1.6022\times10^{-13}\text{ J}}{\text{MeV}} = 8.0 \times 10^{-15}\text{ J} = \frac{hc}{\lambda}$$

$$\lambda = \frac{hc}{8.0\times10^{-15}\text{ J}} = \frac{6.626\times10^{-34}\text{ J}\cdot\text{s}\times3.00\times10^{8}\text{ m/s}}{8.0\times10^{-15}\text{ J}} \times \frac{\text{pm}}{10^{-12}\text{ m}} = 25\text{ pm}$$

75. $\Delta E = (74.9225\text{ u} + 2 \times 1.0087\text{ u} - 74.9216\text{ u} - 2.0140\text{ u})$

$$= 4.3 \times 10^{-3}\text{ u} \times 931.5\frac{\text{MeV}}{\text{u}} = 4.01\text{ MeV}$$

$$= 4.01\text{ MeV} \times \frac{1.6022\times10^{-13}\text{ J}}{\text{MeV}} = 6.42 \times 10^{-13}\text{ J}$$

$E = mv^2$

$$v = \sqrt{\frac{E}{m}} = \sqrt{\frac{6.42\times10^{-13}\text{ J}}{2.0140\text{ u}} \times \frac{\text{u}}{1.6605\times10^{-27}\text{ kg}} \times \frac{\text{kg m}^2\text{ s}^{-2}}{\text{J}}}$$

$$v = 1.39 \times 10^{7}\ \frac{\text{m}}{\text{s}}$$

77. $$\text{activity in 1 mol} = \frac{226.0254\text{ g}}{\text{mol}} \times \frac{3.7 \times 10^{10}\text{ dis}}{\text{g s}} = \frac{8.36\times10^{12}\text{ dis}}{\text{s mol}}$$

$$\lambda = \frac{0.693}{t_{1/2}} = \frac{0.693}{1.60 \times 10^{3}\text{ y}} \times \frac{\text{y}}{365\text{ d}} \times \frac{\text{d}}{24\text{ h}} \times \frac{\text{h}}{3600\text{ s}} = 1.37 \times 10^{-11}\text{ s}^{-1}$$

$$N = \frac{A}{\lambda} = \frac{\frac{8.36 \times 10^{12}\ \text{dis}}{\text{sec mol}}}{1.37 \times 10^{-11}\ \text{s}^{-1}} = 6.1 \times 10^{23}\ \frac{\text{atoms}}{\text{mol}}$$

Because of the fluctuation of activity the curie is defined only to two significant figures which limits the precision and accuracy of the value of N. The crystallography data are more accurate.

79. $\lambda = \frac{0.693}{t_{1/2}} = \frac{0.693}{1.25 \times 10^9\ \text{y}} = 5.54 \times 10^{-10}\ \text{y}^{-1}$

$$A = \lambda N = 5.54 \times 10^{-10}\ \text{y}^{-1} \times 50.0\ \text{mg KMg Cl}_3\cdot 6\text{H}_2\text{O} \times \frac{10^{-3}\ \text{g}}{\text{mg}}$$

$$\times \frac{\text{mol KMgCl}_3\cdot 6\text{H}_2\text{O}}{277.9\ \text{g KMgCl}_3\cdot 6\text{H}_2\text{O}} \times \frac{\text{mol K}}{\text{mol KMgCl}_3\cdot 6\text{H}_2\text{O}} \times \frac{0.0117\ \text{mol K40}}{100\ \text{mol K}}$$

$$\times \frac{6.022 \times 10^{23}\ \text{atoms}}{\text{mol}} \times \frac{\text{y}}{365\ \text{d}} \times \frac{\text{d}}{24\ \text{h}} \times \frac{\text{h}}{60\ \text{min}} = 13.4\ \frac{\text{dis}}{\text{min}}$$

81.

t	cpm	–background of 27 cpm	ln A
0	1784	1757	7.47
1	1232	1205	7.09
2.33	880	853	6.75
3.5	656	629	6.44
4.5	554	527	6.27
5.5	342	315	5.75
6.5	266	239	5.48

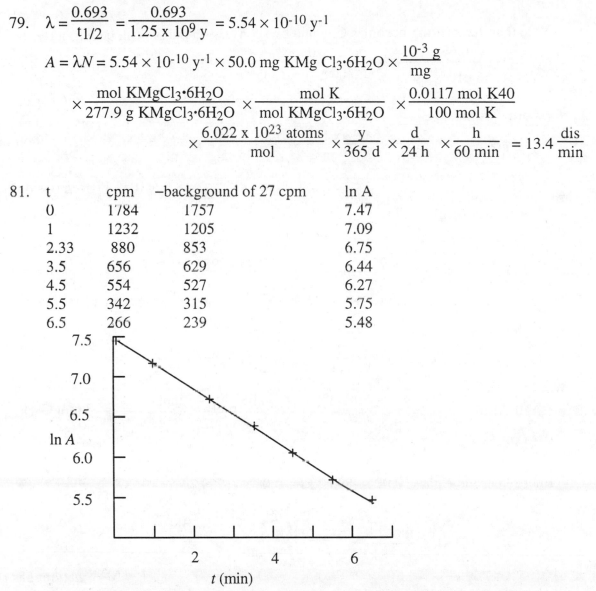

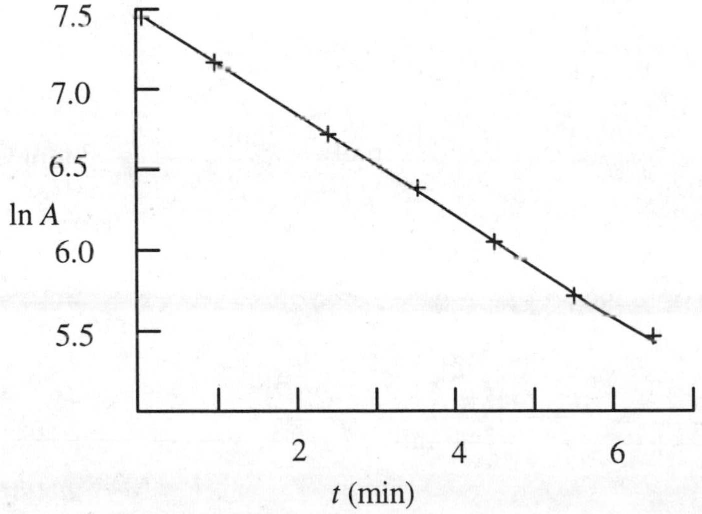

In a research lab, the data would be fed into a computer to give the slope of the best straight time. For this problem, use the third and last points to get the slope.

$$\text{slope} = -\lambda = \frac{5.48 - 6.75}{6.5 - 2.33} = -0.305$$

$\lambda = 0.305\ \text{min}^{-1}$

$$t_{1/2} = \frac{0.693}{0.305\,\text{min}^{-1}} = 2.27\text{ min}$$

Note that from 0 to 2.33 min, the cpm declines by slightly more than half.

83. $$\lambda = \frac{89.8\text{ mCi}}{1.00\text{ mg}} \times \frac{10^{-3}\text{ Ci}}{\text{mCi}} \times \frac{3.7\times10^{10}\text{ dis s}^{-1}}{\text{Ci}} \times \frac{\text{mg}}{10^{-3}\text{ g}} \times \frac{137\text{ g}}{\text{mole}} \times \frac{\text{mole}}{6.022\times10^{23}}$$

$$= 7.56\times10^{-10}\text{ s}^{-1}$$

Note that the mCi/mg becomes Ci/g and two conversion factors can be left out.

$$t_{1/2} = \frac{0.693}{7.56\times10^{-10}\text{ s}^{-1}} \times \frac{\text{h}}{3600\text{ s}} \times \frac{\text{d}}{24\text{ hr}} \times \frac{\text{y}}{365\text{ d}} = 29.1\text{ y}$$

Apply Your Knowledge

85. $$?\text{ kJ} = 3.2\times10^{-11}\frac{\text{J}}{\text{atom}} \times \frac{6.022\text{ x }10^{23}\text{ atom}}{\text{mol}} \times \frac{\text{mol}}{235\text{ g}} \times \frac{10^{-3}\text{ g}}{\text{mg}}$$

$$\times\ 1.00\text{ mg} \times \frac{\text{kJ}}{10^3\text{ J}} = 8.2\times10^4\text{ kJ (from fission)}$$

$CH_4(g) + 2\ O_2(g) \rightarrow CO_2(g) + 2\ H_2O(l)$

$\Delta H_{rxn} = \Sigma\Delta H_f\text{ (products)} - \Sigma\Delta H_f\text{ (reactants)}$

$$\Delta H_{rxn} = 1\text{ mol }CO_2 \times -393.5\frac{\text{kJ}}{\text{mol}} + 2\text{ mol }H_2O \times -285.8\frac{\text{kJ}}{\text{mol}}$$

$$-\ 1\text{ mol }CH_4 \times -74.81\frac{\text{kJ}}{\text{mol}} - 2\text{ mol }O_2 \times 0 = -890.3\text{ kJ/mol }CH_4$$

$2\ C_2H_6(g) + 7\ O_2(g) \rightarrow 4\ CO_2(g) + 6\ H_2O(l)$

$$\Delta H_{rxn} = 4\text{ mol }CO_2 \times -393.5\frac{\text{kJ}}{\text{mol}} + 6\text{ mol }H_2O \times -285.8\frac{\text{kJ}}{\text{mol}}$$

$$-\ 2\text{ mol }C_2H_6 \times -84.68\frac{\text{kJ}}{\text{mol}} - 7\text{ mol }O_2 \times 0 = \frac{-3119.4}{2\text{ mol }C_2H_6}$$

$$\Delta H_{rxn} = -1559.7\ \frac{\text{kJ}}{\text{mol }C_2H_6}$$

$$\text{Energy per mole gas} = 0.92 \times -890.2\text{ kJ} + 0.08 \times -1559.7\text{ kJ} = -943.8\frac{\text{kJ}}{\text{mol gas}}$$

$$?\text{ L} = 8.2\times10^4\text{ kJ} \times \frac{\text{mol gas}}{943.8\text{ kJ}} \times \frac{\frac{62.4\text{ mmHg}}{\text{K mol}}\text{ x }(22^\circ + 273)\text{K}}{744\text{ mmHg}}$$

$$= 2.1\times10^3\text{ L natural gas}$$

87. $2\ C_7H_5N_3O_6 \rightarrow 10\ C + CO + 3\ CO_2 + 3\ N_2 + 5\ H_2O$

$\Delta H_{rxn} = \Sigma\Delta H_f(\text{prod}) - \Sigma\Delta H_f(\text{react})$

$$\Delta H_{rxn} = 10 \text{ mol C} \times \frac{0 \text{ kJ}}{\text{mol}} + 1 \text{ mol CO} \times \frac{-110.0 \text{ kJ}}{\text{mol}} + 3 \text{ mol } CO_2 \times \frac{-393.5 \text{ kJ}}{\text{mol}}$$

$$+ 3 \text{ mol } N_2 \times \frac{0 \text{ kJ}}{\text{mol}} + 5 \text{ mol } H_2O \times \frac{-241.8 \text{ kJ}}{\text{mol}} - 2 \text{ mol } C_7H_5N_3O_6 \times \frac{-65.5 \text{ kJ}}{\text{mole}}$$

$$= 2369 \text{ kJ}$$

$$? \text{ kg } ^{235}U = 20 \text{ kton} \times \frac{10^3 \text{ ton}}{\text{kton}} \times \frac{2000 \text{ lb}}{\text{ton}} \times \frac{454 \text{ g}}{\text{lb}} \times \frac{\text{mol TNT}}{227.14 \text{ g}} \times \frac{2369 \text{ kJ}}{2 \text{ mol}} \times \frac{10^3 \text{ J}}{\text{kJ}}$$

$$\times \frac{\text{MeV}}{1.6022 \times 10^{-13} \text{ J}} \times \frac{\text{atom}}{183.6 \text{ MeV}} \times \frac{\text{mole}}{6.022 \times 10^{23} \text{ atom}} \times \frac{235 \text{ g}}{\text{mole}} \times \frac{\text{kg}}{10^3 \text{ g}}$$

$$= 1.3 \text{ kg } ^{235}U$$

89. Co-57 – $1s^1 2s^2 2p^6 3s^2 3p^6 4s^2 3d^7$
Co-57 - $1s^2 2s^1 2p^6 3s^2 3p^6 4s^2 3d^7$
$\lambda = 0.179$ nm

$$\nu = \frac{3.00 \times 10^8 \text{ m/s}}{0.179 \text{ nm}} \times \frac{\text{nm}}{10^{-9} \text{ m}} = 1.68 \times 10^{18} \text{ s}^{-1}$$

$$E = h\nu = \frac{6.626 \times 10^{-34} \text{ Js}}{\text{atom}} \times 1.68 \times 10^{18} \text{ s}^{-1} \times \frac{6.022 \times 10^{23} \text{ atom}}{\text{mol}} \times \frac{\text{kJ}}{10^3 \text{ J}}$$

$$= \frac{6.70 \times 10^5 \text{ kJ}}{\text{mol}}$$

Chapter 20

The *s*-Block Elements

Exercises

20.1 (a) $2\ NaCl(l) \xrightarrow{\text{electrolysis}} 2\ Na(l) + Cl_2(g)$

$2\ Na(s) + H_2(g) \xrightarrow{\Delta} 2\ NaH(s)$

(b) $2\ NaCl(aq) + 2\ H_2O(l) \xrightarrow{\text{electrolysis}} 2\ NaOH(aq) + Cl_2(g) + H_2(g)$

$2\ NaOH(aq) + Cl_2(g) \rightarrow NaCl(aq) + NaOCl(aq) + H_2O(l)$

20.2 (a) If the sole product were MgO, the mass of the product would be

$$? \text{ g MgO} = 1.000 \text{ g Mg} \times \frac{1 \text{ mol Mg}}{24.305 \text{ g Mg}} \times \frac{2 \text{ mol MgO}}{2 \text{ mol Mg}} \times \frac{40.304 \text{ g MgO}}{1 \text{ mol MgO}}$$

$$= 1.658 \text{ g MgO maximum mass}$$

(b) If the sole product were Mg_3N_2, the mass of the product would be

$$? \text{ g } Mg_3N_2 = 1.000 \text{ g Mg} \times \frac{1 \text{ mol Mg}}{24.305 \text{ g Mg}} \times \frac{1 \text{ mol } Mg_3N_2}{3 \text{ mol Mg}}$$

$$\times \frac{100.93 \text{ g } Mg_3N_2}{1 \text{ mol } Mg_3N_2} = 1.384 \text{ g } Mg_3N_2 \text{ minimum mass}$$

Review Questions

1. Sodium is the most abundant alkali metal, and calcium is the most abundant alkaline earth metal.
   The alkali and alkaline earth metals occur principally as carbonates, chlorides, and sulfates.

4. If only a small quantity of water is electrolyzed, H^+ produced at the anode is removed at the cathode and the concentration of sulfuric acid is essentially unchanged. However, if a larger quantity is electrolyzed, the concentration of the sulfuric acid increases slowly. This slow increase results from the gradual decrease on the amount of water.

12. (a) Portland cement: limestone, clay, and sand
    (b) Soda-lime glass: limestone, sand, and sodium carbonate
    (c) Mortar: slaked lime, sand, and water
    (d) Plaster of Paris: gypsum ($CaSO_4 \cdot 2H_2O$)

14. The anion in temporary hard water is hydrogen carbonate ion. If anions other than hydrogen carbonate are present, the water is said to be permanent hard water. Boiler scale is a mineral deposit produced by ions in hard water. For example, the boiling of temporary hard water produces a deposit of $CaCO_3(s)$.

17. $CH_3(CH_2)_nCOO^- Na^+$. The soap molecule has a long organic region, an anionic carboxylate end, and a cation. If the cation is an alkali metal ion (e.g., Na^+), the soap is soluble. The soaps of more highly charged cations (e.g., Mg^{2+}, Ca^{2+}, and Fe^{2+}) are insoluble. A calcium soap has two long-chain hydrocarbon groups with each Ca^{2+} ion. Sodium soaps have only one hydrocarbon group per Na^+ ion.

Problems

21. Na_2CO_3 is most soluble because 1A salts are more soluble than 2A salts. Li_2CO_3 is next in solubility because lithium is like magnesium in solubility of its salt, even though it is a 1A element. $MgCO_3$ is least soluble because it is a group 2A salt. Interionic attractions are strongest between the small highly charged Mg^{2+} and CO_3^{2-} and are greater for Li_2CO_3 than for Na_2CO_3 because Li^+ is a smaller ion than Na^+.

23. (a) lithium carbonate
    (b) Mg_3N_2
    (c) $CaBr_2{\cdot}6H_2O$
    (d) potassium hydrogen sulfate

25. (a) calcium oxide, CaO
    (b) sodium sulfate decahydrate, $Na_2SO_4{\cdot}10H_2O$
    (c) calcium sulfate dihydrate, $CaSO_4{\cdot}2H_2O$
    (d) $CaCO_3{\cdot}MgCO_3$

27.

$$MgCO_3 \xrightarrow{HCl} MgCl_2(aq) \xrightarrow{\text{evaporation}} MgCl_2(s) \xrightarrow{\text{electrolysis}} Mg(l)$$

$$MgCO_3(s) + 2\,HCl \rightarrow MgCl_2(aq) + H_2O + CO_2(g)$$

$$MgCl_2(s) \xrightarrow{\text{electrolysis}} Mg(l) + Cl_2(g)$$

29. Glauber's salt is sodium sulfate decahydrate $Na_2SO_4{\cdot}10\,H_2O$.

$$H_2SO_4(\text{conc aq}) + 2\,NaCl(s) \xrightarrow{\Delta} Na_2SO_4(s) + 2HCl(g).$$

$$Na_2SO_4(s)\ 10\,H_2O(l) \rightarrow Na_2SO_4{\cdot}10H_2O(s)$$

A volatile acid is produced by heating one of its salts (NaCl) with a nonvolatile acid (H_2SO_4).

31. NaOH(s) has to be prepared from NaCl(aq) by electrolysis. To obtain Na from NaOH requires an additional electrolysis. This is a more expensive route than to electrolyze NaCl(l), despite the requirement of a higher temperature.

33.

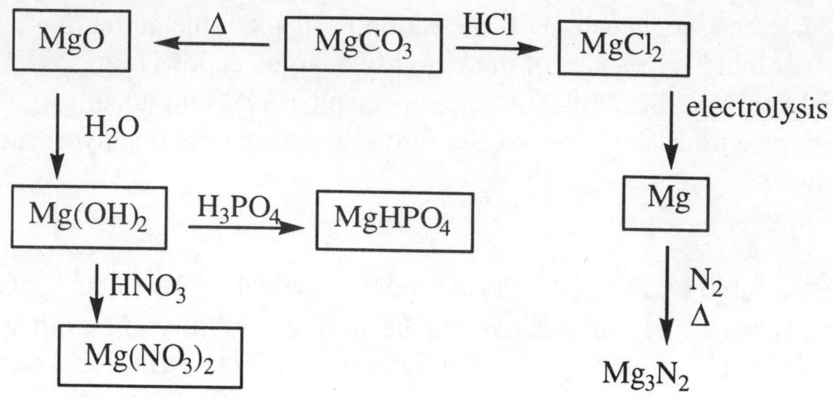

35. (a) $Sr(s) + 2\ H_2O(l) \rightarrow H_2(g) + Sr(OH)_2(aq)$
(b) $LiH(s) + H_2O(l) \rightarrow LiOH(aq) + H_2(g)$
(c) $BaO(s) + H_2O(l) \rightarrow Ba(OH)_2(aq)$
(d) $C(s) + H_2O(g) \xrightarrow{\Delta} CO(g) + H_2(g)$

37. (a) $2\ Al(s) + 6\ HCl(aq) \rightarrow 3\ H_2(g) + 2\ AlCl_3(aq)$
(b) $C_2H_6(g) + 2\ H_2O(g) \rightarrow 2\ CO(g) + 5\ H_2(g)$
(c) $CH_3C{\equiv}CH(g) + 2\ H_2(g) \rightarrow CH_3CH_2CH_3(g)$
(d) $MnO_2(s) + 2\ H_2(g) \rightarrow Mn(s) + 2\ H_2O(l)$

39. (a) $BeF_2(s) + 2\ Na(l) \xrightarrow{\Delta} Be(s) + 2\ NaF(s)$
(b) $Ca(s) + 2\ CH_3COOH(aq) \rightarrow Ca(CH_3COO)_2(aq) + H_2(g)$
(c) $PuO_2(s) + 2\ Ca(s) \xrightarrow{\Delta} Pu(s) + 2\ CaO(s)$
(d) $CaCO_3{\cdot}MgCO_3(s) \xrightarrow{\Delta} CaO(s) + MgO(s) + 2\ CO_2(g)$

41. (a) $Mg(CO)_3(s) + 2\ HCl(aq) \rightarrow MgCl_2(aq) + H_2O(l) + CO_2(g)$
(b) $CO_2(g) + 2\ KOH(aq) \rightarrow K_2CO_3(aq) + H_2O(l)$
(c) $2\ KCl(s) + H_2SO_4(conc.\ aq) \rightarrow K_2SO_4(s) + 2\ HCl(g)$

43. (a) $2\ NaCl(s) + H_2SO_4(conc.\ aq) \xrightarrow{\Delta} Na_2SO_4(s) + 2\ HCl(g)$
(b) $Mg(HCO_3)_2(aq) \xrightarrow{\Delta} MgCO_3(s) + H_2O(l) + CO_2(g)$
$MgCO_3(s) \xrightarrow{\Delta} MgO(s) + CO_2(g)$

45. $CaH_2(s) + 2\ H_2O \rightarrow 2\ H_2 + Ca(OH)_2$
$$n = \frac{PV}{RT}$$

$$? \text{ g } CaH_2 = \frac{746 \text{ mmHg} \times \frac{\text{atm}}{760 \text{ mmHg}} \times 126 \text{ L}}{\frac{0.08206 \text{ L atm}}{\text{K mol } H_2} \times (15 + 273) \text{ K}} \times \frac{\text{mol } CaH_2}{2 \text{ mol } H_2} \times \frac{42.10 \text{ g } CaH_2}{\text{mol } CaH_2}$$

$$= 110 \text{ g } CaH_2$$

47. $Ca(OH)_2 + 2\ HCl \rightarrow CaCl_2 + 2\ H_2O$

$$55.6 \text{ kg} \times \frac{10^3 \text{ g}}{\text{kg}} \times \frac{\text{mol } Ca(OH)_2}{74.10 \text{ g}} \times \frac{2 \text{ mol HCl}}{\text{mol } Ca(OH)_2} \times \frac{\text{L HCl}}{6.00 \text{ mol}} = 2.5 \times 10^2 \text{ L HCl}$$

49. $$1.00 \text{ kg Mg} \times \frac{10^3 \text{ g}}{\text{kg}} \times \frac{\text{ton seawater}}{1270 \text{ g Mg}} \times \frac{2000 \text{ lb}}{\text{ton}} \times \frac{454 \text{ g}}{\text{lb}} \times \frac{\text{mL}}{1.03 \text{ g}} \times \frac{10^{-3} \text{ L}}{\text{mL}}$$

$$= 694 \text{ L seawater}$$

The actual volume required is greater than 694 L because not every step in the overall process has a 100% yield.

51. (a) $Zn(s) + 2\ H^+ \rightarrow Zn^{2+} + H_2(g)$

$$\frac{1 \text{ mole } H_2}{1 \text{ mole Zn}} = \frac{2 \text{ g } H_2}{65 \text{ g Zn}}$$

(b) $2\ Na(s) + 2\ H_2O(l) \rightarrow 2\ NaOH(aq) + H_2(g)$

$$\frac{1 \text{ mole } H_2}{2 \text{ mole Na}} = \frac{2 \text{ g } H_2}{2 \text{ x } 23 \text{ g Na}}$$

(c) $CaH_2(s) + 2\ H_2O(l) \rightarrow Ca(OH)_2(aq) + 2\ H_2(g)$

$$\frac{2 \text{ mole } H_2}{\text{mole } CaH_2} = \frac{2 \text{ x } 2 \text{ g } H_2}{42 \text{ g} CaH_2}$$

The comparison is very simple:

Zn	Na	CaH_2
$\frac{2}{65}$	$\frac{2}{23 \text{ x } 2}$	$\frac{2 \text{ x } 2}{42}$

CaH_2 obviously produces the most H_2 from a given mass of the reactant.

53. The base NH_3 reacts with HCO_3^- to form CO_3^{2-}, which precipitates Mg^{2+}, Ca^{2+} and Fe^{2+}.

$NH_3 + HCO_3^- \rightarrow NH_4^+ + CO_3^{2-}$

$M^{2+} + CO_3^{2-} \rightarrow MCO_3(s)$ $\qquad M^{2+} = Mg^{2+}, Ca^{2+}, Fe^{2+}$

Alternatively, think of the reactions as

$NH_3 + H_2O \rightarrow NH_4^+ + OH^-$

$OH^- + HCO_3^- \rightarrow CO_3^{2-} + H_2O$

$M^{2+} + CO_3^{2-} \rightarrow MCO_3(s)$ $\qquad M^{2+} = Mg^{2+}, Ca^{2+}, Fe^{2+}$

55. $Ca(HCO_3)_2(aq) \rightarrow CaCO_3(s) + H_2O + CO_2(g)$

$$725 \text{ mL water} \times \frac{10^{-3} \text{ L}}{\text{mL}} \times \frac{115 \text{ g } HCO_3^-}{10^3 \text{ L water}} \times \frac{\text{mol } HCO_3^-}{61.02 \text{ g}} \times \frac{\text{mol } CaCO_3}{2 \text{ mol } HCO_3^-}$$
$$\times \frac{100.09 \text{ g } CaCO_3}{\text{mol } CaCO_3} = 0.0684 \text{ g "scale"}$$

57. $Ca(HCO_3)_2$ drips from a cave roof. When the drop evaporates, it leaves $CaCO_3$, which builds up from the floor, making stalagmites. If the drop evaporates before falling, stalactites are produced. Columns are formed where a stalactite meets a stalagmite. The caves themselves are formed when calcium carbonate is dissolved by water charged with $CO_2(g)$.The water is "hard water" after calcium carbonate dissolves in it. That is, the formation of a cave or of "hard water" occurs through the reaction: $CaCO_3(s) + H_2O(l) + CO_2(g) \rightarrow Ca(HCO_3)_2(aq)$. Stalactites, stalagmites, and "boiler scale" (which forms when hard water is boiled) form in the reverse of the reaction.

59. The water is first run through a cation-exchange resin to exchange all cations for H^+. Then an anion-exchange resin will exchange the anions for OH^-. The OH^- and H^+ combine to form water.

Additional Problems

61. (a) $4\ Li(s) + O_2(g) \rightarrow 2\ Li_2O(s)$
(b) $6\ Li(s) + N_2(g) \rightarrow 2\ Li_3N(s)$
(c) $Li_2CO_3(s) \rightarrow Li_2O(s) + CO_2(g)$

63. $CaO(s) + CO_2(g) \rightarrow CaCO_3(s)$. Carbon dioxide reacts with the CaO or water vapor reacts with the CaO.
$CaO(s) + H_2O(g) \rightarrow Ca(OH)_2(s)$ followed by:
$Ca(OH)_2(s) + CO_2(g) \rightarrow CaCO_3(s) + H_2O(g)$.

65. $U + 2\ HNO_3 + 2\ H_2O \rightarrow UO_2(NO_3)_2 + 3\ H_2$

$$?\text{ L } H_2 = 1.00 \text{ kg U} \times \frac{10^3 \text{ g U}}{\text{kg U}} \times \frac{\text{mol U}}{238 \text{ g U}} \times \frac{3 \text{ mol } H_2}{1 \text{ mol U}} \times \frac{62.4 \text{L Torr}}{\text{K mol}} \times$$
$$\times \frac{(25° + 273) \text{ K}}{745 \text{ Torr}} = 315 \text{ L } H_2$$

67. Let X be the mass of Mg that reacts to form MgO and 1.000 - X, the mass of Mg that reacts to form Mg_3N_2.

$$?\text{g MgO} = X \text{ gMg} \times \frac{1 \text{ mol Mg}}{24.31 \text{ g Mg}} \times \frac{2 \text{ mol MgO}}{2 \text{ mol Mg}} \times \frac{40.30 \text{ g MgO}}{1 \text{ mol MgO}} = 1.658\ X \text{ gMgO}$$

$$?\text{g } Mg_3N_2 = (1.000\text{-}X)\text{gMg} \times \frac{1 \text{ mol Mg}}{24.31 \text{ g Mg}} \times \frac{1 \text{ mol } Mg_3N_2}{3 \text{ mol Mg}} \times \frac{101.0 \text{ g } Mg_3N_2}{1 \text{ mol } Mg_3N_2}$$
$$= 1.385 \times (1.000 - X) = (1.385 - 1.385\ X)\text{g}Mg_3N_2$$

The total mass of product is 1.537g. That is,

$1.658\ X + (1.385 - 1.385\ X) = 1.537$
$0.273\ X = 0.152$
$X = 0.557$
Mass of MgO = $1.658\ X = 1.658 \times 0.557 = 0.924$ g

$$\text{Mass \% MgO} = \frac{0.924\text{ g MgO} \times 100\%}{1.537\text{ g products}} = 60.1\%\text{ MgO.}$$

69. $$?\text{ years} = \frac{3.0\text{ V} \times 0.5\text{ A h}}{5\,\mu\text{W}} \times \frac{\mu\text{W}}{10^{-6}\text{ W}} \times \frac{\text{W}}{\text{V A}} \times \frac{\text{d}}{24\text{h}} \times \frac{\text{yr}}{365\text{ d}} = 34\text{ years}$$

$$?\text{ g Li} = 0.5\text{ A h} \times \frac{3600\text{ s}}{\text{h}} \times \frac{\text{C}}{\text{A s}} \times \frac{\text{mol e-}}{96485\text{ C}} \times \frac{\text{mol Li}}{\text{mol e}} \times \frac{6.94\text{ g Li}}{\text{mol Li}} = 0.13\text{ g Li}$$

71. (a) $H_2\text{ (g)} + \frac{1}{2}\ O_2\text{ (g)} \rightarrow H_2O\text{(l)}$ $\qquad \Delta H° = -285.8$ kJ

$\Delta H°_{combustion}$ of H_2 is equal to $\Delta H°_f$ for water.

$C_{12}H_{26} + \frac{37}{2}\ O_2 \rightarrow 12\ CO_2 + 13\ H_2O\text{(l)} \qquad \Delta H°_{comb} = -8087$ kJ

$\Delta H°_{combustion}(C_{12}H_{26}) = \Sigma\Delta H°_f\text{(prod)} - \Sigma\Delta H°_f\text{(react)}$

$$\Delta H°_{comb} = 12\text{ mol } CO_2 \times \frac{-393.5\text{ kJ}}{\text{mol}} + 13\text{ mol } H_2O \times \frac{-285.8\text{ kJ}}{\text{mol}}$$
$$- 1\text{mol } C_{12}H_{26} \times \frac{-350.9\text{ kJ}}{\text{mol}} - \frac{37}{2}\text{ mol } O_2 \times \frac{0\text{ kJ}}{\text{mol}}$$

$\Delta H°_{comb} = -8087$ kJ

$$H_2: \quad \frac{-285.8\text{ kJ}}{\text{mol}} \times \frac{\text{mol}}{2.016\text{ g}} = -141.8\ \frac{\text{kJ}}{\text{g}}$$

$$C_{12}H_{26}: \quad \frac{-8087\text{ kJ}}{\text{mol}} \times \frac{\text{mol}}{170.3\text{ g}} = \frac{-47.5\text{ kJ}}{\text{g } C_{12}H_{26}}$$

(b) $$H_2: \quad \frac{-141.8\text{ kJ}}{\text{g}} \times \frac{0.0708\text{ g}}{\text{mL}} = \frac{-10.0\text{ kJ}}{\text{mL}}$$

$$C_{12}H_{26}: \quad \frac{-47.5\text{ kJ}}{\text{g}} \times \frac{0.749\text{ g}}{\text{mL}} = \frac{-35.6\text{ kJ}}{\text{mL}}$$

73. $2\ CH_3(CH_2)_{14}COO^-Na^+ + Ca^{2+} \rightarrow Ca(CH_3(CH_2)_{14}COO)_2\text{ (s)} + 2\ Na^+$

$$?\text{ g soap} = 5.00\text{ L water} \times \frac{135\text{ g } HCO_3^-}{10^3\text{ L water}} \times \frac{\text{mol } HCO_3^-}{61.02\text{ g}} \times \frac{\text{mol } Ca^{2+}}{2\text{ mol } HCO_3^-}$$
$$\times \frac{2\text{ mol soap}}{\text{mol } Ca^{2+}} \times \frac{278.4\text{ g soap}}{\text{mol soap}} = 3.08\text{ g soap}$$

75. $$\frac{15.17\text{ mL x } \frac{10^{-3}\text{ L}}{\text{mL}} \text{ x } 0.02650\text{ M NaOH}}{100.0\text{ mL solution x } \frac{10^{-3}\text{ L}}{\text{mL}}} \times \frac{1\text{ mol } H^+}{1\text{ mol } OH^-} \times \frac{1\text{ mol } Ca^{2+}}{2\text{ mol } H^+}$$
$$\times \frac{40.08\text{ g } Ca^{2+}}{\text{mol } Ca^{2+}} \times \frac{\text{ppm}}{\text{g}/10^3\text{ L}} = 80.56\text{ ppm } Ca^{2+}$$

(1 ppm = g/10^3 L, 10^3 L = 10^6 g)

77. (a) In a bcc cell, the distance from one corner of the cell to the opposite corner is $4r$. That distance is the hypotenuse of a right triangle. The other two sides of the triangle are the length (l) of the cell (down one corner) and the diagonal of the bottom of the cell. The diagonal distance from the Pythogorian theorem is $d_d = \sqrt{l^2 + l^2} = \sqrt{2l^2}$. The distance from one corner of the cell to the other is $d = \sqrt{2l^2 + l^2}$.

Since $d = 4r$, then

$4r = \sqrt{2l^2 + l^2} = \sqrt{3l^2}$

$4r = \sqrt{3}\, l$

$r_{Na} = 186$ pm $\qquad r_K = 227$ pm $\qquad r_{Rb} = 248$ pm

$l_{Na} = 430$ pm $\qquad l_K = 524$ pm $\qquad l_{Rb} = 573$ pm

(b) $V_{Na} = l^3 = (430\text{ pm})^3 \times \left(\dfrac{10^{-12}\text{ m}}{\text{pm}}\right)^3 \times \left(\dfrac{\text{cm}}{10^{-2}\text{ m}}\right)^3 = 7.95 \times 10^{-23}\text{ cm}^3$

$V_K = 1.44 \times 10^{-22}\text{ cm}^3$

$V_{Rb} = 1.88 \times 10^{-22}\text{ cm}^3$

(c) In a bcc unit there is one atom in the center and $8 \times \frac{1}{8}$ atom in the corners, so 2 atom/unit cell.

$$d_{Na} = \frac{m}{V} = \frac{22.990\text{ g/mol} \times \dfrac{\text{mol}}{6.022 \times 10^{23}\text{ atoms}} \times \dfrac{2\text{ atoms}}{\text{cell}}}{7.95 \times 10^{-23}\text{ cm}^3} = 0.960\text{ g/cm}^3$$

$d_K = 0.901\text{g/cm}^3$

$d_{Rb} = 1.51\text{g/cm}^3$

Although the calculations don't give the actual value, the calculated values are in the same trend of Rb>Na>K.

79. $CaCO_3(s) \rightleftharpoons Ca^{2+}(aq) + CO_3^{2-}(aq)$ $\qquad K_{sp} = 2.8 \times 10^{-9}$

$CO_3^{2-}(aq) + H_3O^+(aq) \rightleftharpoons HCO_3^-(aq) + H_2O(l)$ $\qquad \dfrac{1}{K_{a2}} = \dfrac{1}{4.7\times10^{11}}$

$H_2O(l) + CO_2(g) \rightleftharpoons H_2CO_3(aq)$

$\underline{H_2CO_3(aq) + H_2O(l) \rightleftharpoons H_3O^+(aq) + HCO_3^-(aq)}$ $\qquad K_{a1}\ 4.4 \times 10^{-7}$

$CaCO_3(s) + H_2O(l) + CO_2(g) \rightleftharpoons Ca^{2+}(aq) + 2\,HCO_3^-(aq)$

$$K = \frac{K_{sp} \times K_{a1}}{K_{a2}} = 2.6 \times 10^{-5}$$

CO_3^{2-}, H_3O^+, H_2O, and H_2CO_3 cancel.

81. Spectroscopic methods should reveal whether the emitted radiation is that of hydrogen or of other elements.

Apply Your Knowledge

83. (a) Precipitation $CO_2(s) + Ca^{2+} + 2\ OH^- \rightarrow CaCO_3(s) + H_2O$

When the $Ca(OH)_2$ is used up

Redissolving $CaCO_3(s) + CO_2(g) + H_2O \rightarrow Ca^{2+}(aq) + 2\ HCO_3^-(aq)$

(See also Problem 79.)

(b) $Q_{ip} = [Ca^{2+}][CO_3^{2-}] = 5 \times 10^{-3} \times K_{a2}(H_2CO_3) = 5 \times 10^{-3} \times 4.7 \times 10^{-11}$

$= 2 \times 10^{-13}$

$K_{sp}(CaCO_3) = 2.8 \times 10^{-9}$

$Q_{ip} < K_{sp}$ No precipitate forms.

(c) $Ca^{2+} + CO_3^{2-} \rightarrow CaCO_3(s)$

$CO_2 + H_2O \rightarrow H_2CO_3$

$H_2CO_3 + 2\ OH^- \rightarrow CO_3^{2-} + H_2O$

With a large excess of OH^-, the equilibrium is forced to CO_3^{2-} and H_2O. The excess CO_3^{2-} concentration keeps the precipitate as solid instead of allowing it to dissolve.

85. $? \text{ °C} = 50 \text{ °F} \times \frac{5 \text{ °C}}{9 \text{ °F}} = 28 \text{ °C}$

$? \text{ kJ} = 3.0 \times 10^3 \text{ kg} \times \frac{0.8 \text{ J}}{\text{g °C}} \times 28 \text{ °C} = 7 \times 10^4 \text{ kJ}$

$? \text{ kg} = 7 \times 10^4 \text{ kJ} \times \frac{\text{mole}}{112 \text{ kJ}} \times \frac{322.2 \text{ g}}{\text{mole}} \times \frac{\text{kg}}{10^3 \text{ g}} = 2 \times 10^2 \text{ kg}$

Chapter 21

The *p*-Block Elements

Exercise

21.1 $2\ BCl_3(g) + 3\ H_2(g) \rightarrow 2\ B(s) + 6\ HCl(g)$

21.2 $2\ Br^-(aq) \rightarrow Br_2(l) + 2\ e^-$ $\qquad E^\circ_{Br^2/Br^-} = 1.065\ V$

$SO_4^{2-}(aq) + 4\ H^+(aq) + 2\ e^- \rightarrow 2\ H_2O + SO_2(g)$ $\qquad E^\circ_{SO_4^{2-}/SO_2} = 0.17\ V$

$$E^\circ_{cell} = E^\circ_{cathode} - E^\circ_{anode} = E^\circ_{SO_4^{2-}/SO_2} - E^\circ_{Br^2/Br^-} = -0.90\ V$$

$$\Delta G^\circ = -nFE^\circ_{cell}$$

$$\Delta G^\circ = -2\ \text{mol e}^- \times \frac{96485\ C}{\text{mol e}^-} \times -0.90\ V \times \frac{J}{C\ V} \times \frac{kJ}{1000\ J} = 174\ kJ$$

The positive ΔG° indicates that the reaction is nonspontaneous, just as was concluded in Example 21.2. The ΔG° values do not agree because the states of the reactants and products differ, for example, $H_2SO_4(l)$ and $Br_2(g)$ in Example 21.2 and Br_2 (l) and $[H^+]$ = 1M here.

Review Questions

4. Aluminum is the most active metal, and fluorine is the most active nonmetal in the *p* block of elements.

7. The pellets of NaOH and the granules of Al liberate heat and H_2 (g) when they react in water. The heat helps to melt grease and fat. It also promotes the reaction of NaOH with fat to form soap, which can solubilize the clog. The hydrogen gas helps to push the clog out of a drain.

8. Because aluminum is less dense than copper, the transmission lines can be made lighter than if copper were used. Even if a larger cross section wire is used to equal the current carrying capacity of a copper cable, the aluminum cable is lighter and easier to support.

9. In 1886, Hall and Heroult found an inexpensive method to electrolyze $Al_2O_3(l)$ by dissolving $Al_2O_3(s)$ in molten cryolite. The electrolyic method greatly reduced the cost of aluminum.

16. Copper metal has a positive reduction potential so it cannot be oxidized by H_3O^+ in HCl (aq). It can be oxidized by hot, concentrated H_2SO_4, with the liberation of SO_2 (g).

21. (a) silver azide (b) potassium thiocyanate

(c) astatine oxide (d) telluric acid

22. (a) H_2SeO_4 (b) H_2Te (c) $Pb(N_3)_2$ (d) AgAt

23. bauxite Al_2O_3, boric oxide B_2O_3, corundum Al_2O_3, cyanogen $(CN)_2$, hydrazine NH_2NH_2, silica SiO_2

Problems

25. $3\ Mg(s) + B_2O_3(s) \xrightarrow{\Delta} 2\ B(s) + 3\ MgO(s)$

27. BH_3 is an electron-deficient structure that does not exist as a stable molecule (the stable species is diborane, B_2H_6). The possibility of resonance structures with B-to-F double bonds, and possibly even ionic structures, lead to a resonance hybrid for BF_3 that is a stable molecule.

29. $Na_2B_4O_7{\cdot}10H_2O + H_2SO_4 \rightarrow 4\ B(OH)_3 + Na_2SO_4 + 5\ H_2O$
$2\ B(OH)_3 \xrightarrow{\Delta} B_2O_3 + 3\ H_2O$
$B_2O_3 + 3\ CaF_2 + H_2SO_4 \rightarrow 2\ BF_3 + 3\ CaSO_4 + 3\ H_2O$

31. $Al_2O_3(s) + 3\ H_2SO_4 + 15\ H_2O \xrightarrow{\Delta} Al_2(SO_4)_3 \cdot 18\ H_2O$
$Al_2(SO_4)_3 \cdot 18H_2O + 6\ OH^- \rightarrow 2\ Al(OH)_3(s) + 3\ SO_4^{2-} + 18\ H_2O$
$Al(OH)_3 + 3\ HCl + \rightarrow AlCl_3 + 3\ H_2O$

33. The first key feature is that Al_2O_3 is amphoteric. By adding a strong base, Al_2O_3 can be separated from the impurity Fe_2O_3, which is not amphoteric. The second key is the use of molten cryolite, $Na_3AlF_6(l)$, as a solvent for $Al_2O_3(s)$. Electrolysis can be conducted at a much lower temperature and in a better electrical conductor.

35. $\Delta H^\circ_{rxn} = \Sigma \Delta H^\circ_f \text{ products} - \Sigma \Delta H^\circ_f \text{ reactants}$
$\Delta H^\circ_{rxn} = \Delta H^\circ_f(Al_2O_3) - \Delta H^\circ_f(Fe_2O_3) = -1676 \text{ kJ/mol} - (-824.2 \text{ kJ/mol})$
$\Delta H^\circ_{rxn} \approx -852 \text{ kJ/mol}$
This result is only approximate because it is based on data at 298 K, whereas the reaction occurs at a very high temperature. Also, the estimate assumes $\Delta H^\circ_f = 0$ for Fe(s), even though at the temperature of the reaction the stable form of iron is Fe(l).

37. Aluminum could react with strongly acidic foods to produce H_2 (g); the metal would become pitted. In a strongly basic medium (oven cleaner), aluminum could react to produce $[Al(OH)_4]^-$, and again the metal would become pitted.

39. $Na_2C_2(s) + 2\ H_2O(l) \rightarrow 2\ NaOH(aq) + C_2H_2(g)$

41.

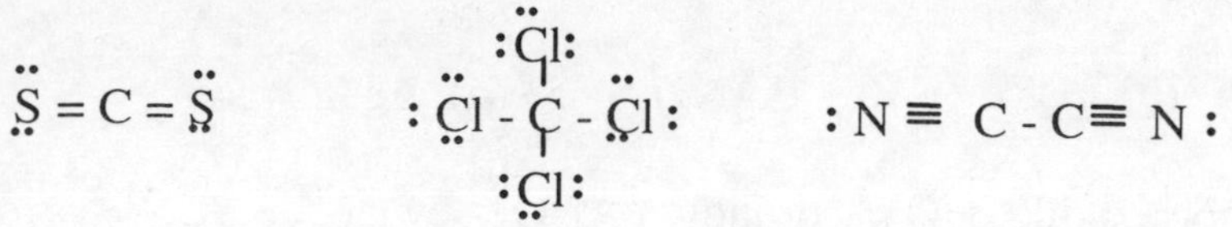

43.

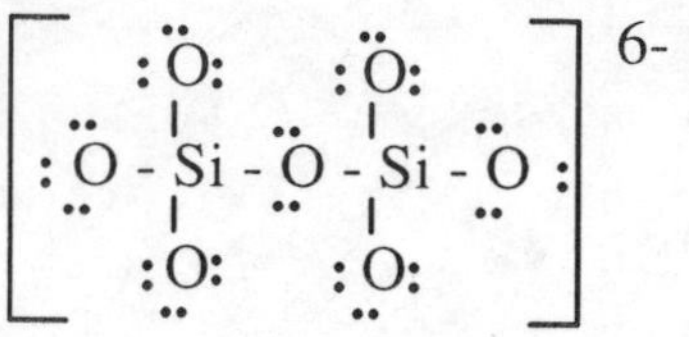

This would be twc tetrahedra with a common corner.

45. $\overset{+2}{Mg_3}(\overset{+4}{Si_2}\overset{-2}{O_5})^{-2}(\overset{-2}{O}\overset{+1}{H})_4^{-1}$ $\quad 3(+2) + (-2) + 4(-1) = 0$

47. (a) $Si(CH_3)_4$ (b) $SiCl_2(CH_3)_2$ (c) $SiH(C_2H_5)_3$

49. $PbO_2 + 4\,H^+ + 2\,e^- \rightarrow Pb^{2+} + 2\,H_2O$ $\quad E°_{PbO_2/Pb^{2+}} = 1.455$ V

(a) $ClO_3^- + H_2O \rightarrow 2\,e^- + 2\,H^+ + ClO_4^-$ $\quad E°_{ClO_4^-/ClO_3^-} = 1.19$ V

Yes: $E°_{cell} = E°_{PbO_2/Pb^{2+}} - E°_{ClO_4^-/ClO_3^-} = 0.27$ V

(b) $H_2O \rightarrow H_2O_2 + 2\,H^+ + 2\,e^-$ $\quad E°_{H_2O_2/H_2O} = 1.763$ V

No: $E°_{cell} = E°_{PbO_2/Pb^{2+}} - E°_{H_2O_2/H_2O} = -0.308$ V

(c) $Ag^+ \rightarrow Ag^{2+} + e^-$ $\quad E°_{Ag^+/Ag} = 1.98$ V

No: $E°_{cell} = E°_{PbO_2/Pb^{2+}} - E°_{Ag^+/Ag} = -0.53$ V

(d) $Sn^{2+} \rightarrow Sn^{4+} + 2\,e^-$ $\quad E°_{Sn^{4+}/Sn^{2+}} = 0.154$ V

Yes: $E°_{cell} = E°_{PbO_2/Pb^{2+}} - E°_{Sn^{4+}/Sn^{2+}} = 1.301$ V

51. (a) $Sn(s) + 2\,HCl(aq) \rightarrow SnCl_2(aq) + H_2(g)$

(b) $SnCl_2 + Cl_2(g) \rightarrow SnCl_4$

(c) $SnCl_4(aq) + 4\,NH_3(aq) + 2\,H_2O \rightarrow SnO_2(s) + 4\,NH_4^+(aq) + 4Cl^-(aq)$

53. (a) $N_2O \quad \frac{2 \times 14}{(2 \times 14) + 16} = \frac{14}{14 + 8}$

(b) $NH_3 \quad \frac{14}{14 + 3}$

(c) $NO \quad \frac{14}{14 + 16}$

(d) $NH_4Cl \quad \frac{14}{14 + 4 + 35}$

(e) $H_2NNH_2 \quad \frac{2 \times 14}{(2 \times 14) + 4} = \frac{14}{14 + 2}$

(f) $(NH_4)_2SO_4 \quad \frac{2 \times 14}{(2 \times 14) + 8 + 32 + 64} = \frac{14}{14 + 4 + 16 + 32}$

The larger the denominator, the smaller the mass percent.
In increaing order f < d < c < a < b < e.

55. (a) $NH_2NH_2 + HCl \rightarrow NH_2NH_3^+ + Cl^-$ followed by
$NH_2NH_3^+ + HCl \rightarrow NH_3NH_3^{2+} + Cl^-$
(b) $3\ Cu(s) + 8\ H^+ + 2\ NO_3^- \rightarrow 3\ Cu^{2+} + 2\ NO + 4\ H_2O$
(c) $2\ NO(g) + O_2(g) \rightarrow 2\ NO_2(g)$

57. $4(Fe^{3+} + e^- \rightarrow Fe^{2+})$ $\quad E^\circ_{Fe^{3+}/Fe^{2+}} = 0.771\ V$
$\underline{NH_2NH_3^+ \rightarrow N_2 + 5\ H^+ + 4\ e^-}$ $\quad E^\circ_{N_2/NH_2NH_3^+} = ?\ V$
$4\ Fe^{3+} + NH_2NH_3^+ \rightarrow N_2 + 4\ Fe^{2+} + 5\ H^+$
$E^\circ_{cell} = E^\circ_{Fe^{3+}/Fe^{2+}} - E^\circ_{N_2/NH_2NH_3^+} = 0.771\ V - E^\circ_{N_2/NH_2NH_3^+} = 1.00\ V$
$E^\circ_{N_2/NH_2NH_3^+} = 0.771\ V - 1.00\ V = -0.23\ V$

59. The principal allotropes of phosphorus are white P and red P, with white phosphorus being the more reactive. The molecular structure of white phosphorus consists of individual P_4 tetrahedra. In red phosphorus, the P_4 tetrahedra are joined into long chains.

61. $P_4(s) + 3\ KOH(aq) + 3\ H_2O \rightarrow 3\ KH_2PO_2(aq) + PH_3(g)$

63. (a) $4\ Al(s) + 3\ O_2(g) \rightarrow 2\ Al_2O_3(s)$
(b) $2\ KClO_3(s) \rightarrow 2\ KCl(s) + 3\ O_2(g)$
(c) $2\ Na_2O_2(s) + 2\ H_2O(l) \rightarrow 4\ NaOH(aq) + O_2(g)$
(d) $Pb^{2+}(aq) + O_3(g) + H_2O(l) \rightarrow PbO_2(s) + O_2(g) + 2\ H^+(aq)$

65. Sulfur will melt at 119 °C. Water is heated under pressure to produce steam at temperatures greater than 119 °C. The super-heated steam melts the sulfur underground. Sulfur neither reacts with hot water nor dissolves in it, so the liquid sulfur can be brought to the surface in a pure condition by compressed air.

67. reduction: $2\ SO_3^{2-} + 3\ H_2O + 4\ e^- \rightarrow S_2O_3^{2-} + 6\ OH^-$
oxidation: $\underline{2\ S(s) + 6\ OH^- \rightarrow S_2O_3^{2-} + 3\ H_2O + 4\ e^-}$
$2\ SO_3^{2-} + 2\ S(s) \rightarrow 2\ S_2O_3^{2-} \rightarrow SO_3^{2-} + S\ (s) \rightarrow S_2O_3^{2-}$

69. $Br_2 + 2\ e^- \rightarrow 2\ Br^-$ $\quad E^\circ = 1.065\ V$
$I_2 + 2\ e^- \rightarrow 2\ I^-$ $\quad E^\circ = 0.535\ V$
$F_2 + 2\ e^- \rightarrow 2\ F^-$ $\quad E^\circ = 2.866\ V$
$Cl_2 + 2\ e^- \rightarrow 2\ Cl^-$ $\quad E^\circ = 1.358\ V$
To displace Br_2 from an aqueous solution of Br^- requires oxidation of Br^-.
$2\ Br^-(aq) \rightarrow Br_2(l) + 2\ e^-$ $\quad E^\circ_{anode} = 1.065\ V$
$E^\circ_{cell} = E^\circ(\text{reduction}) - E^\circ_{Br_2/Br^-} = E^\circ(\text{reduction}) - 1.065\ V$
Only a reduction half-reaction with $E^\circ > 1.065$ will work, and this must be $Cl_2(g) + 2\ e^- \rightarrow 2\ Cl^-(aq)$ $\quad E^\circ = 1.358\ V$. I_2 is too poor an oxidizing agent to work, and I^-, Cl^-, and F^- can only be reducing agents, not oxidizing agents.

71. $I^-(aq) + 3\ Cl_2(g) + 3\ H_2O(l) \rightarrow IO_3^-(aq) + 6\ Cl^-(aq) + 6\ H^+(aq)$

$2\,Br^-(aq) + Cl_2(g) \rightarrow Br_2(l) + 2\,Cl^-(aq)$

When the products of the reaction are treated with $CS_2(l)$, the Br_2 dissolves and the other products remain in the aqueous solution.

73. ICl is an interhalogen compound.

$(CN)_2$ is a pseudohalogen and fits none of the other categories.

$NaCl$ is a halide salt.

$KBrO_3$ is a halate salt.

75. (a) $Cl_2 + H_2O(l) \rightarrow HOCl + H^+ + Cl^-$

(b) $Cl_2 + 2\,NaOH \rightarrow NaOCl + NaCl + H_2O$

(c) $3\,Cl_2 + 6NaOH \xrightarrow{\Delta} 5\,NaCl + NaClO_3 + 3\,H_2O$

77.

(a) $: \ddot{O} - Xe - \ddot{O} :$

$: \ddot{O} :$

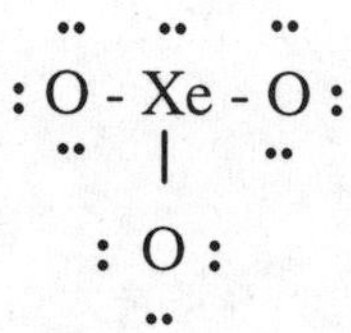

3 bonded atoms

1 L.P.

AX_3E

trigonal pyramidal

(b) $: \ddot{O} :$

$: \ddot{O} - Xe - \ddot{O} :$

$: \ddot{O} :$

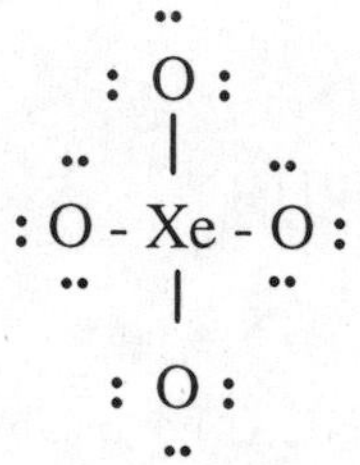

4 bonded atoms

AX_4

tetrahedral

79. The nucleus ^4_2He is very stable and is thus easily formed and emitted as an alpha particle from many nuclei. Argon is not emitted from other nuclei; it is formed only by the nucleus of $^{40}_{19}K$ absorbing an inner shell electron to become argon.

Additional Problems

81.

(a) $: \ddot{F} - \ddot{Br} - \ddot{F} :$

$: \ddot{F} :$

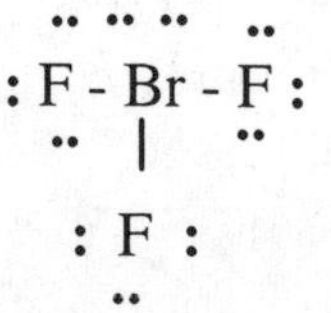

3 bonded atoms

2 L.P.

AX_3E_2

T-shape

(b) $: \ddot{F} :$

$: \ddot{F} - \ddot{I} - \ddot{F} :$

$: \ddot{F} : \quad : \ddot{F} :$

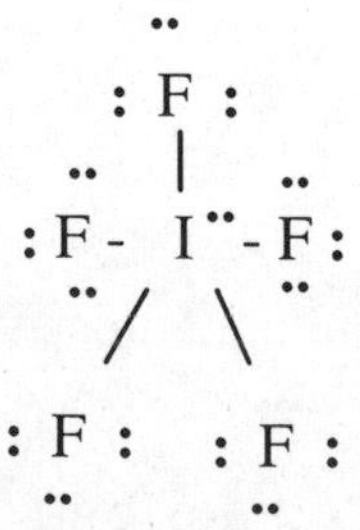

5 bonded atoms

1 L.P.

AX_5E

square pyramidal

83. $Na_2B_4O_7{\cdot}10H_2O + 6\ CaF_2 + 8\ H_2SO_4 \rightarrow 2\ NaHSO_4 + 6\ CaSO_4 + 17\ H_2O + 4\ BF_3$

85. $MnF_6^{2-} + 2\ SbF_5 \rightarrow MnF_4 + 2\ SbF_6^-$

$MnF_4 \rightarrow MnF_2 + F_2\ (g)$

87. $2(XeF_2 + 2\ H^+ + 2\ e^- \rightarrow Xe\ (g) + 2\ HF)$ $\qquad E°_{cathode} = 2.32\ V$

$\underline{2\ H_2O \rightarrow 4\ e^- + 4\ H^+ + O_2}$ $\qquad E°_{anode} = 1.229\ V$

$2\ XeF_2 + 2\ H_2O \rightarrow 2\ Xe\ (g) + 4\ HF + O_2$

$E°_{cell} = E°_{cathode} - E°_{anode} = 2.32\ V - 1.229\ V = 1.09\ V$

By applying Le Chatelier's Principle to the cell equation, it is obvious that HF is neutralized in basic solution and thus the forward reaction should be favored in basic solution.

89. $$2.50\ L \times 1.75\ M\ Na_2SO_3 \times \frac{2\ mol\ H_2S}{mol\ SO_3^{2-}} \times \frac{100\ mol\ air}{1.5\ mol\ H_2S} \times \frac{\frac{62.4\ mm\ Hg\ L}{K\ mol} \times 298\ K}{755\ mm\ Hg} = 1.4 \times 10^4\ L\ air$$

91. The cell is 335 pm on each side.

$$V = l^3 = (335\ pm)^3 \times \left(\frac{10^{-12}\ m}{pm}\right)^3 \times \left(\frac{cm}{10^{-2}\ m}\right)^3 = 3.760 \times 10^{-23}\ cm^3$$

One atom in the unit cell.

$$d = \frac{209\ u}{3.760 \times 10^{-23}\ cm^3} \times \frac{1.661 \times 10^{-24}\ g}{u} = 9.23\ g/cm^3$$

93. $Mg + 2\ H_2O \rightarrow Mg(OH)_2 + 2\ H^+ + 2\ e^-$

$NO_3^- + 9\ H^+ + 8\ e^- \rightarrow NH_3 + 3\ H_2O$

$4\ Mg + 5\ H_2O + NO_3^- + H^+ \rightarrow 4\ Mg(OH)_2 + NH_3$

$4\ Mg + 6\ H_2O + NO_3^- \rightarrow 4\ Mg(OH)_2 + NH_3 + OH^-$

$NH_3 + HCl \rightarrow NH_4Cl$

$HCl + NaOH \rightarrow Na^+ + Cl^- + H_2O$

$$[NO_3^-] = (50.00\ mL \times 0.1500\ M\ HCl - 32.10\ mL \times 0.1000\ M\ NaOH \times \frac{mol\ HCl}{mol\ NaOH}) \times \frac{mol NH_3}{mol HCl} \times \frac{mol\ NO_3}{mol\ NH_3} \times \frac{1}{25.00\ mL} = 0.1716\ M\ NO_3^-$$

95. $2\ CH_4 + S_8 \rightarrow 2\ CS_2 + 4\ H_2S$

$CS_2 + 3\ Cl_2(g) \rightarrow CCl_4 + S_2Cl_2$

$4\ CS_2 + 8\ S_2Cl_2 \rightarrow 4\ CCl_4 + 3\ S_8$

97. (a) $1/2\ Cl_2(g) + 1/2\ F_2(g) \rightarrow ClF(g)$

$\Delta H = BE(react) - BE(prod)$

$$\Delta H = 1/2 \text{ mol } Cl_2 \times \frac{243 \text{ kJ}}{\text{mol}} + 1/2 \text{ mol } F_2 \times \frac{159 \text{ kJ}}{\text{mol}} - 1 \text{ mol ClF} \times \frac{251 \text{ kJ}}{\text{mol}}$$

$$= \frac{-50 \text{ kJ}}{\text{mol}}$$

(b) $1/2\ O_2(g) + F_2(g) \rightarrow OF_2(g)$

$$\Delta H = 1/2 \text{ mol } O_2 \times \frac{498 \text{ kJ}}{\text{mol}} + 1 \text{ mol } F_2 \times \frac{159 \text{ kJ}}{\text{mol}} - 2 \text{ mol OF} \times \frac{187 \text{ kJ}}{\text{mol}} = \frac{34 \text{ kJ}}{\text{mol}}$$

(c) $Cl_2(g) + 1/2\ O_2(g) \rightarrow Cl_2O(g)$

$$\Delta H = 1 \text{ mol } Cl_2 \times \frac{243 \text{ kJ}}{\text{mol}} + 1/2 \text{ mol } O_2 \times \frac{498 \text{ kJ}}{\text{mol}} - 2 \text{ mol ClO} \times \frac{205 \text{ kJ}}{\text{mol}}$$

$$= \frac{82 \text{ kJ}}{\text{mol}}$$

(d) $1/2\ N_2(g) + 3/2\ F_2(g) \rightarrow NF_3(g)$

$$\Delta H = 1/2 \text{ mol } N_2 \times \frac{946 \text{ kJ}}{\text{mol}} + 3/2 \text{ mol } F_2 \times \frac{159 \text{ kJ}}{\text{mol}} - 3 \text{ mol NF} \times \frac{280 \text{ kJ}}{\text{mol}}$$

$$= \frac{-129 \text{ kJ}}{\text{mol}}$$

99. $XeF_4 + 4\ e^- \rightarrow Xe + 4\ F^-$ reduction
$2\ H_2O \rightarrow 4\ e^- + 4\ H^+ + O_2$ oxidation
$XeF_4 + 3\ H_2O \rightarrow XeO_3 + 4\ HF + 2\ H^+ + 2\ e^-$ oxidation
Xe to XeO_3 ratio is 2 to 1. Thus 6 e^- are required from H_2O oxidation.
$XeF_4 + 3\ H_2O \rightarrow XeO_3 + 4\ HF + 2\ H^+ + 2\ e^-$
$\{2\ H_2O \rightarrow 4\ e^- + 4\ H^+ + O_2\} \times 3/2$
$\underline{\{XeF_4 + 4\ e^- \rightarrow Xe + 4\ F^-\} \times 2}$
$3\ XeF_4 + 6\ H_2O \rightarrow 2\ Xe + XeO_3 + 12\ HF + 3/2\ O_2$
OR
$XeF_4 + 4\ H^+ + 4\ e^- \rightarrow Xe + 4\ HF$ reduction
Xe to XeO_3 ratio is 2 to 1.
$\{XeF_4 + 4\ H^+ + 4\ e^- \rightarrow Xe + 4\ HF\} \times 2$
$\underline{XeF_4 + 3\ H_2O \rightarrow XeO_3 + 4\ HF + 2\ H^+ + 2\ e^-}$
$3\ XeF_4 + 6\ H^+ + 6\ e^- + 3\ H_2O \rightarrow 2\ Xe + XeO_3 + 12\ HF$
Thus 6 e^- are required from H_2O oxidation.
$\underline{\{2\ H_2O \rightarrow 4\ e^- + 4\ H^+ + O_2\} \times 3/2}$
$3\ XeF_4 + 6\ H_2O \rightarrow 2\ Xe + XeO_3 + 12\ HF + 3/2\ O_2$

Apply Your Knowledge

101. $2Al + Fe_2O_3 \rightarrow 2Fe + Al_2O_3$
ΔH_f – 824.2 –1676
$\Delta H_{rxn} = -852$ kJ
$4Al + 3MnO_2 \rightarrow 3Mn + 2Al_2O_3$
ΔH_f – 520 –1676
$\Delta H_{rxn} = -1792$ kJ
$2Al + Cr_2O_3 \rightarrow 2Cr + Al_2O_3$

ΔH_f -1135 -1676

$\Delta H_{rxn} = -541$ kJ

$3Si + 2Fe_2O_3 \rightarrow 4Fe + 3SiO_2$

ΔH_f -824.2 -910.9

$\Delta H_{rxn} = -1084$ kJ

$Si + MnO_2 \rightarrow Mn + SiO_2$

ΔH_f -520 -910.9

$\Delta H_{rxn} = -390$ kJ

$3Si + 2Cr_2O_3 \rightarrow 4Cr + 3SiO_2$

ΔH_f -1135 -910.9

$\Delta H_{rxn} = -463$ kJ

The aluminum with MnO_2 should produce the highest temperature.

103. (a) $4\ KO_3 + 2\ H_2O \rightarrow 5\ O_2 + 4\ OH^- + 4\ K^+$

$4\ KO_4 + 2\ H_2O \rightarrow 7\ O_2 + 4\ OH^- + 4\ K^+$

(b) $$\frac{0.35\ \text{atm} \times 10.0\ \text{L}}{\frac{0.08206\ \text{L atm}}{\text{K mol}} \times 298\ \text{K}} \times \frac{4\ \text{mol KO}_4}{7\ \text{mol O}_2} \times \frac{103.1\ \text{g}}{\text{mole KO}_4} = 8.4\ \text{g KO}_4$$

$$\frac{0.35\ \text{atm} \times 10.0\ \text{L}}{\frac{0.08206\ \text{L atm}}{\text{K mol}} \times 298\ \text{K}} \times \frac{4\ \text{mol KO}_3}{5\ \text{mol O}_2} \times \frac{87.1\ \text{g}}{\text{mole KO}_3} = 10\ \text{g KO}_3$$

$4\ KO_2 + 2\ H_2O \rightarrow 3\ O_2 + 4\ OH^- + 4\ K^+$

$$\frac{0.35\ \text{atm} \times 10.0\ \text{L}}{\frac{0.08206\ \text{L atm}}{\text{K mol}} \times 298\ \text{K}} \times \frac{4\ \text{mol KO}_2}{3\ \text{mol O}_2} \times \frac{71.1\ \text{g}}{\text{mole KO}_2} = 14\ \text{g KO}_2$$

$2\ Na_2O_2 + 2\ H_2O \rightarrow 4\ OH^- + O_2 + 4\ Na^+$

$$\frac{0.35\ \text{atm} \times 10.0\ \text{L}}{\frac{0.08206\ \text{L atm}}{\text{K mol}} \times 298\ \text{K}} \times \frac{2\ \text{mole Na}_2\text{O}_2}{\text{mole O}_2} \times \frac{78\ \text{g}}{\text{mole Na}_2\text{O}_2} = 22\ \text{g Na}_2\text{O}_2$$

$4\ LiO_2 + 2\ H_2O \rightarrow 3\ O_2 + 4OH^- + 4\ Li^+$

$$\frac{0.35\ \text{atm} \times 10.0\ \text{L}}{\frac{0.08206\ \text{L atm}}{\text{K mol}} \times 298\ \text{K}} \times \frac{4\ \text{mol LiO}_2}{3\ \text{mol O}_2} \times \frac{38.9\ \text{g}}{\text{mole LiO}_2} = 7.4\ \text{g LiO}_2$$

$LiO_2 < KO_4 < KO_3 < KO_2 < Na_2O_2$

105. $$[Cl_2] = \frac{6.4\ \text{g}}{\text{L}} \times \frac{\text{mole}}{70.90\ \text{g}} = 0.0903\ \text{M}$$

Cl_2	$+ 2\ H_2O \rightarrow$	$HOCl +$	$H_3O^+ +$	Cl^-
0.0903 M				
0.0903 - X		X	X	X

$$K_c = \frac{[HOCl][H_3O^+][Cl^-]}{[Cl_2]} = 4.4 \times 10^{-4}$$

$$4.4 \times 10^{-4} = \frac{X^3}{0.0903 - X}$$

$X^3 + 4.4 \times 10^{-4} X = 3.97 \times 10^{-5}$

Assume $X = 0.010$

$1.0 \times 10^{-6} + 4.4 \times 10^{-6} = 5.4 \times 10^{-6} < 3.97 \times 10^{-5}$

$X \approx 0.020$

$8.0 \times 10^{-6} + 8.8 \times 10^{-6} = 1.68 \times 10^{-5} < 3.97 \times 10^{-5}$

$X \approx 0.030$

$2.7 \times 10^{-5} + 1.3 \times 10^{-5} = 4.0 \times 10^{-5} \approx 3.97 \times 10^{-5}$

$X \approx 0.0290$

$2.44 \times 10^{-5} + 1.28 \times 10^{-5} = 3.72 \times 10^{-5} < 3.97 \times 10^{-5}$

$X = 0.030$

$[Cl_2] = 0.0903 - 0.030 = 0.060$ M

$[HOCl] = 0.030$ M

$[Cl^-] = 0.030$ M

Chapter 22

The *d*-Block Elements and Coordination Chemistry

Exercises

22.1 $4\ FeCr_2O_4 + 16\ NaOH + 7\ O_2 \xrightarrow{\Delta} 8\ Na_2CrO_4 + 4\ Fe(OH)_3 + 2\ H_2O$

22.2 $MnO_4^-(aq) + 2\ H_2O + 3\ e^- \rightarrow MnO_2(s) + 4\ OH^-(aq)$ $\quad E° = 0.60$ V
$MnO_4^-(aq)$ in basic solution will oxidize any species for which $E° < 0.60$ V. This includes $Br^-(aq)$ to $BrO_3^-(aq)$, $Br_2(l)$ to $BrO^-(aq)$, Ag(s) to $Ag_2O(s)$, $NO_2^-(aq)$ to $NO_3^-(aq)$, S(s) to $SO_3^{2-}(aq)$, and so on.

22.3 (a) Coordination number = 6
$ox\#_{Co}$ + anionic charge$_{SO_4{}^{2-}}$ = species charge
$ox\#_{Co} + (-2) = +1$
$ox\#_{Co} = +3$
(b) Coordination number = 6
$ox\#_{Fe} + 6 \times$ anionic charge$_{CN^-}$ = species charge
$ox\#_{Co} + 6 \times (-1) = -4$
$ox\#_{Fe} = +2$

22.4 (a) hexaamminecobalt(II) ion
(b) tetrachloroaurate(III) ion
(c) pentaamminebromocobalt(III) bromide

22.5 (a) $[Co\ (en)_3]^{3+}$ (b) $[CrCl_4(NH_3)_2]^-$ (c) $[PtCl_2(en)_2]SO_4$

22.6A (a) These are isomers. The atom donating electrons is S in the first and N in the second.
(b) These are isomers. The ligands, NH_3, are bonded to Zn in one structure and Cu in the other. Similarly, the Cl^- ligands are bonded to Cu in one structure and Zn in the other.
(c) These are not isomers. There are two NH_3 in one structure and 4 NH_3 in the other.

22.6B (a) Different structural isomers are not possible.
(b) $[Pt(NH_3)_4][CuCl_4]$ is another isomer, as are other combinations of NH_3 and Cl^- on Cu^{2+} and Pt^{2+} such as $[CuCl(NH_3)_3][PtCl_3(NH_3)]$ and $[PtCl(NH_3)_3][CuCl_3(NH_3)]$.

22.7 Assuming that the ethylenediamine molecule can link only to adjacent coordination sites, there is only one form of the molecule and no geometric or optical isomers.

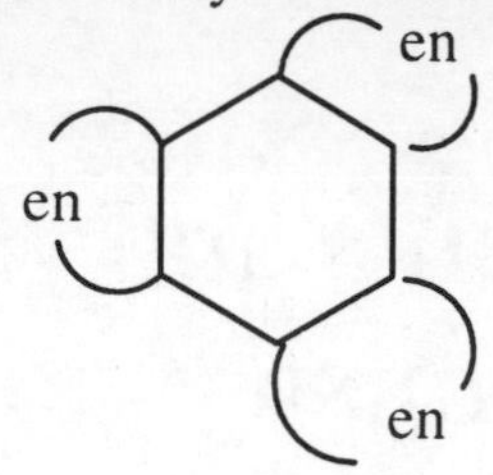

22.8 (a)

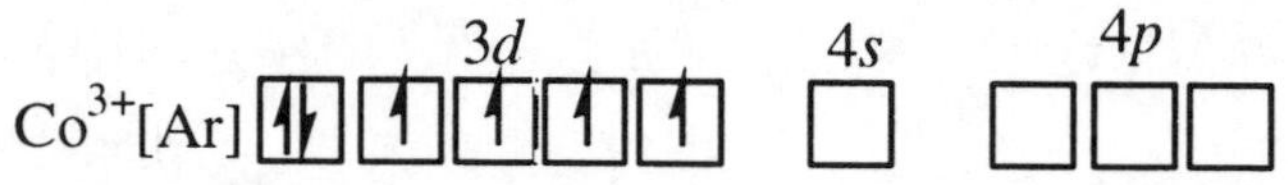

For a strong field ligand (CN^-), the electrons are paired in the lower set of orbitals.

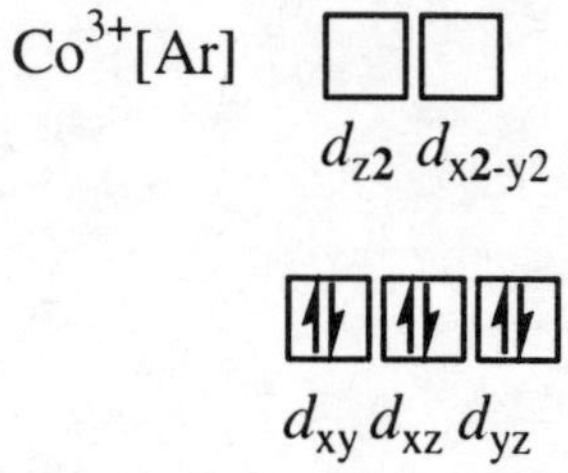

There are no unpaired electrons.

(b)

Cl^- is usually a weak-field ligand, but the distribution of the eight 3*d* electrons would be the same even if it were strong field. There are two unpaired electrons.

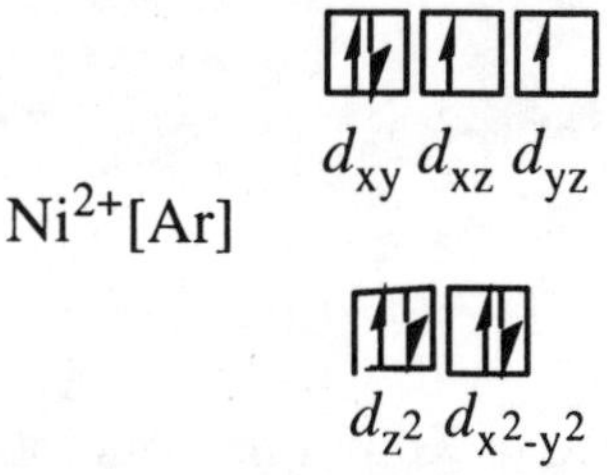

weak field

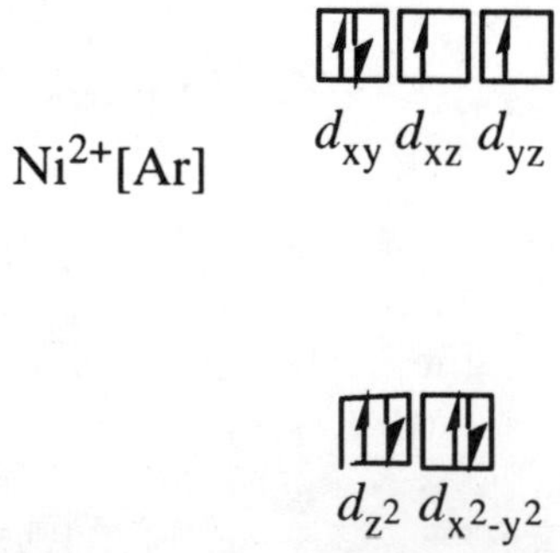

strong field

Review Questions

2. Cobalt is [Ar] $3d^7 4s^2$. It is a transition element as the 3*d* subshell is partially filled. [Kr] $4d^{10} 5s^2 5p^1$ is a representative element, indium. The 4*d* subshell is filled, and the 5*p* subshell is partially filled.

8. (a) scandium chloride (b) iron(II) silicate

(c) sodium manganate (d) chromium (VI) oxide

9. (a) $BaCr_2O_7$ (b) Cr_2O_3 (c) Hg_2Br_2 (d) V_2O_5

13. Only for monodentate ligands is the number of ligands equal to the coordination number. Some ligands are polydentate and occupy more than one coordination site.

17. (a) water (b) chloride ion (c) ammonia (d) ethylenediamine
(e) oxalate ion (f) ethylenediaminetetraacetate ion

Problems

23. In both cases electrons are lost to produce the electron configuration of Ar. With calcium this means the two 4*s* electrons, and the ion Ca^{2+} is formed. With scandium, the 3*d* electron is lost as well, producing Sc^{3+}.

25. (a) The Ca atom is smaller than the K atom because it has a higher nuclear charge (+20 compared to +19), but the same number of electrons in its noble-gas core (18), coupled with the fact that the two 4*s* electrons are not effective in screening one another.
(b) The Mn atom is smaller than the Ca atom because it has a higher nuclear charge (+25 compared to +20) and the same number of valence electrons in the same configuration ($4s^2$).
(c) The Mn and Fe atoms are about the same size because they have about the same nuclear charge (+25 and +26, respectively), the same number of valence electrons ($4s^2$), and inner shell electrons that are about equally effective in shielding the valence electrons from the nucleus. The effective nuclear charges of Mn and Fe are essentially the same and roughly equal to $\{25 - (18 + 5)\}$ and $\{26 - (18 + 6)\}$.

27. (a) $2\ Sc(s) + 6\ HCl(aq) \rightarrow 2\ ScCl_3(aq) + 3\ H_2(g)$
(b) $Sc(OH)_3(s) + 3\ HCl(aq) \rightarrow ScCl_3(aq) + 3\ H_2O(l)$
(c) $Sc(OH)_3(s) + 3\ Na^+(aq) + 3\ OH^-(aq) \rightarrow [Sc(OH)_6]^{3-}(aq) + 3\ Na^+(aq)$

29. (a) $Ba(s) + 2\ H^+(aq) + 2\ Cl^-(aq) \rightarrow Ba^{2+}(aq) + H_2(g) + 2\ Cl^-(aq)$; followed by, $Ba^{2+}(aq) + 2\ Cl^-(aq) + 2\ K^+(aq) + CrO_4^{2-}(aq) \rightarrow$
$BaCrO_4\ (s) + 2\ K^+(aq) + 2\ Cl^-(aq)$
(b) Combine the reduction half-reaction $MnO_4^-(aq) + 2\ H_2O + 3\ e^- \rightarrow MnO_2(s) + 4\ OH^-(aq)$, $E° = 0.60$ V with a half-reaction for which $E° < 0.60$ V. For example, $3\ NO_2^-(aq) + 2\ MnO_4^-(aq) + H_2O \rightarrow 3\ NO_3^-(aq) + 2\ MnO_2(s) + 2OH^-(aq)$, $E°_{cell} = E°_{MnO_4^-/MnO_2} - E°_{NO_3^-/NO_2^-}$
$= 0.60\ V - 0.01\ V = 0.59\ V$

31. $Mn^{2+}(aq) + 4\ H_2O(l) \rightarrow MnO_4^-(aq) + 8\ H^+(aq) + 5\ e^-$
$BiO_3^-(aq) + 6\ H^+(aq) + 2\ e^- \rightarrow Bi^{3+}(aq) + 3\ H_2O(l)$

$2\ Mn^{2+}(aq) + 5\ BiO_3^-(aq) + 14\ H^+(aq) \rightarrow 2\ MnO_4^-(aq) + 5\ Bi^{3+}(aq) + 7\ H_2O(l)$

33. $[CrO_4^{2-}]$ in the reversible reaction
$Cr_2O_7^{2-} + H_2O \rightleftharpoons 2\ CrO_4^{2-} + 2\ H^+$
is large enough that K_{sp} of $PbCrO_4(s)$ is exceeded.

35. Fe(s) is the best reducing agent. The reducing agent is oxidized, and $E° = -0.440$ V for the half-reaction: $Fe^{2+} + 2\ e^- \rightarrow Fe(s)$. For the reaction of $Co^{2+} + 2\ e^- \rightarrow Co(s)$, $E° = -0.277$ V, and for the half reaction, $Fe^{3+}(aq) + e^- \rightarrow Fe^{2+}(aq)$, $E° = 0.771$ V, which means that $Fe^{3+}(aq)$ is readily reduced, making it an oxidizing agent. $Co^{3+}(aq)$ is an oxidizing agent (reduced to $Co^{2+}(aq)$), not a reducing agent.

37. $Fe_2O_3(s) + 10\ OH^-(aq) \rightarrow 2\ FeO_4^{2-}(aq) + 5\ H_2O(l) + 6\ e^-$
$\underline{3(Cl_{2(g)} + 2\ e^- \rightarrow 2\ Cl^-(aq))}$
$Fe_2O_3(s) + 3\ Cl_2(g) + 10\ OH^-(aq) \rightarrow 6\ Cl^-(aq) + 2\ FeO_4^{2-}(aq) + 5\ H_2O(l)$

39. (a) $Cu(s) \rightarrow Cu^{2+} + 2\ e^-$
$\underline{SO_4^{2-} + 4\ H^+ + 2\ e^- \rightarrow SO_2(g) + 2\ H_2O}$
$Cu(s) + SO_4^{2-} + 4\ H^+ \rightarrow SO_2(g) + 2\ H_2O + Cu^{2+}$ net ionic equation
(b) $NO_3^-(aq) + 4\ H^+ + 3\ e^- \rightarrow NO(g) + 2\ H_2O$
$3\ Cu(s) + 2\ NO_3^-(aq) + 8\ H^+(aq) \rightarrow 2\ NO(g) + 4\ H_2O(l) + 3\ Cu^{2+}(aq)$ net ionic equation

41. $NO_3^- + 4\ H^+ + 3\ e^- \rightarrow NO(g) + 2\ H_2O$ $\quad E°_{cathode} = 0.956$ V
$Ag \rightarrow Ag^+ + e^-$ $\quad E°_{anode} = 0.800$ V
$E°_{cell}$ for NO_3^- + Ag is 0.156 V, the Ag reaction will go.
$Au \rightarrow Au^{3+} + 3\ e^-$ $\quad E°_{anode} = 1.52$ V
$E°_{cell}$ for NO_3^- + Au is -0.56 V, the gold reaction will not occur spontaneously.

43. Group 2B elements form complexes like transition elements. The 2A elements don't form many complexes. Group 2B atomic radii are smaller than Group 2A. Compared to Group 2A, the 2B elements have higher ionization energies and $\Delta E°$ values that are less negative.

45. (a) $3(Hg(l) \rightarrow Hg^{2+} + 2\ e^-)$
$\underline{2(NO_3^- + 4\ H^+ + 3\ e^- \rightarrow NO(g) + 2\ H_2O)}$
$3\ Hg(l) + 2\ NO_3^- + 8\ H^+ \rightarrow 2\ NO + 4\ H_2O + 3\ Hg^{2+}$
(b) $ZnO(s) + 2\ CH_3CO_2H \rightarrow Zn^{2+} + 2\ CH_3CO_2^- + H_2O$

47. (a) 6 (b) 4 (c) 4 (d) 6

49. (a) +2 (b) +3 (c) +3 (d) +3

51. (a) tetraamminecopper(II) ion
(b) hexafluoroferrate(III) ion

(c) tetraamminedichloroplatinum(IV) ion
(d) tris(ethylenediamine)chromium(III) ion

53. (a) $[Fe(H_2O)_6]^{2+}$ (b) $[CrBrCl(NH_3)_4]^+$ (c) $[Al(ox)_3]^{3-}$

55. (a) potassium hexacyanochromate(II) (b) potassium trioxalatochromate(III)

57. (a) $Na_2[Zn(OH)_4]$ (b) $[Cr(en)_3]_2(SO_4)_3$ (c) $K_2Na[Co(ONO)_6]$

59. (a) It is an anion and should have an "ate" ending: tetrahydroxozincate(II) ion.
(b) Because the complex ion is an anion, the metal should be named last and given an "ate" ending: hexafluoroferrate(III) ion.

61. (a) Listing the ligands in a different order does not make these structures isomers. These are geometrical isomers only if a pair of ligands such as Cl, NH_3 or H_2O is cis in one structure and trans in the other, but this cannot be indicated by the formula alone.
(b) Yes, these are structural isomers.
(c) No, these are not isomers. They contain different numbers of Cl because the oxidation state of Co is +3 in one ion and +2 in the other.
(d) Yes, these are structural isomers. The NO_2 is bonded through the N atom in one structure and through the O atom in the other.

63. The en can only bond at adjacent coordination sites. Since the other four sites are all filled by Cl^-, there exists only one form. All others can be made by a rotation of the first. *Cis-trans* isomerism is not possible.

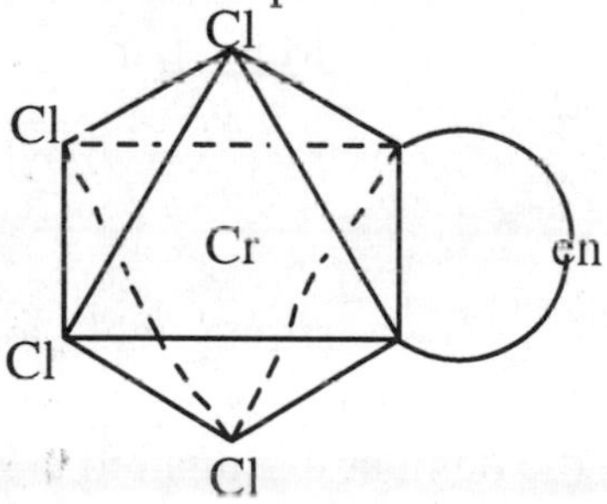

65.

Yes.

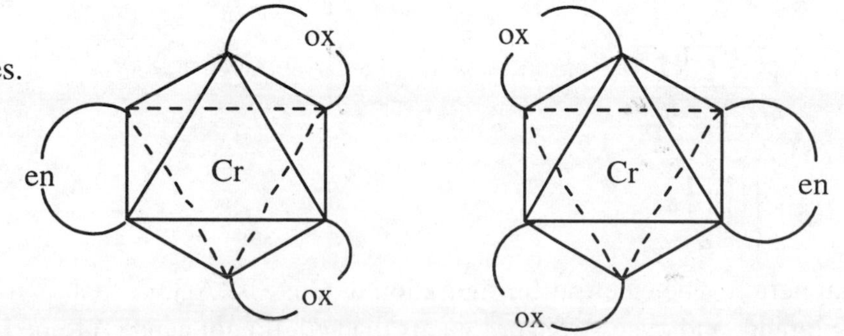

These mirror images are not superimposable.

67. (a)

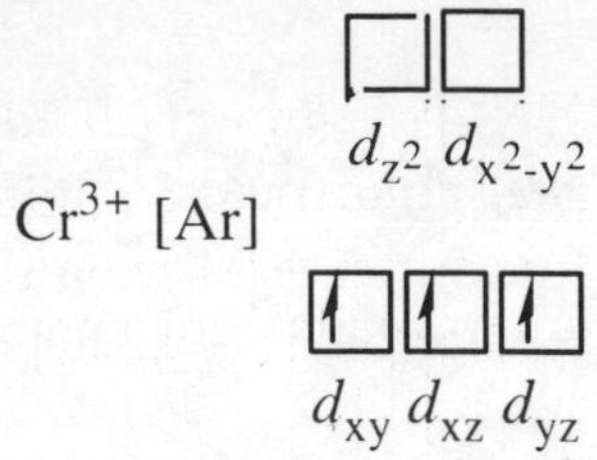

Paramagnetic. Cr^{3+}- [Ar] $3d^3$. with only three electrons, it doesn't matter whether it is low-spin or high-spin; they are the same. There are three unpaired electrons.

(b)

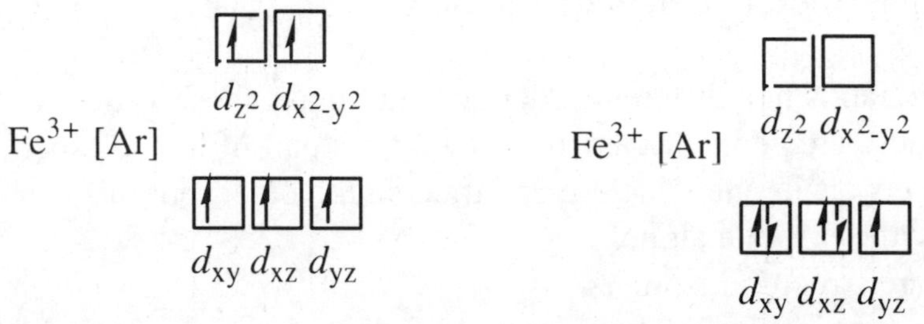

weak field-"high spin" strong field-"low spin"

Paramagnetic. Fe^{3+}-[Ar]$3d^5$. There is an odd number of electrons so it must be paramagnetic. Because Cl^- is a weak-field ligand it is expected that the high spin state will exist with five unpaired electrons.

(c)

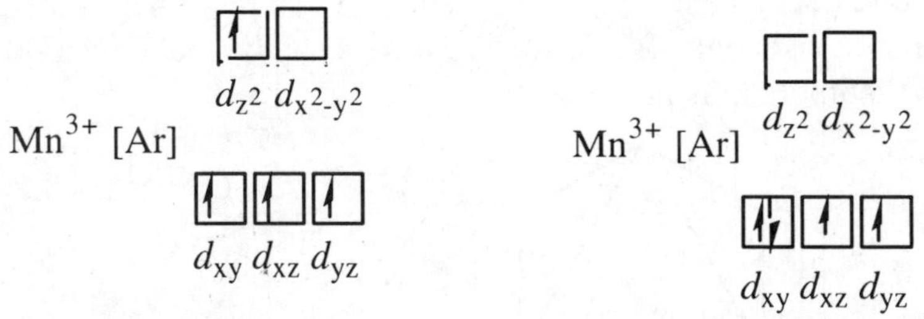

weak field-"high spin" strong field-"low spin"

Paramagnetic. The electron configuraton of Mn^{3+} is [Ar]$3d^4$. If the complex is of the "high spin" type, the number of unpaired electrons is four; for the "low spin" type it is two. Because CN^- is a strong-field ligand, it is expected that the ion is low-spin state with two unpaired e^-.

(d)

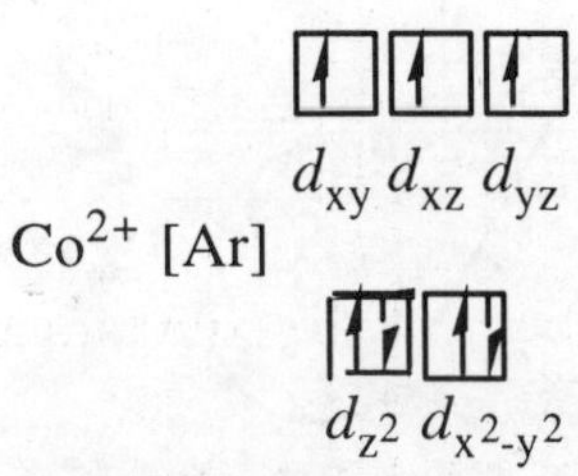

Paramagnetic. The electron configuration of Co^{2+} is [Ar]$3d^7$. Cl^- is usually a weak ligand, and even stronger ligands than Cl^- do not cause Δ to be large enough to force the lower orbitals to be filled first. There are three unpaired electrons.

69. $[Cr(H_2O)_6]^{3+}$ is violet and $[Cr(NH_3)_6]^{3+}$ is yellow. NH_3 is a stronger field ligand, causing a greater Δ value, and requiring light of a higher frequency to stimulate an electronic transition. Light of a higher frequency has a shorter wavelength. Violet is a shorter wavelength than yellow. $[Cr(NH_3)_6]^{3+}$ should absorb violet and transmit yellow light.

Additional Problems

71. A basic solution of chromate ion is not a good oxidizing agent. In HCl, the CrO_4^{2-} is changed to $Cr_2O_7^{2-}$ which is a good oxidizing agent, capable of oxidizing Cl^- to Cl_2 (g). The high concentration of H^+, and the fact that the reaction product Cl_2 escapes, cause the reaction to proceed in the forward direction, despite the slightly negative value of $E°_{cell}$.

73. $MnO_4^{2-} + 8\ H^+ + 4\ e^- \rightarrow Mn^{2+} + 4\ H_2O$
$\underline{\{MnO_4^{2-} \rightarrow MnO_4^- + e^-\} \times 4}$
$5\ MnO_4^{2-} + 8\ H^+ \rightarrow Mn^{2+} + 4\ MnO_4^- + 4\ H_2O$
OR
$MnO_4^{2-} + 4\ H^+ + 2\ e^- \rightarrow MnO_2(s) + 2\ H_2O$
$\underline{\{MnO_4^{2-} \rightarrow MnO_4^- + e^-\} \times 2}$
$3\ MnO_4^{2-} + 4\ H^+ \rightarrow MnO_2(s) + 2\ MnO_4^- + 2\ H_2O$

75. For each reaction, all $E°$ values are for reduction half-reactions and $E°_{cell} = E°_{cathode} - E°_{anode}$.

$Sn^{2+} \rightarrow Sn^{4+} + 2\ e^-$ $\quad E°_{anode} = 0.154$ V
$V^{2+} \rightarrow V^{3+} + e$ $\quad E°_{anode} = -0.255$ V
$Fe^{2+} \rightarrow Fe^{3+} + e^-$ $\quad E°_{anode} = 0.771$ V

(a) $MnO_2(s) + 4\ H^+ + 2\ e^- \rightarrow Mn^{2+} + 2\ H_2O$ $\quad E°_{cathode} = 1.23$ V
Sn^{2+}, V^{2+} and Fe^{2+} are all capable.

(b) $I_2 + 2\ e^- \rightarrow 2\ I^-$ $\quad E°_{cathode} = 0.535$ V
Only V^{2+} and Sn^{2+} are capable.

(c) $2\ H^+ + 2\ e^- \rightarrow H_2(g)$ $\quad E°_{cathode} = 0.0$ V
Only V^{2+} is capable.

77. $CrO_3 + 6\ H^+ + 6\ e^- \rightarrow Cr\ (s) + 3\ H_2O$

$$2.57\ A \times 3.25\ h \times \frac{3600\ s}{h} \times \frac{C}{A\ s} \times \frac{mol\ e^-}{96485\ C} \times \frac{mol\ Cr}{6\ mol\ e^-} \times \frac{52.00\ g\ Cr}{mol\ Cr} = 2.70\ g\ Cr$$

79. The *trans* isomer does not exhibit optical isomerism because the structure and its mirror image are superimposable. The cis isomer's mirror image is not superimposable, and so the cis isomer does exhibit optical isomerism.

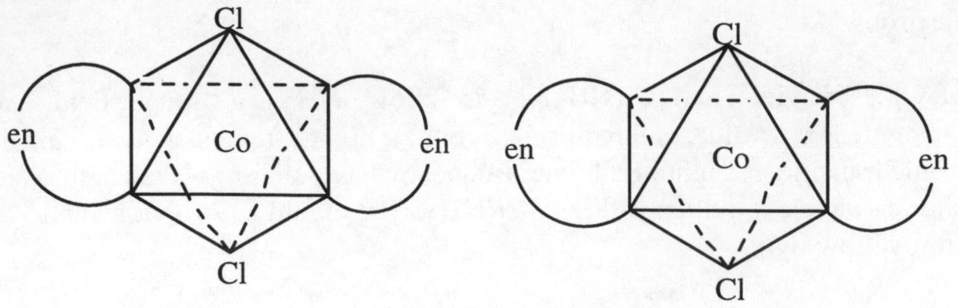

trans isomers

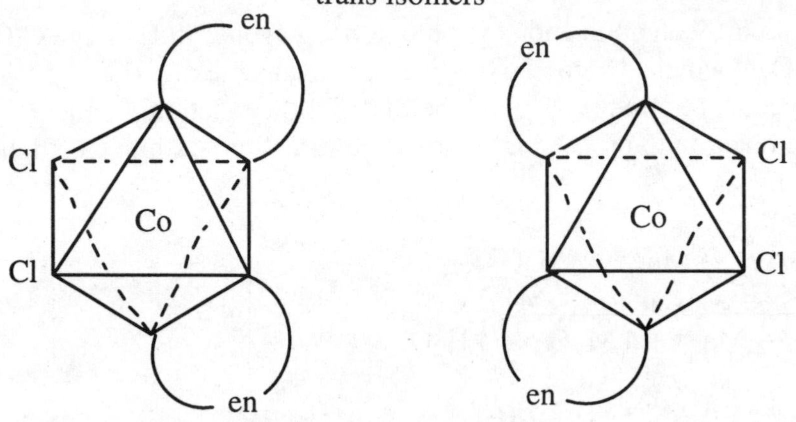

cis isomers

81. $[PtCl_2(NH_3)_2]$ is a square planar complex in which two identical ligands can be either cis (like ligands on adjacent corners) or trans (like ligands on diagonal corners.) $[ZnCl_2(NH_3)_2]$ is a tetrahedral complex, and does not exhibit cis-trans isomerism.

83. Cr^{3+} has an odd number of electrons so it will be paramagnetic with any ligand. Cr^{3+} has the electron configuration $[Ar]3d^3$. The three $3d$ electrons will remain unpaired in the lower-energy set of three $3d$ orbitals, regardless of whether the ligands (L) are strong field or weak field (see also, answer to 67a).

85. Because the complex cation and anion each have Pt^{2+} as the central ion, the formula should be based on a doubling of the empirical formula $2 \times (PtCl_2{\cdot}2NH_3)$: $Pt_2Cl_4{\cdot}4NH_3$. The true formula then is $[Pt(NH_3)_4][PtCl_4]$.

87.

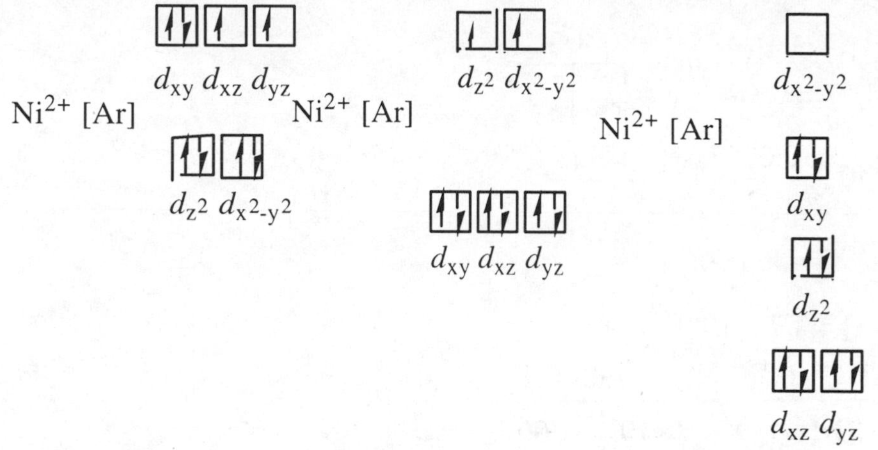

tetrahedral octahedral square planar

The electron configuration of Ni^{2+} is [Ar]$3d^8$. The complex ion $[Ni(CN)_4]^{2-}$ is a strong-field complex in which all the electrons are paired (diamagnetic). The assignment of eight electrons to the tetrahedral splitting diagram would leave two electrons unpaired. The assignment of eight electrons to the octahedral splitting diagram would also leave two electrons unpaired. The assignment of eight electrons to the diagram for a square planar complex would fill the four lowest energy *d* orbitals with electron pairs and leave the highest energy *d* orbital empty. The structure of the complex ion is square planar.

89. By experiment, bis(glycinalto)copper(II) is found to be square planar. The cis form crystallizes as blue needles and the trans form as violet scales.

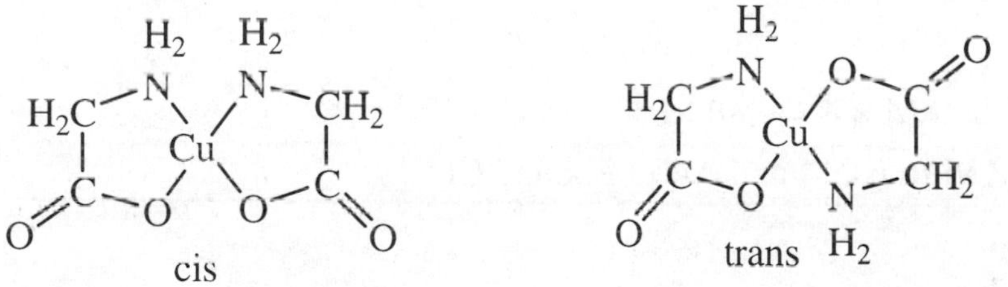

cis trans

Neither the cis form nor the trans form is optically active. A mirror image could be flipped over and would thus superimpose.

Apply Your Knowledge

91. $? \text{ M } Ni^{2+} = (25.0 \text{ mL} \times 0.0200 \text{ M EDTA} - 3.32 \text{ mL} \times 0.0347 \text{ M})$

$$\times \frac{\text{mmol } Ni^{2+}}{\text{mmol EDTA}} \times \frac{1}{25.0 \text{ mL}} = 0.0154 \text{ M } Ni^{2+}$$

The results would have been the same because EDTA reacts with all polyvalent metal ions in a one to one ratio regardless of charge.

93. $Cu \rightarrow Cu^{2+} + 2\,e^-$ $E°_{anode} = -0.340 \text{ V}$

$\underline{2\,H^+ + 2\,e^- \rightarrow H_2}$ $E°_{cathode} = 0.000 \text{ V}$

$Cu + 2\,H^+ \rightarrow Cu^{2+} + H_2$

$E^\circ_{cell} = E^\circ_{cathode} - E^\circ_{anode} = 0.000 - (-0.340\text{ V}) = -0.340$

$$0.08 = -0.340 - \frac{0.0257}{2}\ln\frac{[Cu^{2+}]P_{H_2}}{[H^+]^2}$$

$$0.08 = -0.340 - \frac{0.0257}{2}\ln\frac{[Cu^{2+}]1}{1}$$

$$\frac{0.42 \times 2}{-0.0257} = -33 = \ln [Cu^{2+}]$$

$e^{-33} = [Cu^{2+}] = 4.7 \times 10^{-15}$

$$K_f = \frac{[Cu(NH_3)_4]^{2+}}{[Cu^{2+}][NH_3]_4} = \frac{1.00}{4.7 \times 10^{-15}(1.00)^4} = 2 \times 10^{14}$$

95. $[Cu(H_2O)_4]^{2+} + 4\ Cl^- \rightleftharpoons [CuCl_4]^{2-} + 4\ H_2O \qquad K = 4.2 \times 10^5$

$$\frac{[CuCl_4]^{2-}}{[Cu(H_2O)_4]^{2+}[Cl^-]^4} = 4.2 \times 10^5$$

$$\frac{(0.099)}{(0.001)[Cl^-]^4} = 4.2 \times 10^5$$

$$[Cl^-]^4 = \frac{0.099}{0.001 \times 4.2 \times 10^5}$$

$[Cl^-]^4 = 2.4 \times 10^{-4}$

$[Cl^-] = 0.12$ M

Cl^- in $CuCl_4^-$ =0.099 M × 4 = 0.40 M

Total $[Cl^-]$ = 0.12 M free Cl^- and 0.40 M bound Cl^-

Chapter 23

Chemistry and Life: More on Organic, Biological, and Medicinal Chemistry

Exercises

23.1 (a) Aliphatic. The compound has a conjugated bonding system, but it does not extend completely around the ring. It also has only eight π electrons instead of the 10 $(4 \times 2 + 2)$ π electrons needed.

(b) Aliphatic. The compound has a conjugated bonding system, but it does not extend completely around the ring. It does have the six π electrons $[(4 \times 1) + 2]$ needed to be aromatic.

(c) Aliphatic. The compound has a conjugated bonding system, that does extend completely around the ring. It does not have the six π electrons $[(4 \times 1) + 2]$ or ten π electrons $[(4 \times 2) + 2]$ needed to be aromatic; it has eight π electrons.

(d) Aromatic. The compound has a conjugated bonding system that does extend completely around the ring. It does have the 14 π electrons $[(4 \times 3) + 2]$ needed to be aromatic.

23.2 The sign of the optical rotation cannot be predicted from the structure alone.

(a) Guclose has six carbon atoms and an aldehyde group. In the Fisher projection, the bottom chiral carbon is on the left –the L configuration. It is an L–aldohexose.

(b) Erythrulose has four carbon atoms and a ketone group. In the Fisher projection, the bottom chiral carbon is on the left—the L configuration. It is an L–ketotetrose.

Review Questions

5. (a) Aromatic. The compound has a conjugated π bond system completely around the ring and has six π electrons $[(4 \times 1) +2]$.

(b) Aliphatic. The conjugated bond system does not extend completely around the ring.

(c) Aliphatic. The π bond system is not conjugated.

(d) The ring is aromatic; the side-chain is aliphatic. The side chain has no double bonds. The ring has conjugated π bond system completely around the ring and has six π electrons.

7. Soap molecules and glycerol are formed.

9. The D and L relate to the absolute configurations of groups attached to the chiral carbon atoms adjacent to the CH_2OH end of the chain forms of the sugar molecule. Both sugars have the D configuaration. D and L are not related to the sign of the

optical rotation. Glucose rotates light to the right (+); fructose rotates light to the left (–).

15. DNA controls the process of protein synthesis because single-stranded messenger RNA is synthesized from DNA by a process of complementary base pairing similar to the replication of DNA. This messenger RNA controls protein synthesis by transmitting instructions originally encoded in the DNA.

17. Emission spectroscopy refers to the analysis of the radiation given off (emitted) by an energetically excited species. Absorption spectroscopy refers to the analysis of light absorbed by a sample.

20. UV-spectroscopy uses radiation of the highest energy content. NMR absorbs radiation of the smallest energy content or longest wavelength; radio waves are absorbed.

Problems

33. (a) $CH_3CH{=}CHCHCH_2CH_3$
    |
    CH_3

(b) $CH_3CH_2CHCH_2CH_3$
    |
    CH_3

(c) CH_3
    |
    $CH_3CCH_2CH_2CHCH_3$
    | |
    CH_3 CH_3

(d) $CH_3CH_2CHCHCH_2CH_2CH_2CH_3$
    | |
    H_3C CH_2
    |
    CH_3

(e) $HC{\equiv}CCHCH_2CH_2CH_3$
    |
    $HCCH_3$
    |
    CH_3

(f) $CH_3C{=}CCH_3$
    | |
    H_3C CH_3

35. (a) 3-methylpentane
(b) 3-methylpentane
(c) 2,6-dimethyl-3-heptyne
(d) 2-methyl-2-pentene

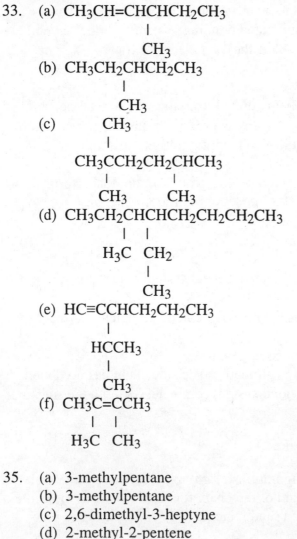

37. (a) 1-hexanol
(b) 2-hexanol
(c) 1,2,4-trichlorobenzene
(d) 2,4-dibromotoluene

39. (a)
```
CH3CH2CHCH2CH2CH3
         |
         O
         H
```
(b)
```
        CH3
         |
   CH3CHCCH3
       |  |
       O CH3
      H
```
(c)
```
CH3CH2CHCHCHCH2CH3
         |  |   |
         O  CH3CH3
         H
```
(d)
```
HOCH2CHCH2CH3
       |
      CH2
       |
      CH3
```
(e)

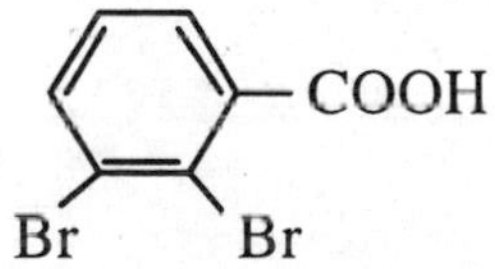

(f)

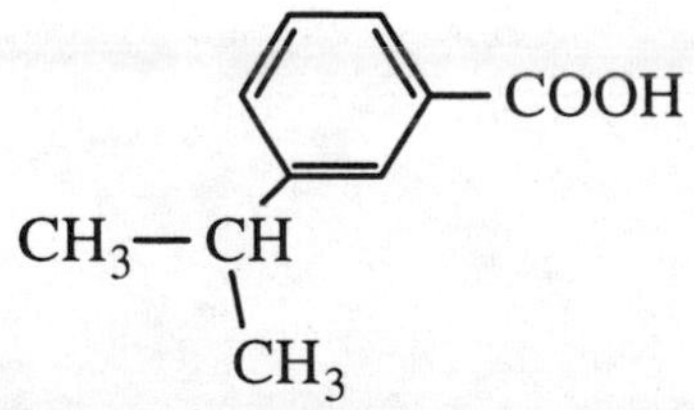

41. (a) 3-chloropentanoic acid, $CH_3CH_2CHClCH_2COOH$
(b) 2-ethyl-1-pentene, $CH_2{=}C(CH_2CH_3)CH_2CH_2CH_3$
(c) 2,4-dibromobenzoic acid, ($2,4\text{-}Br_2C_6H_3COOH$)

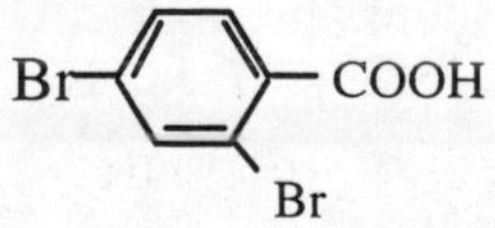

43. (a) oxidation
(b) hydration

45. (a) $(CH_3)_2C-CH_2$
        | |
       Br Br
(b) $CH_3CHCH_2CH_3$
        |
       CH_3

47. (a) $HOCH_2CH_2CH_3 \xrightarrow{H_2SO_4 \text{ and } \Delta} CH_2{=}CHCH_3 + H_2O$
(b) $2\ HOCH_2CH_2CH_3 \xrightarrow{H_2SO_4} CH_3CH_2CH_2OCH_2CH_2CH_3$

49. (a) $CH_2{=}CHCH_3$ (b) $CH_2{=}CCH_3$
        |
       CH_3

51. (a) $CH_3CH_2CH_2COOH + NaHCO_3 \rightarrow CH_3\ CH_2CH_2COO^-Na^+ + H_2O + CO_2$
(b) $CH_3COOCH_2CH_3 + H_2O \xrightarrow{H_2SO_4} CH_3COOH + HOCH_2CH_3$

53. (a) saturated 10C
(b) unsaturated 14C
(c) unsaturated (2 double bonds) 16C

55. (a) $CH_3(CH_3)_{16}COOH$
(b) $CH_3(CH_2)_{12}COO^-K^+$
(c) $CH_2OCO(CH_2)_7CH{=}CH(CH_2)_7CH_3$
|
$CHOCO(CH_2)_7CH{=}CH(CH_2)_7CH_3$
|
$CH_2OCO(CH_2)_7CH{=}CH(CH_2)_7CH_3$
(d) $CH_2OCO(CH_2)_{14}CH_3$
|
$CHOCO(CH_2)_{14}CH_3$
|
$CH_2OCO(CH_2)_{14}CH_3$

57. (a) aldose, triose
(b) aldose, pentose
(c) aldose, pentose
(d) ketose, hexose
(e) aldose, hexose

59. (a) L-sugar

(b) D-sugar
(c) L-sugar

61. (a)

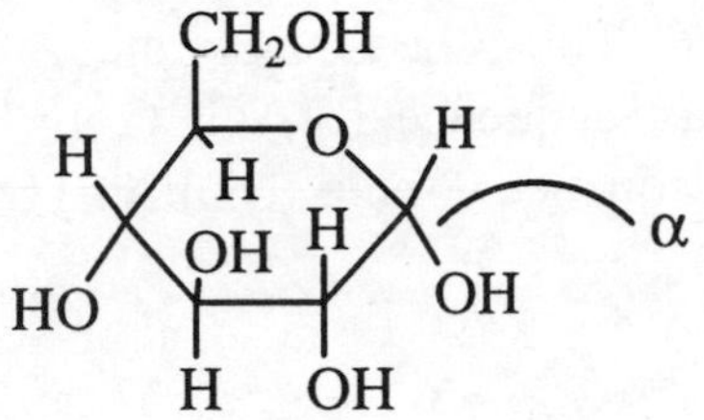

(b)

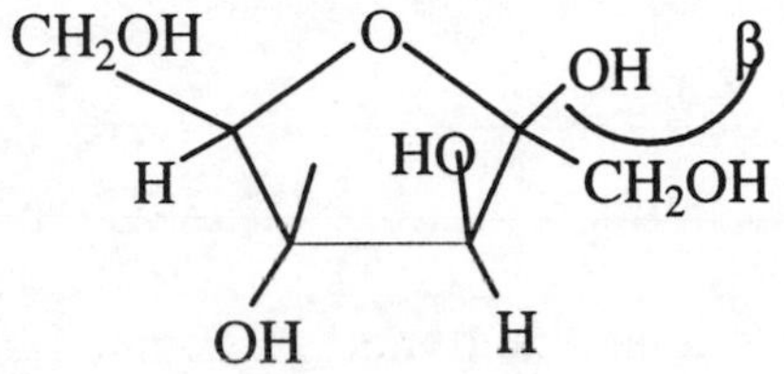

63. (a) β
(b) α

65. (a) $-CH_3$
(b) $-(CH_2)_4NH_2$
(c)

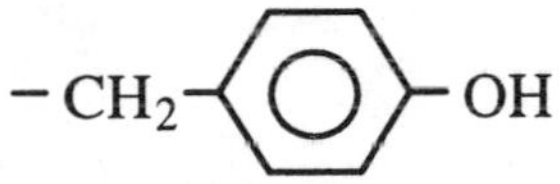

(d) $-\underset{\displaystyle CH_3}{\underset{|}{CH}}CH_2CH_3$

67. There are two amino acids with acidic side chains:

aspartic acid and glutamic acid.

$H_2N\underset{\displaystyle CH_2COOH}{\underset{|}{CH}}COOH$ $\qquad$ $H_2N\underset{\displaystyle CH_2CH_2COOH}{\underset{|}{CH}}COOH$

69. $H_2NCHCOO^-$

71. (a) $pH < pI$
(b) $pH > pI$
(c) at pH equal to the isoelectric point.

73. The sugar is ribose; the heterocyclic base is the pyrimidine uracil.

75. This molecule is not a nucleotide, since it has no phosphate group; rather, it is a nucleoside. The base is a pyrimidine since it has but one ring; it is thymine. Since the sugar is deoxyribose, this compound would be incorporated into DNA.

77. The pairing in DNA is always A to T and G to C. Thus, the DNA strand that is complementary to ATTCGC would have the base sequence TAAGCG. In RNA, uracil takes the place of thymine. The mRNA sequence copied from ATTCGC would have the base sequence UAAGCG.

79. The pairing in RNA is always A to U and G to C.
(a) AAA is the anticodon for UUU.
(b) GUA is the anticodon for CAU.
(c) UCG is the anticodon for AGC.
(d) GGC is the anticodon for CCG.

81. We should be able to distinguish between acetic acid and methyl acetate by detecting an absorption peak for the O—H group of the carboxylic acid in acetic acid in the range 2500 to 3000 cm^{-1}. Methyl acetate does not have a corresponding peak.

83. Yes, we would expect absorption in the infrared because the organic molecule will likely have one or more of the structural features noted in Table 23.6.

85. A substance that appears yellow absorbs the complementary color, violet light with wavelength of about 425 nm (see Table 23.7).

Additional Problems

87. Organic compounds have low melting points and are insoluble in water because the intermolecular forces are of the relatively weak dispersion type. Many organic compounds will burn. This solid is probably organic. Most ionic compounds have moderate to high melting points; many are water soluble; few are combustible.

89. (a) The esterification reaction is reversed (acidic hydrolysis).

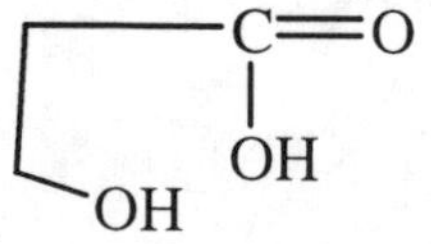

or $HOCH_2CH_2COOH$

(b)

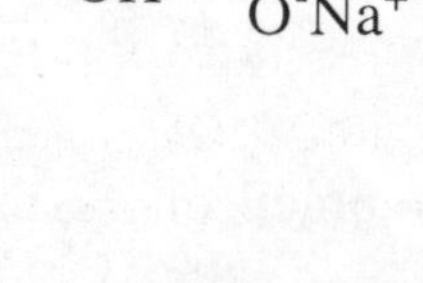

or $HOCH_2CH_2CH_2C{=}O + Na^+$
$\quad\quad\quad\quad\quad\quad |$
$\quad\quad\quad\quad\quad\quad O^-$

91. $? \text{ moles acid} = 125 \text{ ml} \times \dfrac{10^{-3}\text{ L}}{\text{mL}} \times 0.400\text{M NaOH} \times \dfrac{\text{mol acid}}{\text{mol NaOH}}$

$= 5.00 \times 10^{-2}$ moles

$? \text{ molar mass} = \dfrac{5.10\text{ g}}{5.00 \times 10^{-2}\text{ mol}} = 102 \dfrac{\text{g}}{\text{mol}}$

COOH is $12.01 + 2 \times 16.00 + 1.01 = 45.02$

CH_3 is $12.01 + 3 \times 1.01 = 15.04$

CH_2 is $12.01 + 2 \times 1.01 = 14.03$

$102 - 45 - 15 = 42$

$\dfrac{42}{14} = 3$ $\quad CH_3(CH_2)_3COOH$

4 possible structural formulas

$CH_3(CH_2)_3COOH$
pentanoic acid

COOH
|
$CH_3CH_2CHCH_3$
2-methylbutanoic acid

CH_3CHCH_2COOH
|
CH_3 3-methylbutanoic acid

CH_3
|
CH_3CCOOH
|
CH_3
2,2-dimethylpropanoic acid

93.

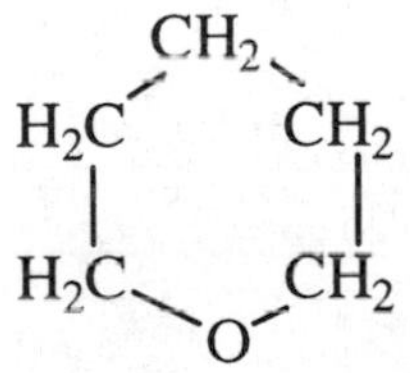

95. $\text{energy} = hc\tilde{\nu} = 6.626 \times 10^{-34}\text{ J}\cdot\text{s} \times \dfrac{2.9979 \times 10^{10}\text{ cm}}{1\text{ sec}} \times 3590\text{ cm}^{-1}$

$\times \dfrac{6.022 \times 10^{23}}{1\text{ mol}} \times \dfrac{1\text{ kJ}}{1000\text{ J}} = 42.9\text{ kJ/mol}$

97. (a) C_6H_5—NH_2 will show aromatic C—H at ≈ 3030 cm^{-1} and N—H at 3300-3500 cm^{-1}. C_6H_5—NH—C(=O)—CH_3 will show those absorptions plus an amide C=O at 1630-1690 cm^{-1} and alkyl C—H at 2850-2960 cm^{-1}.

(b) The alcohol will show alkyl C—H at 2850-2960 cm^{-1} and alcoholic O—H at 3250-3450 cm^{-1}, while the ketone will show alkyl C—H at 2850-2960 cm^{-1} and ketone C=O at 1680-1750 cm^{-1}.

99. (a) Methylbenzene and ethylbenzene should have similar infrared and ultraviolet spectra. Ethylbenzene should have somewhat more peaks in its NMR spectrum (3 peaks versus 2 peaks). But their mass spectra should be truly different; one molecular ion ($C_6H_5CH_3^+$) is smaller than the other ($C_6H_5CH_2CH_3^+$).
(b) One can distinguish between isopropanol and acetone with infrared spectroscopy. Isopropanol will show the alcoholic O—H peak at 3250–3450 cm^{-1}, while acetone shows the ketone C═O peak at 1680–1750 cm^{-1}.

101. "Spin-spin splitting" refers to the interaction of the magnetic field of one H atom with that of another H atom, bonded to an adjacent C atom. This interaction produces a pattern of peaks that enables one to interpret an NMR spectrum.

103. (a) In $CHBr_2CH_3$, the signal from the lone hydrogen is split into a quartet by the three adjacent methyl hydrogens. The methyl hydrogen signal is split into a doublet.
(b) In $CH_3OCH_2CH_2Cl$, the methyl hydrogen signal is not split because there is no H atom on the O atom adjacent to the CH_3 group. The methylene hydrogens (—CH_2—) even on adjacent carbons produce singlets. Spin-spin splitting does not occur when chemically equivalent (the same kind) H atoms are found on the same or adjacent C atoms.

105. $$\text{pH} = \text{p}K\text{a} + \log\left(\frac{[\text{base}]}{[\text{acid}]}\right)$$

$$\text{p}K\text{a} = -\log\left(\frac{K_w}{K_b}\right) = -\log\left(\frac{10^{-14}}{1.05 \times 10^{-6}}\right) = 8.02$$

$$\text{pH} = \text{p}K\text{a} + \log\left(\frac{[C_{10}H_{14}N_2]}{[C_{10}H_{14}N_2H^+]}\right)$$

$$\text{pH} = \text{p}K\text{a} - \log\left(\frac{[C_{10}H_{14}N_2H^+]}{[C_{10}H_{14}N_2]}\right)$$

$$\log\left(\frac{[C_{10}H_{14}N_2H^+]}{[C_{10}H_{14}N_2]}\right) = \text{p}K\text{a} - \text{pH}$$

$$\frac{[C_{10}H_{14}N_2H^+]}{[C_{10}H_{14}N_2]} = 10^{+(\text{p}K\text{a} - \text{pH})}$$

pH 4.5	ratio = 3320
pH 5.0	ratio = 1050
pH 5.5	ratio = 332
pH 6.0	ratio = 105
pH 6.5	ratio = 33.2
pH 7.0	ratio = 10.5
pH 7.5	ratio = 3.32
pH 8.0	ratio = 1.05

The purpose of this problem is to show how much the soubility of nicotine changes with pH. This rough calculation achieved that purpose. It is realized by the authors as stated in the text that the Henderson-Hasselbalch equation is accurate only at ratios between 10 and 1/10. Also at close to neutral pH, the self-ionization of water should also be considered for an accurate calculation.

107. (a) $CH_3CH_2CH_2COOCOCH_2CH_2CH_3 + H_2O \rightarrow 2\ CH_3CH_2CH_2COOH$
$CH_3CH_2CH_2COOCOCH_2CH_2CH_3 + 2\ NH_3 \rightarrow CH_3CH_2CH_2CONH_2 + CH_3CH_2CH_2CO_2^-NH_4^+$

(b) The product from NH_3,that is, ($CH_3CH_2CH_2CO_2^-NH_4^+$) is an ionic compound and would have a higher melting point.

(c) $CH_3CH_2CH_2CO_2^-NH_4^+ \xrightarrow{\Delta} CH_3CH_2CH_2CONH_2 + H_2O$

Apply Your Knowledge

109. $C_6H_5CHO + MnO_4^- \rightarrow C_6H_5CO_2H + MnO_2$

Balance half-equations.

$$\left(C_6H_5CHO + H_2O \rightarrow C_6H_5CO_2H + 2H^+ + 2e^-\right)_{\times 3} \quad \left(MnO_4^- + 4H^+ + 3e^- \rightarrow MnO_2 + 2H_2O\right)_{\times 2}$$

Add half-equations.

$$3\ C_6H_5CHO + 2\ MnO_4^- + 2\ H^+ \rightarrow 3\ C_6H_5CO_2H + 2\ MnO_2 + H_2O$$

Add hydroxides. $+ 2\ OH^-$ $+ 2\ OH^-$

$$3\ C_6H_5CHO + 2\ MnO_4^- + H_2O \rightarrow 3\ C_6H_5CO_2H + 2\ MnO_2 + 2\ OH^-$$

$$10.6\text{ g } C_6H_5CHO \times \frac{\text{mole}}{106.0\text{ g}} \times \frac{1}{3\text{ mole}} = 0.0333$$

$$5.9\text{ g } KMnO_4 \times \frac{\text{mole}}{158.0\text{ g}} \times \frac{1}{2\text{ mole}} = 0.0187 \text{ lesser limiting}$$

$$5.9\text{ g } KMnO_4 \times \frac{\text{mole}}{158.0\text{ g}} \times \frac{3\text{ mole } C_6H_5CO_2H}{2\text{ mole } MnO_4^-} \times \frac{122.0\text{ g } C_6H_5CO_2H}{\text{mole } C_6H_5CO_2H}$$

$= 6.8$ g $C_6H_5CO_2H$ theo yield

$$\%\ \text{Yield} = \frac{6.1\text{ g}}{6.8\text{ g}} \times 100\ \% = 90\ \%$$

111.

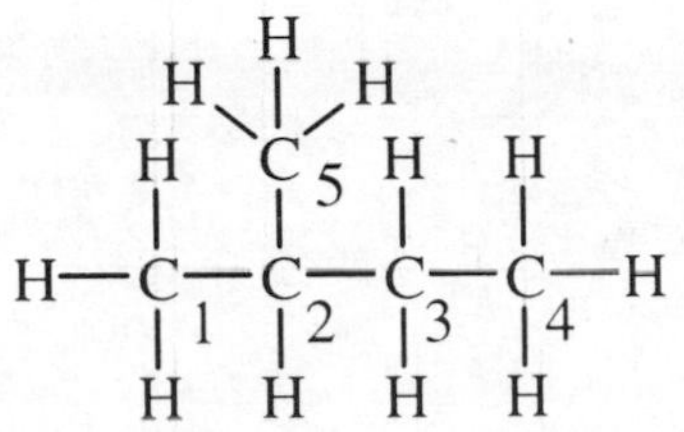

C_1 and C_5 are equivalent and are 1°, C_4 is 1°, C_3 is 2°, and C_2 is 3°.
Chlorine on C_1 or C_5 makes the same isomer 1, $ClCH_2CH(CH_3)CH_2CH_3$
Chlorine on C_4 makes the isomer 2, $CH_3CH(CH_3)CH_2CH_2Cl$

Chlorine on C_2 makes the isomer 3, $CH_3CCl(CH_3)CH_2CH_3$
Chlorine on C_3 makes the isomer 4, $CH_3CH(CH_3)CHClCH_3$
There are four isomers.
There are 12 possible replacement sites (12 H sites). One site on C_2, 2 sites on C_3, and 3 sites on C_1, C_4, and C_5.

	H site	×	reactivity	= # of molecules	
C_2	1	×	4.3	= 4.3	isomer 3
C_3	2	×	3	= 6	isomer 4
C_1	3	×	1	= 3	isomer 1
C_4	3	×	1	= 3	isomer 2
C_5	3	×	1	= 3	isomer 1
				19.3	

$$\text{isomer 1} = \frac{6}{19.3} \times 100\ \% = 31.1\ \%$$

$$\text{isomer 2} = \frac{3}{19.3} \times 100\% = 15.5\ \%$$

$$\text{isomer 3} = \frac{4.3}{19.3}\ 100\ \% = 22.3\ \%$$

$$\text{isomer 4} = \frac{6}{19.3} \times 100\ \% = 31.1\ \%$$

113. Nine amino acids in the peptide.

1 2 3 4 5 6 7 8 9

An Arg on each end

Pro-Phe-Arg

Arg-Pro-Pro Phe-Arg

There are 3 Pro—two at 2 and 3 and one at 7, so Pro-Pro-Gly has to be at 2,3, and 4.

Pro-Pro-Gly

Then Gly-Phe-Ser-Pro must start at 4

and Ser-Pro-Phe

leads to

Arg- Pro- Pro-Gly-Phe-Ser-Pro-Phe-Arg

Chapter 24

Chemistry of Materials: Bronze Age to Space Age

Exercise

24.1 The main reason for using zinc for the displacement reaction is that if a more active metal such as aluminum were used, this metal would displace Zn(s) as well as other, less active metals. This would reduce the yield of Zn(s) obtained in the electrolysis and would also leave other cations in solution that might electrodeposit with Zn(s).

24.2 The error in the given structure is that bonding does not occur through the $-CH_3$ group; the carbon atom of the methyl group would form five bonds—an impossibility for carbon.

24.3 $n\ HOCH_2CH_2COOH \rightarrow \left[O{-}CH_2{-}CH_2{-}\overset{\overset{\displaystyle O}{\|}}{C} \right]_n$

Review Questions

1. Gold, silver, platinum, and copper. The free metals must not be very reactive, (below H_2 in the activity series of Figure 4.15) or they would not be found free (unreacted).

4. Roasting is the process of heating an ore to a high temperature with the purpose of converting metal compounds to oxides. Oxide ores (e.g., Fe_2O_3) do not need to be roasted. Also, some metals can be extracted by hydrometallurgy which does not have a roasting step.

5. Reduction is the process of obtaining the free metal from its oxide. Thus, this is the process of removing oxygen from a metal compound. The preferred or most widely used metallurgical reducing agent is carbon, as coke. Often the CO(g) produced in a furnace is the actual reducing agent. Carbon and CO are especially effective in reducing metal oxides. There are a few metals that can be extracted by heating an ore to a high temperature and thus do not require a reducing agent.

14. An electric current requires a net flow of electrons in one direction. This does not occur through random motion. The effect of imposing an electric field is to create a preferred direction of motion of the electrons of the sea.

PROBLEMS

25. An ore containing a high percentage of a metal may not be the best industrial source of the metal because the metal may be in a compound that is chemically difficult to

reduce. For example, many clays contain high percentages of aluminum, but the metal is difficult to extract from them. Another difficulty can be the by-products produced during the extraction process. If these by-products complicate the purification process or are hazardous or difficult to dispose of, their formation will make the use of the ore uneconomical.

27. It is reasonable to find cadmium as an impurity since cadmium and zinc are quite similar chemically, as indicated by their adjacent positions in Group 2B of the periodic table. Cadmium can be removed either by fractional distillation of liquid zinc or by adding powdered zinc to displace the less active cadmium before refining the zinc electrolytically.

29. When HgS(s) is roasted, any HgO(s) that would form would immediately decompose to the elements: $2\ HgO(s) \xrightarrow{\Delta} 2\ Hg(l) + O_2(g)$. Decomposition of HgO(s) is one of the ways that Joseph Priestly, the discoverer of $O_2(g)$, prepared the element.

31. $2\ [Ag(CN)_2]^-(aq) + Zn(s) \rightarrow 2\ Ag(s) + [Zn(CN)_4]^{2-}(aq)$ summarizes the displacement of silver from its cyano complex by zinc.

33. Dissolving: $ZnO(s) + H_2SO_4(aq) \rightarrow ZnSO_4(aq) + H_2O(l)$
Displacement: $Zn(s) + CdSO_4(aq) \rightarrow Cd(s) + ZnSO_4(aq)$
Electrolysis: $2\ ZnSO_4(aq) + 2\ H_2O(l) \xrightarrow{\text{electrolysis}} 2\ Zn(s) + 2\ H_2SO_4(aq) + O_2(g)$

$Zn^{2+} + 2e^- \rightarrow Zn(s)$ cathode
$2H_2O \rightarrow O_2 + 4e^- + 4H^+$ anode

The $H_2SO_4(aq)$ solution is all that remains after electrolysis, and it then can be reused to dissolve more ZnO(s).

35. Chlorination: $Sn(s) + 2\ Cl_2(g) \rightarrow SnCl_4(l)$
Hydrolysis: $SnCl_4(l) + (2+x)H_2O(l) \rightarrow SnO_2 \cdot xH_2O(s) + 4\ HCl(aq)$
Dehydration: $SnO_2 \cdot xH_2O(s) \xrightarrow{\Delta} SnO_2(s) + x\ H_2O(g)$
Reduction: $SnO_2(s) + 2\ C(s) \xrightarrow{\Delta} Sn(l) + 2\ CO(g)$

37. $$\text{mass of ore} = 1 \times 10^7 \text{ kg pig iron} \times \frac{95 \text{ kg Fe}}{100 \text{ kg pig iron}} \times \frac{1 \text{ kmol Fe}}{55.85 \text{ kg Fe}} \times \frac{1 \text{ kmol } Fe_2O_3 \text{(hematite)}}{2 \text{ kmol Fe}} \times \frac{159.7 \text{ kg } Fe_2O_3}{1 \text{ kmol } Fe_2O_3} \times \frac{100 \text{ kg ore}}{82 \text{ kg hematite}} = 1.7 \times 10^7 \text{ kg ore}$$

39. Calcium is a better conductor of electricity than potassium because calcium has twice as many valence electrons. These valence electrons are free to move through

the 4*p* band in calcium, which overlaps the 4*s* band. Potassium metal has half as many electrons, mostly confined to the 4*s* band.

41. The lustrous appearance of metals results from their reflecting all incident light. Since there are energy levels at all spacings in a band, photons of all energies can be absorbed. When they are re-emitted, we see lustre. Because of the greater number of transitions of some energies, colored metals reflect some wavelengths of light to a greater extent than others.

43. $$? \text{ energy levels} = 35.0 \text{ mg Na} \times \frac{10^{-3} \text{ g}}{\text{mg}} \times \frac{\text{mol Na}}{22.99 \text{ g Na}} \times \frac{6.022 \times 10^{23} \text{ atoms}}{\text{mol Na}} \times \frac{1 \text{ energy level}}{\text{atom}} = 9.17 \times 10^{20} \text{ energy levels}$$

$$? \text{ e}^- = 35.0 \text{ mg Na} \times \frac{10^{-3} \text{ g}}{\text{mg}} \times \frac{\text{mol Na}}{22.99 \text{ g Na}} \times \frac{6.022 \times 10^{23} \text{ atoms}}{\text{mol Na}} \times \frac{1 \text{ e}^-}{\text{atom}} = 9.17 \times 10^{20} \text{ e}^-$$

45. In a semiconductor, there is an energy gap between the valence and conduction bands. In a metallic conductor, either the valence band is itself a conduction band or it overlaps one; there is no energy gap.

47. (a) *n*-type
(b) *p*-type

49. In *n*-type semiconductors, conduction electrons and positive holes are present in the same number, but the positive holes are not able to move, since they are in the atomic orbitals of the dopant, such as in P^+ in phosphorus-doped silicon. (The dopant's electrons have moved into the conduction band and are the freely moving conduction electrons.) In *p*-type semiconductors, the number of holes equals the number of electrons that are now trapped in the valence orbitals of the dopant, such as in Al^- in aluminum-doped silicon. But these electrons are not conduction electrons; they are not free to move throughout the metal. Positive holes predominate as charge carriers.

51. Metals become superconductors near 0 K because all resistance to electron flow disappears. Semiconductors usually do not become superconductors at low temperatures as the band gap remains as an energy barrier to be overcome, and the ability of electrons to jump from the valence to the conduction band *decreases* as the temperature is lowered.

53. We obtain the number of valence electrons (v.e.), outer-shell electrons, from the electron configuration of each element.
(a) [Cu] = [Ar] $3d^{10}\,4s^1$ 1 v.e. [S] = [Ne] $3s^2\,3p^4$ 6 v.e.
average = 3.5 v.e.

(b) [Zn] = [Ar] $3d^{10}\,4s^2$ 2 v.e. [Se] = [Ar] $3d^{10}\,4s^2\,4p^4$ 6 v.e.
average = 4 v.e.
(c) [Pb] = [Xe] $4f^{14}\,5d^{10}\,6s^2\,6p^2$ 4 v.e. [O] = $1s^2\,2s^2\,2p^4$ 6 v.e.
average = 5 v.e.
(d) [Ga] = [Ar] $3d^{10}\,4s^2\,4p^1$ 3 v.e [P] = [Ne] $3s^2\,3p^3$ 5 v.e.
average = 4 v.e.
ZnSe and GaP meet the average number of valence electrons criterion for being semiconductors.

55. (a) Cellulose nitrate is simply cellulose (e.g., cotton) in which —OH groups have been replaced by $—ONO_2$ groups. It is made by treating cellulose with nitric acid.
(b) Rayon is cellulose in which first the —OH groups are replaced with $—O—CS_2^-\ Na^+$ groups by treatment with NaOH and CS_2 to form cellulose xanthate, which is then converted back to cellulose (regenerated cellulose) in an acid bath.
(c) HDPE, high-density polyethylene, has the formula $-\!\!\!\!+ CH_2—CH_2 +\!\!\!\!-_n$. It is made by the free-radical polymerization of ethylene, $CH_2{=}CH_2$. Its chains are long and pack well together.

57. Rubber is elastic because its coiled polymer chains are straightened out during stretching but return to their coiled state when relaxed. Vulcanization forms crosslinks between chains which more readily pull the chains back into their original shape after stretching.

59.

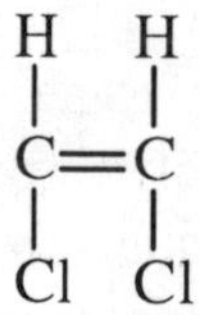

$-\!\!\!\!+ CHCl—CHCl +\!\!\!\!-_n$ is the condensed formula for polydichloroethene,

or 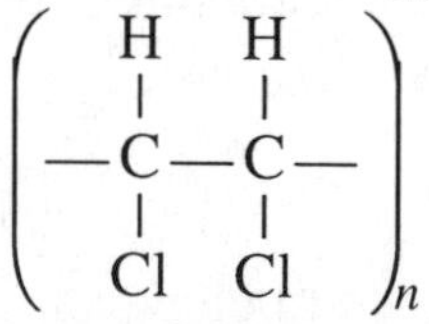

61. (a) polyacrylonitrile

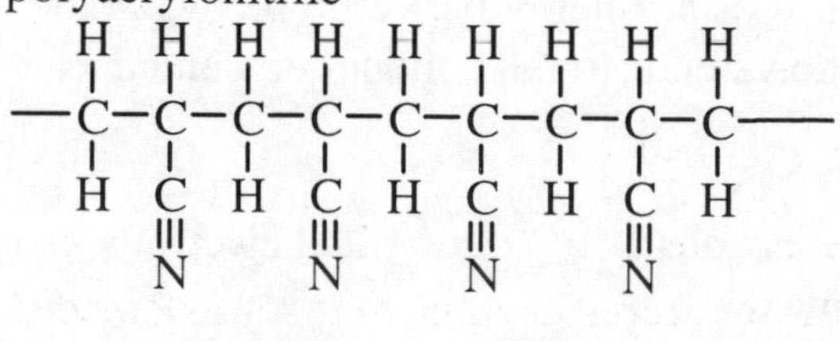

(b) poly(vinyl acetate)

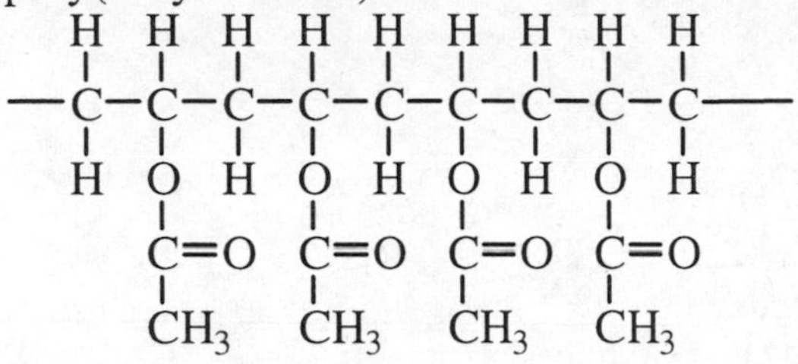

63. (a) polyglycolic acid

$$-O-CH_2-\overset{\overset{O}{\|}}{C}-O-CH_2-\overset{\overset{O}{\|}}{C}-O-CH_2-\overset{\overset{O}{\|}}{C}-O-CH_2-\overset{\overset{O}{\|}}{C}-$$

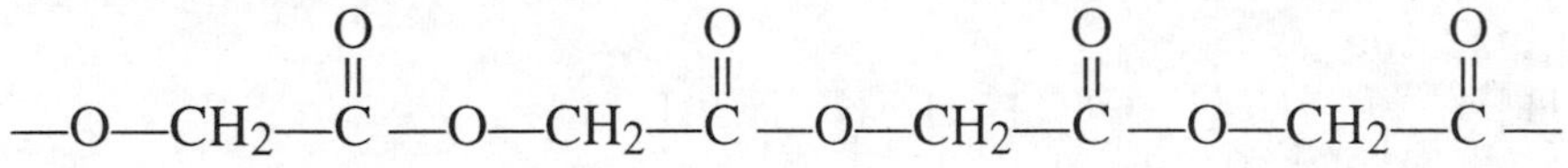

(b) Kevlar has the structure

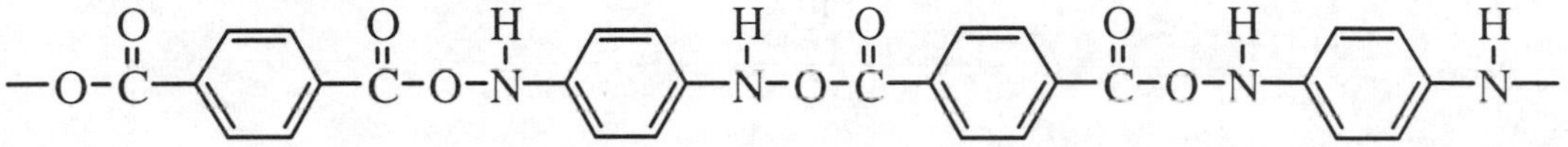

65. (a)

$$H-O-\overset{\overset{O}{\|}}{C}-(CH_2)_6-\overset{\overset{O}{\|}}{C}-O-H \quad \text{and} \quad NH_2-\bigcirc-CH_2-\bigcirc-NH_2$$

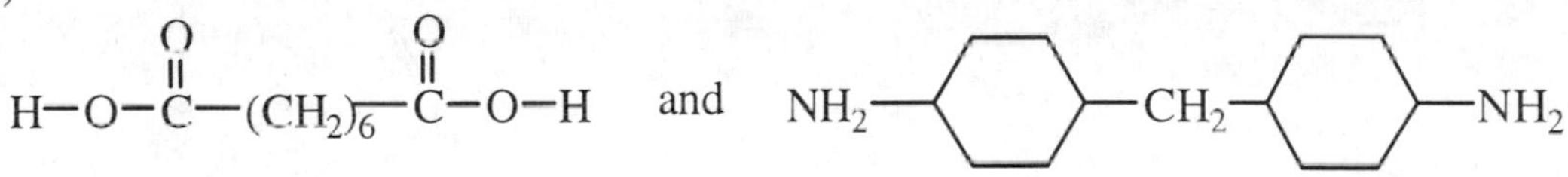

(b) polyamide

Additional Problems

67. Determine $[Cr_2O_7{}^{2-}]$ in the 10.00-mL aliquot.

$$\text{mol } Cr_2O_7{}^{2-} = 0.1387 \text{ g BaCrO}_4 \times \frac{1 \text{ mol BaCrO}_4}{253.34 \text{ g BaCrO}_4} \times \frac{1 \text{ mol CrO}_4{}^{2-}}{1 \text{ mol BaCrO}_4}$$

$$\times \frac{1 \text{ mol Cr}_2O_7{}^{2-}}{2 \text{ mol CrO}_4{}^2} = 2.737 \times 10^{-4} \text{ mol Cr}_2O_7{}^{2-}$$

$$[Cr_2O_7{}^{2-}] = \frac{2.737 \times 10^{-4} \text{ mol Cr}_2O_7{}^{2-}}{10.00 \text{ mL} \times \dfrac{1L}{1000 \text{ mL}}} = 0.02737 \text{ M}$$

Then determine the mass of Cr in the 250.0 mL of solution.

$$\text{Cr mass} = 250.0 \text{ mL} \times \frac{1 \text{ L}}{1000 \text{ mL}} \times \frac{0.02737 \text{ mol Cr}_2O_7{}^{2-}}{1 \text{ L}} \times \frac{2 \text{ mol Cr}}{1 \text{ mol Cr}_2O_7{}^{2-}}$$

$$\times \frac{51.996 \text{ g Cr}}{1 \text{ mol Cr}} = 0.7116 \text{ g Cr}$$

And finally determine the % Cr in the stainless steel.

$$\% \text{ Cr} = \frac{0.7116 \text{ g Cr}}{5.000 \text{ g steel}} \times 100\% = 14.23\% \text{ Cr}$$

69. The addition of both donor and acceptor atoms in equal number to a semiconductor means that the holes in the valence band that are created by electrons moving into the atomic orbitals of acceptor atoms will be filled by electrons moving from the

atomic orbitals of the donor atoms. The net result is no holes and no conduction electrons.

71. polyisobutylene

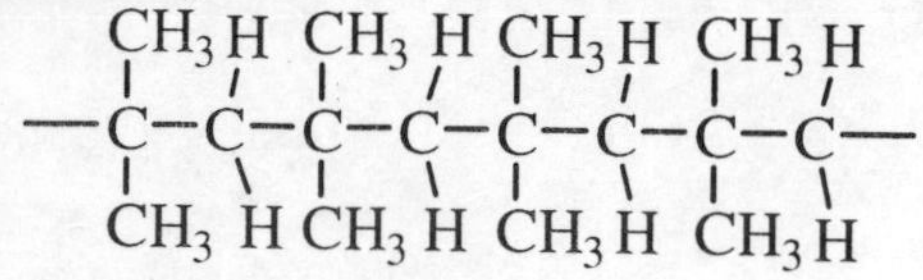

copolymer of isobutylene and 1,3-butadiene

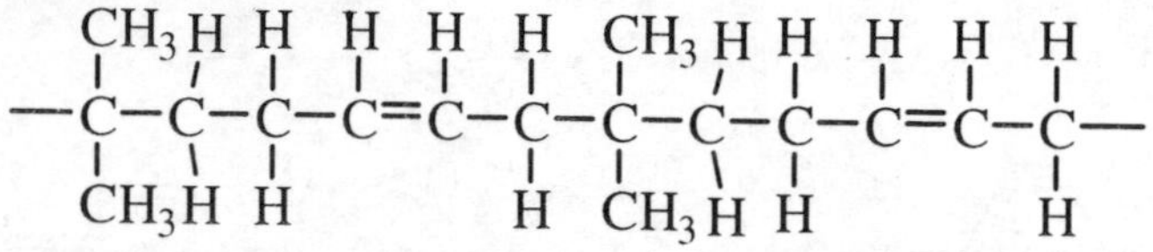

73. Kodel

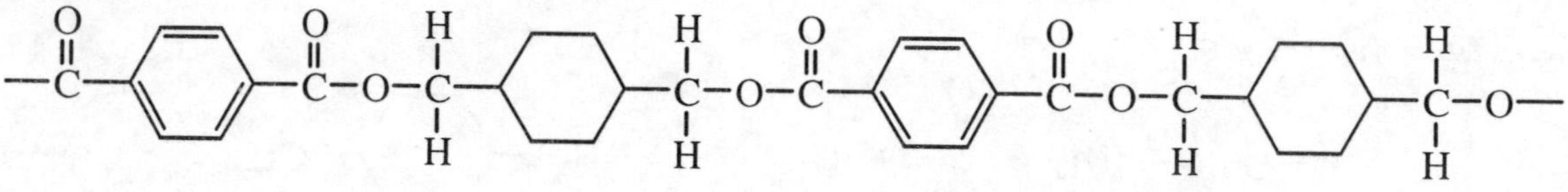

75. 1,2-ethanediol has two functional groups (—OH) but 1,2,3-propanetriol has <u>three</u> functional groups (also —OH). The polymer of 1,2,3-propanetriol and phthalic acid has extensive cross-linking and is a more rigid polymer.

77. number average molecular mass = (0.28 × 786 u) + (0.25 × 702 u) + (0.15 × 814 u) + (0.32 × 758 u) = 760 u

Apply Your Knowledge

79. $$? \text{ g Ni} = 0.0906 \text{ g} \times \frac{\text{mole}}{288.9 \text{ g}} \times \frac{\text{mol Ni}}{\text{mole complex}} \times \frac{58.69 \text{ g}}{\text{mole Ni}} \times \frac{250 \text{ mL}}{25 \text{ mL}} = 0.184 \text{ g Ni}$$

$$? \text{ \% Ni} = \frac{0.184 \text{ g Ni}}{5.108 \text{ g sample}} \times 100\% = 3.60 \text{ \% Ni}$$

81. Boat resin cures by a free-radical process. A reduction in the amount of peroxide slows the curing reaction because there are fewer free radicals, but as long as there are any free readicals, the mixture will eventually cure. The polyurethane reaction is clearly an addition reaction. A large change in the amount of one of the two monomers will make for an excess of the other monomer; that excess does not react. The mixture will thicken, but cure will not complete.

Chapter 25

Environmental Chemistry

Exercise

25.1 $\frac{38.5\%}{100\%} \times 17.5 \text{ mmHg} = 6.74 \text{ mmHg}$

Review Questions

1. (a) troposphere
   (b) stratosphere

2. N_2 78 mol %, O_2 21 mol %, Ar 1 mol %

5. (a) PAN - peroxyacetyl nitrate
   (b) SO_X - sulfur oxides SO_2 and SO_3
   (c) Fly Ash - mineral matter in coal that doesn't burn and that may be carried on the wind as small particles.

6. (a) carbon monoxide and nitrogen monoxide
   (b) methane, ozone, nitrous oxide, CFCs
   (c) nitric acid, and sulfuric acid

10. A fuel mixture that is rich in fuel and lean in oxygen will produce more carbon monoxide.
    Carbon monoxide ties up the iron atom in hemoglobin, preventing the transport of oxygen to the cells of the body.

14. (a) Nitrogen monoxide participates in reactions with hydrocarbons that lead to ozone, PAN, and other smog components.
    (b) Carbon monoxide is not important in formation of photochemical smog.
    (c) Hydrocarbon vapors react with oxygen atoms to make free radicals, ozone, and PAN.
    (d) Sulfur oxide is not important in photochemical smog. It is important in industrial smog.

21. Na^+, K^+, Ca^{2+}, Mg^{2+}, Fe^{2+}, Cl^-, SO_4^{2-}, HCO_3^-

32. (a) Aluminum sulfate and slaked lime will produce a gelatinous precipitate of $Al(OH)_3(s)$ that will trap contaminants in the water.
    (b) Aeration will remove dissolved gases (for example, chloronated hydrocarbons) from the water. Also, CO_2 and O_2 dissolved in the water, will improve its taste.
    (c) Chlorine is added to kill disease-causing microorganisms.

40. Cotinine is less toxic than nicotine and is more water-soluble, and thus, more likely to be excreted.

Problems

47. If 99% of the mass is within 30 km, then only 1% is outside 30 km, and this can support a mercury column that is only about 1% × 760 mmHg = 7.6 mmHg or ≈ 10 mmHg.

49. $\text{mole \%} = \dfrac{2.00\text{ mmHg}}{760\text{ mmHg}} \times 100\% = 0.263\text{ mole \% } H_2O$

$$\frac{0.263\text{ mol }H_2O}{100\text{ mol air}} \times \frac{10^4}{10^4} = \frac{2630\text{ molecules }H_2O}{10^6\text{ molecules air}} = 2.63 \times 10^3\text{ ppm }H_2O$$

51. $\dfrac{75.5\%}{100\%} \times 17.5\text{ mmHg} = 13.2\text{ mmHg}$

53. The dewpoint is about 20.4 °C, the temperature at which the V.P. of water is 18.0 mmHg.

55. Combustion of a hydrocarbon is an oxidation-reduction reaction in which the oxidation state of carbon atoms can increase to either +2 (CO) or +4 (CO_2). The reaction of an acid with a metal carbonate is an acid-base reaction. Only CO_2 can form. The oxidation state of C is +4 in the metal carbonate and remains the same in CO_2. A reduction would be required to produce CO, but there is no accompanying oxidation.

57. $2\ C_6H_{14} + 19\ O_2 \rightarrow 12\ CO_2 + 14\ H_2O$

It is impossible because no one can know the ratio of CO to CO_2 produced at any one time. The ratio can even change during the reaction.

59. (a) $CH_4 + 2\ O_2 \rightarrow CO_2 + 2\ H_2O$

$$?\text{ metric tons }CH_4 = 19.8\ t \times \frac{10^3\text{ kg}}{1\ t} \times \frac{10^3\text{ g}}{1\text{ kg}} \times \frac{\text{mol }CO_2}{44.01\text{ g }CO_2} \times \frac{\text{mol }CH_4}{\text{mol }CO_2} \times \frac{16.04\text{ g }CH_4}{\text{mol }CH_4} \times \frac{10^{-3}\text{ kg}}{1\text{ g}} \times \frac{10^{-3}\ t}{1\text{ kg}} = 7.22\text{ metric tons }CH_4$$

(b) $2\ C_8H_{18} + 25\ O_2 \rightarrow 16\ CO_2 + 18\ H_2O$

$$?\text{ metric tons }C_8H_{18} = 19.8\ t \times \frac{10^3\text{ kg}}{1\ t} \times \frac{10^3\text{ g}}{1\text{ kg}} \times \frac{\text{mol }CO_2}{44.01\text{ g }CO_2} \times \frac{2\text{ mol }C_8H_{18}}{16\text{ mol }CO_2} \times \frac{114.22\text{ g }C_8H_{18}}{\text{mol }C_8H_{18}} \times \frac{10^{-3}\text{ kg}}{1\text{ g}} \times \frac{10^{-3}\ t}{1\text{ kg}} = 6.42\text{ metric tons }C_8H_{18}$$

(c) $C + O_2 \rightarrow CO_2$

$$?\text{ metric tons coal} = 19.8\text{ t}\ \frac{10^3\text{ kg}}{1\ t} \times \frac{10^3\text{ g}}{1\text{ kg}} \times \frac{\text{mol }CO_2}{44.01\text{ g }CO_2} \times \frac{\text{mol C}}{\text{mol }CO_2} \times \frac{12.01\text{ g C}}{\text{mol C}} \times \frac{100\text{ g coal}}{94.1\text{ g C}} \times \frac{10^{-3}\text{ kg}}{1\text{ g}} \times \frac{10^{-3}\ t}{1\text{ kg}} = 5.74\text{ metric tons coal}$$

61. (a) $S(s) + O_2(g) \rightarrow SO_2(g)$
(b) $2\ ZnS(s) + 3\ O_2(g) \rightarrow 2\ ZnO(s) + 2\ SO_2(g)$
(c) $2\ SO_2(g) + O_2(g) \rightarrow 2\ SO_3(g)$
(d) $SO_3(g) + H_2O(l) \rightarrow H_2SO_4(aq)$
(e) $H_2SO_4(aq) + 2\ NH_3(aq) \rightarrow (NH_4)_2SO_4(aq)$

63. (a) Wave action causes droplets of salt water to be sprayed into the air. When the water evaporates, a salt particle is left in the air.
(b) $S(\text{in coal}) + O_2(g) \rightarrow SO_2(g)$
$2\ SO_2(g) + O_2(g) \rightarrow 2\ SO_3(g)$
$SO_3(g) + H_2O(l) \rightarrow H_2SO_4(aq)$
$2\ NH_3(g) + H_2SO_4(aq) \rightarrow (NH_4)_2SO_4(s)$

65. $CaCO_3 + 2H^+ \rightarrow Ca^{2+} + H_2CO_3$
$H_2CO_3 \rightarrow H_2O + CO_2$

Additional Problems

67. $$V_{\text{particle}} = \frac{4}{3}\pi r^3 = \frac{4}{3}\pi(0.5\ \mu\text{m})^3 \times \left(\frac{10^{-6}\ \text{m}}{\mu\text{m}}\right)^3 \times \left(\frac{\text{cm}}{10^{-2}\ \text{m}}\right)^3 = \frac{4}{3}\pi(0.5 \times 10^{-4}\ \text{cm})^3$$
$$= 0.52 \times 10^{-13}\ \text{cm}^3/\text{particle}$$

$$?\ \text{particles/cm}^3\ \text{air} = \frac{100\ \mu\text{g particles}}{\text{m}^3\ \text{air}} \times \left(\frac{10^{-2}\ \text{m}}{\text{cm}}\right)^3 \times \frac{10^{-6}\ \text{g particle}}{\mu\text{g particle}}$$
$$\times \frac{\text{cm}^3\ \text{particle}}{1\text{g particle}} \times \frac{\text{particle}}{0.52\times10^{-13}\ \text{cm}^3} = 2 \times 10^2\ \text{particles/cm}^3\ \text{air}$$

69. $$\text{R.H.} = \frac{5.67\ \text{mmHg}}{17.5\ \text{mmHg}} \times 100\% = 32.4\%$$
It does take some moisture out of the air, but only if the initial relative humidity is greater than 32.4%.

71. $C + O_2 \rightarrow CO_2$ $\Delta H_f = -393.5$ kJ /mol
$CH_4 + 2O_2 \rightarrow CO_2 + H_2O$ $\Delta H_{comb} = -890$ kJ /mol
$C_4H_{10} + \frac{13}{2}O_2 \rightarrow 4CO_2 + 5H_2O$ $\Delta H_{comb} = -2879$ kJ /mol

(a) $$1\ \text{g C} \times \frac{\text{mol}}{12.01\ \text{g}} \times \frac{\text{mol } CO_2}{1\ \text{mol C}} \times \frac{44.01\ \text{g}}{\text{mol } CO_2} = 3.66\ \text{g } CO_2$$
$$1\ \text{g } CH_4 \times \frac{\text{mol}}{16.04\ \text{g}} \times \frac{\text{mol } CO_2}{1\ \text{mol } CH_4} \times \frac{44.01\ \text{g}}{\text{mol } CO_2} = 2.74\ \text{g } CO_2$$
$$1\ \text{g } CH_4H_{10} \times \frac{\text{mol}}{58.12\ \text{g}} \times \frac{4\ \text{mol } CO_2}{1\ \text{mol } CH_4H_{10}} \times \frac{44.01\ \text{g}}{\text{mol } CO_2} = 3.03\ \text{g } CO_2$$
CH_4 is smallest per gram fuel.

(b) C: $1\text{ kJ} \times \frac{\text{mol}}{393.5\text{ kJ}} \times \frac{\text{mol } CO_2}{\text{mol C}} \times \frac{44.01\text{ g}}{\text{mol } CO_2} = 0.111\text{ g } CO_2$

CH_4: $1\text{ kJ} \times \frac{\text{mol}}{890\text{ kJ}} \times \frac{\text{mol } CO_2}{\text{mol } CH_4} \times \frac{44.01\text{ g}}{\text{mol } CO_2} = 0.0494\text{ g } CO_2$

C_4H_{10}: $1\text{ kJ} \times \frac{\text{mol}}{2879\text{ kJ}} \times \frac{4\text{ mol } CO_2}{\text{mol } C_4H_{10}} \times \frac{44.01\text{ g}}{\text{mol } CO_2} = 0.0611\text{ g } CO_2$

CH_4 is smallest per kJ heat involved.

73. $C + O_2 \rightarrow CO_2$ $\Delta H_f = -393.5\text{ kJ/mol}$

$$8.7 \times 10^8\text{ ton} \times \frac{2000\text{ lb}}{\text{ton}} \times \frac{454\text{ g}}{\text{lb}} \times \frac{\text{mol}}{12.01\text{ g}} \times \frac{393.5\text{ kJ}}{\text{mol}} \times \frac{2\text{ mg } SO_2}{\text{kJ}} \times \frac{10^{-3}\text{ g}}{\text{mg}}$$
$$\times \frac{\text{mol } SO_2}{64.07\text{ g}} \times \frac{\text{mol } H_2SO_4}{\text{mol } SO_2} \times \frac{98.09\text{ g}}{\text{mol } H_2SO_4} \times \frac{\text{lb}}{454\text{ g}} = 1.75 \times 10^{11}\text{ lb } H_2SO_4.$$

This mass of H_2SO_4 is about twice the typical U.S. annual production.

75. $1\text{ L} \times \frac{\text{mL}}{10^{-3}\text{ L}} \times \frac{0.80\text{ g}}{\text{mL}} \times \frac{\text{mol}}{114.22\text{ g}} \times \frac{8\text{ mol CO}}{\text{mol } C_8H_{18}} \approx 56\text{ mol CO}$

C_8H_{18} is less dense than water; that is, $d < 1.00$ g/mL. Let's assume about 0.80 g/mL.

$$95\text{ m} \times 38\text{ m} \times 16\text{ m} \times \left(\frac{\text{cm}}{10^{-2}\text{ m}}\right)^3 \times \frac{\text{mL}}{\text{cm}^3} \times \frac{10^{-3}\text{ L}}{\text{mL}} \times \frac{1\text{ mol air}}{25\text{ L air}} \approx 2.3 \times 10^6\text{ mol}$$

Assume a molar volume of air of about 25 L/mol at the prevailing T and P.

$$\frac{56\text{ mol CO}}{2.3 \text{ x } 10^6\text{ mol air}} \approx 24\text{ ppm}$$

The limit of 35 ppm would not be exceeded.

77. The rate = $k[Cl][O_3]$, then doubling [Cl] would double the rate.

79. $$?\frac{\text{nmol}}{\text{L}} = 34\text{ ppb} \times \frac{\text{g}}{\text{ppb } 10^9\text{ g}} \times \frac{1\text{ g } H_2O}{\text{mL } H_2O} \times \frac{\text{mL}}{10^{-3}\text{L}} \times \frac{\text{mole}}{133.4\text{ g } CH_3CCl_3}$$
$$\times \frac{\text{nmol}}{10^{-9}\text{ mol}} = 2.5 \times \frac{10^2\text{ nmol}}{\text{L}}$$

81. $C_3H_8O + \frac{9}{2} O_2 \rightarrow 3\ CO_2 + 4\ H_2O$

$$? = \frac{\text{mg } O_2}{\text{L}} = \frac{875\text{ kg } C_3H_8O}{1.8 \times 10^8\text{ L}} \times \frac{10^3\text{ g}}{\text{kg}} \times \frac{\text{mol } C_3H_8O}{60.09\text{ g } C_3H_8O} \times \frac{9\text{ mol } O_2}{2\text{ mol } C_3H_8O}$$
$$\times \frac{32.00\text{ g } O_2}{\text{mol } O_2} \times \frac{\text{mg}}{10^{-3}\text{ g}} = 12\frac{\text{mg } O_2}{\text{L}} = \text{BOD}$$

83. (a) $OCl^- + HSO_3^- + OH^- \rightarrow Cl^- + SO_4^{2-} + H_2O$
(b) $OCl^- + H_2O_2 \rightarrow Cl^- + O_2 + H_2O$

85. $[PO_4^{3-}] = 10\ ppm \times \frac{1\ g\ PO_4^{3-}}{ppm\ 10^6\ g\ solution} \times \frac{1\ g}{mL} \times \frac{ml}{10^{-3}\ L} \times \frac{mol\ PO_4^{3-}}{94.97\ g}$

$= 1.1 \times 10^{-4}$ M

(a) $Fe^{3+} + PO_4^{3-} \rightleftharpoons FePO_4(s)$

$K_{sp} = [Fe^{3+}][PO_4^{3-}]$

$[Fe^{3+}] = \frac{K_{sp}}{[PO_4^{3-}]} = \frac{1.3\times10^{-22}}{1.1\times10^{-4}} = 1.2 \times 10^{-18}$ M

(b) $Al^{3+} + PO_4^{3-} \rightleftharpoons AlPO_4(s)$

$K_{sp} = [Al^{3+}][PO_4^{3-}] = 6.3 \times 10^{-19}$

$[Al^{3+}] = \frac{K_{sp}}{[PO_4^{3-}]} = \frac{6.3 \times 10^{-19}}{1.1 \times 10^{-4}} = 5.7 \times 10^{-15}$ M

(c) $3\ Ca^{2+} + 2\ PO_4^{3-} \rightleftharpoons Ca_3(PO_4)_2$

$Ksp = [Ca^{2+}]^3[PO_4^{3-}]^2$

$[Ca^{2+}] = \sqrt[3]{\frac{Ksp}{([PO_4^{3-}])^2}} = \sqrt[3]{\frac{2.0\times10^{-29}}{(1.1\times10^{-4})^2}} = 1.2 \times 10^{-7}$ M

CaO is insoluble in water; $Ca(OH)_2$ is moderately soluble.

$CaO(s) + H_2O(l) \rightarrow Ca(OH)_2(s)$

$Ca(OH)_2(s) \rightleftharpoons Ca^{2+}(aq) + 2\ OH^-(aq)$

87. $5.2 \times 10^{15}\ metric\ ton \times \frac{1000\ kg}{metric\ ton} \times \frac{10^3\ g}{kg} \times \frac{mol\ air}{28.97\ g} \times \frac{368\ mole\ CO_2}{10^6\ mol\ air}$

$\times \frac{44.01\ g\ CO_2}{mol\ CO_2} = 2.91 \times 10^{18}\ g\ CO_2$

$\frac{4.6\times10^{16}\ g\ CO_2}{2.91\times10^{18}\ g\ CO_2} = 0.016$

molar mass of air = $28.0134 \times 0.78084 + 32.000 \times 0.20946 + 39.948 \times 0.00934$

$+44.009 \times 0.000356 = 28.966$ g/mol

Apply Your Knowledge

89. $2\ C_8H_{18} + 25\ O_2 \rightarrow 16\ O_2 + 18\ H_2O$

N_2	28.02	×	0.78084	=	21.88
O_2	32.00	×	0.20946	=	6.70
Ar	39.95	×	0.00934	=	0.37
CO_2	44.01	×	0.00036	=	0.02
					28.97

? moles air = 1 moles $C_8H_{18} \times \frac{25\ mol\ O_2}{2\ mol\ C_8H_{18}} \times \frac{100.00\ mol\ air}{20.946\ mol\ O_2}$

= 59.68 moles air for 1 mol C_8H_{18}

$$? \text{ g } C_8H_{18} = 1 \text{ mol } C_8H_{18} \times \frac{114 \text{ g}}{\text{mol}} = 114 \text{ g } C_8H_{18}$$

$$? \text{ g air} = 59.68 \text{ mol air} \times \frac{28.97\text{g}}{\text{mol}} = 1.73 \times 10^3 \text{ g air}$$

$$\text{mass ratio air to fuel} = \frac{1.73\times10^3 \text{ g air}}{114 \text{ g } C_8H_{18}} = 15$$

91. (a) $? \frac{\text{atoms C}}{\text{L}} = \frac{505 \text{ mg}}{\text{L}} \times \frac{10^{-3} \text{ g}}{\text{mg}} \times \frac{\text{mol}}{12.01 \text{ g}} \times \frac{6.022\times10^{23} \text{ atoms}}{\text{mol}}$

$$= \frac{2.53\times10^{22} \text{ atoms C}}{\text{L}}$$

$$? \frac{\text{atoms N}}{\text{L}} = \frac{92 \text{ mg}}{\text{L}} \times \frac{10^{-3} \text{ g}}{\text{mg}} \times \frac{\text{mol}}{14.01 \text{ g}} \times \frac{6.022\times10^{23} \text{ atoms}}{\text{mol}}$$

$$= \frac{3.95\times10^{21} \text{ atoms N}}{\text{L}}$$

$$? \frac{\text{atoms P}}{\text{L}} = \frac{14 \text{ mg}}{\text{L}} \times \frac{10^{-3} \text{ g}}{\text{mg}} \times \frac{\text{mol}}{30.97 \text{ g}} \times \frac{6.022\times10^{23} \text{ atoms}}{\text{mol}}$$

$$= \frac{2.72\times10^{20} \text{ atoms P}}{\text{L}}$$

$\frac{2.53\times10^{22}}{2.72\times10^{20}} = 93.0$ $\quad$ $\frac{3.95\times10^{21}}{2.72\times10^{20}} = 14.5$

ratio 93.0 C : 14.5 N : 1 P

$\frac{93.0}{14.5} = 6.4$ $\quad$ $\frac{106}{16} = 6.6$

C is the limiting nutrient.

(b) Begin with either $(NH_4)_2HPO_4$ or $NH_4H_2PO_4$. If $(NH_4)_2HPO_4$ is chosen, it provides the P and 2 moles N. Add 2.5 moles $CO(NH_2)_2$ for the N and 13.83 moles $CH_3CHOHCOOH$ for the rest of the C.

$1 \text{ mole } (NH_4)_2HPO_4 \times \frac{132.06 \text{ g}}{\text{mole}} = 132 \text{ g}$

$2.5 \text{ mole } CO(NH_2)_2 \times \frac{60.05 \text{ g}}{\text{mole}} = 150 \text{ g}$

$13.83 \text{ mole } CH_3CHOHCOOH \times \frac{90.08 \text{ g}}{\text{mole}} = 1246$

$\frac{150 \text{ g}}{132 \text{ g}} = 1.136$ $\quad$ 1528 g

$\frac{1246 \text{ g}}{132 \text{ g}} = 9.439$

The mass ratio of $CH_3CHOCOOH$ to $CO(NH_2)_2$ to $(NH_4)_2HPO_4$ is 9.439:1.136:1.000.

If instead one begins with 1 mole $NH_4H_2PO_4$. That provides the P and one mole of N; add 3 moles of $CO(NH_2)_2$ for the N and 3 C and 13.66 moles of $CH_3CHOHCOOH$ for the rest of the C.

$$1 \text{ mole } NH_4H_2PO_4 \times \frac{115.09 \text{ g}}{\text{mole}} = 115$$

$$3 \text{ mole } CO(NH_2)_2 \times \frac{60.05 \text{ g}}{\text{mole}} = 180$$

$$13.66 \text{ moles } CH_3CHOHCOOH \times \frac{90.08 \text{ g}}{\text{mole}} = \underline{1230}$$

$$\frac{180 \text{ g}}{115 \text{ g}} = 1.565 \qquad 1526 \text{ g}$$

$$\frac{1231}{115} = 10.70$$

The mass ratio of $CH_3CHOHCOOH$ to $CO(NH_2)_2$ to $NH_4H_2PO_4$ is 10.70:1.565:1.0000

$NH_4NO_3$was not used. It would provide the N but not the C and would thus increase the total mass.

93. (a) $? \text{ L in pond} = 7.91 \text{ acre} \times \frac{(\text{mile})^2}{640 \text{ acre}} \times \left(\frac{5280 \text{ ft}}{\text{mile}}\right)^2 \times 5.25 \text{ ft} \times \left(\frac{12 \text{ in}}{\text{ft}}\right)^3$

$$\times \left(\frac{2.54 \text{ cm}}{\text{in}}\right)^3 \times \frac{\text{ml}}{\text{cm}^3} \times \frac{10^{-3} \text{ L}}{\text{mL}} = 5.12 \times 10^7 \text{ L}$$

$$? \text{ molecules} = 2.50 \times 10^2 \text{ kg} \times \frac{10^3 \text{ g}}{\text{kg}} \times \frac{\text{mole}}{172.0 \text{ g}} \times \frac{6.022 \times 10^{23} \text{ molecules}}{\text{moles}}$$
$$= 8.75 \times 10^{26} \text{ molecules}$$

$$? \frac{\text{molecules}}{\text{L}} = \frac{8.75 \times 10^{26} \text{ molecules}}{5.12 \times 10^7 \text{ L}} = 1.71 \times 10^{19} \text{ molecules/L}$$

$$? \text{ ppm} = \frac{2.50 \times 10^5 \text{ g}}{5.12 \times 10^7 \text{ L}} \times \frac{10^{-3} \text{ L}}{\text{mL}} \times \frac{\text{mL}}{1 \text{ g}} \times 10^6 \text{ ppm} = 4.9 \text{ ppm}$$

(b) $\ln \frac{N}{N_o} = -\lambda t \qquad \lambda = \frac{0.693}{t_{1/2}} = \frac{0.693}{1 \text{ y}} = 0.693$

$$t = \frac{\ln \frac{1.00 \text{ ppm}}{4.9 \text{ ppm}}}{-0.693} = 2.29 \text{ y}$$

(c) $\ln \frac{N}{N_o} = -\lambda t$

before 2nd application

$$\ln \frac{N}{4.9\text{ ppm}} = -0.693(1/2)$$

$$\frac{N}{4.9\text{ ppm}} = e^{-\frac{0.693}{2}} = 0.707$$

$N = 3.5$ ppm

before 3rd application

$$\ln \frac{N}{8.4\text{ ppm}} = -0.693(1/2)$$

$N = 0.707 \times 8.4$ ppm $= 5.9$ ppm

before 4th application

$$\ln \frac{N}{10.8\text{ ppm}} = -0.693(1/2)$$

$N = 0.707 \times 10.8$ ppm $= 7.7$ ppm

before 5th application

$$\ln \frac{N}{12.6\text{ ppm}} = -0.693(1/2)$$

$N = 0.707 \times 12.6$ ppm $= 8.9$ ppm

before 6th application

$$\ln \frac{N}{13.8\text{ ppm}} = -0.693(1/2)$$

$N = 0.707 \times 13.8 = 9.8$ ppm

(d) The increase becomes smaller each time, so that the solubility limit is probably never reached. Each time, the increase is 71% of the previous increase. It only requires 12 times before the increase is less than 0.1. At that point the concentration is only about 9.6 ppm, less than the saturation limit.

Increases 2.4 + 1.8 + 1.2 + 0.9 + 0.6 + 0.5 + 0.3 + 0.2 + 0.2 + 0.1 + 0.1 + 0.1 = 6.1.

Total = 6.1 + 3.5 =9.6 before an application, and 11.0 just after the application.